The Routledge Handbook of Postsecularity

The Routledge Handbook of Postsecularity offers an internationally significant and comprehensive interdisciplinary collection which provides a series of critical reviews of the current state of the art and future trends in philosophical, theoretical, and conceptual terms. The volume likewise presents a range of empirical knowledges and engagements with postsecularity. A critical yet sympathetic dialogue across disciplinary divides in an international context ensures that the volume covers a wide and interrelated intellectual and geographical scope.

The editor's introduction with Klaus Eder offers a robust foundation for the volume, setting out the central aims and objectives, the rationale for the contributions, and an outline of the structure. Thorny issues of normativity and empirical challenges are highlighted for the reader. The handbook comprises four interrelated sections. Part I: *Philosophical meditations* discusses postsecularity from philosophical standpoints, and Part II: *Theological perspectives* presents contributions from a variety of theological viewpoints. Part III: *Theory, space, social relations* contains pieces from geography, planning, sociology, and religious studies that delve into theoretically informed empirical implications of postsecularity. Part IV: *Political and social engagement* offers chapters that emphasize the political and social implications of the debate. In the Afterword, Eduardo Mendieta joins the editor to reflect on the notion of *reflexive secularization* across the volume as a whole, alluding to new lines of inquiry.

The handbook is an invaluable guide for graduate and advanced undergraduate teaching, and a key reference for students and scholars of human geography, sociology, political science, applied philosophy, urban and public theology, planning, and urban studies.

Justin Beaumont is Honorary Research Fellow at the School of Geography, Queen Mary University of London, UK.

The Routledge Handbook of Postsecularity

Edited by Justin Beaumont

Routledge
Taylor & Francis Group
LONDON AND NEW YORK

First published 2019
by Routledge
2 Park Square, Milton Park, Abingdon, Oxon OX14 4RN

and by Routledge
605 Third Avenue, New York, NY 10017

First issued in paperback 2020

Routledge is an imprint of the Taylor & Francis Group, an informa business

British Library Cataloguing-in-Publication Data
A catalogue record for this book is available from the British Library

Library of Congress Cataloguing-in-Publication Data
Names: Beaumont, Justin, editor.
Title: The Routledge handbook of postsecularity / edited by
Justin Beaumont.
Description: 1 [edition]. | New York: Routledge, 2018. |
Includes bibliographical references and index.
Identifiers: LCCN 2018030624 | ISBN 9781138234147
(hardback: alk. paper) | ISBN 9781315307831 (ebook)
Subjects: LCSH: Postsecularism.
Classification: LCC BL2747.8 .R68 2018 | DDC 306.6—dc23
LC record available at https://lccn.loc.gov/2018030624

ISBN 13: 978-0-367-73264-6 (pbk)
ISBN 13: 978-1-138-23414-7 (hbk)
ISBN 13: 978-1-315-30783-1 (ebk)

Typeset in Bembo
by codeMantra

For Beryl and Grigory

Contents

Illustrations

Figures

Tables

Contributors

Christopher Baker is William Temple Professor of Religion and Public Life in the Department of Social, Therapeutic and Community Studies at Goldsmiths, University of London, where he co-directs the Faiths and Civil Society Unit. He is also Director of Research for the William Temple Foundation. His interdisciplinary research and publications engage theology and religious studies with sociology, sociology of religion, human geography, planning, and policy studies. His edited volume *Postsecular Cities: space, theory and practice* (with J. Beaumont; 2011) is a much-cited and authoritative text in this field. His latest book publication is *Re-imagining Religion and Belief for 21st Century Policy and Practice* (with B. R. Crisp and A. Dinham; 2018). A forthcoming title is *Geographies of Postsecularity: re-envisioning politics, subjectivity and ethics* (with Cloke et al.; 2019).

Justin Beaumont is Honorary Research Fellow at the School of Geography, Queen Mary University of London (UK). He has a PhD in human geography from Durham University (UK) and has held positions in the UK and the Netherlands. His books include: *Exploring the Postsecular* (co-edited with A. L. Molendijk and C. Jedan; 2010); *Postsecular Cities* (co-edited with C. Baker; 2011); *Faith-Based Organizations and Exclusion in European Cities* (co-edited with P. Cloke; 2012); and *Working Faith* (co-edited with P. Cloke and A. Williams; 2013). His research blends human geography with social justice, critical theory, and postsecularity. He's currently working on a single-authored monograph *Enlightened City* and an edited volume on Russian intellectual history and its impact on geographical scholarship.

Matt Bullimore has worked in West and South Yorkshire as a parish priest, a theological educator and a former domestic chaplain of the Bishop of Wakefield. He studied theology at the Universities of Cambridge, Harvard, and Manchester. He currently works in outreach and widening participation for Churchill College, University of Cambridge in the UK. He edited *Graced Life: the writings of John Hughes (1979–2014)* (2016).

Giuseppe Carta is PhD candidate at the School of Geographical Sciences, University of Bristol in the UK. His research is entitled *Religious pluralism and imagination: towards a post-secular city*. Across 2016 and 2017, he has organized, coordinated, and led the project *Re-thinking the mosque: sacred space and the city*, consisting of two series of public workshops, held in Bologna during February and March 2016, with the name 'Una moschea per Bologna?' and in Rome during May 2017 'Una moschea per Roma?' Thanks to the collaboration of a number of researchers and artists, these practice-as-research workshops assisted conflict resolution and management around conflicts over mosques.

Dritëro Demjaha is doctoral candidate in the Faculty of Theology at the University of Oxford in the UK. His research focuses on thought and paradox in the philosophy of G. W. F. Hegel, as well as more generally on questions about the secular and postsecular. He is also the editor of the *Radical Orthodoxy Journal.*

Klaus Eder is retired professor of Comparative Sociology at Humboldt-Universität zu Berlin, Germany. He has written extensively on sociological theory, social evolution and learning, public sphere, symbolic power, collective memory, and social movements. His numerous books include: *New Politics of Class* (1993); *The Social Construction of Nature* (1996); and *Collective Identities in Action: a sociological approach to ethnicity* (with B. Giesen, O. Schmidtke and D. Tambini; 2003). He has co-edited: *European Citizenship: national legacies and transnational projects* (with B. Giesen; 2001); *Environmental Politics in Southern Europe* (with M. Kousis; 2001); *Citizenship, Markets, and the State* (with C. Crouch and D. Tambini; 2001); and *Collective Memory and European Identity: the effects of integration and enlargement* (with W. Spohn; 2005).

Mikhail Epstein is Samuel Candler Dobbs Professor of Cultural Theory and Russian Literature at Emory University, Atlanta, GA (US). Between 2012 and 2015, he was professor of Russian and Cultural Theory and founding director of the Centre for Humanities Innovation at Durham University (UK). His research interests include new directions in the humanities and methods of intellectual creativity, contemporary philosophy, postmodernism, Russian literature, and philosophy and religion of the twentieth and the twenty-first centuries. He has authored 33 books and more than 700 articles and essays; his work has been translated into 23 languages. His recent books include: *The Irony of the Ideal* (2017); *The Transformative Humanities* (2012); and he has edited *Russian Postmodernism* (with A. Genis and S. Vladiv-Glover; 2016).

Francis Schüssler Fiorenza is Charles Chauncey Stillman Professor of Roman Catholic Theological Studies at Harvard Divinity School (USA). He has served as president of the American Theological Society (2013–2014) and as President of the Catholic Theological Society of America. His books include: *Foundational Theology: Jesus and the church* (1984); and *Modern Christian Thought vol. 2* (with J. Livingston; 2006). He has edited: *Political Theology: contemporary challenges and future directions* (with M. Welker and K. Tanner; 2013); *Systematic Theology: Roman Catholic perspectives* (with J. Galvin; 2011); and *Habermas, Modernity and Public Theology* (with D. Browning; 1992).

Robbie B. H. Goh is professor in the Department of English Language and Literature at the National University of Singapore. He works on Christianity in Asia and the diaspora, popular culture, and Indian Anglophone literature. Some of his books include: *Protestant Christianity in the Indian Diaspora* (2018); *Christianity in Southeast Asia* (2005); and *Contours of Culture: space and social difference in Singapore* (2005). His articles appear in *Material Religion, Culture and Religion, Asian Studies Review,* and other journals and edited volumes.

Warren S. Goldstein is executive director of the Center for Critical Research on Religion, in Newton, Massachusetts (USA). His PhD in sociology is from the New School for Social Research in New York. He is co-editor of *Critical Research on Religion* (SAGE Publications), book series editor of 'Studies in Critical Research on Religion' (Brill Academic Publishers and Haymarket Books), and co-chair of the American Academy of Religion Sociology of

Religion Group. While his research aims to develop a critical sociology of religion as a 'new paradigm' in the sociology of religion, he is more broadly interested in the development of a critical paradigm in the study of religion as a whole.

Elaine Graham is Grosvenor Research Professor of Practical Theology at the University of Chester (UK). She was educated at the Universities of Bristol and Manchester and has held academic posts in Sheffield, Manchester, and Chester, where she has worked since 2009. She is the author of several books, including: *Transforming Practice* (1996); *Representations of the Post/Human* (2002); *Words Made Flesh* (2009); *Theological Reflection: methods* (with H. Walton and F. Ward; 2005); *Invitation to Practical Theology Research* (with Z. Bennett, S. Pattison and H. Walton; 2018). Her most recent work considers public theology as a form of Christian apologetics: *Between a Rock and a Hard Place: public theology in a post-secular age* (2013) and *Apologetics without Apology* (2017).

Robert Joustra is associate professor of politics & international studies and founding director of the Centre for Christian Scholarship at Redeemer University College, Toronto (Canada). He is editor of *Public Justice Review*, with the D.C. think tank the *Center for Public Justice*, and an editorial fellow with *The Review of Faith & International Affairs*. He is author, co-author, and co-editor of several books, most recently *The Religious Problem with Religious Freedom: why foreign policy needs political theology* (2017).

Lily Kong is Lee Kong Chian Chair Professor of Social Sciences at the Singapore Management University (SMU). She has written widely on urban transformations, and social and cultural change in Asia. In particular, she has published a large body of work on religion, cultural policy and creative economy, urban heritage and conservation, and national identity. Recent publications include: *Arts, Culture and the Making of Global Cities: creating new urban landscapes in Asia* (with C. Chia-Ho and C. Tse-Lung; 2015); *Food, Foodways and Foodscapes: culture, community and consumption in post-colonial Singapore* (co-edited with V. Sinha; 2015); and *Religion and Space: competition, conflict and violence in the contemporary world* (with O. Woods; 2016).

Hagar Lahav is senior lecturer in the Communication Department at Sapir College, Israel, and an associated scholar at the Hadassah-Brandeis Institute, Brandeis University in the USA. She specializes in sociology of faith and gender studies. Her recent studies on Jewish-Israeli secular-believer women were published in *Journal of Contemporary Religion*, *Israel Studies*, and *Israel Studies Review*.

Bruno Latour is Emeritus Professor at Sciences Po, Paris, France, associated with the médialab and the program in political arts (SPEAP). Since January 2018, he is fellow at the *Zentrum fur Media Kunst* (ZKM) and professor at the HfG both in Karlsruhe. Member of several academies and recipient of six honorary doctorate, in 2013, he received the Holberg Prize. He received his PhD in philosophy from the Université François-Rabelais/Université de Tours and after field studies in Africa and California, he specialized in the analysis of scientists and engineers at work. In addition to work in philosophy, history, sociology, and anthropology of science, he has collaborated on many studies in science policy and research management. His many influential books include: *We Have Never Been Modern* (1993); *On the Modern Cult of the Factish Gods* (2009); and *Facing Gaia* (2017).

Michiel Leezenberg is associate professor of philosophy at the University of Amsterdam, where he teaches in the departments of Philosophy and Religious Studies. After graduating in classical languages, philosophy, and general linguistics, he received his PhD in 1995, and spent some time in the Middle East. On his return he published the Socrates-exchange-winning *Islamitische Filosofie* ('Islamic Philosophy') (2001). His research focuses on the intellectual history of the Islamic world, the history and philosophy of the humanities, and culture and society of the Kurds. He started writing for the Dutch national newspaper, *NRC Handelsblad*, in 2002.

David Lewin is lecturer in philosophy of education at the University of Strathclyde in the UK. His work draws together diverse interests in philosophy of education, philosophy of religion, and philosophy of technology. He has published numerous articles, two books, and three co-edited collections. His work has appeared in a range of journal publications including: *Journal of Philosophy of Education*; *Journal of Curriculum Studies*; *Studies in Philosophy and Education*; *Journal of Contemporary Religion*; and *Techne: Research in Philosophy and Technology*. His most recent book is *Educational Philosophy for a Postsecular Age* (2016).

Rafael Luciani is Professor Extraordinarius at Boston College School of Theology and Ministry in the USA, Full Professor at Universidad Católica Andrés Bello in Caracas, Venezuela, and Senior Adviser to CELAM (Latin American Bishop's Council) in Bogotá, Colombia. He holds degrees of Doctor in Theology and Licenciate in Dogmatic Theology from the Pontifical Gregorian University of Rome; Baccalaureatum in Philosophy and Baccalaureatum in Theology from the Pontifical Salesian University of Rome; and Licenciate in Education (with mention in Philosophy) earned from the Jesuit`s Catholic University Andrés Bello in Caracas. He has also been engaged in postdoctoral activities at the Julius-Maximilians Universität in Würzburg, Germany. His books include: *Regresar a Jesús de Nazaret* (2014); *Al Estilo de Jesús* (2015); *Francisco y la Teología del Pueblo* (2016); *Retornar a Jesus de Nazaré* (2016); and *Francis and the Theology of the People* (2017).

Martin Beck Matuštík is Lincoln Professor of Ethics and Religion at Arizona State University, joining the College of Interdisciplinary Arts and Sciences at ASUW in fall 2008. After earning his PhD from Fordham University in 1991, he has been on the faculty in the Department of Philosophy at Purdue University. His research and teaching specialties range from critical theory, continental philosophy, phenomenology, and existentialism, to post-Holocaust and reparative ethics and East-Central European thought. His interests integrate in a genuinely transdisciplinary form the fields of literature, philosophy, social theory, and spirituality. He published seven single author books, edited two collections, and co-edited *New Critical Theory*, a book series at Rowman and Littlefield Publishers. Among his publications are *Jürgen Habermas: a philosophical-political profile* (2001); *Radical Evil and the Scarcity of Hope: postsecular meditations* (2008); *Out of Silence: repair across generations* (2015); *Kierkegaard in Post/Modernity* (co-edited with M. Westphal; 1995); and *Calving O. Schrag and The Task of Philosophy after Postmodernity* (co-edited with William L. McBride; 2002).

Gregor McLennan is professor of sociology at the University of Bristol (UK). Following postgraduate degrees at the renowned Birmingham Centre for Contemporary Cultural, he worked in social sciences at the Open University in the 1980s, going on to head the Department of Sociology at Massey University, New Zealand, from 1991 to 1997. He is the author of *Marxism and the Methodologies of History* (1981); *Marxism, Pluralism and Beyond* (1989); *Pluralism* (1995);

Sociological Cultural Studies: reflexivity and positivity in the human sciences (2006); and *Story of Sociology* (2011), and is co-editor of several themed collections, including *The Idea of the Modern State* (with D. Held and S. Hall; 1985). Always concerned with central issues in the philosophy of the social sciences, over the past decade he has been engaged with the challenges posed by postcolonial and postsecularist thought to the radical enlightenment tradition and its epistemologies. His selection and extensive framing of Stuart Hall's writings on *Questions of Marxism and Post-Marxism* will appear next year (2019).

Eduardo Mendieta is professor of philosophy, associate director of the Rock Ethics Institute, affiliated faculty at the School of International Affairs and at the Bioethics Program at Penn State University (USA). He is the author of *The Adventures of Transcendental Philosophy* (2002) and *Global Fragments: globalizations, Latinamericanisms, and critical theory* (2007). He is also co-editor of *The Power of Religion in the Public Sphere* (with J. VanAntwerpen; 2011); *Habermas and Religion* (with C. Calhoun and J. VanAntwerpen; 2013); and *Reading Kant's Geography* (with Stuart Elden; 2011). He recently finished a book titled *The Philosophical Animal,* which will be published by SUNY Press in 2018. He was in 2017 the recipient of the Frantz Fanon Outstanding Achievements Award.

Kasia Narkowicz is lecturer in sociology at the University of Gloucestershire. She holds a PhD in human geography from the University of Sheffield. Kasia's research interests include religion, race/ethnicity and gender, and the exclusion of Others. Her publications include: 'Refugees not welcome here: state, church and civil society responses to the refugee crisis in Poland' (*International Journal of Politics, Culture, and Society*), 'Unmaking citizens: passport removals and the reorientation of colonial governmentalities' (*Ethnic and Racial Studies*, with Nisha Kapoor) and 'Saving and fearing Muslim women in a post-communist context: troubling Catholic and secular anti-Muslim narratives in Poland' (*Gender, Place and Culture*, with Konrad Pedziwiatr).

Krzysztof Nawratek is senior lecturer in humanities and design architecture at the University of Sheffield (UK). Before joining Sheffield, he was an associate professor in architecture and MArch and MA in architecture programme leader at the Plymouth University (UK). He has worked as an architect and urban designer in Poland, Latvia, and Ireland. He was a Visiting Professor at the Geography Department at the University of Latvia, and a researcher at National Institute for Regional and Spatial Analysis, Maynooth, Ireland. His books include *City as a Political Idea* (2011) and *Holes in the Whole: introduction to the urban revolutions* (2012). His edited volumes include *Radical Inclusivity: architecture and urbanism* (2015) and *Urban Re-Industrialization* (2017). A forthcoming title is *Total Urban Mobilisation: Ernst Jünger and the post-capitalist city* (2019). He has authored several papers and chapters in edited volumes.

Peter Nynäs is professor of comparative religion at Åbo Akademi University (Finland) and director for the Centres of Excellences in Research on 'Post-Secular Culture and a Changing Religious Landscape in Finland' (2010–2014) and 'Young Adults and Religion in a Global Perspective' (2015–2018). He has combined his interest in both psychology and sociology of religion into research on religious change and contemporary religious subjectivities, alongside his interest in place and space, gender and sexuality, and methodology in the study of religion. He has edited several volumes: *On the Outskirts of "the Church": diversities, fluidities and new Spaces of religion in Finland* (with R. Illman and T. Martikainen; 2015); *Religion, Gender and Sexuality in Everyday Life* (with A. Yip; 2012); *Post-Secular Society* (with

M. Lassander and T. Utriainen; 2012); and *Transforming Otherness* (with Jason Finch; 2011). A forthcoming collection is *Sensitizing Religious' Variety in a Global Perspective: between universalism and particularism* (with R. Illman, N. Novis and R. Fernandez; 2019).

Umut Parmaksız is assistant professor of sociology at TED University, Ankara (Turkey). He received his PhD in sociology from the University of Bristol (UK) in 2015. His thesis examines the cogency, import, and utility of the concept of the postsecular for social and political theory. Before starting work at TED University, he worked as an assistant teacher at the University of Bristol and as a part-time instructor at Bilkent University in Turkey. His research lies at the intersection of social and political theory and he is interested in topics such as postsecularity, secularization, truth, religion in the public sphere, and secular ways of life.

Richard Phillips is professor of human geography at the University of Sheffield (UK). He is the author of numerous books including: *Mapping Men and Empire* (1997); *Sex, Politics and Empire* (2006); and *Muslim Spaces of Hope* (2008). His research covers: the world after empire, including Muslim geographies and postcolonial cities; sexuality, space and power, focusing on constructions and contestations of sexual identities; and curiosity and adventure, dealing with children's books to health and well-being policies. He has a masters in geography from the University of California, Santa Barbara (1988) (USA) and a PhD at the University of British Columbia (1994) in Vancouver (Canada). He taught at the Universities of Aberystwyth, Salford, and Liverpool before taking up a chair in human geography at the University of Sheffield in 2012. He teaches social and cultural geography and postcolonial criticism.

Junxi Qian is assistant professor in the Department of Geography, The University of Hong Kong (China). His works are situated at the intersection of geography, urban studies, and cultural studies. He holds a PhD in human geography from University of Edinburgh, Scotland, UK. His recent research focuses on urban public space, small cities in China, religion and China's ethnic frontiers. He is the author of *Re-visioning the Public in Post-Reform Urban China: poetics and politics in Guangzhou* (2018) and has authored numerous journal articles and book chapters in both Chinese and English languages.

Manav Ratti is associate professor of English at Salisbury University in the University of Maryland (USA) system and author of *The Postsecular Imagination: postcolonialism, religion, and literature* (Routledge 2013; pbk. 2014; South Asian edition, Routledge 2018). His multidisciplinary work has appeared in journals such as *The Journal of Commonwealth Literature*, *Journal of American Studies*, *ARIEL*, and *Sikh Formations*. Since completing his doctorate at Oxford University, he has served as a Fulbright Scholar at New York University, a TEDx speaker, and has held fellowships at the Australian National University, the Gandhi Ashram in Ahmedabad, Jawaharlal Nehru University, Queen's University Belfast, and the University of Toronto.

Christopher Rowland is Dean Ireland's Professor of the Exegesis of Holy Scripture Emeritus, University of Oxford (UK). He taught at the Universities of Newcastle upon Tyne, Cambridge, and Oxford. He has written on the New Testament, with special reference to the nature of apocalypticism and eschatology, and also on liberation theology. His most recent books are: *The Cambridge Companion to Liberation Theology* (2007); *Blake and the Bible* (with Zoë Bennett; 2010); *In a Glass Darkly: the Bible, reflection and everyday life* (with Zoë Bennett; 2016); and *Radical Prophet: the mystics, subversives and visionaries who strove for heaven on earth* (2017).

Chris Shannahan is Research Fellow in Faith and Peaceful Relations at the Centre for Trust, Peace, and Social Relations, Coventry University (UK). Prior to joining Coventry University in 2015, Chris Shannahan was the head of Religious Education in a large East London Secondary school; a youth worker in the East End of London and Trenchtown, Jamaica; a Methodist Minister in inner-city London and Birmingham; and a community organizer. He has held academic positions at the Universities of Birmingham and Manchester. His first monograph, *Voices from the Borderland* (2010), is a set text at Universities and Theological Colleges in the UK, the USA, and Australia, and his second, *A Theology of Community Organizing* (2014), provided the first systematic theological analysis of broad-based community organizing. At present he is working on his third book, provisionally entitled *Contextual Theology and the Politics of Faith* and the development a cross-cultural and interfaith contextual theology project entitled the 'Faith and Peaceful Relations Forum'. He is now beginning a major new three-year ESRC urban theology project 'Life on the Breadline: Christianity, politics and poverty in the twenty-first century city'.

Roger Speare MBE is a social entrepreneur based in Bolton in the UK. He read natural sciences and economics at the University of Cambridge *en route* to industry and commerce. He has worked at CEO level in both large and small companies making batteries, scientific instruments, and tensile structures, later switching to social enterprise and founding Emmaus Bolton. He is currently planting two new Emmaus communities in Stoke-on-Trent and South Manchester to help combat poverty and homelessness in these localities.

Timothy Stacey is currently postdoctoral researcher on the Religion and Diversity Project, Department of Classics and Religious Studies, University of Ottawa, Canada, where he is undertaking ethnographic inquiries into the values that motivate secular citizens to campaign for social and economic justice. His research explores the role of religion and belief in renewing a sense of solidarity in culturally and economically fragmented spaces. In particular, he explores what he calls 'myths', the stories of great events and characters that point towards an ideal way of living. He uses the term to highlight the parity between religious and secular ways of inhabiting the world. His monograph, *Myth and Solidarity in the Modern World: beyond religious and political division*, was released with Routledge in 2018. He is a member of several research and policy networks and co-founded the LivedReligionProject.com, an online campaign that challenges prejudiced understandings of religion as dogma by highlighting the complexity and humour with which randomly selected individuals navigate their religious and non-religious lives.

Kristina Stoeckl is associate professor at the University of Innsbruck in Austria where she teaches sociology of religion and social and political thought within the Department of Sociology. Her research interests cover religion and politics, Orthodox Christianity, and church-state relations in Russia. She is the author of *The Russian Orthodox Church and Human Rights* (2014) and principal investigator of the five-year research project 'Postsecular Conflicts' (ERC-STG-2015-676804).

Øyvind Strømmen is a Norwegian journalist. His master's thesis focused on the anti-Catholicism of the Norwegian author and theologist Marta Steinsvik in the 1920s and 1930s. He has written three books on extremism: *Det Mørke Nettet* ('The Dark Net') (2011); *Den Sorte Tråden* ('The Black Thread') (2013); and *I Hatets Fotspor* ('In the Footsteps of the Hat') (2014). He has also published a biography on the founder of the Norwegian right-wing populist party known as the Progress Party ('Fremskrittspartiet'): *Anders Lange: en norsk historie* ('Anders Lange: a Norwegian history') (2016) and a book on Hungarian history and politics: *Ungarn: en fortelling om Europa* ('Hungary: a story about Europe') (2018).

Phra Nicholas Thanissaro is Associate Fellow of Warwick Religions and Education Research Unit (WRERU) at the University of Warwick's Centre for Education Studies. He is currently reworking his PhD thesis entitled 'Templegoing Teens: the religiosity and identity of Buddhists growing up in Britain' for publication. A Buddhist monk affiliated with the Dhammakâya Foundation, he is also a qualified school teacher and MBTI practitioner.

Dmitry Uzlaner holds a PhD in philosophy and religious studies (M.V. Lomonosov Moscow State University) and currently is Research Fellow at both the Moscow School of Social and Economic Sciences (MSSES) and the Russian Presidential Academy of National Economy and Public Administration (RANEPA). He is editor-in-chief of the journal *Gosudarstvo, Religiia, Tserkov v Rossii i za Rubezhom* ('State, Religion, Church in Russia and Abroad'). In 2016, he joined Kristina Stoeckl's five-year international research project 'Postsecular Conflicts' at the University of Innsbruck (Austria). He has published numerous articles on theoretical aspects of secularization and postsecularity both in Russian and global contexts.

Guido Vanheeswijck is full professor at the Department of Philosophy of the University Antwerp and part-time professor at the Institute of Philosophy of the Catholic University Louvain in Belgium. His research areas are metaphysics, philosophy of culture, and philosophy of religion. He is best known as a defender of active pluralism and diffuser of the thinking of Robin Collingwood, René Girard, and Charles Taylor in the Low Countries. His books include: *Dionysus of de Gekruisigde: Friedrich Nietzsche versus René Girard over de kern van het christendom* ('Dionysus or the Crucified: Friedrich Nietzsche versus René Girard about the core of Christianity') (with S. Latré; 2001); *Voorbij het Onbehagen: ressentiment en christendom* ('Beyond Discomfort: resentment and christianity') (2002; awarded with the Max Wildiers Essay Prize 2002); *Verlossend Inzicht: filosofie en christendom* ('Redemptive Insight: philosophy and christianity') (with W. van Herck; 2005); and *Tolerantie en Actief Pluralisme: de afgewezen erfenis van More, Erasmus en Gillis* ('Tolerance and Active Pluralism: the rejected legacy of More, Erasmus and Gillis') (2008).

Herbert De Vriese is assistant professor at the Center for European Philosophy, University of Antwerp, Belgium. He obtained his PhD in philosophy from the University of Antwerp in 2011 with a thesis on Young Hegelianism. His main field of expertise is nineteenth-century German philosophy, with special attention to the 'revolutionary rupture' between Hegel and Nietzsche. Critique of religion and secularization belong to the heart of his research interests. In the co-edited volume *Rethinking Secularization: philosophy and the prophecy of a secular age* (with G. Gabor; 2009) and a number of other publications, he discusses the philosophical repercussions of the end of classical secularization theory.

Edward Wigley is post-doctoral research associate at the Department of Geography of the Open University (UK). He completed his PhD entitled *Everyday Mobilities, Place and Spirituality* at the University of the West England, Bristol (UK). He is currently working on the ESRC-funded project *Smart Cities in the Making: learning from Milton Keynes* based at the Open University in the UK. His work has recently been published in *Social and Cultural Geography* and *Mobilities*.

Acknowledgements

To the reader who is unfamiliar with my work, a word of explanation as part of a larger acknowledgement maybe be useful.

Around 20 years ago I first encountered enlightened faith-based action at *The Cedarwood Trust* on the Meadow Well council estate in North Tyneside, UK. I was undertaking fieldwork for my thesis at Durham at the time. While millions had been spent to try and heal the wounds from the riot on the estate in 1991, The Cedarwood Trust community development charity worked—and continues to work—to raise confidence, aspirations, and hope among people of a community that are often still hindered by debilitating levels of poverty and income deprivation.

Experience showed that these tireless efforts against deep-seated structural inequalities and injustices were about far more than serving God. To try and better understand those innate motivations—how one might be faith-oriented or not yet work together so those powerful innate human energies can contribute to a flourishing of a better world for everyone—is, to my mind, what postsecularity is all about. My first thanks, then, go to the people of Cedarwood and North Tyneside more generally who helped me experience and confirm these essential truths.

Second, I would like to thank the years of personal and professional support from Paul Cloke. Paul is one of the very few geographers who really 'gets' it. What he and I have called, perhaps esoterically 'postsecular rapprochement in the city', reflects a line of inquiry that moved into theoretical and philosophical engagements with Jürgen Habermas and Klaus Eder. These ideas, among others, helped to try and make better sense of our previous empirical investigations into faith-based organizations and urban social justice. The spirit of Paul's oeuvre is very much present in the volume.

Third, the time I spent as a volunteer at *Emmaus St. Albans* in 2017 after leaving Groningen was genuinely inspirational on many levels, and I cannot thank the companions and workers there enough. At a particularly challenging time, the people there—and they know perfectly well who they are—helped restore belief in our innate capacities and desire to work together and help each other to help ourselves. My time at Emmaus brought me into contact with Roger Speare. Roger's contribution to this volume means a great deal, in symbolic as well as substantive terms.

Fourth, I would like to thank Andrew Mould and Egle Zigaite at Routledge who supported the project from inception. Their flexibility and understanding throughout has been vital to the completion of the volume, and I thank you wholeheartedly for that.

Fifth, thanks to all contributors for their diligence, commitment, and professionalism. Special thanks go to Warren Goldstein for his sharp editor's eye, as well to Klaus Eder and

Eduardo Mendieta, both who came in late but whose involvement was just what I was seeking during the final push.

Finally, and most importantly, my warmest and most heartfelt appreciation goes to my family, without whom this volume would never have come about. They can see and feel so much of me in this project, as I can them. Thank you.

J.R.B.
June 2018

Introduction

1

Concepts, processes, and antagonisms of postsecularity

Justin Beaumont and Klaus Eder

... let us unite, not in spite of our differences, but through them. For differences can never be wiped away, and life would be so much the poorer without them. Let all human races keep their own personalities, and yet come together, not in a uniformity that is dead, but in a unity that is living.

Rabindranath Tagore[1]

Neutralization of culture—the words have the ring of a philosophical concept. They posit as a more or less general reflection that intellectual constructs have forfeited their intrinsic meanings because they have lost any possible relation to social praxis ... Every now and then ... it is possible to name a work in which the neutralization of culture has expressed itself most strikingly ... while it remains enigmatically incomprehensible.

Theodor W. Adorno[2]

Message of *The Handbook*

The Routledge Handbook of Postsecularity (*The Handbook* hereafter) comes at a crucial time. First conceived in 2015, *The Handbook* presents 35 original chapters from philosophers, theologians, geographers, and sociologists who, from a variety of standpoints, critically embrace the notion of *postsecularity*.[3] It is our pleasure to introduce the message of the volume, present an initial framing of the collection of articles, and deal briefly with postsecularity as a complementarity of discourses and as a confrontation of normativities. While the recent Zuckerman and Shook (2017) compendium includes voices on liberal democratic constitutionalism and multiculturalism, pointing to the interplay of the secular and postsecular in theoretical and normative terms, the contributions tend to veer, implicitly or otherwise, towards a defence of secularism. *The Handbook* is quite different and intentionally so. The volume advocates an ethos, sensibility, or consciousness of postsecularity across a purposively diverse disciplinary range and opens a genuinely radically pluralist, inclusive, and reflexive space for intellectual and political engagement. This platform concerns dialogue across the broad spectrum of postsecular thought. The interchange in it occurs on an equal footing—perhaps as Jürgen Habermas would applaud—welcoming atheists, believers, and agonistics, indeed any position or worldview that values this reflexive opportunity.

The Handbook builds on a growing yet hugely contested, diverse, and fragmented set of literatures on the persistence, resurgence, or reformulation of religious belief and practices in relation to secular ideas, modes of existence, and governance. The assumption at the core is that the concept of postsecularity—the generic term covering 'the postsecular', 'postsecularism', and 'postsecular society' together—offers a path-breaking and innovative means for grappling with (at least to address in a new and hitherto overlooked or underexplored way) the most pressing intellectual, political, and human concerns of our time. The claim is not that postsecularity in its various incarnations offers in any way an all-encompassing panacea to problems of the human predicament. Deeper engagement with, and a clearer understanding of, the concept across disciplines, however, are required to further debates in social sciences and the humanities in valuable and meaningful directions that might, one way or another, assist in social transformation. The collection offers a valiant and heart-felt attempt to better understand postsecularity in theoretical, empirical, and normative terms. It does not fall prey to one ideological imperative over another. The aim is to understand and guide, not to dictate or prescribe.

Mikhail Nesterov's *The Philosophers* (1917) adorning the front of this collection depicts philosopher-theologians Pavel Florensky and Sergei Bulgakov on the eve of the Great October Revolution. Both figures were radical thinkers who thought of the world and all living creatures including humans as a single organism. Bulgakov, who started out as a Marxist but turned to Idealism after reading Soloviev, Tolstoy, and Dostoyevsky, showed tense foreboding of what lay ahead. Florensky was an Orthodox theologian, polymath, and prominent member, like Nesterov, of the Russian symbolist movement of the time, and conveyed calm and humble acceptance of his impending tragedy.[4] The ferment of conservative, progressive, and radical ideas just prior and leading up to the events of 1917 is arguably somewhat overlooked in the West while also engaging and awe-inspiring (cf. Epstein 2011). This intensely creative and tumultuous period, linked to the Silver Age in Russian poetry and literature—for example, Soloviev, Berdyaev, Florovsky, Frank, among others—has been invoked as 'postsecular' and evidence for the failure or at least unfinished project of modernity (Mrowczynski-Van Allen et al. 2016) (see Figure 1.1).[5]

That coming together—or complementarity of discourses—as a whole, a totality, and a potentially higher perspective is what is sought in *The Handbook*. History has a peculiar way of repeating itself. We have in mind here the eternal recurrence or return—in a Nietzschean sense, the bigger scheme of things despite the passage of time and the range of individual and sometimes collective, impassioned efforts and dreams for social transformation and a better world.[6] Religion in a putative secular society has played and will continue to play its part, in conjunction with the range of competing and sometimes also coalescing ideas, inspirations, and motivations. Critical engagement with diverse scholars over the notion of postsecularity, with origins in the 'religious turn' in continental philosophy and the advent of radical orthodoxy (RO) (Blond 1997; de Vries 1999; Žižek 2000, 2009; Rorty and Vattimo 2007; Žižek and Milbank 2011), has the potential to convey at least part of people's progressive spiritual essences, interior to themselves and externally in relation to landscape and in harmony with their social and spatial environmental contexts.

Such an optimistic view may not sit easily with some observers. We fully acknowledge that tensions exist between differing epistemological and normative positions in the debate. The creativities that emerge from this confrontation are precisely the point and worthy of deeper, critical engagement. The aim is not to smooth over these disputes; rather, the purpose in the spirit of the contextual modernism and universalism of Rabindranath Tagore, the Bengali philosopher, poet, and polymath, is to sense and experience a unity in our

Figure 1.1 Portrait of philosopher Vladimir Soloviev by Ivan Kramsky (1885)
Source: Wikipedia Commons.

intrinsic human diversity (Tagore 1961; Sarkar and Keipeng 2017) (see Figure 1.2). A blindly normative quest for integration, assimilation, and solidarity that eschews difference would achieve little and play into the hands of the trite and all too familiar critiques of universalist thinking. In this introduction we are sensitive to 'normative optimistic' and 'normative pessimistic' readings of postsecularity and the human condition more generally.

Adorno's (1976) critique of Beethoven's Mass in D major, Op. 123, *Missa Solemnis*, captures this sentiment and anxiety (see Figure 1.3). While recognizing that composers from Bach to Schoenberg typically have dealt with the form of themes and variations on these themes, the tensions of which are ultimately resolved, Adorno feels uncomfortable with the tendency to non-resolution and aesthetics for itself in Beethoven's choral masterpiece (Adorno 2002). One can allude to the logical refinement, pristine secular constructivism, and religious piety in Bach's Suites for Piano/Harpsichord. The spiritual awakening in Beethoven's later life was quite a different matter. Beethoven, the radical freethinker and individualist of the Enlightenment, sought the transcendental and 'a longing for a totally other' (Siebert 2005) as the ultimate human goal. Less clear is whether the work was motivated by his own feelings of the celestial or an attempt to inspire those feelings in listeners.

Figure 1.2 Tagore: thinker and poet of universal values, Germany 1931
Source: Wikipedia Commons: photo: German Federal Archive.

Figure 1.3 Beethoven's walk in nature by Julius Schmid
Source: Wikipedia Commons: image: Michael Martin Sypniewski.

Either way, that mutual imbrication between human and earthly immanence and the other-worldly quest for the transcendental strikes at the heart of critical inquiries into postsecularity. The verge, edge, or liminal space between the immanent and transcendent, the metaphysical and postmetaphysical, as well as the material and spiritual (see Chapters 4–6 and 20) captures a certain intellectual urgency but also sense of ambiguity that typifies the discussion—enchanting and frustrating proponents and opponents in more or less equal measure. The several contributions in this volume speak in multiple, contrasting, and sometimes overlapping approaches to the concepts, notions, and dilemmas of postsecularity. At this stage we would like to draw attention to at least three claims:

1 Critical inquiries into postsecularity unsettle both debates on the postsecular as well as the secular and the complementarities between these two discourses. The outworking of this disturbance tends towards transcendence, philosophical monism, and pantheism.
2 Central to this engagement is a revival of certain strands of critical theory of the Frankfurt School, and in particular the negative dialectics of Adorno and Horkheimer. Alongside the postmetaphysical intersubjectivity of Habermas, these currents bring distinctive normativities to the fore.
3 Steps from postmetaphysical immanence that reach outwards to transcendental modes raise questions about the universal and authentic human condition that eclipse East and West, while paying respect to context-specific postsecularisms outside the Occident including Russia, India, and Asia.

Framing of the volume

We face another decisive moment in European history, with ramifications worldwide. Unprecedented conflicts, uncertainties, and insecurities unleashed by post-9/11 radical Islamism, the Arab Spring and ongoing Syrian Civil War, Europe's refugee and humanitarian crises, tightening punitive politics of securitization and surveillance amid terror attacks in several European cities, the rise of ethno-national and far-right extremist movements and political parties, an escalating ecological crisis, and, perhaps most importantly, the global financial calamity and geopolitical *précarité* unleashed in the USA that soon spread to Eurozone countries like austerity-ridden Greece—all these events and processes manifest themselves as deep human injustices felt in urban areas and society at large. Notwithstanding the withdrawal of the UK from the EU, and Trump's showman geopolitics and authoritarian populism in the USA, these grievances challenge us to develop new frameworks of understanding in order to conceive of and aim for a more socially just, humane, and equitable world.

Despite little agreement over the meaning of the term, at its most general level, postsecularity might refer to the persistence, reformulation, or resurgence of religion in the public sphere. Charles Taylor's (2007) *A Secular Age* is widely considered one of the most important critiques of secularism. In this *magnum opus* Taylor pays attention to the cultural conditions of secularity, in which unbelief in religion is thought to be a viable option among several others, and religious and secular ideas coexist on equal terms. Most commentators attribute the surge of popularity in the concept of the postsecular to *The Dialectics of Secularization* (Habermas and Ratzinger 2007) and in particular Habermas' ideas on the alleged rise of postsecular society (Habermas 2008) (see Chapters 4, 5, and 9) (see Figure 1.4). When taken alongside José Casanova's (1994) *Public Religions in the Modern World* with reference to the deprivatization of religion, we have at least three of the main intellectual cornerstones of what has become the conceptually interrelated narratives of postsecularity.

Figure 1.4 Jürgen Habermas during a discussion in the *Munich School of Philosophy*
Source: Wikipedia Commons: photo: Wolfram Huke.

Other important works include Talal Asad's (2003) Nietzschean-inspired Foucaldian anthropology of religion that shows how European and North American presecularisms have mutated into exclusionary forms of secularism today (cf. Scott and Hirschkind 2006). The much-vaunted *The Power of Religion in the Public Sphere* (Butler et al. 2011) presents Judith Butler's critical engagement with state violence in Israel-Palestine. An outspoken critic of Zionism, Butler addresses the potential role of religion in revitalizing critical thought and progressive politics (see Figure 1.5). Work in feminist theology (e.g., Plaskow 1991; Kien 2000; Raphael 2003; Ross 2004) holds implications for the relationship between gender and postsecularity (Greed 2011; Graham 2012; see Chapter 13) and points to a possible relationship between postsecularity and a growing feminist movement (Braidotti 2008; Butler 2008; Vasilaki 2016). The forthcoming *Cambridge Habermas Lexicon* (Allen and Mendieta 2019) draws attention to how decolonial thought and Frankfurt School critical theory help us understand postsecularity and also the relations between world society, the global public sphere, and postsecularism (see Chapters 4, 5, and 9). All in all, these intellectual developments sharpen our awareness of the relations between religious, secular, and humanist forces, previously viewed in isolation, which are now placed together in the public sphere. At the forefront of the analysis are empirical developments on the ground, political advances for social transformation, and theoretical explanations of macrosocial developments at large.

Building on four previous volumes, (1) *Exploring the Postecular* (Molendijk et al. 2010);[7] (2) *Postsecular Cities* (Beaumont and Baker 2011); (3) *Faith-based Organizations and Exclusion in European Cities* (Beaumont and Cloke 2012); and (4) *Working Faith* (Cloke et al. 2013), the objective in this *Handbook* is to deliver the much needed theoretical and conceptual clarity and coherence across disciplines. The strength of *Postsecular Cities* (Beaumont and Baker 2011) was the deepening of dialogue between geographers and theologians. This *Handbook* takes a concerted step forward from these previous volumes towards coherence and importantly deals directly with the thorny issue of normativity invoked in debates. Work on postsecularity tends towards a Western- and especially Euro-centrism, and so reaching out to critical scholarship in Russia, India, and Asia on decolonial thought beyond a global development perspective is an important new feature (see Chapters 4, 6, 7, 9, 21, 22, 24, and 29). So is

Figure 1.5 Judith Butler receives the Theodor W. Adorno Award in 2012
Source: Wikipedia Commons.

the linking to debates on the crises of secularism and multiculturalism (see Chapter 20) and the inclusion of voices from non-Christian theistic traditions such as Islam, Judaism, and Buddhism (see Qian 2018; Qian and Kong 2018; see also Chapters 8, 13, 21, and 30).

While the volumes alluded to earlier herald a timely and unprecedented exchange between geographers and several other disciplines, the sought-after crossover narratives remain limited to a few protagonists (Cloke and Beaumont 2013; Williams 2014; Cloke et al. 2019), while the more general reception to the concept remains sceptical and at times even hostile (Joas 2008; Kong 2010; Reder 2010; Wilford 2010; Ley 2011; Beckford 2012). The defiance rests, one might argue, on perceived tensions between otherwise problematic binaries of essentialism and anti-essentialism, universalism and particularism, as well as controversies over religion in a secular age, conflict and consensus in political theory, and the perennial issue of normativity in social science more generally. For instance, one needs to deal directly with tensions between 'postsecularity as stance' versus 'postsecularity as debate'—in other words, to handle the confrontation of normativities in a meaningful way. *The Handbook* critically addresses these tensions and offers pathways out of the impasse for new scholarship.

The new scholarship is beginning to challenge this deadlock, and *The Handbook* dovetails with this growing corpus of work. We now see important scholarship that deal with debates on postsecularity with reference to, for instance, spaces of postsecular rapprochement (Cloke and Beaumont 2013; cf. Bosetti and Eder 2006), Christian youth in Scotland (Olson et al. 2013), postsecular geographies (Williams 2014; Gökarıksel and Secor 2015; Cloke et al. 2019), competition, conflict, and violence in the contemporary world (Kong and Woods 2016), and new cultural anthropological forays into entrepreneurial religion and urban space (Lanz and Oosterbaan 2016). More recent work has called for 'infrasecular geographies' (della Dora 2018), an 'epistemology of religiosity' in urban development (Qian and Kong 2018) and the 'enlightened city' (Beaumont 2018). Now that the tide has started to turn, the time is right for an all-encompassing volume that captures postsecularity philosophically, theoretically, and empirically for all concerned and provides the basis for a new era of human geographical and urban scholarship in the years to come.

Complementarity of discourses and all-unity[8]

From this initial framing of the volume, it would seem, then, that the concept of postsecularity, particularly, but not exclusively, in the Russian context, characterizes one of the main trends of our time. The notion reflects a complementarity of discourses and the admirable but often-misplaced quest for all-unity (see Gutner 2016; see also Chapter 22). Understanding the current events in terms of the dichotomy 'secular-postsecular' initially holds some promise but is problematic on several grounds including the variety of interpretations of postsecularity. The concept of secularity also hardly enjoys a generally accepted interpretation (Calhoun et al. 2011; Zuckerman and Shook 2017; see Chapter 20). We propose, as an initial presupposition, to adhere to the triadic scheme, *presecularity*, *secularity*, *postsecularity*, which has certain heuristic value in the European historical context but is problematic to see this development as a unilinear progression. The discussion is confined to Christianity but provides the basis for exploring the decolonization and all-unity implications of postsecular thought.

The presecular epoch forms the idea of the integral universe, which includes both natural and supernatural realities. The integrality of the universe is determined by the fact that it is created and permanently supported by God and the unity of the universe is perfectly expressed in the image of the cosmic liturgy.[9] Each created thing is a participant of the liturgy because it participates in mystical action of God. The important implication is that this applies to each aspect of human activity. Cognition, social and private life, and political activity are ways to participate in God's action. So, on this basis and tallying with ideas of RO (see Chapters 11–13), science, ethics, and politics are derived from and are implicated in theological principles. Characteristic in this sense is the scientific paradigm adopted during the High and Late Middle Ages and based on Aristotelian metaphysics, which makes the basis of all existence a movement towards the goal (Aristotle 1999). Teleology permeates scientific knowledge. To understand anything in the world means to answer the 'for what?' question. Such an approach presupposes a close connection between science and theology, and the initial assumption of any scientific investigation is the idea of the Divine Mind, within which all the goals lie (see also the section 'Theological perspectives' in this volume). Thomas Aquinas (1989) in *Summa Theologiae* points out that the comprehension of goals in nature is the way from the mind to God. In this sense, science turns out to be a 'servant of theology' since scientific knowledge is a precondition for knowledge of God. For this reason, the study in the theological faculties of the medieval universities was preceded by the study of the liberal arts and philosophy. According to this argument, these domains are only approaches to theology and thus did not have value in their own right.

In continuation of this line of thought, political power was also understood in the context of the integrality of the universe. The monarch receives sanction from God and therefore must protect and support the Christian faith. Secular authority sits in relation to the spiritual in the same way that science does in relation to theology. This authority enjoys a certain independence and solves its own tasks, but ultimately these tasks are derived from others: from those who face the church. The maintenance of order and the protection of law are important primarily because they contribute to the mission of the church. Secularization is an emancipation of these, as well as other, spheres of life. In the course of the seventeenth through eighteenth centuries, science increasingly was the pioneer of the autonomous and self-sufficient development of these multiple spheres of life in what became known as the Age of Enlightenment (Hampson 1990; Israel 2001, 2006, 2011; Edelstein 2010). The Aristotelian paradigm, which had hitherto dominated, presupposed the teleological explanation of things: understanding anything implied knowing the aim (telos) of its existence. Science depends on theology because aims ultimately reside in the Divine Mind; but the new scientific paradigm, however, refused teleological explanation. The understanding of phenomena in this paradigm means knowing aims, not causes. Causal explanation implied understanding nature without appealing to a transcendent creator. After the scientific revolution of the seventeenth century, scientific research became independent from Christian faith, with the latter being the private matter of the scientist. Science, on the contrary, becomes universal in the sense that its principles are the same for Christians, Muslims, Jews, atheists, and other worldviews and belief systems.

Another important sphere of secularization is policy and political philosophy. The first significant step in this direction was made by Thomas Hobbes (2010): in *Leviathan*, he presupposed that the state appeared as a result of a social contract. So, political power—and science for that matter—have human, not divine, origins and as a result one has no or little need for theological justification that is independent of any religion. The state cannot be Christian, Islamic, Judaic, atheistic, and so on, as religious affiliation is the private matter of every citizen, including politicians and state officials. The state, however, is common to everyone.

The same argument applies to morality. This implication might seem strange, not least because morality is usually taken as an important element of religion in the form of moral commandments and religious doctrine. Pierre Bayle, the French philosopher and Calvinist Protestant (Huguenot) of the seventeenth century, tried to demonstrate, in his celebrated *Historical and Critical Dictionary* written in Rotterdam, the Netherlands, and first published in 1697 (Bayle 1991), that morality is founded on the principle of reason and therefore it was unnecessary to include any justification by means of Holy Scripture or religious tradition. As an early advocate of reason and toleration of divergent beliefs, Bayle, a forerunner of the *Encyclopédistes* of the Enlightenment, influenced Immanuel Kant's (1998) *Religion within the Boundaries of Mere Reason* that showed morality cannot be deduced from religion. The autonomy of moral will in Kant's ethic is correlated with the universality of moral law, and it must be the same for both religious and non-religious people (see Chapter 5).

The same could be said about other spheres of human life such as art, culture, economics, and professional activities of various kinds (Burkholder et al. 2006). Most of these spheres, although not all, became autonomous and had their justification only in reason. Such Enlightenment reasoning reduced religion to a specific and private sphere, existing on equal or even less significant terms than other spheres. The integral nature of the world became increasingly under threat, and consequently waned and disappeared. Previously the lives of medieval human beings were united by their faith because all aspects of their lives had an essential religious component. The lives of human beings during the secular epoch became divided between several autonomous spheres. The Enlightenment made an attempt to find a new wholeness or integration of these spheres. The foundation of wholeness, according to the Enlightenment philosophers, must be reason. All aspects of human life must be based on clear and distinctive principles, evident for each reasonable human being to comprehend. The problem was that it was notoriously difficult to ultimately formulate such principles: not even philosophers could agree about the nature and origin of such principles, let alone of their definitive content.

The search for wholeness took an essential place in the minds of Europeans during the nineteenth century and even the first half of the twentieth century. Hope was placed in science and projects developed in a frame of positivism that considered human beings, as well as secular religion, as part of nature (Comte 2009a, 2009b; see also Giddens 1974; Rorty 1982; Petit 2016). As physics, chemistry, and biology provide us objective knowledge of nature, so psychology and sociology would be able to obtain objective knowledge of human life. Ultimately, some unified scientific image of the world was expected to appear. All human problems, it was claimed, would be solved by means of the scientific method. It was against this prevailing positivism, 'scientific socialism', and anarchist political activity of the mid- to late nineteenth century that Raskolnikov of *Crime and Punishment* (Dostoyevsky 1964) 'stepped over' humanity in a proto-Nietzschean, and, for Dostoyevsky, ultimately flawed way.[10] This virulent critique of positivism, its denial of spontaneous emotionality, and the frailties of the human condition exemplifies the ironies and paradoxes of magnificent ideals turned cruelly into evil and self-destructive forces (Epstein 2017).

The failure of these positivist hopes became particularly apparent after the Great War (1914–1918) and the Russian Revolution (1917). The crisis of European civilization that ensued reflected, for the Russian philosopher Nikolai Berdyaev (2009), the failed projects of modernity—and their closely related humanism—marked the beginning of a new Middle Age. This epoch was one of darkness and the return to the mysterious life of the spirit. The transcendental phenomenologist Edmund Husserl described the period as the crisis of European science (Husserl 1970). New attempts to find wholeness resulted in monstrous consequences: totalitarian regimes in Germany and the Soviet Union. These movements tried to create absolute wholeness (totality) on the basis of ideology, the latter a peculiar hybrid of scientific theory and religious doctrine. Nevertheless, it appeared quite effective for uniting the masses, at least for a time, by means of Nazism and Stalinism.[11] The disastrous injustices and evils of totalitarianism in these countries, as elsewhere, are evidence of the limits of secular attempts to create a measure of wholeness. History ultimately demonstrated that the price of wholeness was the destruction of human beings and everything human. The fact that the origin of totalitarianism lies in the Enlightenment project—in other words, in the attempt of secular reason to constitute its own universal integrity—was first disclosed by Theodor Adorno and Max Horkheimer (2002) (see Figure 1.6). Their analysis of the idea of Enlightenment demonstrates that its imperialism, Eurocentrism, and scientism contained the seeds of its demise. Other commentators disagree that Enlightenment equates with totalitarianism (Bronner 2006). Later, Jean-François Lyotard and other postmodernists described the limits of the Enlightenment in quite different terms and without apparent reference to totalitarianism (Lyotard 1984). He described the end of metanarratives as the core idea of the new postmodern epoch. A metanarrative, according to Lyotard, can be interpreted as an attempt by secular reason to legitimate scientific knowledge and ultimately itself. According to Lyotard, the failure of this attempt characterizes the postmodern epoch. For Habermas, postmodern thought in turn totalizes and thus fails to differentiate phenomena in modern society (Ritzer and Stepnisky 2013). Postmodern thought has retreated since the 1990s with

Figure 1.6 Heidelberg, April 1964, at the Max Weber-Soziologentag. Horkheimer is front left, Adorno, front right

Source: Wikipedia Commons: photo: Jeremy J. Shapiro.

the ascendency of counter-tendencies such trans-postmodernism (Epstein et al. 1999) and metamodernism (Vermeulen and van den Akker 2010).

The presupposition that the postmodern implies or at least implicates postsecularity is quite understandable. These concepts are, of course, different and not synonymous. No secular image of universal integrity, such as a unified scientific worldview, totalitarian ideology, or metanarrative, can now be adopted, at least not as before. In such a situation, the meaning of religion must change in comparison with the previous epoch. The previous era was a secular or modern one, but the changing meaning and value of religion cannot reflect an uncritical return to a presecular situation devoid of normative ambitions. The advent of postsecularity, at least of the debate around the alleged challenges to the hegemony of modes of secularity, reflects this unhinging of secular universal integrity and the quest for new accounts of differentiated and radically diverse, yet globally interconnected, expressions or discourses of human integrity. One can either continue the path of absolute heterogeneity as set out by the postmodern challenge, or one can attempt a new global interconnectivity, unity in diversity, or 'radical difference' across all outlooks, value systems, and social identities (Beaumont 2018; see also Chapter 27).

The postsecular attitude is not initially inherent in the Christian or any other religious belief system or worldview, or in the scientific community. The outlook or sensibility of the postsecular develops gradually in the course of benevolent and mutually beneficial interaction, with mutual interest, goodwill, and a desire for authentic collectivity at stake. In this way, the complementarity of discourses emerges. Postsecularity is, to a certain degree, paradoxical. On the one hand, it reflects the reality of our time and clearly replaces or certainly unsettles modes of secularity, just as postmodernity did over questions of modernity. The impossibility of metanarratives clearly implies postsecularity (Mrowczynski-Van Allen et al. 2016; see also Antonov 2016; Gutner 2016). On the other hand, postsecular sensibility does not emerge spontaneously. Intriguingly reminiscent of Habermas, one could say that it requires the political and ethical will of diverse communities and bearers of otherwise opposing discourses in the search for genuine equality of interaction, intersubjective dialogue, and complementarity.

Confrontation of normativities and empirical challenge

We have never been modern, as Latour once said (see also Chapter 2) (see Figure 1.7). That we have never been secular, holds equally. Yet, the notion of postmodernity as well as that of postsecularity is premised on the assumption that we have gone through something like modernity with respect to secularity. But one day we might become modern and secular. Modernity and secularity are 'unfinished projects'; postmodernity and postsecularity can be seen intermediate steps towards finally achieving these goals. These intermediate phases produce situations in which modernizers and anti-modernizers have to coexist, where secularists and non-secularists have to relate to each other and maybe even learn from each other. Yet, in the long run, the modern and the secular will win over the non-modern and the non-secular. The reciprocal recognition and tolerance of secularists and non-secularists is, in this story and arguably to its detriment, biased towards the modern and the secular. It is an asymmetric relation that in this modernist perspective will dissolve over the course of time.

This narrative is one which not only the disciples of the social sciences have believed for two centuries, but also the people who consider themselves modern and secular. This depiction is so well established that it can be said to belong to the background beliefs of the modern world. There has been much criticism that argues that this story is a Western one,

Figure 1.7 Bruno Latour, French anthropologist and philosopher, at a conference on the Anthropocene

Source: Wikipedia Commons: photo: Gérald Garitan.

and that non-Western cultures do not share such background beliefs. This criticism has been answered by claiming there are different paths towards modernity (Eisenstadt 2000)—and therefore also, by implication, different paths to secularity ('multiple secularities') such as through Muslim secularity. Revitalized religion in modern society, on this view, reflects a European exceptionalism and common reality elsewhere in the world (see Eder 2002; Wohlrab-Sahr and Burchardt 2012; Burchardt and Wohlrab-Sahr 2013; Burchardt et al. 2015; see also Chapters 4, 6, 7, 8, 9, 13, 21, 22, 24, and 29). Yet, what tends to come out ahead in the end, time and time again, are modernity and secularism; the telos of the story.

One way to undermine such narratives is to be empirical and check what secularism is about on the ground. Secularism has institutional aspects: for example, the appropriation of church property by the state and the dismantling of church power. Secularism also has cognitive aspects: emphasizing the autonomy of people from religious prescriptions, thus freeing the critical capacities of people and turning them into autonomous individuals. Secularism finally has a site where the critical capacities of human beings can be practised: in other words, the public sphere of critical debate. Empirically, none of the processes have come to an end. Church property not only survived in Western nation-states, it even outlived decades of communism in the Soviet Union. Religious values and norms still dominate the lives of the majority of people, even those living in allegedly the most secularized countries such as those of Scandinavia. The public sphere is permeated by religious symbols, whether in state rituals or in using religious symbols in public buildings. Secularism seems to be an unfinishable project. It is cognitively resisted by the majority of people who still look for the religious specialist in the critical situations of life: birth, marriage, and death. Secularism also lives in the public rituals of handing over power to others—to open a new parliament, to mourn those losing their lives from terrorist violence. The list goes on.

Herein lies the empirical basis to the thesis of postsecularism. The simple version is that religion returns to the public sphere of secular society and that the project of secularism, based on the rule of law and on a universalist and rationalist ideology, continues to wane. Religion returns, as growing Pentecostalist, Islamist, and Buddhist movements show (Kepel 2002; Jondhale and Beltz 2004; Miller and Yamamori 2007; Aslan 2009). It also returns in the form of new religious wars and in the form of a growing variety of privatized forms of religion, fostering the diversification of sectarian movements into increasingly smaller units of believers. Religion also returns as a renaissance of traditional religious symbols and institutions. The recent order by the predominately Catholic German state of Bavaria shows how this 'renaissance' can invoke controversy. All government buildings, not just public schools and courtrooms, must now, as a matter of law, adorn their entrances with Christian crucifixes. The state law which undermines multicultural values has generated allegations of identity politics against far-right and Islamophobic parties in the run-up to federal state elections in October 2018 (BBC News 2018; Bennhold 2018; Posener 2018). Even national states such as Russia re-enthrone traditional religion, Orthodoxy, as a pillar of Russian society (see Chapters 6 and 22). Some conservative interpreters have taken these observations as signs of an end of secularism, an end that opens the door to the revival of presecular society. On this view, the postsecularism thesis turns into a 'presecularism thesis'. Yet, there is another and more radical and progressive version of the postsecularism thesis based on these and other empirical observations.

This radical version claims that a social world emerges where differences between people increase that belief in increasingly different things (see Chapter 27)—'radical difference', as one of us calls it (Beaumont 2018). People do so in a changing spatial context: at home, in the workplace, in the public sphere, in everyday life beyond the home, and in social networks.

They do so over time in a way that makes discontinuity normal or at least in a way that gives the appearance of normalcy. The notion of postsecularism reflects a world that produces many different people with dissimilar beliefs spread out over space and time and united in a higher perspective or reality. This leads to the central question provoked by the postsecularism thesis: how actually do people with different beliefs, values, and worldviews relate to each other in space and time?

Spatially, the question is where we find or can locate the sites of postsecular practices. The classic argument has been that the city is the realm of the secular, and the countryside the realm of the religious. It is in fact easier to live a secular life in the city than it is in the countryside. Yet, this binary no longer works, especially when taken in light of new developments in urban theory that challenge this Wirthian separation (Brenner 2016; Brenner and Katsikis 2018). The city and the countryside offer equally the tension underlying the secular living together with the non-secular: the countryside as the quasi-religious locus of salvation for urban dwellers and the urban village in the city quarters as a quasi-religious community, raising the issues of secular individualism and the romantic search for belonging to an authentic community. Adding the global aspect, the spatial dimension multiplies the sites that cross-cut established notions of being spatially situated. *Temporally*, secular and religious beliefs are no longer stable beliefs that can be attributed to an individual or a group. People change their beliefs over the course of their life. What people do depends on whether they are in the early or late phases of their life course—whether they are part of a generation that marks discontinuities with former generations. We have to take into account sequential effects that undermine the notion of a path towards a telos (be it religious renaissance or secular culture). Contrary to the triadic heuristic referred to earlier, postsecularity is more cyclical or dialectical than linear.

Postsecular society—our first conclusion—is a social world in which differences and boundaries between beliefs and values abound but critically also where attempts are made at their reconciliation. Analyzing this multiplicity, the traditional conceptual tools for making differences visible turn out to be inadequate. Nations, ethnic groups, religions, urban and rural inhabitants, classes, and so on no longer suffice for grasping the multiple differences and boundaries across sites and across time. Even concepts like scaling up or down a boundary (scaling up a region to a nation and then to a civilization, or scaling down a nation to a region and then to a city and finally to a neighbourhood) are no longer analytically sufficient strategies for grasping the diversity of sites. We have to abstract from such taken-for-granted units ('things'). Scrutinizing the structure of social relations emerging from the multiplicity of sites (ranging from individuals to social networks) and the multiplicity of points in time needs devices that make the network structure of social relations visible. These relations transcend the 'natural' boundaries of ethnic groups, classes, states, or families. The structures of social relations can no longer be read from such 'naturalized' social entities. A non-naturalistic way to make them visible is to analyze them as relations between individuals that occupy structurally equivalent positions in networks (see Smith 2007; White 2008; Eder 2011; Forchtner et al. 2018).

This interpretation of postsecularism requires the relinquishing of the idea that there are entities like 'modern' or 'secular' society. These entities exist as concepts, and as concepts they even circulate well in networks of social relations, certainly among Western liberal intellectuals, but also among people being proud of having cut the bonds with religious authorities and beliefs. Thinking about postsecular society forces us to give up not only the idea of natural groups but also to renounce the habit of attributing to these groups meanings that are either secular or non-secular and religious. Not only the variation of social structures, but also the variation of the meanings of secular and non-secular requires rethinking society as one with diversifying networks through which these differentiating beliefs circulate. What follows is whether the secular self-understanding—and its antithesis, the religious re-enchantment of the world—is just one of the meanings that increasingly circulate through networks of social relations. What happens to secular individualism anchored in liberal society and to the national community that served as its container? Does the unsettling of the secular and the postsecular produce a society of 'disillusioned' and 'insecure' people? Does it produce a society with disintegrating and moreover inauthentic communities? Or, is postsecular society a catchword for evolving networks of social relations through which secular and religious (and other) stories or narratives coalesce and circulate simultaneously?

Some prominent thinkers of postsecularity have already taken up the idea of the simultaneous circulation of different beliefs in society but with different normative orientations. 'Democratic secularism', for the German constitutional judge and liberal conservative political theorist Ernst-Wolgang Böckenförde (1976), presents obstacles for the development of social capital. He questions the ethical shaping force of the secular constitutional state. Jürgen Habermas (2008), as we have seen in detail earlier (see Chapters 4, 5, 8, and 9), has focused on 'postsecular society' as reciprocal learning processes between believers and non-believers. Both reveal versions of a secular position in public debate under the stress of the 'return of religion'. The public sphere is the haven for secularity repaired by debating with non-secular believers, and the secular state is its guarantee. This account, however, provides only half the story. Modern secular society has a second site, the private sphere, which is the haven for religious and non-religious sentiments alike and at least relativizes the exclusiveness of a secular principle of judging the common good. The two spheres, the private and the public, follow two different principles of judgement, a secular one (in the public sphere) and a non-secular one (in the private). This separation guaranteed the dominance of the secular principle as long as there was a clear separation and hierarchical organization of these spheres—which in fact was never the case and increasingly no longer is. If two principles of judgement of the common good coexist in modern society, we cannot exclude the possibility of further principles, such as principles of judgement that can be derived from the market, the arts, or the industrial and postindustrial worlds of work. Postmodern society, we would claim, is a society where the structure of social relations allows the circulation of different principles of the common good.

Postsecular society, then, is a subcase of this observation: a society where secular and non-secular principles—beyond others—circulate simultaneously through social relations. This observation creates a particular opportunity and a particular risk for postsecular society. The opportunity is that people learn to switch between different principles of judgement when no consensus on a dispute relative to a principle of judgment is possible. By switching sites—not only from public to private, but also to new emerging sites such as social networks—people can move the dispute to other principles of judgement. If this does not work, then there is the choice of tolerating reciprocal criticism and waiting for further debate; it is the solution of permanent dialogue. Another answer would be to compromise between competing principles, for example, secular and religious criteria, a situation that runs the risk of being denounced as betraying both sides in the dispute. When there is no agreement or compromise, two solutions remain: forgiving and forgetting, or war and violence. Whatever happens and how traumatic collective experience might become, social processes will continue and produce situations in which people have to restart disputes and re-enter processes of reconciliation and negotiation with each other.

There is a second conclusion from the postsecularism thesis: the issue of taking a 'postsecular perspective' as a social scientist observing this postsecular world. The postsecular observer has a critical stance towards this world, and she can do that from a secular or a non-secular perspective. If things go wrong in the secular or religious world, this observer criticizes what is going wrong as any other social actor does. In the case of strong and unbridgeable dissent, observers of social science normally do not forgive nor do they fight. Instead, they are specialized in fostering critical capacities. This means to expose the social scientists not only to the critical capacities of their colleagues, but also to the critical capacities of the many others who inhabit postsecular society. The postsecular position forces us to recognize that there is no ultimate solution, final consensus, or telos whether defined in religious or secular terms. Postsecularism dethrones any claim of normative superiority by religious or secular criteria in judging the common good. It enthrones permanent and critical debate as the only solution left.

Overview of *The Handbook*

Following this Introduction, *The Handbook* is organized around four thematic sections: (I) Philosophical meditations;[12] (II) Theological perspectives; (III) Theory, space, social relations; and finally, (IV) Politics and social engagement. An Afterword completes the collection. Contributing authors include annotated further reading suggestions as well as an extensive list of references for the avid reader to explore further.

PART I: PHILOSOPHICAL MEDITATIONS presents nine chapters that discuss postsecularity from a philosophical standpoint. *Bruno Latour* critiques social explanations of religion, calling for attention to a plurality of modes of existence as an alternative to violence. *Martin Beck Matuštík* addresses the difficulty of unforgiving acts of evil and experiences of trauma from the perspective of redemptive critical theory and the killing fields of Cambodia. *Eduardo Mendieta* discusses Jürgen Habermas' oeuvre, particularly the postsecular condition and the genealogy of postmetaphysical thinking. *Warren S. Goldstein* continues in this vein showing that Habermas engages a soft, redemptive criticism of religion, one that falls short of evaluation and negation typically associated with critique. *Mikhail Epstein* embraces postatheism and minimal religion in Russia, locating the postsecular in liminal spaces that transcend both worldly immanence and all religious dogmas, rituals, and affiliations. *Herbert de Vriese* and *Guido Vanheeswijck* discuss the performative force of the postsecular, revealing deeper motives and arguments behind the rhetorical gesture of the term. *Michiel Leezenberg* argues that Habermas and Asad deny that recent forms of violent radicalism, like salafi-jihadism, have anything to do with religion, religious claims, or traditions as reasonable. *Manav Ratti* presents arguments for five interrelated facets of postsecularism, drawing upon examples primarily from postcolonial and literary studies. Finally, *David Lewin* discusses formations of the postsecular in education, with particular reference to the challenges posed to value-neutrality of secularism.

PART II: THEOLOGICAL PERSPECTIVES contains nine contributions from a variety of theological perspectives on postsecularity. *Matt Bullimore* introduces the theological sensibility of RO, its understanding of the postsecular, and the perceived dangers and opportunities including a vision of a transformative Christian socialism. Equally concerned with RO, *Dritëro Demjaha* argues that the postsecular also entails the postmodern and the postliberal with political implications. *Hagar Lahav* explores postsecular theologies, through post-Christian and Jewish theologies, as fluid and dynamic re-enchantments of the secular world and warns against a backlash on humanistic values. *Francis Schüssler Fiorenza* explores the challenges of postsecularity for political theologies, namely privatization and detraditionalization, emergence of new spaces, and religious diversity and pluralism. *Robert Joustra* engages with the work of Abraham Kuyper and Herbert Butterfield, two Augustinian thinkers, providing a causal and constitutive framework for global politics. *Christopher Rowland* recovers the secular in the biblical tradition through the secular theologies of Gerrard Winstanley and William Blake, pointing to affinities with liberation theology. *Rafael Luciani* addresses key concepts of the theology of the people, a branch of Latin American liberation theology, to help understand Pope Francis' theological and pastoral options. *Elaine Graham* interrogates the postsecular, showing how its unprecedented nature signals the contradictions inherent in the renewed presence of faith in public life, alongside continuing opposition to religion as a source of legitimate public discourse. Finally, *Chris Shannahan* argues for a more holistic approach to postsecularity based on urban liberation theology, cases of urban youth exclusion, and faith-based community organizing.

PART III: THEORY, SPACE, SOCIAL RELATIONS presents 12 chapters from geography, planning, sociology, and religious studies that delve into theoretically informed empirical implications of postsecularity. *Gregor McLennan* critically engages with what he calls expansive secularism via Ledowitz, Braidotti, Critchley, and Unger, drawing attention to their philosophical naturalism and historicism. *Lily Kong* and *Junxi Qian* critique Western- and Euro-centric discourses on postsecularity, arguing for temporal and spatial specificities through cases in China, Singapore, and Hong Kong. *Kristina Stoeckl* and *Dmitry Uzlaner* identify four genealogies of postsecularity: a sociological, normative, postmodern, and theological genealogy. *Umut Parmaksız* identifies thin normativities around solidarity within postsecular scholarship, arguing instead for an alternative hermeneutical approach that can lead to a more substantive affirmation of religion's truth-content. *Robbie Goh* examines the effect of diaspora and transnationalism on Christianity in Asia, focusing on the Indian diaspora that reinforces certain evangelical and fundamental traits. *Christopher Baker* shows how the postsecular describes an operant transcendent tradition that subverts and complexifies the normative, often neo-Marxist denial of the role of religion and belief in modernity. *Kryzsztof Nawatrek* proposes radicalized postsecularism as an ontology of contemporary architecture, shaped by the fundamental tension between spatial hierarchy, Bill Hillier's configurational theory, and emancipatory and egalitarian ambitions. *Giuseppe Carta* contends that postsecular urbanism should conceive the urban as the multiplicity of incommensurable specificities, apophatic principles, and imagination. *Peter Nynäs* argues when dealing with postsecularity from a Habermasian or other perspective, one needs to critically rethink how we conceptualize religious identity and subjectivity. *Kasia Narkowicz* and *Richard Phillips* problematize the postsecular with reference to British Muslims of Pakistani heritage and Polish Muslims with mixed heritage where religion never privatized. *Phra Nicholas Thanissaro* tests the notion of postsecularity in empirical terms with reference to Buddhist teenagers in Britain, alluding to the relationship of postsecularity to Buddhism in the West. Finally, *Edward Wigley* focuses on how mobilities, flow, and movement contribute to a personal sense of religious experience, identity, and subjective spiritual geographies in Western urban secular contexts.

PART IV: POLITICAL AND SOCIAL ENGAGEMENT contains three contributions that emphasize the political and social implications of the debate. *Timothy Stacey* takes a normative stance on the potential benefits of postsecularity for re-enchanting and thus reinvigorating secular people and politics through 'performative postsecularism'. *Øyvind Strømmen* examines the use of religious imagery—Christian and Pagan—by extra-parliamentary far-right groups in Nordic countries, invoking conceptions of Self and Other in debates on postsecularity. Finally, *Roger Speare* argues that Abbé Pierre was a prophet and pioneer of postsecular thinking and social action when he founded Emmaus International, the solidarity movement to combat poverty and homelessness.

In the Afterword, *Eduardo Mendieta* and *Justin Beaumont* reflect on the notion of reflexive secularization across the volume as a whole, alluding to new lines of inquiry.

Conclusion

The Handbook advocates interdisciplinary and reflexive engagement that embraces debates on the postsecular openly and critically. The collection acknowledges antagonistic processes unleashed by the secularization of the state, on the one hand, and respect for the right to one's faith on the other. Conflicts as well as solidarities emerge from this complementarity of discourses, the inevitable confrontation of normativities, and new empirical challenges. We do not argue that we should all become 'postsecularists' in a card-carrying or ideological sense, but rather for the value in adopting postsecular sensibility, consciousness, or ethos. At stake is the question whether postsecularity is really any different from 'complex secularity'. Addressing that question requires reasoned, balanced, and critical intellectual debate. *The Handbook* offers a powerful set of arguments from a variety of standpoints to do so.

Notes

1 See Adorno (1976: 113).
2 See Tagore (1994; 2008: 712).
3 We are grateful to Warren S. Goldstein, Gregor McLennan, Christopher Rowland, Dmitry Ulanzer, and Kristina Stoeckl who commented on an earlier draft of this introduction. The responsibility for the final version, or course, remains our own.
4 For more on Nesterov's life and work, see Mikhailov (1958) and Nikonova (1962). See Louth (2015) for biographical sketches on a range of Orthodox thinkers and their intellectual legacies.
5 The chapter by Konstantin Antonov (2016) discusses the legacy of Vladimir Sergaevitch Soloviev, the philosopher, theologian, and poet who inspired Dostoyevsky's *The Brothers Karamazov* and other classics. He authored *A Story of the Anti-Christ* (Soloviev 1900, 2012). Note that recent scholarship on religion and literature draws attention to the postsecular (see McClure 2007; Branch 2016; Knight 2016).
6 Nietzsche is infamous for his misogynistic dislike for women, his scathing contempt for the impoverished (in his words, the 'bungled and the botched') in favour of the aristocratism of the superman, and his bitter critique of Christianity. We do not invoke him in this context. Rather, we share Adorno and Horkheimer's interest in Nietzsche's totality of happiness and critique of the entire bourgeois culture of society. Beneath the veneer of misanthropic torment, they see in Nietzsche—and concur—with his disdain for passive, self-avoiding, and conformist acquiescence in the *status quo*.
7 The Brill volume was the outcome of a conference one of us (Justin Beaumont) organized at the University of Groningen, the Netherlands. The conference, *Religion, politics, and the postsecular city*, was held at the *Doopsgezindekerk Groningen* ('Groningen Mennonite Church') in November 2008, and brought together philosophers, theologians, sociologists, and geographers, including world-leading scholars Paul Cloke, Gregor McLennan, and Ed Soja. The world-renowned sociology of religion scholar, Jim Beckford presented at the conference and subsequently had published a searing critique of the postsecular (Beckford 2012). Arie L. Molendijk, a theologian and co-organizer of the conference, reviews the uses and misuses of the 'postecular', noting that religious actors claim a role for religion in public and secular domains (Molendijk 2015).
8 This section draws upon the first draft of a contribution from Grigory Gutner (1960–2018), formerly of the Institute of Philosophy, Russian Academy of Sciences, and Saint Philaret Christian Orthodox Institute (SFI), the first independent theology school in Soviet Russia. Grigory sadly passed away in February 2018 during the completion of *The Handbook*. The volume is dedicated to his memory. See also Gutner (2016).
9 'Cosmic liturgy' refers to theology of synthesis in terms of Rome and Byzantium, of antiquity and the Middle Ages, for example, with Maximus the Confessor, the monk, theologian, and scholar in Early Christianity (see von Balthasar 2003).

10 Raskolnikov, an impoverished ex-law student in nineteenth-century St. Petersburg, hatches a plot—which he also carries out—to murder a mean and unscrupulous pawnbroker in the name of the greater good of humanity as a whole. He believes that certain crimes are admissible, if not in actual law then within evolutionary law, when committed by certain 'higher types'. Powerful historical figures like Napoleon embody such higher morality, he claims, rather than the ordinary masses. The split between an instinctive moral universe (Russian Orthodox, for Dostoyevsky) and the principles of rationality, utilitarianism, and ultimately nihilism imported from the West allows the protagonist to commit the deadly deed, but at the price of deep, authentic, and spiritually rooted humanity.

11 Dostoyevsky was mysteriously prophetic of the advent of twentieth-century totalitarianisms. *The Legend of the Grand Inquisitor*, a 'poem' on human nature and freedom in *The Brothers Karamazov* (Dostoyevsky 1982), has Jesus visit Seville, Spain, at the time of the Inquisition; he is immediately picked up and sentenced to burn at the stake the next day. The narrative concerns a 'dialogue' between the captive Jesus and the Inquisitor. The latter explains that in offering humanity freedom, Jesus had to be executed because of his obstruction to the Church's historic mission.

12 Martin Beck Matuštík's (2008) *Radical Evil and Scarcity of Hope* inspired the title of this section. Drawing from this work, Matuštík writes passionately in this volume about the difficulty of unforgiving.

References

Aquinas, T. (1989) *Summa Theologiae: a concise translation*, edited by Timothy McDermott, London: Eyre and Spottiswoode.

Adorno, T. W. (1976) 'Alienated masterpiece: the *Missa Solemnis* (1959)', *Telos*, 28(Summer): 113–24.

——— (2002) *Essays on Music*, edited by Richard Leppert, trans. Susan H. Gillespie, Berkeley: University of California Press.

Adorno T. W. and M. Horkheimer (2002) *Dialectic of Enlightenment*, Stanford, CA: Stanford University Press.

Allen, A. and E. Mendieta (eds.) (2019) *The Cambridge Habermas Lexicon*, Cambridge: Cambridge University Press.

Antonov, K. (2016) '"Secularisation" and "post-secular" in Russian religious thought: main features'. In Mrowczynski-Van Allen, A., Obolevitch, T. and P. Rojek (eds.) *Beyond Modernity: Russian religious philosophy and post-secularism*, Eugene, OR: Wipf and Stock, pp. 25–38.

Aristotle (1999) *Aristotle's Metaphysics*, a new translation by Joe Sachs, Santa Fe, NM: Green Lion Press.

Asad, T. (2003) *Formations of the Secular: Christianity, Islam, modernity*, Stanford, CA: Stanford University Press.

Aslan, R. (2009) *Global Jihadism: a transnational social movement*, Saarbrucken: VDM Verlag.

Balthasar, H. U. von (2003) *Cosmic Liturgy: the universe according to Maximus the Confessor*, San Francisco, CA: Ignatius Press.

Bayle, P. (1991) *Historical and Critical Dictionary: selections*, trans. Richard H. Popkin, Indianapolis, IN: Hackett.

Beaumont, J. (2018) 'The enlightened city', submitted to *Political Geography*.

Beaumont, J. and C. Baker (eds.) (2011) *Postsecular Cities: space, theory and practice*, London: Continuum.

Beaumont, J. and P. Cloke (eds.) (2012) *Faith-based Organizations and Exclusion in European Cities*, Bristol: The Policy Press.

Beckford, J. A. (2012) 'Public religions and the postsecular: critical reflections', *Journal for the Scientific Study of Religion*, 51(1): 1–19.

Bennhold, K. (2018) 'Crosses go up in public offices: it's culture, Bavaria says, not religion', *New York Times*, 30-05-2018, accessed online 09-06-2018.

Berdyaev, N. (2009) *The End of Our Time*, English edition, trans. Donald Attwater with a foreword by Boris Jakim, New York, NY: Semantron Press.

Blond, P. (1997) *Post-Secular Philosophy: between philosophy and theology*, London/New York: Routledge.

Braidotti, R. (2008) 'In spite of the times: the postsecular turn in feminism'. *Theory, Culture & Society*, 25(6): 1–24.

Branch, L. (2016) 'Postsecular studies'. In Knight, M. (ed.) *The Routledge Companion to Literature and Religion*, London/New York, NY: Routledge.
Brenner, N. (2016) *Critique of Urbanization*, Basel: Birkhäuser Verlag.
Brenner, N. and N. Katsikis (eds.) (2018) *Is the World Urban?* Barcelona: Actar.
BBC News (2018) 'Germany's Bavaria orders Christian crosses in all state buildings', *BBC News*, 25-04-2018, accessed online 09-06-2018.
Böckenförde, E.-W. (1976) *Staat, Gesellschaft, Freiheit*, Frankfurt: Suhrkamp.
Bosetti, G. and K. Eder (2006) 'Post-secularism: a return to the public sphere', interview with G. Bosetti, *Eurozine*, 17 August. www.eurozine.com/post-secularism-a-return-to-the-public-sphere
Bronner, S. E. (2006) *Reclaiming the Enlightenment: toward a politics of radical engagement*, New York, NY: Columbia University Press.
Burchardt, M. and M. Wohlrab-Sahr (2013) 'Multiple secularities: religion and modernity in the Global Age—Introduction', *International Sociology*, 28(6): 605–11.
Burchardt, M., Wohlrab-Sahr, M. and M. Middell (eds.) (2015) *Multiple Secularities Beyond the West: religion and modernity in the global age*, Boston, MA: de Gruyter.
Burkholder, J. P., Grout, D. J. and C. V. Palisca (2006) *A History of Western Music,* 7th Edition, New York, NY: W.W. Norton & Company, Inc.
Butler, J. (2008) 'Sexual politics, torture, and secular time', *British Journal of Sociology*, 59(1): 1–23.
Butler, J., Habermas, J., Taylor, C. and C. West (2011) *The Power of Religion in the Public Sphere*, edited by Eduardo Mendieta and Jonathan VanAntwerpen, afterword by Craig Calhoun, New York, NY: Columbia University Press.
Calhoun, C., Juergensmeyer, M. and J. VanAntwerpen (eds.) (2011) *Rethinking Secularism*, 1st edition, Oxford: Oxford University Press.
Casanova, J. (1994) *Public Religions in the Modern World*, Chicago, IL: University of Chicago Press.
Cloke, P. and J. Beaumont (2013) 'Geographies of postsecular rapprochement in the city', *Progress in Human Geography*, 37(1): 27–51.
Cloke, P., Beaumont, J. and A. Williams (eds.) (2013) *Working Faith: faith-based organizations and urban social justice*, Milton Keynes/Carlisle: Paternoster Press.
Cloke, P., Baker, C., Sutherland, C. and A. Williams (2019) *Geographies of Postsecularity: re-envisioning politics, subjectivity and ethics*, London/New York: Routledge.
Comte, A. (1865; 2009a) *A General View of Positivism*, trans. J. H. Bridges, Cambridge: Cambridge University Press.
——— (1891; 2009b) *The Catechism of Positive Religion*, trans. R. Congrev, Cambridge: Cambridge University Press.
Dora, della V. (2018) 'Infrasecular geographies: making, unmaking and remaking sacred space', *Progress in Human Geography*, 42(1): 44–71.
Dostoyevsky, F. M. (1964) *Crime and Punishment*, trans. Jessie Coulson, including background and sources, essays in criticism, edited by George Gibian, 3rd edition, New York, NY/London: W. W. Norton & Company.
——— (1982) *The Brothers Karamazov*, translated with an introduction by David Magarshack, London: Penguin Books.
Edelstein, D. (2010) *The Enlightenment: a genealogy*, Chicago, IL: University of Chicago Press.
Eder, K. (2002) 'Europäische säkularisierung: ein sonderweg in die postsäkulare sesellschaft?', *Berliner Journal für Soziologie*, 12: 331–44.
——— (2011) 'Europe as a narrative network: taking the social embeddedness of identity constructions seriously'. In Lucarelli, S., Cerutti, F. and V. A. Schmidt (eds.) *Debating Political Identity and Legitimacy in the European Union: interdisciplinary views*, London: Routledge, pp. 38–54.
Epstein, M. (2011) *Russian Spirituality and the Secularization of Culture*, New York, NY: FrancTireur-USA.
——— (2017) *The Irony of the Ideal: paradoxes of Russian literature*, Boston, MA: Academic Studies Press.
Epstein, M., Genis, A. and S. Vladiv-Glover (1999) *Russian Postmodernism: new perspectives on post-Soviet culture*, New York, NY: Berghahn Books.
Eisenstadt, S. N. (2000) 'Multiple modernities', *Daedalus*, 129(1): 1–29.
Forchtner, B., Engelken-Jorge, M. and K. Eder (2018) *Towards a Revised Theory of Collective Learning Processes: argumentation, narrative and the making of the social bond*, unpulished manuscript, Leeds-Bilbao-Berlin.
Giddens, A. (1974) *Positivism and Sociology*, London: Heinemann.
Gokarıksel, B. and A. J. Secor (2015) 'Postsecular geographies and the problem of pluralism: religion and everyday life in Istanbul, Turkey', *Political Geography*, 46: 21–30.

Graham, E. (2012) 'What's missing? Gender, reason and the post-secular', *Political Theology*, 13(2): 233–45.
Greed, C. (2011) 'A feminist critique of the post-secular city: god and gender'. In Beaumont, J. and C. Baker (eds.) *Postsecular Cities: space, theory and practice*, London: Continuum.
Gutner, G. (2016) 'Post-secularity vs. all-unity'. In Mrowczynski-Van Allen, A., Obolevitch, T. and P. Rojek (eds.) *Beyond Modernity: Russian religious philosophy and post-secularism*, Eugene, OR: Wipf and Stock, pp. 39–47.
Habermas, J. (2008) 'Secularism's crisis of faith: notes on post-secular society', *New Perspectives Quarterly*, 25: 17–29.
Habermas, J. and J. Ratzinger (2007) *The Dialectics of Secularization: on reason and religion*, San Franscisco, CA: Ignatius Press.
Hampson, N. (1990) *The Enlightenment*, new edition, London: Penguin.
Hobbes, T. (2010) *Leviathan: or the matter, forme, and power of a common-wealth ecclesiasticall and civill*, edited by Ian Shapiro, New Haven, CT: Yale University Press.
Husserl, E. (1970) *The Crisis of European Sciences and Transcendental Phenomenology: an introduction to phenomenological philosophy*, Evanston, IL: Northwestern University Press.
Israel, J. I. (2001) *Radical Enlightenment: philosophy and the making of modernity 1650–1750*, new edition, Oxford: Oxford University Press.
——— (2006) *Enlightenment Contested*, Oxford: Oxford University Press.
——— (2011) *Democratic Enlightenment: philosophy, revolution, and Human Rights 1750–1790*, Oxford: Oxford University Press.
Joas, H. (2008) *Do We Need Religion?*, trans. A. Skinner, Boulder, CO: Paradigm.
Jondhale, S. and J. Beltz (eds.) (2004) *Reconstructing the World: B.R. Ambedkar and Buddhism in India*, Oxford: Oxford University Press
Kant, I. (1998) *Religion within the Boundaries of Mere Reason: and other writings*, trans. Allen Wood and George di Giovanni, with an introduction by Robert Merrihew Adams, Cambridge: Cambridge University Press.
Kepel, G. (2002) *Jihad: the trail of political Islam*, trans. Anthony F. Roberts, 1st edition, Cambridge, MA: Belknap Press of Harvard University Press.
Kien, J. (2000) *Reinstating the Divine Woman in Judaism*, Parkland, FL: Universal Publishers.
Knight, M. (ed.) (2016) *The Routledge Companion to Literature and Religion*, London/New York, NY: Routledge.
Kong, L. (2010) 'Global shifts, theoretical shifts: changing geographies of religion', *Progress in Human Geography*, 34(6): 755–76.
Kong, L. and O. Woods (2016) *Religion and Space: competition, conflict and violence in the contemporary world*, London/New York, NY: Bloomsbury.
Lanz, S. and M. Oosterbaan (2016) 'Symposium: entrepreneurial religion in the age of neoliberal urbanism', *International Journal of Urban and Regional Research*, 40(3): 487–506.
Ley, D. (2011) 'Preface: towards the postsecular city?'. In Beaumont, J. and C. Baker (eds.) *Postsecular Cities: space, theory and practice*, London/New York, NY: Continuum.
Louth, A. (2015) *Modern Orthodox Thinkers: from the philokalia to the present*, Downers Grove, IL: IVP Academic.
Lyotard, J.-F. (1984) *The Postmodern Condition: a report on knowledge*, Minneapolis, MN: University of Minnesota Press.
Matuštík, M. B. (2008) *Radical Evil and the Scarcity of Hope: postsecular meditations*, Bloomington: Indiana University Press.
McClure, J. (2007) *Partial Faiths: postsecular fiction in the age of Pynchon and Morrison*, Athens: University of Georgia Press.
Mikhailov, A. (1958) *Mikhail Vasilievich Nesterov: life and work* ("Mikhail Vasil'evich Nesterov: zhizn' i tvorchestvo"), first edition, Moscow: Sovetskii Khudozhnik.
Miller, D. E. and T. Yamamori (2007) *Global Pentecostalism: the new face of Christian social engagement*, Berkeley: University of California Press.
Molendijk, A. L. (2015) 'In pursuit of the postsecular', *International Journal of Philosophy and Theology*, 76(2): 100–15.
Molendijk, A. L., Beaumont, J. and C. Jedan (2010) *Exploring the Postsecular: the religious, the political and the urban*, Leiden/Boston, MA: Brill.
Mrowczynski-Van Allen, A., Obolevitch, T. and P. Rojek (eds.) (2016) *Beyond Modernity: Russian religious philosophy and post-secularism*, Eugene, OR: Wipf and Stock.
Nikonova, I. (1962) *M. V. Nesterov*, first edition, Moscow: Sovetskii Khudozhnik.

Olson, E., Hopkins, P., Pain, R. and G. Vincett (2013) 'Retheorizing the postsecular present: embodiment, spatial transcendence, and challenges to authenticity among young Christians in Glasgow, Scotland', *Annals of the Association of American Geographers*, 103(6): 1421–36.

Petit, A. (2016) *Le Système d'Auguste Comte: de la science à la religion par la philosophie*, Paris: Vrin.

Plaskow, J. (1991) *Standing Again at Sinai: Judaism from a feminist perspective*, New York, NY: HarperOne.

Posener, A. (2018) 'Can Catholic crosses keep the far right at bay in multicultural Bavaria?', *The Guardian*, 01-06-2018, accessed online 09-06-2018.

Qian, J. X. (2018) 'Redeeming the Chinese modernity? Zen Buddhism, culture-led development and local governance in Xinxing County, China', *Environment and Planning A*. doi: 10.1177/0308518X16687555

Qian J. and L. Kong (2018) 'Buddhism Co. Ltd? Epistemology of religiosity, and the reinvention of a Buddhist monastery in Hong Kong', *Environment and Planning D: society and space*, 36(1): 159–77.

Raphael, M. (2003) *The Female Face of God in Auschwitz: a Jewish feminist theology of the Holocaust*, London: Routledge.

Reder, M. (2010) 'How far can faith and reason be distinguished?'. In Habermas, J., Brieskorn, N., Reder, M., Ricken, F. and J. Schmidt (eds.) *An Awareness of What Is Missing: faith and reason in a postsecular age*, Cambridge, MA: Polity Press, pp. 36–50.

Ritzer, G. and J. Stepnisky (2013) *Sociological Theory*, 9th edition, New York, NY: McGraw-Hill Higher Education.

Rorty, R. (1982) *Consequences of Pragmatism: essays 1972–1980*, Minneapolis, MN: University of Minnesota Press.

Rorty, R. and G. Vattimo (2007) *The Future of Religion*, edited by Santiago Zabala, trans. William McCuaig, New York, NY: Columbia University Press.

Ross, R. (2004) *Expending the Palace of Torah: orthodoxy and feminism*, Waltham, MA: Brandeis University Press.

Sarkar, S. and S. Keipeng (2017) *Tagore's Thoughts & Ideas on Education*, Düsseldorf: LAP Lambert Academic Publishing.

Scott, D. and C. Hirschkind (eds.) (2006) *Powers of the Secular Modern: Talal Asad and his interlocutors*, Stanford, CA: Stanford University Press.

Siebert, R. J. (2005) 'The critical theory of society: the longing for the totally other', *Critical Sociology*, 31(1–2): 57–113.

Smith, T. (2007) 'Narrative boundaries and the dynamics of ethnic conflict and conciliation', *Poetics*, 35(1): 22–46.

Soloviev, V. S. (1900; 2012) *A Story of Anti-Christ*, Fredericksburg, TX: Kassock Bros. Publishing Co.

Tagore, R. (1961) *Towards Universal Man*, New York, NY: Asia Publishing House.

——— (1994; 2008) *The English Writings of Rabindranath Tagore: a miscellany*, New Delhi: Sahitya Akademi.

Taylor, C. (2007) *A Secular Age*, Cambridge, MA: Harvard University Press.

Vasilaki, R. (2016) 'The politics of postsecular feminism', *Theory, Culture & Society*, 33(2): 103–23.

Vermeulen, T. and R. van den Akker (2010) 'Notes on metamodernism', *Journal of Aesthetics and Culture*, 2(1): 5677.

Vries, H. de (1999) *Philosophy and the Turn to Religion*, Baltimore, MD: Johns Hopkins University Press.

White, H. C. (2008) *Identity and Control: how social formations emerge*, 2nd edition, Princeton, NJ: Princeton University Press.

Wilford, J. (2010) 'Sacred archipelagos: geographies of secularization', *Progress in Human Geography*, 34(3): 328–48.

Williams, A. (2014) 'Postsecular geographies: theo-ethics, rapprochement and neoliberal governance in a faith-based drug programme', *Transactions of the Institute of British Geographers*, 40(2): 192–208.

Wohlrab-Sahr, M. and M. Burchardt (2012) 'Multiple secularities: toward a cultural sociology of secular modernities', *Comparative Sociology*, 11(6): 875–909.

Žižek, S. (2000; 2009) *The Fragile Absolute: or, why the Christian legacy is worth fighting for*, new edition, London: Verso.

Žižek, S. and J. Milbank (2011) *The Monstrosity of Christ: paradox or dialectic?*, edited by Creston Davis, paperback edition, Cambridge, MA: The MIT Press.

Zuckerman, P. and J. R. Shook (eds.) (2017) *The Oxford Handbook of Secularism*, Oxford: Oxford University Press.

Part I
Philosophical meditations

2

Beyond belief

Religion as the 'dynamite of the people'

Bruno Latour

Introduction

In my contribution to the debate on postsecularity I wish to revisit my long fascination with the dichotomy between knowledge and belief based on a keynote lecture I gave in Groningen, the Netherlands, in 2014 (see Figure 2.1).[1] I'll posit the idea that a plurality of templates to measure and understand the world could be conducive to a new public space that would allow respect towards religion as much as politics without mixing the two.

As Jan Assmann (2009a) has suggested in a recent book on 'religion and violence', how much we regret the time when religion was the 'opium of the people'. Now, it is rather the 'dynamite of the people'! From a drug putting the damned of the world into somnolence instead of doing revolution, religion has become the spear of revolutionary changes, and not always for the better. Religious studies have become an entry into the most misunderstood source of extreme violence and radical politics, and this is true not only of Islam, but is everywhere visible, from India to the evangelical church of North America, all the way to Russian orthodoxy, without forgetting the violent act of destruction of idols and fetishes that keep accompanying so much of the missionary work. While the state of the planet leaves everybody cold, the destruction of someone else's cult brings vast masses into action immediately. While modernism had long been defined by 'secularization', it seems that we are witnessing a reinforcement of modernist violence through new type of what should be called religious wars.

But far from being a 'remnant of the past' or an 'archaic return to the past', this metamorphosis of opium into dynamite proves that religion has to be taken as a fully *modernist* attitude. Specialists of religious studies should be ideally capable of probing this odd novelty—and if there is one place where all the tension of religion with modernity is being open to inquiry, it is in Europe, with its long history of simultaneously pluralism and most recently the hard testing of the extreme fragility of tolerance. So, what I want to do in this contribution to the volume is to follow the metaphor of the drug but to add to it what biochemists would call the study of its *potentialization*. It used to be a drug that put people to sleep—opium—and now it makes them active to the point

Figure 2.1 Protestant Nieuwekerk, Groningen: venue for the keynote lecture
Source: Wikipedia Commons: photo: Edi Weissmann.

of frenzy: it has been 'potentialised'. We have to discover what chemists call the action principle of this drug that explains such metamorphosis. I'll undertake this task in three parts, trying to find out why is it that this drug has become so strange. One part is, very quickly, about social explanation, the other about belief in belief, and the third one about politics. I will show that the three are actually combined together, which might explain some of the difficulty we have in understanding this contemporary emergence of religious wars.

The limit of social explanation of religion

So, let me start with the first one. We are not much helped in this search by the sociological principle—most clearly articulated, to take a classic case, by Emile Durkheim—that religion is made of the rites and words put in place to hide and reveal the existence of what he called 'society'. Durkheim (1947[1915]), as everybody knows, initiated a long set of studies that try to replace the enigmatic nature of religion by an even more enigmatic set of entities called society or social relations. As any sociologist will tell you, Durkheim claimed that the impersonal force of society was the only reality behind the vast mythical elaboration of religion. But what is not as often underlined is how strange, how active, and how enigmatic was the so-called impersonal force claimed by Durkheim to be the reality behind the enigma. One example:

> Society could not *abandon* the categories to the free choice of individual [...]. For this reason, society uses *all its authority* upon its members to *forestall* such dissidences [...] it is frequently *rude* to individuals; it is constantly *doing violence* to our natural appetites.
>
> (my emphasis) *(cited in Latour 2014)*

That's a lot of action for something that is supposed to be impersonal.

It is not too complicated to divine behind the impersonal agent implied by Durkheim (and sociology of religion after him)—the very personal agent implied by monotheistic religions. It's hard not to see in those 'social explanations' of religion, the mere replication of the being that Western religions invoke at the origin of their social life. The notion of 'society' is the 'one God, one people' of tradition. To put it bluntly, 'society' is the name given to a barely secularized 'Yavhé' (Karsenti 2017).

So, secularization has always been an attempt at reinforcing the 'one God, one society' argument. The obsession of sociology for explaining the obscure by appealing to what is more obscure is based on the denegation that there is *something* that makes people act, something whose agency has to be carefully scrutinized on its own term and for which the umbrella term 'religion' is terribly inadequate and which is not 'society'. In other words, it is not society that is behind religion; on the contrary, society is made in part by connections made by people with highly specific types of beings. This reversal in the direction of explanation is essential if we want to understand and avoid the 'one people, one God' argument. Society is what is to be explained, not what brings any explanation, especially not when by 'society' scholars of a Durkheimian persuasion mean, in effect, the God of Israel and Christianity. Religion, just like science or law, is not what is to be explained by alluding to social ties but includes some of the ingredients, making the social ties hold. At least this is the general principle of actor-network theory of the social order, a principle especially forgotten when religion is 'explained' away by sociologists (Latour 2005). If we consider how religious people define the beings that they encounter, it seems that a better definition would be that there are agents on which they have limited control and whose disappearance will make them die. Let's call them, for this reason, beings of salvation and try to get at them without using the sociological notion of belief.

Belief as a category mistake

This brings me to the second problem that renders the potentialization of opium into dynamite difficult to follow. This time it is not due to the explanation that appeals to the society instead of explaining the religious contribution to the solidity of social ties. The problem is due—and often on the part of those who pride themselves in being 'religious'—to what makes them act in competition with science. By science I mean at least information to render the idea of a totally utopian space where things, argument, people, and goods could be transported without being transformed. Transportation without transformation has always been my personal nemesis. This is what I call double-click information (Latour 2013b).

My thesis is that it is the spread of double-click information that is at the origin of the invention of obscurantism in matters of religion, that is, the idea articulated by opponents as well as by proponents of religion that there is something 'occult' in its rituals and practices.

The very use of the word 'religion' has come to mean what is inexplicable, irrational, what requires an appeal to an extraordinary set of drives (for the analyst), or what requires *supra-natural* entities for those who are called 'believers' who are forced to accept belief as what accounts for their faith. This requires some explanation.

I claim that there is nothing obvious in this link of religion with the strange, the occult, the supernatural, nor, to use the main notion that rocks any understanding of the question, with 'belief'. The idea of belief is the result—and an unhappy one—of interrogating a mode of existence by using another mode. I want to try to propose that belief is always the result of an unfortunate crossover between two modes of existence. The use of the notion of belief proves that there has been a conflict during an interchange in the templates we should use to define an entity on its own terms. This is what I call a category mistake (Latour 2013a).

Those category mistakes are banal, but very often they don't have the huge consequences we are witnessing in the potentialization of the opium into explosive. For instance, if, after a judge has rendered her verdict, you, the plaintiff, keep saying 'I don't feel appeased by this judgement', your lawyer will be right to say psychological peace of mind is not what law is about—a verdict has its own logic and nobody hearing it would conclude that law is irrational, occult, or obscure. You might keep complaining against the formalism of law, but most probably you will not conclude that law is 'irrational'. You most probably conclude that law has *its own strange and painful* way of being right. Thus, law seems to resist the accusation of being 'just about irrational belief' (Latour 2009).

Ideally, we should be able to say the same thing when registering any crossing between two incompatible templates. Such is the principle of an inquiry into modes of existence: double-click is not the universal template for every encounter. Faced with a judgement of law, you simply recognized that as far as psychological appeasement is concerned, legal vectors are found wanting. No more, no less.

So now we can ask ourselves how come that the same thing does not happen when you ask the carriers of religious salvation—Bible, angel, sermon, or icons—how come they are not producing accurate information about a certain state of affairs? Why don't we simply conclude: 'well religious vectors are simply not good at transferring facts because they do something else that facts are not asked to do: namely to *transform* those who are addressed by religious beings'. Imagine the Virgin Mary asking Gabriel what information he is carrying. He should obviously reply: 'I'm not carrying information, I'm transforming you!' Information content: zero, transformation content: maximum, that is, the birth of the Son of God! The idea of some occult kind of message would only be produced if, by mistake, the answer was: 'there is a message (that is, an information), but it is encoded in some mysterious language'. At this point, the transformative (by opposition to the informative) mode would be lost for good.

The difference between modes has been well demonstrated by Louis Marin commenting on the famous 'Annunciation' by Piero della Francesca (Marin 1989; Marin 1991) (see Figure 2.2). Piero painted an annunciation, and he did it very beautifully so that the angel is actually *hidden by the pillar*; there is no way for the Virgin in the newly invented perspective space to see Gabriel! Piero della Francesca was amazingly careful in his disposition of objects in space—after all, he invented this new optical regime!—and that's why he made it absolutely clear that the Virgin *should not* see Gabriel in that space to indicate as clearly as possible to the viewer that Gabriel was not a carrier of information but a completely different type of vector. Marin comments that to make sure the difference of the two modes is understood, perspective logic is used to render the protagonists invisible to one another. But it has nothing to do with the obfuscation of a message that could be clarified by painting Gabriel facing Mary straight on.

Figure 2.2 *The Annunciation,* Piero della Francesca, 1460
Source: WikiArt.

The notion that religion is about the irrational is thus the result of an embarrassment. Gabriel would be embarrassed at being asked the wrong question, at being interrogated in the wrong key: 'What information (meaning exact information) do you bring to me?' Poor Gabriel would not know what to say. But you would agree that it would be worse if we concluded from his unease that he has something to hide, another more esoteric and less rational message. He has nothing to hide, he does something else. He brings a total transformation of Mary.

Belief arises when we have two exit routes left. One is to withdraw into a rather shameful 'yes, I believe in strange things but I won't tell anybody', and the other is, on the contrary, to assert that 'yes, indeed, there is a world that belief can access *just as much* as information can access the world of common sense, except it's a supra natural world of beyond to which you have no access'. This means that you are *not transformed* by the message but left simply hanging eyes looking up. Belief has eaten up the originality of religion. There is a totally invented competition between the double-click messages transporting information about the natural world and double-click information transporting information about the supra natural world.

At this point, the poison comes in when belief that started as a misunderstanding on the part of the interrogator is *accepted* by the interrogated as what he or she *has to hold* in order to be respected. This is where the difference between religion and law is most striking. The lawyer will never say, 'law is exactly as information transfer except it is much more esoteric'. He or she will say, 'law's job is not to carry information nor is it to cure psychological miseries. *Dura lex sed lex*'. But in matter of religion, religious people themselves have accepted to submit to the power of double-click when they begin to confess: 'Yes, I believe in what *cannot* be explained by normal means but you are right that it is a message'.

What started as absurdity on the part of interrogator, not using the correct template, becomes now what the wrongly understood soul begins to hold *most dearly* to. That sits at the heart of the question because now there is a deep *lack of authenticity* in accepting to be a believer yourself because of the way you have been requested to bear witness for the beings who make you act. The potentialization of opium into dynamite comes, in my view, from this operation by which the imputation of belief by the outside observer has been *interiorized* by the agent as the only way to understand what makes him or her act. 'Yes, you're right, after all I believe in occult, irrational, supernatural sort of things'. Except this cannot be true. Belief is *not*, and cannot be, the sincere and authentic way in which you are acted by the being activating you. Belief is always a mistake, whether it is imputed from the outside or accepted as inside as the only definition of the situation.

Since it is what I call a category mistake, it *deprives* the now entrenched believer of any possibility of rearticulating what makes him or her act. Now the believer is poisoned from the inside by this imputation of believing into something strange that does *not* correspond at all on how he or she is acted upon by the beings coming to make him or her saved. I claim that *this* is the source of the modernist form of fundamentalism—a fully modern extension of the poisonous notion of belief, coming at first from the outside as a category mistake on what it is to be acted upon and then *interiorized* as the only positive way to *assert* oneself in the face of a confrontation by people who don't understand what you are. At this point, violence is the only solution.

I am not suggesting an old, premodern, archaic violence. Assmann, in *The Price of Monotheism* (2009b), is really interesting in this regard. He classifies the different sources of religious violence—rather, a fully modernist violence the intensity of which, as we see more or less every day in the press, increases with the violence of the others and the slow disappearance of any alternative template to define action by other beings. The result is fanatism; there is no other way; this reflects the contemporary situation. Formerly religious souls, now convinced that what they have to die for, or to kill for, is made in the name of *their beliefs*. There is a built-in inauthenticity in believing in belief, which means that the meeting with religious beings can no longer be articulated.

Not surprisingly, it is this lack of authenticity that creates the conflagration. Violence ensues immediately. Believers, by which I now designate people who are *forced to believe* (in belief), have to believe in something in a way that would not otherwise require the notion. Ashamed of himself, Gabriel now withdraws, convinced that his secret coded message is a fully undecodable esoteric mystery! Believers are deprived of any *reasonable* way of following their own beings. When believers encounter other believers, who are just as deprived of their resource, what happens is necessary violence. There is no other way.

Let me be clear in my argument. I'm not saying that fight among believers is a remnant of a past or a new relapse into archaism. Rather, I am suggesting that it is the incorporation into a vast array of modes of existence of one template—belief—itself due to the competition of another template: information or communication or double-click (often conflated erroneously with science). If asked 'what information do you carry?' bearers of religious salvation have to say 'none'—to which the interrogator replies, 'then you are not part of our common world'. To which the carriers may reply, 'ok, we're out, can you leave us a small place?'—that's the opium phase—but they also may conceivably reply, 'we will fight to the death for our belief', and that ushers in the dynamite phase. The opium has become dynamite because no one seems to understand any more what religion was about. Gabriel goes away without delivering its message—which no one understands anymore was *not* a message.

Notice that there is nothing in the original situation that should have connected religion with 'strange' or 'irrational' or even with belief. In other words, religion could be treated like law. If the judge gives me a verdict that is not satisfactory on psychological ground, it is perfectly possible for the judge accused of not being sensitive to the 'peace of mind' of the plaintiff to explain why law has its own logic that gives a quality to the common world but would have been lost if she had accepted to transform law into a solution to psychological trauma. So, both disagree, but they now *reasonably* disagree because they accept to use *two* templates, one for law and another one for trauma, and because the plaintiff has no way to accuse the judge of 'believing' irrationally in law. Make the test by yourselves: no one would as a judge: 'do you believe in law?' The judge or the lawyer would say: 'yes, of course, this is completely material. Why would you ask me about belief? I'm a lawyer, I pursue the specific type of being I meet'. The law resists the interiorization of the accusation of being an object of mere belief. This is my point. Why do we not do the same for what is called 'meeting with carriers of salvation'?

Let me take my favourite example; suppose your lover asks you (a classic question): 'do you love me?' and you answer: 'I've already told you hundreds of time, yes, I love you!' The first lover could legitimately ask: 'thus, you don't love me anymore'. If the second pushed the button of the tape recorder and said 'hear, is this not a proof? Listen, this is what I have said last year, don't you *believe* now that I love you?'—it would be reasonable to understand that the first lover is right in throwing the tape recorder at the face of a man. How *absurd* to believe in making the original question—'do you love me?'—a matter of *information* double-click.

If you don't *hear* the question as a matter of *transformation,* obviously no amount of proof will ever convince the interrogator. The difference with religion in this case is that the lover, because he or she *loves*, has a robust reticence to any interpretation of the interchange as being about transfer of information. So, by resisting belief, he or she keeps open the possibility of *converting* his or her lover to a reasonable—not a rational—change of repertoire. Imagine the scene a few days later: 'how could you be so dumb, so insensitive to push the button of your tape recorder'. To which he or she could answer, probably in tears, 'yes, I was so stupid'. A different outcome was possible because the authenticity of the shift from information to transformation was actually not erased by the notion of information, and thus by belief. A different outcome than violence is possible in this case because values may be rearticulated from the inside.

This is what is completely impossible when the *extraneous* template of belief has been made to be only the definition of how we stick to one's values, because if we have only one template, it's impossible to articulate any statement and the guy will keep saying 'I was transmitting information, is this not what you ask? You ask me if I love you, I said yes'. To which the reply should be: 'you understood nothing, it's another mode of existence, you have to answer in another tone'. Everything is different, you completely misunderstood, but if you have only one template, you cannot even hear the accusation of misunderstanding. You have become inauthentic to yourself. It is for this reason that I think we should modify the meaning of agnosticism and that it should be used to mean a way to *abstain* from using the notion of 'belief' when considering anything having to do with religious matters.

When violence moves from religion to politics

Moving to the third part, the question now becomes: how come the meeting with carriers of religious salvation is not met with the same resistance and robust opposition as those who carry legal means or the transformative aspect of love talks? Conversely, how come that the modernist toxic influence of information double-click has not been able to intoxicate those other sources as well? In other words, why is it that we find ourselves again thrown into *religious wars* infinitely crueller and longer lasting than those of the early modern period? Of course, in the same way as the post-Reformation wars were a product of the time—and not at all a remnant of the past—the new religious wars are brand new and totally modern. We have to turn for this question not to a competition with information double-click, but with *politics*, which is the third mode of existence I wish to consider.

I know this is an immense topic. But the principle of method I'm following in the inquiry of mode of existence could help account for the *potentialization* I'm after. If the conflagration of believers unable to rearticulate the attachment to their gods is so strong, it is not only because of the expansion of information as the only template acceptable everywhere. It is also because it seems to have become impossible to deploy politics as its own *independent template* just as law or love has theirs. The disappearance of politics has its own logic, and truth conditions make, of course, the assertion of belief the ideal *shortcut* to define what it is to hold values and to stick together in a group that shares some identity.

It is not that religions have been politicized. On the contrary, it is that a religious state of mind—itself, as I showed before, entirely corrupted, if you want, into belief in belief—has been unable to respect this other completely original and strange *sui generis* mode of existence that is called politics. Compared to the certainties requested now by religious believers, politics always appears too uncertain, tentative, even irrational. The great irony of our time, which is a great tragedy, is that it's religious characters—themselves transformed by belief in belief into inauthentic souls—claim now to pass judgement on politicians for being *not* rational enough, not certain enough, too *wavering*, too disposed to compromise. And amazingly, contrary to lawyers and lovers, politicians seem to agree; they confess, they apologize, and they attempt to be more virtuous, *principled*, rational!

These days it is especially strange to hear, for instance in America, Trump supporters accusing scientists of irrationality because they 'believe' in climate change! As if science was a matter of belief. We have come full circle: religious people believing in belief now accuse climate science to be a mere belief and those who believe in it to be irrational, as if they themselves were the last rational persons left on Earth. What a mess! The competition of values which at first pushes the religious souls to believe in belief—that is fundamentalism—is now pushing one step further politics itself into being about belief, values, and absolute certainty: this might be called 'value politics'.

As I have tried to show recently, *what is wrongly called* 'populism' is the expansion one more time of a religion corrupted by belief in belief and now attacking politics for its lack of 'values' (Latour 2018). The idea is that political personnel lack 'values' and 'identities'—exactly what religion has *lost* itself in competing with information. We have lost religion transformed into belief, and now it is the belief in belief as the only key that gets into politics and in turn that makes us lose politics. A cascade of catastrophe ensues that brings us to the present state of 'alternative facts' and 'post truth' society.

In a sense we are back to the time of Spinoza: the answer should not support 'value politics' but, on the contrary, *value politics for itself*. And yet it's not with Spinoza that I wish to conclude this contribution but rather with a lesser known figure, that of Eric Voegelin, whose diagnosis of 'gnosticism'—although made in the 1950s—fits exactly the present situation I wish to describe (Voegelin 1952 [1987]).

Gnosticism, in Voegelin's interpretation, is the extension of Christian religion into wanting the full realization of the kingdom of God. Nothing wrong with that, except if it means the kingdom of God *in this world*. I'm referring to what he calls, for this reason, 'immanentisation'—the rendering of the heavenly ideal in the here and now. The key point is that modernity is in no way understood as a break away from Christian religion; Voegelin, as is well known, is completely against the notion of secularization. Rather, modernity concerns the utopian ideal of realizing the kingdom of God here, now, for good, and together. This understanding is almost the same as expecting the kingdom of God, but the small difference is that in the traditional, premodern case, it is God who does the work; in the new gnostic one, humans take on themselves the task of God. That's what Voegelin calls 'immanentisation'.

Secularization, thus, is Christian through and through, but a Christianity *deprived* of a robust, basic, essential instinct that the Kingdom of God is not *of* this Earth (hence the word immanentization of transcendence, not to be confused with immanence which is a healthy and perfectly normal sentiment referring to the spiritual imbrication with the material). Secularization in Voegelin's interpretation, like belief, is a highly poisonous attitude that renders the adept just as unable to respect immanence as to respect transcendence. The psychosocial portrait of a militant drawn by Voegelin from the first Puritan all the way to the communists of his time—and that could be easily expanded to the Christian or Islamic fundamentalists of today—depends crucially on this inability to respect politics as a *mundane*, immanent, extremely fragile mode of existence.

The history of political theology in the Western world depends on the articulation of three religious dimensions that Voegelin calls unabashedly 'truths'. He refers to those as cosmological, anthropological, and soteriological (*Ibid.*: 149–50). The first is really civic religion of the early Empires such as the Roman gods; the second relates to the notion of cosmopolitic, basically the philosophical achievement of the Athenian political culture, encompassing the entirety of the psyche and human affairs; soteriology refers to the Christian type of religious truths. The problem of political theology, according to Voegelin, is that it is never possible to hold the three together in what gets called, very interestingly, the 'complication of symbolism'.

The tendency of secularization therefore is not to abandon Christianity. Rather it is to simplify the number of templates necessary to build the order of society. For instance, Augustine will be just as unable to understand the power and legitimacy of cosmological truth—the Roman gods—as Hobbes would to understand the power and legitimacy of soteriological truth. Augustine will build the *City of God* without the indispensable and healthy presence of the *civic gods*, while Hobbes will write his *Leviathan* without the balancing power of *salvation* (*ibid.*: 159). The secularized gnostic will build modernity as the worst possible solution. The result is gods without gods, belief and disbelief mixed up, without any one of them being able to articulate the multiplicity of templates necessary to have a reasonable—not a rational—political theology.

So, for Voegelin, the idea that we live in a time of *pluralism* is just patently absurd. *D*edifferentiation is exactly what has happened. It has become almost impossible to rearticulate the *multiplicity* of templates, to extract religion from irrationality—and still less from belief in belief—it is even impossible to break down the religious political amalgam around the social values and identities. In the same way as secularization has nothing to do with the abandon of religion—witness Durkheim who is just replacing a personal god by another personal god—what is called pluralism today is the *end* of a plurality of templates and the sad realization that the multiplicity of *clashes* about belief and values is not only *inevitable* but without any other result than violence.

Secular people often pride themselves in the idea that they live in a pluralist society and that this pluralism should be conserved. However, plurality of templates, if you have followed my argument, is the scarcest commodity of all. Social sciences have long attempted to erase entirely the originality of religion by the three ways I've reviewed in this contribution.

They claimed that religion was nothing—or at least, nothing but society; they insisted that it was about belief and largely succeeded in convincing adepts to be also believers; and, finally, they have decided that religion had been distorted, polluted, and kidnapped by politics, when it is exactly the opposite movement that we are witnessing today. Politics has become impossible to articulate precisely because of an inarticulable definition of religion as belief.

Conclusion

The only solution to the present religious wars indeed is in insisting on plurality, but not plurality of knowledge—we rather need more unity of knowledge!—but plurality of *templates* with which to measure the beings that are making us act and are thus holding us—be they law, love, politics, religion, or many others. The idea of plurality or multiplicity of templates is how I embrace the notion of postsecularity. Rest assured I am far from comfortable with another term using 'post'. But I can accept it when one places it in the quest for new ways of conceptualizing the multiplicity of modes of existence. What we cannot afford is to have a war of gods just at the time when we have to also deal with the war of the world imposed by the intrusion of this strangest goddess of all, Gaia. In this way, pluralism is still something that lies way ahead of us.

Note

1 An earlier version of this paper 'Beyond belief: on the forms of knowledge proper to religious beings' was presented *On the Forms of Knowledge Proper to Religious Beings*, keynote lecture, 400th anniversary of Groningen University, EASR meeting, The Netherlands, 12-05-2014.

Further reading

Latour, B. (1993) *We Have Never Been Modern*, trans. Catherine Porter, Cambridge, MA: Harvard University Press.

This book is a classic in anthropology of science that questions the modernist dualism of nature and society. Instead of postmodern and anti-modern movements, Latour argues for the hybrid interrelations of natural and social phenomena with discourse.

Latour, B. (2010) *On the Modern Cult of the Factish Gods*, Durham, NC: Duke University Press.

Latour redescribed the Enlightenment of universal scientific truth with the argument that there are no facts inseparable from their fabrication. The argument in this book extends to religious fetishes—objects invested with mythical powers—to show that the objectivity of facts and the power of fetishes ('factishes') are both truth-making.

Latour, B. (2017) *Facing Gaia: eight lectures on the new climatic regime*, Cambridge, MA: Polity Press.

Based on a series of lectures on 'natural religion', Latour invokes Lovelock's 'Gaia' to interrogate the Anthropocene in ethical, political, theological, and scientific terms. He calls for a new collaboration between scientists, theologians, activists, and artists to embrace the challenges of the 'new climatic regime'.

Lovelock, J. (2016) *Gaia: a new look at life on earth*, Oxford Landmark Science edition, Oxford: Oxford University Press.

In this new edition, Jim Lovelock advances the hotly debated idea that the Earth functions as a single living organism to self-regulate and keep itself a place fit for life.

References

Assmann, J. (2009a) *Violence et Monothéisme*, Paris: Bayard.

——— (2009b) *The Price of Monotheism*, trans. Robert Savage, Stanford, CA: Stanford University Press.

Durkheim, E. (1947 [1915]) *The Elementary Forms of the Religious Life*, trans. Joseph Ward Swain. New York, NY: Free Press.

Karsenti, B. (2017) *La Question Juive des Modernes: philosophie de l'émancipation*, Paris: PUF.

Latour, B. (2005) *Reassembling the Social: an introduction to actor-network theory*, Oxford: Oxford University Press.

——— (2009) *The Making of Law: an ethnography of the conseil d'état*, trans. Marina Brilman and Alain Pottage, Cambridge, MA: Polity Press.

——— (2013a) *An Inquiry into Modes of Existence: an anthropology of the moderns,* trans. Catherine Porter, Cambridge, MA: Harvard University Press.

——— (2013b) *Rejoicing: the torments of religious speech*, trans. Julie Rose, Cambridge, MA: Polity Press.

——— (2014) 'Formes élémentaires de la sociologie: formes avancées de la théologie', *Archives de Sciences Sociales des Religions,* 167 (juillet-septembre): 255–77.

——— (2018) *Down to Earth: politics in the new climatic regime*, trans. Cathy Porter, Cambridge, MA: Polity Press.

Marin, L. (1989) *Opacité de la Peinture: essais sur la représentation au Quattrocento*, Paris: Usher.

——— (1991) 'Stating a mysterious figure'. In Bogue, R. (ed.) *Mimesis in Contemporary Theory: an interdisciplinary approach, vol. II: mimesis, semiosis and power*, Philadelphia, PA: John Benjamins Publishing Company, pp. 45–68.

Voegelin, E. (1952 [1987]) *The New Science of Politics*, new Foreword by Dante Germino, Chicago, IL: University of Chicago Press.

3

The difficulty of unforgiving

Martin Beck Matuštík

I shiver when I think how you can forgive on behalf of others. I can't forgive. I don't have the permission.

(Jona Laks in Kor 2006)

Is it time to pardon, or at least to forget?... [I]n relation to the infinite all finite magnitudes tend to equal one another.... What happened is inexpiable.... Forgiveness died in the death camps.... Father, do not forgive them, for they know precisely what they do.

(Jankélevitch 1996: 554, 558, 567, 564)

You can't cover an elephant with a rice basket.

(Kav Savuth's Cambodian defense of Kaing Guek Eav in Mydan 2009c)

My current plea is that I would like you to please leave an open window for me to seek forgiveness.

(Kaing Guek Eav in Mydan 2009a)

Introduction

Much has been written on the gift, giving, forgiving. Even if we consider all giving to be a conditional act, rather than a pure gift, forgiving poses a difficult task for finite human persons. There is nothing wholly secular about *unforgiving*. Unforgiving is an excessive, saturated phenomenon that engenders this postsecular meditation (Matuštík 2008). I use the locution *postsecular* in a twofold sense: one, in the West, secularization refers to the historical Enlightenment whereby the vanishing point of modernity is envisioned to yield a full rationalization of religious claims. Return of the religious after the process of rationalization of religious claims (their translation without a remainder into one of the cultural spheres of modernity, i.e., sciences, law, and aesthetics) would give birth to postsecular religiosity. Two, the post- is not the same as anti-, and we must admit the possibility of new forms of understanding, such as spirituality without religion, postmetaphysical religiosity, or new enlightenment that includes both rationalization of traditional religious contents and spiritual mindfulness.

In the postsecular age, the binary secular-sacred has been emptied of signification it held during the process of rationalization of the sacred.

This is why there is nothing wholly secular about unforgiving. Forgetting the past requires no sustained effort, while forgiving trauma needs an ongoing effort at remembering something as otherwise than it was. If I cannot forgive, why not rather forget the whole thing? Why would the secular age want to hold on to a postsecular excess of unforgiving? If I could induce amnesia, modify my brain proteins that record a scene of trauma, why would I choose to cohabit my present with my unforgiving self? Unforgiveness separates the scene of trauma one recalls from forgetting oneself at that scene altogether. Walking away from or working through the scene differs from fleeing oneself from the scene. Forgetting does not require my whole self for its work; unforgiving consumes the self that one has been. A new self, albeit recognizable in remembering, arises in coming to terms with the traumatic past.

The scene of trauma choreographs three basic scenarios: the what, who, and how. The what—*the deed*—requires repair, repayment, sentence, punishment, and memorialization. The who—*the doer*—refers to the agent who harmed and *the sufferers* who were injured. The agent and the sufferer are pinned down by their unforgiving. The how—*the intra—or impersonal address* of unforgiving holds the doer and the sufferer bound to the lost possibility of oneself or another. The first scenario emerges in courts, moral repair, and struggles over memory. The second, interpersonal scenario implies that one never accepts, gives, or withholds gifts, even to oneself, wholly alone. In solitude, one never addresses oneself without addressing another. The third scenario brings out the curious insight that my capacity to give or refuse forgiveness does not come from another human or because of trauma. Unforgivingness reveals a vocative dimension of life—there is always a cosmic address to which one addresses this difficulty. Memory retraces the deed (what), hatred holds the doer and the recipient (who) in the difficulty, and the lost possibility of oneself or another (how) addresses unforgiving to *intra- or impersonal address.*

Postsecular meditations

Meditating on the difficulty of unforgiving as an interpersonal affair, the *who* in the second scenario, one apprehends the self and alters as either bound or released at the scene of trauma. Without *postsecular unforgiveness* surviving in the secular age, the death sentence would lose its repulsive sympathy and attractive antipathy. Are there conditions that allow for, though do not legislate, the possibility of forgiving? What does unforgiving require of perpetrators and sufferers?

Meditating on the lost possibility of oneself and another, the *how* in the third scenario, the unforgiving, appeals to an intra- or impersonal address. Even if one meets the moral, legal, and economic obligations of the first scenario, can one prosecute the irreparable? What does the remainder of unforgiveness reveal at the heart of justice?

In this meditation on the postsecular excess of unforgiving, I draw on the trial of Kaing Guek Eav—'Duch', who directed the Tuol Sleng prison during the Khmer Rouge reign. I focus in particular on the witness testimony of Vann Nath, one of the seven Tuol Sleng survivors.

Forgiving

Forgiving seems to be more difficult than not forgiving. Human persons generally believe in human renewal, but sometimes one becomes unable or unwilling to forgive. Forgiving requires self-change. There is no normative or economic shortcut from traumatized self to self-renewal. There is no foreseeable shortcut from doer of unforgivable deeds to person capable of renewal. Where forgiveness does not arise for some deeds, the punishment of the doer and sustained horror at the deeds do not settle the question of whether or not their originator is in principle forgivable.

Are crimes against humanity unforgivable? Unforgiving is a recognized dimension of such crimes against humanity—yet is it possible to subject this dimension to legal action; is it indeed punishable? Truth commissions can articulate political apology, ritualize pardon, declare amnesty, require reparations. But can anyone legislate forgiveness of crimes against humanity? Courts place no time limit on punishing certain crimes. But can anyone demand that they be forgiven in *due* time?

Walter Benjamin (1999: N7a8, p. 471) asked Max Horkheimer in their letter exchange of 16 March 1937, whether past suffering is closed to later generations. Are victims safe from posthumous victories of their oppressors if history of the past is complete? Rituals of political apology address intergenerational liability where interpersonal forgiveness is not available. Moral and political restoration may require the third party or intergenerational political apology. Even when we just remember the humanly unforgivable, we forbid ourselves conceiving of history as fundamentally closed. The categorical imperative of anamnestic solidarity with victims of history transmits this 'messianic hope' as a dangerous memory of those who have expected our coming (Benjamin 1968: 254). I am raising now new questions—is unforgiving transmitted through time that will not pass away, is there memory that can resist its erasure, are there repairs without measure?

Humanly forgivable. To admit that an offender may be humanly forgivable (Griswold 2007: 115) establishes three baseline conditions for forgiveness: if the injury is humanly forgivable, then both the victim and the offender must be willing to reframe their mutual self- and other relations. The wrongdoer and the victim, each, must meet certain conditions in order for forgiveness to take place. Those conditions must be met from each side. Whether one faces the difficulty from the side of the perpetrator or the harmed, one must retain and yet reframe the self at the scene of trauma. The victim's readiness for reframing of oneself, as well as of the perpetrator, exhibits a supererogatory (morally good but not strictly required) capacity for self-modification—a disposition and virtue of 'forgivingness' (17). In greater detail, Griswold (2007: 50–4, 149f., 174) enlists six conditions that, ideally and in a paradigmatic case of forgiving, must be met by an offender and six by the injured for forgiveness to take place.

A wrongdoer must enable but cannot require forgiving:

- Own up
- Repudiate the deed and oneself as its source;
- Regret the deed and oneself as its originator;
- Commit to lasting change in words and deeds;
- Take a sympathetic understanding of harm;
- From the victim's perspective offer a narrative account of one's own wrongdoing and future reframing.

And a victim must satisfy conditions of granting forgiving:

- Forswear revenge;
- Moderate resentment;
- Commit to giving up resentment;
- Reframe the self of the wrongdoer;

- Reframe oneself as a victim;
- Communicate forgiveness to the offender.

In real life, these ideal requirements fall short. This deficit does not mean that one can take shortcuts. Griswold (2007: 118–29) describes imperfect approximations of the ideal-typical forgiveness. A third party may offer or receive forgiveness on behalf of someone else, but only indirectly. One must be related to the dead or to the one who cannot meet the conditions for some legitimate reason. A victim may forgive an unrepentant offender without waiting for all enabling conditions to be met.

Even when both sets of conditions are achieved, there is still the task of self-forgiveness. A victim need not wait on the perpetrator's conditions to be satisfied. A victim's path to freedom from trauma cannot be hostage to the perpetrator. Still self-forgiving requires that one foregoes a great deal of the resentment. Griswold's virtue of 'forgivingness' (2007: 17) resembles interpersonal love in that one cannot forgive or be forgiven without relating to another. Neither self-love nor self-forgiving would make full sense without an interpersonal address. Even in the singularity of self-forgiving, one calls out aloneness not only interpersonally but also intra- or impersonally (appealing to an ideal other, deep self, a cosmic address).

Unforgiving seems humanly easier to harness than fulfilling Griswold's conditions. Even half-baked forgiving and its weaker forms, apology, or reconciliation require a degree of self-modification. In a truly secular age, why should we prolong unhappy unforgiving and rather strive for happy forgetting? The secular mind suspects forgiveness of superstition, while it embraces hatred as an ordinary worldly attitude. If the secular ear is tone-deaf to spiritual music, how can its mindset raise a faith claim whereby unforgiving endures now and forever (*saecula saeculorum*)? When victims fail to reframe the self of a wrongdoer, how is it their weighty memory is not performing spiritual underwriting? Withholding forgiving, rejecting its offer, hardening one's heart—why remembering these very states of mind should not require a degree of self-modification? Indeed, sidestepping oneself while persisting in unforgivingness, this attempt as erasure of traumatic memory traces is a self-contradictory performance. Un/forgivingness is a Janus-faced phenomenon, the vocative *who*, that un/binds the selves of the wrongdoer and the victim. This double difficulty attests to the interpersonal requirement of self-modification.

The difficulty of unforgiving presents us with three distinct outcomes of trauma:

- Forgiving is achieved when the full double set of six conditions (the self- and other relations between the perpetrator and sufferer) is met; or
- Only unforgiving is humanly possible (the self- and other relation remains unchanged); or
- Self- and other relations no longer pose the difficulty of forgiving (one has forgotten the scene of trauma or become otherwise than human).

I either struggle with myself and another, the *who*, at the scene of trauma; or I modify the *what*, sidestepping the *who* and the *how*. Skipping myself and another, I win but a pyrrhic victory.

The high stakes of unforgiving

Imprescriptible means forever punishable. There is a shared normative intuition that only transgressions that admit punishment can be repaired. The twentieth century set aside a special category of misdeeds we call the crimes against humanity. These crimes are in legal terms imprescriptible, that is, there is no time limit on punishing their perpetrators.

Imprescriptibility carries two peculiar outcomes. *First,* there is no time limit on punishment as long as perpetrators live. It is as if infinity immigrated into imprescriptibility. A suprahuman temporal weight sinks into crimes against what it means to be human. However, this expanded secular time limit, boasting an imagined infinity, exceeds only the passing away of finite time. Imprescriptibility does not inscribe human law into the messianic now-time of redemption. The imprescriptible is saturated by another excess—the unforgivable—though it is not identical with it. I use the locution *unforgivable* to infinitize temporality of unforgiving. Only crimes that are unforgivable in principle can also come under the rubric of imprescriptible. Punishment does not require forgiving; the latter cannot be legislated.

Second, victims of history can be memorialized in public dramatizations of reconciliation, healing, and coming to terms with the past. Yet, legally punished misdeeds do not restore shattered human solidarity. There is no human-type contrition that by itself expiates evil deeds. Their postsecular excess tone-deaf ears cannot hear, but phantom limbs of the secular age suffer it even if all dastardliness could be punished and morally repented in time.

Both outcomes of imprescriptibility bring us to the threshold of unforgivingness. Crimes against humanity have become legally imprescriptible: There is no time limit on their prosecution. In a truly secular age, why should we prolong unhappy unforgiving and rather than striving for happy forgetting? Imprescriptibility is saturated by what is humanly unforgivable. Evil crimes must be in principle always already punishable, though not necessarily forgivable. Forgiving them introduces the possibility of *redemptive critical theory.*

The imprescriptible is not the same as the unforgivable. If a criminal deed does not expire with time, we know what it means to call it *in principle* imprescriptible and punishable. These are the high stakes of un/forgiving such crimes. Punishable imprescriptibility of crimes does not provide any warrants that they may be forgivable in time. Punishability of imprescriptible crimes cannot require or legislate their forgiveness *in principle.* The imprescriptibility of crimes, at best, means that their perpetrators must never be granted a posthumous victory (Fackenheim 1996). The judgement on this type of crime (the *what*) never expires with time, the deed must never be forgiven. Indeed, sidestepping oneself while persisting in unforgivingness, this attempt at erasure of traumatic memory traces is a self-contradictory performance.

Critical theory can articulate reparations, moral restoration, public performances of amnesty, pardon, and reconciliation. We can identify negatively the *publicly recognizable features of unforgiving:* It breaks the bounds of mere reasonableness, it exceeds the bounds of the legally imprescriptible, and it negatively saturates the bounds of economic, legal, political, and moral repair. Redemptive critical theory points to what lies outside of the secular 20/20 vision. What are the sources of healing that cannot be legislated? What is forgiving that is not morally categorical? From where is hope derived that is assumed in public settings yet whose promise has been lost in time?

There are two questions in play. Critical theory asks about the *what*: the deed to be punished and its effects to be repaired. Redemptive critical theory asks about the *who* and the *how*: self-modifications of the doer and sufferer. The imprescriptible crime describes the 'what' of the unforgivable deed and in that sense assigns punishment to its agent. The unforgiven address concerns the selves of the perpetrator and sufferer. Must doers of imprescriptible crimes who are punished yet unforgiven in their lifetime become *in principle unforgivable?* This 'in principle' tweets the secular age with a postsecular hashtag.

Evidence from the difficult freedom of unforgiving

The '*who*'—case file 002/14-08-2006 filed in Extraordinary Chambers in the Courts of Cambodia—contains crimes against humanity and grave breaches of the Geneva Conventions of 12 August 1949. The accused is Kaing Guek Eav alias 'Duch' *who* directed the Tuol Sleng Prison under the Khmer Rouge in Cambodia (1975–1979). From 1975 to 1979, between 17,000 and 20,000 people were imprisoned at Tuol Sleng (these are estimates). From the records of 14,000 prisoners who passed through Tuol Sleng, there were twelve known survivors: seven adults and five children. There is no time limit on prosecuting the living doer who has been in hiding for 20 years under another name (1979–1999). 'Duch' was arrested in May 1999, and indicted on 8 August 2008. His trial concluded on 27 November 2009, with the prison term of 35 years, of which he had already served 16 (Mydans 2010a).

The '*who*' calls out the addressee of unforgivingness that attaches to the verdict. A juridifiable address is the material case. A non-juridifiable address is the 'who'. The latter address keeps unforgiving alive. In this imprescriptible case, some mail might have been forwarded from one address to the other. In February 2008, 'Duch' was taken, with his consent, to the scenes of trauma. Witnesses were shocked by an extra-judicial act, choked by great emotions: 'I ask for your forgiveness—I know that you cannot forgive me but I ask you to leave me the hope that you might', 'Duch' cried out, then collapsed in tears on the shoulder of one of his guards. A survivor of Tuol Sleng exclaimed back: 'Here are the words that I've longed to hear for 30 years!' (Bizot 2009).

In April 2009, 'Duch' offered an 18 minute non juridifiable court plea: 'I would like you to please leave an open window [possibility 1] for me to seek forgiveness [possibility 2]. I would like to express my regret and heartfelt sorrow' (Mydans 2009a). He presented the court with a pencil sketch of men at desks and piles of skulls that, he said, explained the workings of the regime's hierarchy. Nic Dunlop, a biographer of this case, discovered 'Duch' living incognito 10 years before that moment. He said that even had it been merely tactical, this apology would be significant. 'Duch's' cooperation and truth-telling would offer some of the historical clarification that Cambodians have been seeking. At a news conference, the defence lawyer, Martine Jacquin, noted: 'the most important thing is that he spoke today and expressed regret, remorse, and sought forgiveness, which was something the civil parties have been waiting for [for] a long time'. This defence suggested that 'Duch's' apology addressed traumas that had not been publicly acknowledged for three decades. But Robert Petit, one of the prosecutors, objected that 'Duch's' contrition had come too late.

> The accused was knowingly and intentionally in control of the entire Tuol Sleng criminal enterprise. Rather than a victim of fear, he was the one who created fear. Fifteen years later the accused would supervise the torture and execution of Chay Kim Huor, teacher and mentor [who recruited him into the Communist Party in 1964]. That single fact I submit as highly revealing.
>
> *(Mydans 2009a)*

But then 'Duch's' non-juridifiable address becomes the elephant in the room. *Who* is 'Duch'?

> I wanted to be a well-disciplined boy who respected the teachers and did good deeds. I never believed that the confessions I received told the truth. At most, they were about 40 percent true. Whoever was sent to S-21 was considered to be already dead. The work expanded, people were arrested illegally, right or wrong. I considered it evil eating evil eating evil.
>
> *(Mydans 2009c)*

In April, 'Duch' pleaded for the possibility: (1) that a window may be left open for the possibility; and (2) for him seeking forgiveness. Two days before the end of the trial,

Kav Savuth pursued for 'Duch' the Albert Speer defence, contradicting the plea (Cambodian Defense 2009):

> You can't cover an elephant with a rice basket, [but] Duch acted under duress obeying superior orders. We cannot ask for acquittal of the accused as well as enter a guilty plea for him. We are in the Albert Speer defense now. At Nuremberg, Speer did not plead guilty but he acknowledged his responsibility... Can 'Duch' still be useful to humanity?

Next day, the international defence counsel, François Roux, gave a witness account of 'Duch's' death and apparent resurrection. Because he will never be forgiven, 'Duch' should be sent home:

> But can we look 'Duch' in the eye and see him for the human he is? Will you bring "Duch" back into the fold of humanity? "Duch" is dead. Today his name is Kaing Guek Eav. He is no longer the "Duch" of the Revolution.
>
> *(Mydans 2009c)*

If one does not die, can one live? The difficulty of resurrecting the living dead resurfaced on the last day of the trial when Kav Savuth challenged the imprescriptibility statute that expired in 1989: 'This is like a person dying and then Resurrecting a dead body—that is impossible'. But it was the international counsel who proclaimed on a previous day an impossible death of this same agent who sought acquittal and resurrection in forgiveness. Negating the time limit on punishing crimes against humanity (their imprescriptibility) inscribes certain temporal infinity into the law. The capital punishment of genocidal offender betrays an already-vanishing limit of the imprescriptible law. Just as there is no positive time limit on achieving conditions for forgiving, so there is no secular 'quickie' resurrection or cheap grace at a non-juridifiable address, especially for a doer who is an elephant trying to hide under the rice basket.

Roux pressed on with precisely this theology of resurrecting the 'who' hidden under the rice-basket. He stressed the impact of dehumanization on the living person. Before dehumanizing their victims, executioners dehumanize themselves. Roux's claim echoes deeper existential truth: 'No one is born an executioner, one becomes so'. I am not a Christian, I am trying to become one (Kierkegaard); I am trying to become an atheist (Sartre); one is not born a woman, one becomes one (de Beauvoir). Roux reminded the court that he had declared 'Duch' dead on Thursday. Why should that unaccounted for dying be enough to underwrite 'Duch's' new life now? The court faced suddenly a former math teacher, who was not a torturer, Roux insisted. The co-prosecutor, William Smith, underscored the two self-contraindicating defence strategies advanced *uno tenore*. 'If his request is for an acquittal, that undermines his pleas of remorse and invites a longer sentence ... He comes to court but he is not facing up to what he was!' The perpetrator is hiding under the rice basket, and that is how the secular age blushes in embarrassment at hidden postsecular sensibility suffusing its rationalized sacred contents. 'Duch' did not challenge the material case of 12,380 deaths at S-21. 'I am responsible for crimes without any denial. I'm responsible for crimes as part of a criminal party [CPK]. I acknowledge that these people died at S-21'. But he conceived of his redemption without dying to his criminal self: 'I have learned from the psychiatrists that I need to be restored into the ambit of Humankind ... My ability to analyze is limited to what I can report. I would like the Chamber to release me' (Mydans 2009c). Can one live if one is not there (non-juridifiable address is lacking) to die?

Humanly unforgivable or the unforgivable in principle?

I would like to make the following claims:

- *Critique*: There is nothing redemptive or holy in conducting a war on the unforgivable in principle. The unforgivable deeds are not committed by moral monsters.
- *Moral restoration and punishment*: There is nothing wholly secular about irreparable violence and inexpiable destruction of hope. The infinite immigrates into the imprescriptible dimension of crimes against humanity.
- *The humanly unforgivable*: Even as the humanly irreparable exceeds moral repair and punishment, its painful face remains recognizably human, and it is this struggle with trauma that cannot remain unforgivable in principle.
- *Redemptive critique as a margin of hope*: There is nothing humanly satisfying in self-modifications that would no longer grapple with unforgiving.

There are radically evil deeds. Humans live and die unforgiven or unforgiving. Is unforgivingness same as the unforgivable in principle? 'Forgiveness forgives only the unforgivable'. 'Forgiveness must announce itself as impossibility itself' (Derrida 2001: 32, 33). Unforgiving calls one to struggle with rather than flee or forget the scene of trauma. Forgiving the humanly unforgivable is beyond ordinary human capacity, yet a wrongdoer, to be recognisable as a human doer of crimes against humans, cannot be regarded as an address that is unforgivable in principle. Such a doer beyond the reach of un/forgiving could not be faced as a *human* transgressor. She or he would be a member of a devilish race from another planet. Induced amnesia, willed forgetting, ideological modifications of historical transmission of trauma, management of brain proteins—these options for removing trauma do not satisfy the post-secular hope living at the address to which un/forgiving appeals. Struggling with one's and another's unforgivingness, whether it is therapeutic and spiritual change awaited at the scene of crime, invokes the dangerous memory of an agent with a human face.

Why does forgiving seem not only paradoxically impossible but also exceedingly difficult? The possibility of redemptive critical theory emerges when we articulate the publicly recognizable, yet untranslatable, asymmetries of unforgiving.

The first asymmetry opens between the depth of the wounding and the height of forgiving required for its healing. Healing cannot be redressed by a repayment schedule. There is no economy of exchange and no measurable moral debt that can be repaid to purchase healing once and for all. Healing can neither be commanded politically nor translated into publicly accessible procedural language. There is no juridifiable economy between crime against humanity and forgiveness, albeit there is a juridified economy between crime and punishment.

The second asymmetry lies between punishment that repays the imprescriptible deeds, morally and economically repairs the harm (the 'what')—and forgiveness. The latter unbinds the criminal and the victim (the 'who') from despair over the unforgivable. The matter of unforgiving lives like a remainder within the harm even when the deed is otherwise repaired.

The third asymmetry lies between the wounded moral, political, and legal space and hope lost in time. *The irreparable and the imprescriptible pertain to the deeds; the unforgivable addresses the persons.* The fault, its bearing, and admission as well as absolution speak to the unforgivable which consumes the self.

Redemptive critical theory articulates the boundaries between the publicly recognizable features of unforgiving and untranslatable asymmetries of transitive and intransitive hope. We require and assume hope as a necessary condition of our practical intent. But we cannot require forgiveness categorically or legislate it politically. How do we respond to the unforgivable—how is one responsible—without invoking theodicy or committing idolatry? One must neither translate hope into critique (I must resist idolatry), nor translate moral and legal repair into hope (I must not underwrite remembrance with theodicy). Redemptive criticism resists hopelessness as well as the destruction of hope. This critical awareness is postsecular in that after the crisis of modernity we must articulate hope after hope has been lost.

The difficulty of unforgiving reveals itself through excessive human cruelty with no tangible material economic, political, legal, or moral purpose (the 'what'). The works of cruelty sow gratuitous destruction of hope in oneself and another (the 'who'). This difficulty testifies to a very human breed of spiritual violence. Making humans superfluous, as Hannah Arendt says, must be registered at non-juridifiable addresses of the crime scene. If *unforgiveness* is not definable as mere absence of forgiveness, then human cruelty must not be ascribed to some metaphysical principle. Being like an elephant hiding under a rice basket, a self-envisioned as something 'in principle', one could never suffer difficulty of withholding or granting forgiveness. Un/forgiving always resides at some human address.

What do we learn from the Albert Speer defence strategy? Can a perpetrator claim resurrection without dying to the criminal self? The question of forgiving cannot be settled at a juridifiable, that is, purely secular, address. If the claim is nonetheless made within the legal proceedings (e.g., I am guilty and yet I merit forgiveness), then I cannot resurrect myself while under the rice basket. The one 'who' wills to be oneself in this manner does so in defiance, that is, wills to be oneself cruelly (Kierkegaard 1983). Cruelty is a positively enacted, that is, willed and self-accountable act. But then, at this cruel address, one finds no fast track to forgiving via the presumed Speer defence. There is no address at which one could cheaply resurrect the living dead. The difficulty of granting or withholding forgiveness is a human-all-too-human phenomenon. Even if there were no forgiving forthcoming or possible, the very act of unforgiving is a postsecular, that is, legally imprescriptible yet spiritually non-juridifiable phenomenon. The 'who' of trauma calls for more than moral or political repair, more than legal or economic reparation, more than a metaphysical deduction of the first principle. Un/forgiving is not something one catches or cures as flu. Or as Kierkegaard shows, the opposite of despair is not moral virtue (or reparations), but faith. Yet faith in this sense is a qualification of the self, and so faith must be distinguished from beliefs. Faith is a postsecular modality of existence, and beliefs are sacred contents that in the West have been rationalized (secularized) in one of the validity spheres of culture.

What does unforgiving do that other repair cannot?

Vann Nath (1998, 2002), one of the adult survivors of Tuol Sleng prison, served as a monk from the age of seven to twenty-one. Before enrolling in a private painting school in 1965, he became interested in painting while still attending school. 'I became very attracted to painting when I went into the pagoda and I saw people painting a picture on the side of the wall of a temple'. Vann Nath was arrested on 7 January 1978, while working in a rice field. The Khmer Rouge took him to Wat Kandal, the Buddhist temple used as a concentration camp. After a week of torture, he was sent to Tuol Sleng prison. 'Duch' made Vann Nath paint pictures of propaganda for the Khmer Rouge. Among them were several portraits of Pol Pot, the Khmer leader of Cambodia between 1975 and 1979, who died on 15 April 1998. Vann Nath was released from Tuol Sleng one year after his arrest, on 7 January 1979. He was on the list of the dead, but 'Duch' wrote next to his name, 'keep the painter'.

Another survivor, Bou Meng, was singled out from a row of shackled prisoners to produce portraits of Pol Pot. Chum Mey was spared because the torturers needed him to repair typewriters used to record the confessions that they extracted from prisoners like himself. Nhem En was spared as a photographer. He would say to the newly arrived prisoners as he removed their blindfolds and adjusted the angles of their heads: 'I'm just a photographer [one of six]; I don't know anything'. Every one of them would be killed. 'I had to clean, develop and dry the pictures on my own and take them to Duch by my own hand. I couldn't make a mistake. If one of the pictures was lost I would be killed' ('Conscience of Nhem En', 2009).

Vann Nath has been working after his release from the Tuol Sleng as a painter. He began painting scenes from the Tuol Sleng prison in Cambodia in late 1979 and early 1980. One of several torture scenes by Khmer Rouge shows waterboarding. Another scene recalls a month in the *D room* on the third floor where about 50 people were shackled together. Nath was forced to lie down and could only sit if he asked the guard for permission: 'It was like hell I can't describe it. At that time, in that condition, the hope that I had earlier had disappeared'. Many people died, and at night guards would come to remove the bodies. When Tuol Sleng prison became a Genocide Museum, Vann Nath wanted people to know what happened there: 'What I saw while we were in the prison was those people who used to scream for help. But we could not help them. I would like their souls to get something from what I paint'. The other two survivors, Bou Meng and Chum Mey, have continued to return to the scene of their imprisonment and torture as if their souls remained trapped there together with the souls of the dead. Some of Vann Nath's paintings hang now in the Tuol Sleng's *D room* as part of the historical evidence.

> Some believe and some do not believe, but they can come and look at my paintings as evidence, and make up their own minds. They can look at my pictures as testament to the atrocities carried out at Tuol Sleng.
>
> *(Vann Nath 1998)*

The Museum houses Nhem En's photograph portraits of the Tuol Sleng victims and other documents kept by the Khmer Rouge of the tortured confessions obtained by them from the prisoners before they were executed.

Un/forgiving defines the dual possibilities of bondage and releasement. I may be bound by despair, suffering, loneliness, and cruelty. I may be released by healing, mending the world, soundings of silence, and unconditional love. Four Hebrew words express such modalities of hope. *Teshuvah* unbinds the humanly unforgivable when I know that I have but one self-defence, 'I am more responsible than all the others' (Levinas 1981: 146 with reference to Dostoevsky 2003: 374, 386). *Tikkun olam* mends the unjust world. *Chasmal* (chash-silent, mal-to speak) addresses silence of unspeakable suffering. *Berakhah/Barakah* restores with unconditional love those who were wounded by wanton cruelty.

On 1 July 2008, Vann Nath opened his gallery and exhibition space at his *Kith Eng Restaurant* in Phnom Penh, displaying paintings documenting his time at Tuol Sleng (S-21) prison. The gallery is a permanent installation at the *Kith Eng Restaurant at 33B Street 169.* Nath hoped to build a retirement centre in Battambang where people who survived the Khmer Rouge regime can spend their last days in peace.

Sentencing the irreparable/punishing the immeasurable

'Duch's' case started with hearing on February 17 and 18, lasted from 30 March to 27 November 2009, and concluded with sentencing on 26 July 2010. The court heard 9 expert witnesses, 17 fact witnesses, 7 character witnesses, and 22 Civil Parties. More than 31,000 people followed the proceedings at the court building (see Mydans 2009b, 2017b). After the sentencing, Vann Nath's words choked in disbelief:

> It was just like a shock when I go there to the court and see him. When I tell them the truth they doubt me, ask me a lot of questions. I don't feel the trust when I tell them, and that makes me feel bad. It seems like the accused person has more rights than the civil parties do, and I'm really not satisfied with that. The verdict should be balancing what Duch has done, how many people he killed and how many he caused suffering. *For me, I can't forgive.*
>
> (*Cambodia's Khmer Rouge Trial 2010,* my emphasis)

At the conclusion of the trial on 27 November 2009, 'Duch' asked for an acquittal and release. With no death penalty in Cambodia, the prosecutors sought 40 years. 'Duch' was sentenced on 26 July 2010, and found guilty of crimes against humanity and war crimes. His sentence amounts to 35 years, of which 16 years were deducted from his sentence for the time already served. Should one outlive one's time in prison for imprescriptible crimes (Mydans 2010a)? The survivors were disheartened by the numerical value of years in prison for crimes that were not only imprescriptible but also beyond any satisfying temporal metric to punish proportionately (Mydans 2010b). 'Duch' and prosecutors appealed the sentence, the former for harshness, the latter for leniency. 'Duch's' sentence was changed to life imprisonment (McDonald 2012).

Moral remainders are the postsecular phenomena that remain in the secular age. Multiple layers of inverse evidence from the difficulty of unforgiving accumulate at the heart of justice and mercy. We confront the firewall of unyielding moral remainders just as we try to sentence the irreparable and punish the immeasurable. Running up against the fury of these moral remainders, why does human imaginary produce theologies of eternal punishment? Must there be hells worse than any life sentence? Should there be? Must there be dying more difficult than death? Should such living deaths last for eternity? In these imaginary constructions, we wrestle with difficult forgiveness. How are justice and mercy dispensed? How should they be?

Insofar as humans remember shame, they will have remembered hope. Aristotle praises shame as a quasi-virtue that safeguards our humanness. As long as humans struggle with the past, they retain umbilical ties to what Bernard Williams named 'moral remainders'. Williams would insist that if not all 'moral conflicts are systematically avoidable', then not all moral problems are 'soluble without remainder' (1973: 179). Card (2002: chap. 6) joins Aristotle's quasi-virtues of moral guilt with Williams' moral remainders. Victims, survivors, and their descendants embed the supra-locutionary power of moral remainders.

In *punishing the immeasurable,* I am offended by the heart of justice that fails to produce temporal payment schedules that could satisfy human retributive scales. 'No sentence could measure up to the atrocities Duch committed', said one eye witness. 'To come up with a number doesn't seem to make sense'. 'I'm not sure how you come up with a number'. Justice is not only blind, but is also late and lame (Mydans 2010b, 2017a).

In *sentencing the irreparable,* one is shocked by *infinite woundings at the heart of justice.* The heart of justice is wounded by infinite remainders of mercy. Can there be a satisfying substitution of another's suffering for my own to balance injustice? Can I extract such suffering for satisfying a just punishment? May the other offer sufferings to warrant redemptive repair? 'Even if we chop him ["Duch"] up into two million pieces', will justice be free of woundings?

In *confronting my unforgiving,* one shudders with *infinite woundings of the heart of mercy*. The heart of mercy is wounded with every cry for mending and healing. Is there an inward or therapeutic measure for healing wounded justice and suffering mercy? Is there a definitive measure how to address calls for mending and healing? 'I'm not sure how I should feel [after the sentencing of "Duch"]. I'm not happy, not sad, just kind of numb' (Cambodia's Khmer Rouge Trial 2010).

Conclusion

Withholding forgiveness requires equal if not greater strength than forgiving. When forgiveness seems performatively impossible, when perpetrators and sufferers live and die without it, then unforgiving is a curiously worthwhile difficulty to have even without imprescriptibility. Unlike Griswold (2007: 92f., 212f.), I do not hold that everything is humanly forgivable. The criminal self is not necessarily renewed in innocence when Griswold's (2007: 94) baseline conditions for forgiveness are met from each side. Must victims be culpable if they refuse to forgive after asked to do so three times? One may still suffer forgiveness as something humanly impossible.

Yet I do not claim that persons can ever become in principle unforgivable or that they would have to be consigned to the inner circles of eternal hell. To accept the possibility of lasting human unforgiving is human-all-too-human. Not withholding forgiving, but rather undoing its difficulty by other means (e.g., erasure of trauma from the brain paths), runs the greater danger of producing a posthumanist monster (Griswold 2007: 72–6, 94). We must, therefore, contend with a traumatic infinity, a postsecular reminder of something that may not be solvable by moral and legal means. Still from the dark side of unforgiving we glean this uncanny hope of the ultimate frontier that saves the human possibility in us.

Further reading

Dostoyevsky, F. (2003) *The Brothers Karamazov,* trans. David McDuff, New York, NY: Penguin Books.

In his last novel, the author articulates radical vision of forgiving and loving unconditionally. This work greatly impacted Levinas's supra-ethical response to the Holocaust to be compared with those given by Jankélevich, Derrida, and Hannah Arendt.

Kor, E. M. (2006) *Forgiving Dr. Mengele,* directed and produced by Bob Hercules and Cheri Pugh, Documentary film, First Run Features.

Eva Mozes Kor underwent in Auschwitz Josef Mengele's medical experiments on twins. Surviving the camp, she later forgives Mengele in order to heal herself.

Wiesenthal, S. (1998) *The Sunflower: on the possibilities and limits of forgiveness,* New York, NY: Shocken Books.

The author ponders his wartime encounter with a Nazi soldier who asked him for forgiveness on behalf of all Jews. Later, he poses a question to moral and religious leaders whether or not he should forgive.

References

Benjamin, W. (1968) *Illuminations,* trans. Harry Zohn, New York, NY: Shocken Books.

——— (1999) *The Arcades Project,* Cambridge, MA: Harvard University Press.

Bizot, F. (2009) 'My savior, their killer', *New York Times,* 27 February.

Cambodian Defense (2009) *Cambodia Trial Monitor,* Center for International Human Rights, NWU School of Law.

Cambodia's Khmer Rouge Trial (2010) *Religion and Ethics,* Video recording, 14 May, www.pbs.org/wnet/religionandethics/episodes/may-14-010/cambodias-khmer-rouge-trial/6280/

Card, C. (2002) *The Atrocity Paradigm: a theory of evil,* Oxford: Oxford University Press.
Conscience of Nhem En (2008) Director Steven Okazaki, USA, HBO documentary film.
Derrida, J. (2001) *On Cosmopolitanism and Forgiveness,* London: Routledge.
Fackenheim, E. L. (1996) *Jewish Philosophers and Jewish Philosophy,* edited by M. L. Morgan, Bloomington: Indiana University Press.
Griswold, C. L. (2007) *Forgiveness: a philosophical exploration,* Cambridge: Cambridge University Press.
Jankélevitch, V. (1996) 'Should we pardon them?', *Critical Inquiry,* 22(3): 552–72.
Kierkegaard, S. (1983) *The Sickness Unto Death,* edited by Edna H. Hong and Howard V. Hong, Princeton, NJ: Princeton University Press.
Lévinas, E. (1981) *Otherwise than Being or Beyond Essence,* trans. A. Lingis, The Hague: Martinus Nijhoff.
Matuštík, M. B. (2008) *Radical Evil and the Scarcity of Hope: postsecular meditations,* Bloomington: Indiana University Press.
McDonald, M. (2012) 'A life sentence for 14,000 deaths', *New York Times,* 6 February.
Mydans, S. (2009a) 'Khmer Rouge defendant apologizes for atrocities', *New York Times,* 01 April.
——— (2009b) 'Survivors shed light on dark days of Khmer Rouge', *New York Times,* 17 May.
——— (2009c) 'Legal strategy fails to hide torturer's pride', *New York Times,* 21 June.
——— (2010a) 'Khmer Rouge figure is found guilty of war crimes', *New York Times,* 25 July.
——— (2010b) 'Anger in Cambodia over Khmer Rouge sentence', *New York Times,* 26 July.
——— (2017a) '11 Years, $300 million and 3 convictions: was the Khmer Rouge tribunal worth It?', *New York Times,* 10 April.
——— (2017b) 'Khmer Rouge trial, perhaps the last, nears end in Cambodia', *New York Times,* 23 June.
Nath, V. (1998) *A Cambodian Prison Portrait: one tear in the Khmer Rouge's S-21,* Bangkok: White Lotus Press.
——— (2002) 'The survivor of Tuol Sleng', *CBC News: Sunday,* Broadcast 27 October, Website Credits Copyright ©CBC.
Williams, B. (1973) *Problems of Self: philosophical papers 1956–1972,* Cambridge: Cambridge University Press, pp. 166–86.

4

The postsecular condition and the genealogy of postmetaphysical thinking

Eduardo Mendieta

German and European public intellectual

Habermas has been engaged in the German and European public sphere for more than half a century (Matuštík, 2001; Specter 2010; Müller-Doohm 2016). He was engaged in the debates in the late 60s with the youth movement around questions of the use of violence; he held a public debate with Niklas Luhmann on the conservative dimensions of systems theory; he debated Gadamer on similar grounds on the lack of self-reflexivity of certain versions of hermeneutics; he was a vociferous critic of German historians who advocated for a 'normalization' of German history that in Habermas' view would lead to a normalization of the holocaust; he advocated for a revision of the German law on naturalization and path to citizenship of non-Germans (Holub 1991); over the last two decades, furthermore, he has advocated the constitutionalization of the EU, and, in tandem, he has been a severe critic of nationalistic, xenophobic, and racist discourses that aim to close borders and exclude allegedly non-whites and non-Europeans from the EU. It is against Habermas' position on constitutionalizing the EU, the modification of the citizenship laws in Germany, and his call for a postconventional moral position that takes the form of what he calls 'constitutional patriotism' that his adoption of the term postsecular must be contextualized (Müller-Doohm 2008).

In this narrow political sense, postsecular for Habermas means that European civil society and public sphere must see itself as impartial towards all forms of religious belief and practice. While the assumption and general belief is that Europe has become secular, the fact is that this secularism masks an unspoken and all too evident form of religiosity, namely, Christian religion that suffuses institutions, practices, beliefs, and assumptions. The European social imaginary is suffused by a Christian/Jewish religious imaginary. The dominant form of Christian secularism that configures European civil society in fact acts as a double shield: it protects an unspoken Christianity, while rejecting and excluding other forms of religion: Islam. In fact, since World War II (WWII), Europe has become a continent of immigrants, many coming from former European colonies, and also many from Islamic countries in which Europe was and remains entangled for colonial and imperial reasons. These immigrants have brought their religions and cultural practices, creating a new *Kulturkampf*, a new cultural war. It is in this context that the term postsecular is meant to unmask Christian secularism, a secularism by means of which Christianity maintains its grip on civil society, while pointing to the social fact of the resilience of both Christian and non-Christian religious beliefs and practices in European and also non-European societies (see Butler 2011).

Influences on Habermas' postsecular

Dialogue is not simply a metonymic term that encapsulates the alpha and omega of Habermas' thinking; it is also a lived practice that is the germinal of his thinking (see Wingert and Günther 2001; Habermas 2008). Over the span of his intellectual career, Habermas has become known for his engagement with other well-known figures: Luhmann, Gadamer, Foucault, Derrida, Rorty, Vattimo, Benhabib, Fraser, Taylor. There are shelves of volumes devoted to critical engagement with Habermas' thinking, many closing with a generous and extended reply. Habermas is truly a philosopher of dialogue. His corpus bears the imprint of other thinkers' affirmation, agreements, but also disagreements and dissent. As I have indicated in several essays (see my introduction to Habermas 2002; Mendieta 2013, forthcoming), over his long intellectual career, Habermas has paid close attention to religion, from different angles and through different lenses. Arguably, he began thinking about the relation between religion, philosophy, history, and society as early as his 1954 doctoral dissertation on Schelling. Still, given German and German-Jewish preoccupation with religion, it is not by accident that Habermas also has had to deal with the debate not simply between Athens and Jerusalem, but also those between Athens and Rome (and the Vatican), Frankfurt, and Freiburg. Close to his intellectual home, the Frankfurt School, several thinkers in the tradition have had a lot to say about religion, faith, secularization, and critique that have also been a reservoir for his own thinking. Quite evidently, there is the work of Walter Benjamin and Ernst Bloch, who never shielded their theological and religious concerns. Theodor W. Adorno and Max Horkheimer, who had a more guarded relationship to their Jewish background, also did not keep these questions at bay. Instead of wondering why a rationalist and Enlightenment figure like Habermas would take up a term like postsecular, we ought to wonder what if he had resolutely refused to engage with these questions and his own immediate tradition. To a large extent, Habermas' engagement with religion, theology, and in general the philosophy of religion has been substantive, long-lasting, and, in many ways, unsurprising and *de rigueur.*

Now, with respect to Habermas' adoption of the term postsecular, four immediate figures appear decisive. First, there is his long-term collaborator and colleague Klaus Eder, who collaborated with him during Habermas' tenure as the director of the Max Planck Institute for the Study of the Scientific-Technical World, at Starnberg (1971–1983). Eder was a key interlocutor, especially with respect to Habermas' adoption of the terms of postconventional societies. Eder is a sociologist who has focused on the evolution and transformation of social systems. In the early 1990s, as part of his work on the emergence of postnational societies, Eder began to use the term postsecular precisely to refer to the resilience of religious beliefs and practices in societies across the world (see Eder 2002; Bosetti and Eder 2006; see also Chapter 1).

Another key interlocutor has been José Casanova, who, already in the early 1990s, with his book *Public Religions in the Modern World* (1994) challenged the European myth of secularization by pointing to the global importance of 'public religions' in the Europe. He noted, for instance, the key role Catholicism played in the opening up of some then-Soviet countries, and the role of Protestant and Catholic denominations in the USA and Latin

America. He also drew attention to the opposite movements of evangelicals in the south of the USA, and the liberation theology movements in Latin America. Casanova is the one sociologist who has contributed most extensively to articulating the relationship between religion, civil society, and the public sphere. As a sympathetic critic, Habermas has paid close attention to Casanova's articulation of the resurgence of religion in the global public sphere (see Habermas 2017).

The other obvious point of reference is, of course, Cardinal Ratzinger, with whom Habermas had a very public dialogue on 19 January 2004. Ratzinger would be named pope in 2005, taking the name of Benedict XVI. This public dialogue received world-view attention precisely because, on the one side, you had a major representative of the Enlightenment, secular tradition, and, on the other, a key representative of the Catholic Church. Ratzinger, in fact, was the prefect of the Catholic Church's Congregation for the Doctrine of the Faith, the role of which was to attend to the orthodoxy of all Catholic missions. This was indeed a most unique dialogue: on the one side, an advocate of inclusiveness and the power of deliberative reason, on the other, the guardian of faith and orthodoxy (Habermas and Ratzinger 2006; Habermas 2017).

Habermas' incisive and provocative contribution to the dialogue, titled 'Pre-political Foundations of the Constitutional State?', begins with the unsuspecting question whether constitutional democracies presuppose motivational resources which themselves can neither proffer nor create. This is an unsuspecting question as Habermas seems to imply that there are both moral and ethical orientations that antecede and are constitutive of a democratic temper that democratic practices cannot and do not generate. It is here where religious beliefs and practices became significant for Habermas. In this text, Habermas also talks about 'secularization' as a twofold and complementary 'learning process'. It is twofold because non-believing citizens learn to articulate their 'secular' reasons within a plural public sphere, and believing citizens learn to translate their beliefs into reasonable arguments in a non-homogeneous civil society and public sphere. This ability to be polyglot citizens about reason-giving and reason-taking, as a 'learning process', is what Habermas calls 'postsecular' consciousness (Habermas and Ratzinger 2006: 21–39).

Perhaps the most explicit source for Habermas' adoption of the term postsecular is John Rawls. Although Habermas had already engaged Rawls work, as early as the early 1970s (Habermas 1985a, 1985b), it was during the early 1990s that Habermas began a dialogue in earnest with Rawls' work, and in particular *Political Liberalism* and the *Laws of Peoples* (Habermas 1998). Among some of the keys under contestation was Rawls' idea that well-ordered societies aim to bring about an overlapping consensus on the basic ideals that would orient the pursuit of the public good. This means that 'substantive theories' or ideas with substantive commitments that may appeal only to specific faiths or religious beliefs would have to be bracketed and cordoned off behind a firewall, insulating them from the political debate determining a specific common good. Habermas, in contrast, thought that the substantive beliefs of religious citizens should be among the many ideas and perspectives to inform a robust public debate. Excluding, bracketing, or censuring the beliefs and commitments of religious citizens from the outset favoured secular or non-religious citizens while demeaning and discriminating against religious citizens. For Habermas, this was not the basis for establishing a well-ordered, just, and inclusive democratic society, one that furthermore nourished itself from the plurality of beliefs that nonetheless form part of their respective lifeworld. Unfortunately, due to Rawls stroke and early death, their debate could not be continued. Still, Rawls remains the key point of reference for Habermas' thinking on religion and the postsecular. This latter term is pivotal for a self-reflexive deliberative democracy that aims to guide debates on the common good in accordance with what Habermas, in *Between Facts and Norms* (1996), called the D principle:

> A rule of action or choice is justified, and thus valid, only if all those affected by the rule or choice could accept it in a reasonable discourse.
>
> *(Habermas 1996: 107)*

'A reasonable discourse', here, would be one that would not from the outset invidiously label the beliefs of religious citizens as either irrational or unacceptable in the public sphere. Thus, while non-religious citizens cannot from the outset discriminate against religious reason in the public sphere, religious citizens must engage in a process of translation and reason-giving that aims to make their commitments relevant and worthy of consideration for all. In later versions of Habermas' articulation of the importance of religious beliefs in the public sphere, he refers to, on the one hand, the 'semantic contents' that lay as if inexhaustible and ever renewable in religious beliefs and practices, and to the need for 'translation', of these contents into reasons that could or would be acceptable to all citizens after a 'reasonable' discourse. This translation, however, goes both ways: religious and non-religious citizens must translate their reasons into reasons that could meet the reasonable assent of all citizens, regardless of their commitments to substantive comprehensive theories of the common good. In a deliberative, constitutional democracy, there are no reasons that cannot be part of the public debate about what may orient and inform the deliberation about the basic principles that ensure the dignified treatment of all members of the polity.

How Habermas deploys the postsecular

Over the past decade, in light of his debates with Rawls and Ratzinger, as well as dialogues with Taylor, Butler, West, as well as Bellah, Assmann, and Joas, Habermas has undertaken to write a major treatise on what, at first, he called 'An Essay on Faith and Reason'. The manuscript in its latest iterations has turned into a massive manuscript on what he now calls 'The genealogy of post-metaphysical thinking: with a focus on the leitmotif of the dialogue of faith and reason'. Parts of this manuscript have been presented as lectures in several venues, and some abbreviated versions of chapters have been published, for instance, in the five-volume edition of what could be called 'Habermas' Greatest Philosophical Hits', published by Suhrkamp as a Student Edition to celebrate his 80th Birthday in 2009 (Habermas 2009b). Some of the essays have also been published in *Postmetaphysicial Thinking II* (Habermas 2017). In general, however, the unpublished manuscript is an analogue of Habermas' other *magnus opus*, *The Theory of Communicative Action* (Habermas 1984/1987). If this latter aims to lay the foundation for a socio-theoretical account of communicative action, the unpublished manuscript aims to provide a genealogy of enlightenment of reason out of the very sources of religious debates about what is faith and what is reasonable. What can be stated at this moment is that Habermas offers a breathtaking historical, reconstructive account of the emergence of what he calls postmetaphysical thinking (Habermas 1992). This undertaking by Habermas is one that few, even those who have followed his work closely, myself included, could have anticipated or foretold in any way.

For my purpose, this unfinished monumental manuscript contains extensive discussion on the notion of secularization. In these discussions, which are part of the book length introduction, Habermas makes a distinction between secularization and secularism. The latter refers to a particular ideological formation that conceals and masks the evident non-secularization of the lifeworld. Secularism is a self-deceiving ideological distortion that aims to claim superiority for allegedly secular societies, in opposition to non-secularized societies, generally associated with non-Western societies. Secularism is an inauthentic version of secularization. Secularism, as a form of ideology in the pejorative sense, has been promoted by certain sociological theories of social development, the most famous being those articulated by Max Weber and Emile Durkheim, for whom modernization and secularization went hand in hand. In light of Eder, Taylor, and Casanova's objections and critiques, Habermas now recognizes that a sociological account of the modernization of society does not demand the coupling of secularization to modernization processes. There can be different paths to modernity that do not involve the rejection or exclusion of religion from either civil society or the lifeworld of democratic citizens. In this manuscript, thus, Habermas calls for the uncoupling of the process of secularization of political power and institutions, from the modernization of different institutions and cognitive orientation of modern subjects.

For these reasons, then, secularization refers to the process of the domestication of political power and institutions by their uncoupling from specific religious beliefs and institutions (such as churches). The secularization of political power and institutions is a decisive achievement of both Western and non-Western societies, one that should be neither undermined nor rejected. In this sense, secularization is part and parcel of what Habermas calls 'learning processes', which are achievements but, for that very reason, are also fragile and unstable. The highest achievement of the secularization of political power, in this analysis, is the constitutionalization of political power, a constitutionalization that establishes the co-originarity of popular, that is, national, sovereignty, and the inalienable, rights of citizens.

What secularization cannot do and should not do is thoroughly 'disenchant' the lifeworld that is the indispensable background against which, and from within which, citizens and political subjects develop their own cognitive and moral orientations. The lifeworld, most importantly, is not simply the praxical and hermeneutical background that informs the communicative action of social agents; it is also the ensemble of practices, habits, and activities within which social agents form their moral and political commitments. The lifeworld is the reservoir that nourishes social agents' deep commitments to certain beliefs and practices that enrich their deliberations about the common good. If we were to use Habermas language from *The Theory of Communicative Action*, we could talk about the colonization of the lifeworld by the secularization of state power, one that would result in the impoverishment of the motivational resources of social agents. By doing so, in fact, and as Habermas already elucidated in his debate with Ratzinger, constitutional democracies would deprive themselves of major resources that both antecede and are indispensable for their health and inclusiveness. A thoroughly secularized lifeworld would lead to a moral and ethical desert, one that would eventually erode the very foundations of constitutional democracies.

Using Habermas' language from *Between Facts and Norms* (1996), democratic deliberations in the public sphere within constitutional rule of law act as 'conveyor' belts that translate ethical commitments and insights, tapping into the semantic contents of religious beliefs, into the constitutionalization of procedural norms that are oriented towards the standpoint of justice—the fair and equal treatment of all citizens. Every democratic iteration is nourished by substantive, overarching, and comprehensive doctrines that enact translation that then become institutionalized into formal and procedural rules of democratic legislation (see Habermas 2009a: 59–78).

As a synopsis, it could be said that the motivations for Habermas' use of the postsecular have to with sociological, political, philosophical, and moral reasons. Postsecular names a sociological condition, and perhaps even a social factum: that religion has neither been extinguished and that it remains a vital dimension of social life in a form that cannot be simply reduced to negative reasons. Faith, in particular, and religion, in general, are not simply atavistic remnants from earlier stages of social evolution, but co-evolving and integral dimensions of it. As Habermas has argued, religion was integral in the linguistification of the sacred, and thus, in the linguistification of social interaction. To the extent that the semantic reservoirs of faith and its related religious rituals remain unexhausted, they continue to catalyze the processes of the linguistification of social interaction.

Postsecular is also meant to deal with the *factum* of religious pluralism within rule of law democracies. In this narrow sense, postsecular directs our attention to the imperative to respect the religious beliefs of each and every citizen, even when these beliefs may seem alien and may not have roots in the dominant traditions. We could say that Habermas uses the term postsecular to name a form of *democratic enlightenment*, one in which citizens's faith commitments are respected and regarded as contributions to the ethical pluralism of the community, which in turn is seen as a resource not only of dignity but of moral insight for the entire political community.

The postsecular also names a particular philosopheme, namely, the perennial question of philosophy's relationship to faith. In fact, this philosopheme was captured in St. Anselm's felicitous motto *fides quaerems intellectum*. Postsecular is the latest iteration of 'faith seeking understanding' in the aftermath of the failures of secularist self-understanding of modern society and the still ongoing processes of secularization.

Finally, postsecular also refers to a moral imperative, to respect the moral autonomy and integrity of moral agents. As is well known, at the core of Habermas' theory of communicative action are the core validity claims that every speech act raises: truth, rightness, and truthfulness (or authenticity). Morality is not simply the orientation towards justice, as the postconventional moral stage demands, but also towards the right kind of motivation. It is not enough to aim towards what is just, but one must do so for the right reasons and with the right conviction. Morality also has to do with the depth of one own's subjectivity, the theatre of one's inner moral life. This is where truthfulness enters. Postsecular as a term that has some moral valences points precisely in this direction: to the depth of subjective life, the degree of authenticity with which one lives one's moral life. The postsecular further expands and deepens the topography of moral subjectivity, one that nonetheless remains on the same plane as having to give accounts of ourselves to others through rational discourses.

Conclusion

A final and concluding remark is in order. In an interview I conducted which first appeared in the *Immanent Frame* (Habermas 2010), Habermas made an important clarification, which was a response to some confusion that his use of the term 'postsecular' instigated (Habermas 2017: 59–76). Habermas makes it clear that postsecular refers to both a societal condition and a self-reflexive orientation of postconventional, postnational democracies. The resilience and renewal of religious orientations and practices are an unquestionable fact of modern societies, even of those that could be said to be secular. Postsecular, in this context, means an enlightened orientation of plural societies that are part of what Habermas now calls, following Niklas Luhmann, world society. In short, postsecular consciousness is the self-awareness of societies that now inescapably see themselves as part of a world society that also shares elements of a world public sphere.

Postsecular, notes Habermas, does not and cannot refer to a new version or articulation of reason. There is postmetaphysical thinking, or a reason that has turned itself to its own self-clarification and reflexivity. Reason, *tout court*, is neither secular nor postsecular. What there is, however, is postmetaphysical thinking, that is, a particular way to understand reason: one that is non-identity thinking, non-substantive—which means it does not rely on metaphysical principles—and not hubristic—which means that philosophy has surrendered its kingly throne to an interdisciplinary research agenda pursued under the watchful eye of fallibilism. Postmetaphysical thinking is enlightened reason, reason that understands itself as historically located, procedural, and fallibalistic. Postsecularism, then, could be understood as further manifestation of postmetaphysical thinking or its fulfilment and realization, in the Left-Hegelian sense of the actualization of a concept's potentiality. Postsecular consciousness, then, is what results when postmetaphysical thinking opens reason to reasons that itself can neither dismiss nor extinguish, but on which it still can nourish itself.

Further reading

Forst, R. (2017) *Normativity and Power: analyzing social orders of justification*, trans. C. Cronin, Oxford: Oxford University Press.

The first English translation in English of Forst's influential *Normativität und Macht* (2015), this book addresses the key normative questions current in social theory and political philosophy. Reflecting the latest developments in Frankfurt style critical theory, the work discusses the relevance of classic authors (Bayle, Kant, Marx, Habermas) to contemporary debate on topics including the idea of power, progress, justice, toleration, transnationalism, and democracy.

Habermas, J. (2018) *Philosophical Introductions: five approaches to communicative reason*, trans. C. Cronin, Cambridge, MA: Polity Press.

An indispensable text for students and scholars in philosophy and critical theory today, this book provides a unique and comprehensive overview of Habermas' philosophy—including the challenge posed by religion in a secular age—in his own words.

Mendieta, E. (2011) 'Rationalization, modernity, and secularization'. In Fultner, B. (ed.) *Habermas: key concepts*, Durham: Acumen, pp. 222–38.

In this contribution to a compendium on Habermas, Mendieta addresses the question of religion at the centre of not just sociology and philosophy, but also political theory. There has been no philosopher, Habermas included, who has not grappled with the question of the dependence of philosophy on religious notions. The question has been posed in terms of the possibility of a dialogue, or subterranean co-dependence, between Athens and Jerusalem, where each is a metonym for reason and faith, respectively.

References

Butler, J. (2011) 'Is Judaism Zionism?'. In Mendieta, E. and J. VanAntwerpen (eds.) *The Power of Religion in the Public Sphere*, New York, NY: Columbia University Press, pp. 70–91.

Casanova, J. (1994) *Public Religions in the Modern World*, Chicago, IL: Chicago University Press.

Bosetti, G. and K. Eder (2006) 'Post-secularism: a return to the public sphere', interview with G. Bosetti, *Eurozine*, 17 August. www.eurozine.com/post-secularism-a-return-to-the-public-sphere/

Eder, K. (2002) 'Europäische säkularisierung: ein sonderweg in die postsäkulare gesellschaft', *Berliner Journal für Soziologie*, 12(5): 331–44.

Habermas, J. (1984/1987) *The Theory of Communicative Action*, 2 vols, trans. T. McCarthy, Boston, MA: Beacon Press. [*Theorie des kommunikativen Handelns*, 2 vols, Frankfurt: Suhrkamp, 1981)]

——— (1985a) *Die Neue Unübersichtlichkeit: kleine politische schriften V*, Frankfurt: Suhrkamp.

——— (1985b) 'Civil disobedience: litmus test for the democratic constitutional state', trans. John Torpey, *Berkeley Journal of Sociology*, 30: 95–116.

——— (1992) *Postmetaphysical Thinking: philosophical essays*, trans. W. Hohengarten, Cambridge, MA: The MIT Press, [*Nachmetaphysisches Denken* (Frankfurt: Suhrkamp, 1988)]

——— (1996) *Between Facts and Norms: contributions to a discourse theory of law and democracy*, trans. W. Rehg, Cambridge, MA: The MIT Press. [*Faktizität und Geltung: beiträge zur diskurstheorie des rechts und des demokratischen rechtsstaats* (Frankfurt: Suhrkamp, 1992)]

——— (1998) *The Inclusion of the Other: studies in political theory*, trans. C. Cronin, Cambridge, MA: MIT Press. [*Die Einbeziehung des Anderen: Studien zur politischen Theorie* (Frankfurt: Suhrkamp, 1996)]

——— (2002) *Religion and Rationality: essays on reason, god and modernity*, Cambridge, MA: The MIT Press.

——— (2003) *The Future of Human Nature*, trans. H. Beister, W. Rehg and M. Pensky, Cambridge, MA: Polity Press. [*Die Zukunft der menschlichen Natur: Auf dem Weg zu einer liberalen Eugenik?* (Frankfurt: Suhrkamp, 2001)]

——— (2008) *Between Naturalism and Religion: philosophical essays*, trans. C. Cronin, Cambridge, MA: Polity Press. [*Zwischen Naturalismus und Religion: Philosophische Aufsätze* (Frankfurt: Suhrkamp, 2005)]

——— (2009a) *Europe: the faltering project*, trans. C. Cronin, Cambridge, MA: Polity Press. [*Ach, Europa* (Frankfurt: Suhrkamp, 2008)].

——— (2009b) *Philosophische Texte: studienausgabe in fünf Bänden*, Frankfurt: Suhrkamp.

——— (2010) 'A postsecular world society? On the philosophical significance of postsecular consciousness and the multicultural world society', an interview with Jürgen Habermas by E. Mendieta, in SSRC, *The Immanent Frame* 1. http://blogs.ssrc.org/tif/wp-content/uploads/2010/02/A-Postsecular-World-Society-TIF.pdf.

——— (2017) *Postmetaphysical Thinking II: essays and replies*, Cambridge, MA: Polity Press.

Habermas, J. and J. Ratzinger (2006) *The Dialectics of Secularization: on reason and religion*, San Francisco, CA: Ignatius Press.

Holub, R. C. (1991) *Jürgen Habermas: critic in the public sphere*, New York, NY: Routledge.

Lutz-Bachmann, M. (Forthcoming) 'Enlightenment'. In Allen, A. and E. Mendieta (eds.) *The Cambridge Habermas Lexicon*, Cambridge: Cambridge University Press.

Matuštík, M. B. (2001) *Jürgen Habermas: a philosophical-political profile*, London: Rowman and Littlefield.

Mendieta, E. (2013) 'Appendix: religion in Habermas' work'. In Calhoun, C., Mendieta, E. and J. VanAntwerpen (eds.) *Habermas and Religion*, Cambridge, MA: Polity Press, pp. 391–407.

——— (Forthcoming) 'Religion'. In Allen, A. and E. Mendieta (eds.) *The Cambridge Habermas Lexicon*, Cambridge: Cambridge University Press.

Müller-Doohm, S. (2008) *Jürgen Habermas (Suhrkamp BasisBiographien)*, Frankfurt am Main: Suhrkamp Verlag.

——— (2016) *Habermas: a biography*, Cambridge, MA: Polity Press.

Specter, M. (2010) *Habermas: an intellectual biography*, Cambridge: Cambridge University Press.

Wingert, L. and K. Günther (eds.) (2001) *Die Öffenlichkeit der Vernunft und die Vernunft der Öffenlichkeit: festschrift für Jürgen Habermas*, Frankfurt am Main: Suhrkamp Verlag.

5

Redemptive criticism or the critique of religion

Warren S. Goldstein

Introduction

We live in an age of the 'posts', declared Steven Best and Douglas Kellner (1997). One such post is postmodernity, marked by the rise of religious fundamentalism; it represents a break from modernity, which was thought to go hand in hand with secularization (Best and Kellner 1997: 30). However, Habermas (1987a) was not a fan of postmodernism, and due to the resurgence of religion, he felt a need to develop his own terms to capture the relation of religion to modernity: the postmetaphysical, and later on the postsecular, which is the 'most influential and widely debated version of the concept that has been developed thus far' (Moberg et al. 2012: 4). Habermas (2017) admits that distinguishing new phenomena by adding the 'preposition "post"' is a 'fashion' that suffers from a lack of precision. Nevertheless, many distinguished scholars have jumped onto the postsecular bandwagon; it has become the topic of numerous books, edited volumes, and journal articles. Habermas' iconic status has contributed to this phenomenon.

Religion was previously off Habermas' radar. His interest in it has occurred only lately—particularly since the events of 11 September 2001 (Mendieta and VanAntwerpen 2011; Calhoun et al. 2013). This latest resurgence of religion, which can be traced back to the rise of the religious right in the USA and the Iranian Revolution of 1979, escaped his attention. His embrace of the term postsecular has been driven by the ongoing resurgence of religious activity (Habermas 2017). According to him, religion has not only 'remained an effective historical force', but its influence worldwide is growing (*ibid.*: 124–5, 214). Conservative and orthodox religious groups are advancing everywhere. Habermas has given repeated disclaimers about his research on religion. He admits his lack of knowledge in secularization theory (Habermas 2017: 78), biblical studies (*ibid.*: 98), theology (*ibid.*: 155), or even the practices of any religious community (*ibid.*: 122). Most of his earlier works only peripherally engaged with religion. His writings on religion can best be described as philosophy of religion, which has little engagement with empirical research on religion.

When Habermas employs the term 'religion', he does so in a general and unspecific way by treating it 'as a monolithic and reified phenomenon' (Dillon 2012: 252). Despite the debates that have raged within religious studies, he employs religion as a universal concept that can be applied globally. He writes, 'every religion is originally a "*worldview*" or comprehensive doctrine'; 'it claims the authority to structure a form of life *as a whole*' (Habermas 2008: 111). However, as he admits, religion no longer exercises a monopolistic position when it comes to worldviews. Due to functional differentiation, it is simply one sphere among many others. Discussing varying forms of religiosity in a nuanced way like paying attention to denominational differences is not something that Habermas does. Rather than describing a broad spectrum of religiosity (see, for example, Roof and McKinney 1987), he identifies only two 'modern forms of religious consciousness'. The first is fundamentalism 'that either withdraws from the modern world or turns aggressively against it'. The second is reflective and attempts to have a positive relationship with other religions; it respects science and human rights (Habermas 2017). If one were to treat these as ideal types and discuss the variations in between, this would be a more accurate description. However, Habermas does not do this.

Critique of religion

While claiming to carry the mantle of critical theory, the question this article asks is how critical is Habermas when it comes to religion? Habermas explicitly refrains from a strong critique of religion. His approach to religion more closely follows the soft redemptive criticism of the Frankfurt School, which, reflecting the historical circumstances in which its members lived (during which secularization was on the advance), was more concerned with the dangers imposed by Fascism, anti-Semitism, or advanced industrial capitalism. If anything, the Frankfurt School saw the emancipatory potential within the messianic tradition and its secularization in Marxism. However, Habermas is writing during a different era, during which there has been a global resurgence of religion, in particular fundamentalism, most notably Christian and Islamic. Despite this, for the most part, he refrains from a critique of religion.

While critical theory engages in a soft redemptive criticism, the foundations of it are in the much stronger critique of religion (Kant, the Left Hegelians, Marx, Nietzsche, and Freud). Critical theory views 'itself as the "successor" to theology' and is caught in a paradox because the morality upon which it rests ultimately stems from it (Habermas 2002: 99). While Habermas (2017) concedes that the militant rejection of religion of the Enlightenment was once justified due to alignment of the church with absolute monarchies, he thinks that in contemporary Western Europe, this position no longer has a place. While admitting that the critique of religion is regrettably topical due to the rise of religious fundamentalism, Habermas confesses that his interests lie elsewhere. This is due to his location in Europe, which in comparison to the rest of the world, is highly secularized. Habermas declares that the 'conflict between anthropocentric and theocentric' worldviews is 'yesterday's battle'. He is more interested in 'incorporating central contents of the Bible into a rational faith' than in fighting against 'priestcraft and obscurantism' (Habermas 2008: 211). He accuses those who engage in a critique of religion as merely confirming their own beliefs about the religious other who they do not understand. Rather he suggests 'trying to comprehend religion in a self-critical way' (Habermas 2017: 144). His critique is not directed outward, but rather inward. The critique of religion should not come from those who are secular, but from those who are religious:

> A 'modernizing' self-enlightenment of religious consciousness is only able to be successful when it is undertaken by each religious tradition from the inside, because in the end

> the community of believers must decide for itself whether the 'reformed' faith that has become reflexive is still the 'true' faith.
>
> *(Habermas 2017: 133)*

Despite Habermas' attempt to refrain from a critique of religion, periodically it slips through. He accuses religious people of being uncritical. He writes that while scientists are 'open to criticism', religious people hold on to a revealed truth that they are unwilling to revise (Habermas 2008: 260). Theology is not able to engage in 'critical self-revision' as long as it promises salvation and people believe in it (Habermas 2017: 114–5). Those who are secular are more likely to treat other perspectives equal to their own. In contrast, those who are religious are more likely to be close-minded. For the believer, other ethics and lifestyles are 'not only different but *mistaken*'; treating them equally is an imposition (Habermas 2008: 263). The strongest critique of religion of Habermas' is when he discusses fundamentalism. It is one of 'the fastest-growing religious movements' and it poses a challenge to secularized European societies (Habermas 2017: viii). Habermas (2002: 151) defines fundamentalism as a return to 'premodern religious attitudes'. It provides a 'false answer' in modern secular society (*ibid.*: 151, 2017: 71). Among those groups he considers to be fundamentalist are not only radical Muslims but also Pentecostals (Habermas 2017: 212). Habermas is harshest on Islamic Fundamentalism, which he says is able to unleash the 'violence innate in religion' (*ibid.*: 213). But beyond this, he spends scant little time analyzing it.

Critique of secularism and modernity

Rather than engaging in a critique of religion, Habermas prefers to direct his critique towards secularists and secularism, and along with them towards modernity. He distinguishes between those who are 'secular' and those who are 'secularist'. Someone who is secular does not believe; they take an agnostic position towards the validity claims of religion. Secularists, on the other hand, take 'a polemical stance' towards religious traditions, doctrines, and beliefs, even though these claims cannot be scientifically validated one way or the other (Habermas 2017).

Habermas (2017) calls extreme secularists, 'Enlightenment fundamentalists'. They have a rigid conception of secularization, see religion as an 'obsolete "intellectual formation"' (*ibid.*: 221), and are fixated on criticizing it. Habermas (2008: 140, 143) thinks that secularists should engage in 'self-reflection' and calls for a 'critical overcoming' of narrow secularist mindsets. While religion has gone through its own reformation, Habermas (2017) questions whether dogmatic secularists are capable of doing the same. Religion for Habermas provides normative content, which secularism and modernity have destroyed. The threat to these norms comes from the outside from reactionary anti-modern fundamentalists and from the inside from the process of modernization itself (Habermas 2008).

Secularization

Habermas' position on the postmetaphysical and postsecular needs to be seen in the context of Western European society, in which secularization has advanced the furthest but is confounded by the immigration of religious Muslims. In order for postmetaphysical thinking to gain a self-understanding, it needs to start with the question of secularization (Habermas 2017). Although familiar with the secularization debate, Habermas is not deeply immersed in it. He recognizes Europe as the exceptional case, and the USA as the norm. Habermas sides with those who defend the secularization thesis. Like José Casanova (1994), he thinks that the aspect of secularization that has occurred is differentiation—that is, a separation of church and state. As a result of the process of differentiation, religion as a subsystem now only engages in 'pastoral care'; it has lost its other functions (Habermas 2017: 213).

John Rawls, who thought that secularization and modernization do not go hand in hand, is a significant influence on Habermas. Rawls observed that although many parts of the world have become modernized, secularization has not always accompanied it. Based on this, he did not think that religion was destined to disappear (Habermas 2017). Even though he agrees with Rawls that religion is not in immanent decline, Habermas does not think that there can be a reversal of the secularization of state power. In other words, he rejects secularization as a decline in religion while accepting the differentiation thesis (see Casanova 1994). Habermas' socio-evolutionary view that there is the release of semantic potentials contained in sacred complexes does not mean that religion is disappearing (Habermas 2017). For him, it is an open question whether religion remains a viable option.

Even though Habermas' exchange with Joseph Ratzinger is given the title *The Dialectics of Secularization*, there is, in fact, no dialectic of secularization described in it. If there were, there would have been a better explanation of the relationship between the postsecular and the secular. Such an understanding of secularization is based on the recognition of it as a process of secular and religious movement and countermovement (Goldstein 2016). Habermas (2008: 310) does have a 'dialectical understanding of cultural secularization', which separates secularization from modernization but sounds more like a process of religious rationalization. He writes that non-European societies will only be able to resist the process of secularization if they discover alternative paths to modernity and if their religious consciousness becomes receptive 'to modernization *from within*' (*ibid.*: 311). After the 9/11 attacks, Habermas came to see the process of secularization in the Occident as incomplete (Mendieta 2013). If secularization is an ongoing process, the question is in what pattern it occurs. One of these patterns, observed by a critical theory of religion, is a dialectical one (see Goldstein 2016).

The secular state

Since Habermas thinks that secularization is a process of differentiation, one of his primary concerns is the relationship that secular and religious citizens should have with each other in a secular state. In pluralistic societies, the secular state takes a neutral stance towards both religious and ideological groups (Habermas 2008, 2017). At the same time, in such societies, religions must renounce their claim to govern life entirely (Habermas 2008). Those who are religious must adapt to the secular state and assimilate its 'principles under the premises of their own faith' (Habermas 2017: 222). The secular state expects the support of religious minorities since their freedom depends on it. The state should not open itself up to the conflict between different religious communities. Otherwise, a religious majority could use it to impose its will on a minority. However, just because the state is neutral towards competing religions, it does not mean that religion should be banished from the public sphere.

Methodical atheism and agnosticism

In his earlier writings on the postmetaphysical, Habermas (2002) adopted the position of methodical or methodological atheism, which he saw as a form of demythologization. Methodological atheism forbids importing theology (Habermas 2017). Later on, Habermas shifts his position, arguing that the postmetaphysical thinking is agnostic: it has an 'ambivalent attitude' towards religion, is unable to decide which aspects of it are rational, and refrains from making judgements about its truth claims. While Habermas (2017: 97) is not willing to criticize 'the dogmatic foundations' of the beliefs of those who are religious but who he considers to be enlightened, by accusing them of being dogmatic, he is in fact criticizing them. To the contrary, he claims that postmetaphysical thinking, in its agnostic position, is willing to engage in a dialogue with theology; it is willing 'to learn from religion' and has an open mind about its fate (Habermas 2008: 143).

Communicative action

Habermas' (1987b) desire to have dialogue between those who are religious and those who are secular is based on the ideals he expressed in his *The Theory of Communicative Action* (TCA). Postmetaphysical thinking, which embraces these ideals, is derived from the idea of communicative reason expressed in TCA (Habermas 2017). Habermas' (1992: 44) theory of communication action marked his 'linguistic turn', which led out of an impasse 'between metaphysical and antimetaphysical thinking' (i.e., idealism and materialism) to a philosophy of language. Key to this turn is that actors are able to communicate with each other linguistically and are thus able to come to an understanding with each other by taking into consideration each other's validity claims with the hope of achieving consensus and coordinating action (Habermas 1987b, 2017). While having this hope, which is idealistic, Habermas (1992) is aware of the paradox in retaining it. When Habermas calls for dialogue between those who are religious and those who are secular, the real dialogue that he engages in is mostly with theologians (Habermas 2017). However, he is realistic when it comes to the possibility of achieving consensus, and therefore expects 'continuing disagreement' (*ibid.*: 152).

Kant

The origins of the postmetaphysical are contained in the writings of Kant who Habermas (2008, 2017) considers to be the first 'postmetaphysical' thinker and who, for him, provides the foundation for postmetaphysical thinking. Kant engaged in a 'self-criticism of reason'. He used theoretical or pure reason (*Vernunft*) to critique the metaphysical tradition, and practical reason (*Verstand*) to critique Christian doctrine (Habermas 2008: 210). Kant's critique of the limitations of both pure and practical reason was supposed to protect against 'metaphysical claims to knowledge' and the 'religious truths of faith' (*ibid.*: 243). Kant, according to Habermas, did not want to 'reduce the philosophy of religion to the critique of religion' (*ibid.*: 215). His priority was on the critique of metaphysics rather than on the critique of religion. Through his critique of pure reason, he was able to dispel its illusions leaving room for practical reason to deal with more scientific empirical tasks and the construction of a morality freed from religion. Kant did this to guard against two types of dogmatism: the orthodoxy of the Church and the unbelief imposed by the Enlightenment.

Kant, who Habermas uses to justify his own position, wanted to 'rescue the contents of faith', which could be justified by reason (Habermas 2008: 211). Habermas is interested in using Kant's philosophy of religion to assimilate the semantic content of religious traditions while maintaining the boundary between faith and knowledge. For Kant, preserving the rational content of religion is achieved by the 'critical and self-critical differentiation between' faith and knowledge (*ibid.*: 228). Habermas (2017: 124) is concerned 'with *the limits of postmetaphysical thinking*, which' can be clarified using Kant's premises and can help determine the relationship between faith and knowledge (Habermas 2008: 243). Kant critically appropriates 'religious contents on a rational basis', but does not always do so in a critical manner (Habermas 2008: 212, 2017: 126). For example, he appropriates aspects of religious traditions such as the biblical idea of the 'kingdom of God on earth', which he translates into the metaphysical concept of the 'highest good' without acknowledging its origin. He did this, according to Habermas, because he wanted to replace a positive religion with a rational one (Habermas 2008: 216, 228, 2017: 124). The project marking the boundary between reason and religion 'in a self-critical way' therefore needs to proceed in a 'hermeneutically more cautious' manner (Habermas 2017: 124).

While Kant engaged in a critique of religion, Habermas for the most part does not. For Kant, religion is both a heritage and an opponent; it is simultaneously a source of morality and a place of 'obscurantism and zealotry'. Habermas (2008: 227) thinks that Kant's goals of 'assimilating religious contents' and engaging in a critique of religion conflict with each other—that one cannot simultaneously engage in a critique of religion while at the same time appropriate some of its contents. Here Habermas is mistaken and forgets the very purpose of critique is to evaluate both the positive and negative aspects of any given religious tradition (Goldstein et al. 2013). It is only through the critique of religion that one can rescue its rational contents.

Hegel and beyond

Habermas' embrace of the postmetaphysical involves him moving beyond his earlier Hegelian view of religion, in which he thought that religions would 'be dialectically superseded in the modern world' (Habermas 2017: 143). Since then, his perspective on religion has changed, as it has not only been able to survive in modernity but has even flourished (Habermas 2017). He has modified his position and now describes himself as 'as a student of Hegel who thinks in postmetaphysical terms' (*ibid.*: 97–98). Hegel was the last thinker who was able to maintain a synthesis between metaphysics and philosophy (Habermas 2002). Hegel, Schleiermacher, and Kierkegaard thought that Kant, 'as a critic of religion', was a prisoner of the Enlightenment who 'stripped religious traditions' of their substance (Habermas 2008: 229). While Kant, Hegel, and Marx translated the content of religion into secular form, Schleiermacher and Kierkegaard, who Habermas considers 'Christian and postmetaphysical thinkers' (*ibid.*: 232), gave religion and philosophy equal status.

The Left Hegelians (Feuerbach and Marx) radicalized the Kantian critique of religion in a materialistic direction by prioritizing 'practical over theoretical reason' (Habermas 2008: 231). Their critique of religion replaced 'the idea of the kingdom of God on earth' in a secular and revolutionary form in a vision of an emancipated society (*ibid.*: 231). The Left Hegelians and Kierkegaard initiated a rupture between the metaphysical and the postmetaphysical (Habermas 1992, 2002). There are two different approaches to religion that Habermas distinguishes between: a rationalist and a dialogical one. A rationalist approach, which comes out of the Hegelian tradition, subsumes (*aufhebt*) 'the substance of faith into the philosophical concept', while a dialogical approach, which comes from Karl Jaspers, adopts 'a critical attitude toward religious traditions while' '*learning* from them' (Habermas 2008: 245).

The Frankfurt School: Adorno and Benjamin

Whereas Habermas rejects the strong secularist critique of religion, found in figures like Kant, the Left Hegelians, Marx, Nietzsche, and Freud, his own approach is more influenced by the soft critique of the Frankfurt School, in particular Theodor Adorno and Walter Benjamin. Habermas (2017) considers Benjamin as a religious writer and sees figures in Western Marxism like Benjamin, Adorno, and Ernst Bloch as engaging in an 'atheistic assimilation of religious contents' (Habermas 2008: 232).

Habermas interprets Adorno along the lines of his own postsecularist thinking by repeating the same quotation of Adorno who wrote: 'nothing of theological content will persist without being transformed; every content will have to put itself to the test of migrating into the realm of secular, the profane' (Habermas 2017: 82, 87, 147). This insight, according to Habermas, raises the question whether the semantic contents of religion can be detached from religious tradition and achieve validity in modern secular society. Adorno, like Kant, thought that this 'critical appropriation' of the 'unutilized potentials' of religion was the secular equivalent of realizing 'the kingdom of God on earth' (*ibid*.: 147). Along the same lines, Habermas conjectures that the only messianic hope that Adorno retained was in the history of philosophy.

Rather than engage in a critique of religion, Habermas (1979, 2017) follows the approach of redemptive criticism that he characterized Walter Benjamin as having. He sees redemptive criticism as a form of consciousness-raising, which preserves 'endangered semantic potential' (Habermas 1979: 50). It seeks to recover images from myth and rescue the lost utopian contents of religious tradition (Habermas 1979, 2017). Under the influence of Benjamin, Habermas (2017) is interested in the secular emancipation of semantic potentials from the messianic tradition. Benjamin along with the young Bloch, Derrida, and Levinas teaches us that philosophy should have 'a receptive and dialogical attitude toward *all* religious traditions' (*ibid*.: 203).

The postmetaphysical

Prior to Habermas' embrace of the postsecular was his development of the term postmetaphysical. Habermas uses it in two senses: the first is methodological; it 'concerns procedures and conceptual means'. The second is substantial; it describes agnostic positions, which distinguish 'between belief and knowledge' without giving validity to any particular religion but, at the same time, not denying their content (Habermas 2008: 245). 'Postmetaphysical thinking is a secular form of thought', which 'coexists with religion' (Habermas 2017: 64, 149, 204). It does not fit within an 'immanent frame' that lacks transcendence (*ibid*.: 82). Postmetaphysical thinking is not able to replace or repress religion as long as its content is inspiring. While the postmetaphysical is a break with the metaphysical, religion continues to coexist alongside it because it escapes explanation; it resists translation into more rational discourses (Habermas 1992: 51).

Prior to the development of postmetaphysical thought, all cultural systems developed within 'a sacred framework' (Habermas 2017: xii). The rise of empirical sciences forced the transition from the metaphysical to the postmetaphysical (Habermas 1992: 37). Postmetaphysical thinking's goal was to deconstruct religious perspectives (Habermas 2017: 96). The postmetaphysical, however, refrains from a 'critique of metaphysics' because it is supportive of 'the autonomy of self-conscious subjects' (Habermas 2008: 278–9). Postmetaphysical thinking does not resort to metaphysics; it 'recovers the meaning of the unconditional without recourse to God or an Absolute' (Habermas 2002: 108). Postmetaphysical philosophy is unable to provide any consolation for human suffering. It is not indifferent to morality but uses it without employing God. Rather, in the spirit of communicative action, it defers moral authority to 'the tribunal of justificatory discourse' (Habermas 2017: 145).

Postmetaphysical thinking has a dialogical relationship to all religious traditions—one that is 'open to learning' (Habermas 2017: 64). Religious teachings are equal to other worldviews. Religious interpretations are part of the 'the legitimate discourses of modernity' just as much as postmetaphysical thinking, so long as they are modernized (*ibid.*: 78). Secular and religious citizens should meet each other 'on an equal footing'; the contributions of both are equally relevant (*ibid.*: 173, 223). Postmetaphysical thought and the major world religions have a shared genealogy (*ibid.*: 78–9). In this genealogy, a process of differentiation took place. Secular thinking and religious consciousness split 'into polar opposites', and metaphysical claims lost their validity (*ibid.*: 63, 78). Postmetaphysical thinking accepts this differentiation, albeit with 'a critical intent' (Habermas 2008: 246); it enables this genealogy to be 'critically overcome' (Habermas 2017: 67). Postmetaphysical thinking is groundless unless it borrows from some traditional authority, which is why it retains a connection to the 'semantic potentials of religion' (*ibid.*: 93, 147). On the other hand, postmetaphysical thought engages in a critique of reason enabling us to see that it too is fallible and has limits.

Translation and linguistification of the sacred

In order to borrow the semantic content of religion, postmetaphysical thinking must translate it into secular language. Postmetaphysical philosophy is not able use religious language of any specific tradition, but it must translate it into a universal discourse, which is secular:

> Under the conditions of postmetaphysical thinking whoever puts forth a truth claim today must, nevertheless, translate experiences that have their home in religious discourse into the language of a scientific expert culture—and from this language retranslate them back into praxis.
>
> *(76)*

Postmetaphysical thinking seeks to discover the 'semantic potentials in religious traditions' and transfer their meaning into a generally accessible everyday language for those who do not belong to that tradition. It facilitates the migration of religious content from the realm of the sacred to that of the profane (Habermas 2017: 126). Habermas (2017: xii, 2013: 398) refers to this as 'the "linguistification of the sacred"', the result of which is disenchantment.

The model of this translation is critical theology, exemplified by figures like Johann Bapist Metz and Francis Schüssler Fiorenza (see Chapter 14), who engaged in a 'theological protest' against rigid and reductionist forms of metaphysics (Habermas 2002: 69–71, 74, 76). Standing on the secular side, Habermas shies away from engaging in a critique of religion himself, but would rather see this done on the religious side from within through self-criticism as exemplified by critical theology. Both postmetaphysical thinking and religious thinking are capable of self-critique. The reformation of both religious consciousness and a postmetaphysical thinking that engages in a 'critique of reason' is the answer to the challenges posed by the Enlightenment. 'Postmetaphysical thinking and 'reformed' religious consciousness' are capable of learning from each other (Habermas 2017: 81, 94).

Today, religious traditions are confronted with a plurality of other traditions, both religious and secular. In this situation, they are forced to self-reflect back on the validity of their own beliefs and the rationality of their own practices. Habermas (2002: 150) argues that criticism must come from within: only when faith engages in 'self-criticism' can it be included in a discourse that is shared with other religious traditions and is limited by secular knowledge. One of the questions Habermas (2008: 138) poses regarding translation is whether there is an '*assymetrical* burden' placed upon religious citizens in comparison with those who are secular. In order for religious citizens to have a state that is neutral to competing religions and worldviews and in order for religious communities to achieve their moral goals, they must accept the 'translation proviso' (Habermas 2008: 130, 136, 2017: 172–3). They must translate their own religious reasons into a generally accessible secular language before they can be implemented. In the secular state, all political decisions must be made in an equally accessible language and justified by 'universally acceptable reasons' (Habermas 2008: 131, 134, 139, 2017: 71, 102, 204).

Habermas places an equal burden on those who are secular. They must give up the prejudices that they hold for those who are religious. They should not deny the possible truth content of their beliefs but rather take their claims seriously. Neither should they deny those who are religious the right to express themselves in religious language. Rather, they have the obligation to translate the relevant contributions of their fellow religious citizens from religious language into a publicly accessible secular language (Habermas 1992: 4, 2008: 113, 310, 2017: 204–5). Religious contents have a 'rational core' that is capable of being translated. To get at it, one must separate the wheat from the chaff (Habermas 1992, 2008, 2017). Religious convictions should not merely be seen as 'irrational from the perspective of secular knowledge' (Habermas 2008: 112). Secular citizens have an obligation to open their minds to the 'possible truth content' of those who are religious (*ibid.*: 132). At the same time, they ought to be 'self-reflective' about their own secularist assumptions (*ibid.*: 138). Postmetaphysical philosophy is open to the rationality of religious traditions and willing to learn from them through the process of translation (*ibid.*).

The conclusion that Habermas (2008, 2017) comes to is that the burdens placed upon both sides are not asymmetrical even though the translation is only one way—from the religious to the secular. He does not want those who are secular to criticize those who are religious (nor does he want to do this himself). Instead, he calls for a self-criticism on the part of those who are secular as well as those who are religious. Critical theology takes this role within religion, but those who are secular also need to be self-critical of their own secularist assumptions. The secular state should not impose an unreasonable burden upon its religious citizens. They should be allowed to express themselves in religious language and take part in political decision-making even if they cannot find secular translations for their ideas. Religious citizens need to realize that their religious principles are compatible with the secular state and society, in which religious pluralism and 'the authority of science' prevail (Habermas 2008: 134).

Postmetaphysical philosophy is based on the principle of tolerance not only between different religious groups but also between those who are religious and those who are secular. However, there is not an equal burden of tolerance for secular and religious citizens. Unlike for those who are religious, for the secular citizen, a pluralism of worldviews and lifestyles does not create 'any cognitive dissonance' (Habermas 2008: 263).

The translation of each religious community's core convictions into democratic principles can help to find 'a shared language' which can bridge 'irreconcilable religious differences'. Through the use of reason shared by both secular and religious citizens, both sides are capable of adopting 'a common standpoint' and transcending their differences (Habermas 2017: 105). However, Habermas (2008) thinks that it is reasonable for there to be continued disagreement between the two sides. The act of translation does have a critical component. Habermas' intention in translating is aimed 'at critically appropriating certain articles of faith' (Habermas 2017: 139). Reason is critical of religion; its goal is to redeem 'religious contents through translation' (Habermas 2008: 228). At the same time, secular reason has its own limits, which can be obtained through a self-critical assessment (*ibid.*: 139–40).

Morality

In a secularizing act, Habermas (2008) wants to appropriate the moral content of religion, while separating morality from it. Historically, morality has been embedded in religious traditions. The onset of the process of secularization associated with modernity raises the potential for the loss of morality (Habermas 2017). Habermas questions whether modernity can make up for this loss through the generation of its own norms. To do this, it must look to religious traditions. Religions have a moral content and provide motivation which secular perspectives lack. Habermas (*ibid.*) questions whether the appropriation of religious contents by philosophy can serve as a counterbalance to the lack of morality in modernity. However, he concludes that anthropocentric and theological perspectives cannot be easily converted into one another because it is not 'a zero-sum game'; religious experiences cannot be translated into secular discourse without there being a remainder (Habermas 2017: 149). In other words, in order to translate religious into secular content, one must also discard.

Faith and knowledge

Postmetaphysical thinking emerges out of the potentially symbiotic relationship between faith and knowledge (Habermas 2017). There is a difference of what it means that something is true in the realm of faith as opposed to that of knowledge. For postmetaphysical thinking, faith and knowledge are two different ways of considering something to be true. Faith and knowledge are 'complementary intellectual formations' which both 'belong to the genealogy of postmetaphysical thinking' (*ibid.*: 123). They have a shared rational basis, which became severed because philosophy lost its connection with theology since it wanted to keep up with science. Scientific knowledge became emancipated from metaphysics, and subsequently faith and knowledge have become polarized.

The Enlightenment conception of the relationship between faith and knowledge does not do justice to traditional religious teachings because it claims to have a greater knowledge about religion than it does (Habermas 2017: 133). Rather, philosophy based on the Enlightenment has its limits and needs to be 'self-reflexive'. 'The relation between faith and knowledge' needs to be determined 'self-critically from the perspective of secular knowledge' (Habermas 2008: 112). Both the religious and secular should have an understanding of the relationship between 'faith and knowledge that enables them to live together in a self-reflective manner' (Habermas 2017: 224). Here I think Habermas overplays the parallels between faith and knowledge. The fact that they became separated from each other only indicates that one lagged behind. Following an Enlightenment perspective, I would argue that acquiring knowledge is rational, while belief, which rests upon faith, is not. Habermas in his refrain from a critique of religion steers away from a critique of faith.

Philosophy between religion and science

The same lines of tension that exist between faith and knowledge also exist between religion and science. Jose Casanova (2013: 47–8) accuses Habermas (2013) of having a 'consciousness of stages' in which philosophy sits in an intermediary stage between religion and science. Habermas' answer to this is that he does not see religion as being devalued in postmetaphysical or postsecular thought. Philosophy, according to Habermas, is caught between religion and science, but can play a moderating role between the two. Philosophy and religion have a common genealogy that has a critical function (Habermas 2017). Philosophy has 'appropriated semantic contents from the Judeo-Christian tradition' (*ibid.*: 173). It can grasp the rational content of religion only if it recognizes that the imposition of religious orthodoxy upon it is unacceptable (Habermas 2008).

Philosophy is secular; it severed its relationship to religion, renouncing 'metaphysical claims to knowledge' and gave up competing with it (Habermas 2008: 272, 2017: 5, 14, 203). Yet, philosophy is not able to replace religion because, unlike religion, it is unable to provide consolation for human suffering (Mendieta 2013). Philosophy and religion have become differentiated but at the same time have developed a complementary relation to each other (Habermas 2017). Philosophy and religion have 'a dialogical relationship' to each other; the two sides can learn from each other through dialogue (Habermas 2017: xiv, 94). Philosophy's relationship to religion is also dialectical; it incorporates some of the contents of religion as its own in a modified secularized form (Habermas 2008). While philosophy draws on a 'knowledge-based tradition', religion relies on theology (Habermas 2017: 98). Theology, which was once connected with science (e.g., Aristotle), has lost its relationship to it (Habermas 2017: 203). Theology is 'the guardian of faith' and rests on it. Unlike philosophy, it cannot afford to expose its 'validity claims to criticism' (*ibid.*: 141).

Postmetaphysical thinking mediates between religion and the sciences. Philosophy developed in conjunction with 'the modern empirical sciences' and has contributed to a disenchanted world (Habermas 2017: 14). 'Philosophy is not a science' but secularism misleads it into thinking so (*ibid.*: xiii, 67). In contrast to the sciences, which attempt to be objective, philosophy 'shares with religious and metaphysical "worldviews"' a 'self-reflexive attitude' (*ibid.*: xiii).

The postsecular

Habermas' notion of the postmetaphysical culminates in that of the postsecular, which he develops later. The postmetaphysical is a genealogical perspective used to achieve self-understanding that occurs in the realm of philosophy, whereas the postsecular is 'a sociological predicate' that occurs from the observer's perspective and refers to a type of society (Habermas 2008: 140, 119, 2017: 63, 215). While the two are related, Habermas does not mean to equate them with each other. Postmetaphysical thinking is secular even 'in a situation depicted as post-secular; but, in this different situation, it may become aware of a secularistic self-misunderstanding' (Habermas 2017: 62). (Note that Habermas use the hyphen 'postsecular' in his writings.) The misunderstanding is the misinterpretation of the theory of secularization that religion was destined to disappear. The emergence of postsecular society represents a 'conservative turn' (Habermas 2008: 111); it corresponds with the rise of the religious right. Among the societies Habermas (2017) considers to be postsecular are Western Europe, Canada, and Australia, which are fairly secularized. Noticeably, he excludes the USA from this categorization.

When Habermas uses the predicate postsecular, it does not mean a reversal of the process of secularization, but rather a shift in the consciousness of a secularized society—ones which have had to adjust to the fact that religious groups continue to flourish and be quite vocal in the public sphere despite ongoing secularization (Habermas 2008, 2013, 2017; Casanova 2013). A postsecular consciousness, in contrast to a secularist attitude, leads to a change in attitude towards religion (Habermas 2017: 64). It places into question a linear theory of secularization 'toward a nonreligious future' (Wickström and Illman 2012: 219). While the limitation of the postsecular to consciousness may be true of most of Habermas' writings on the concept, it is not consistently true in the way in which he uses it and is not logically consistent. It does not hold true for his exchange with Joseph Ratzinger (Habermas and Ratzinger 2005), in which he refers to society itself as being postsecular. Nor does it make logical sense. How can there be a change in consciousness (beliefs) without a change in the practices in society? The very notion of the postsecular is triggered not only by the persistence of religion in modern society but by in its resurgence. Nevertheless, for the most part, Habermas uses the term postsecular in a prescriptive way—as a way for those who are secular to deal with those who are not, and vice versa.

As Charles Taylor (2007) has done in *A Secular Age*, the more interesting question would be whether we have entered a postsecular age. Indeed, the postsecular implies 'that we have left the secular era behind' (Gorski et al. 2012: 1). If there had not been a reversal of the process of secularization with the rise of religious fundamentalism, even with the persistence of religion, would there be a need to use the term postsecular? By Habermas' (2017: 210) own admission, one can describe a society as 'post-secular' only if it has been secular in the past. However, if the postsecular referred to an age, Habermas would not view it as such. He writes, 'as measured by the usual sociological indicators of the religious beliefs and practices the' population has not 'changed to such an extent as to justify labelling these societies post-secular' (*ibid.*: 211). But he does not provide evidence to support this claim. Which sociological indicators? Is it church participation, belief in God, or the many other ways one can measure religiosity in surveys? In which countries? Over which time period? If he were more deeply engaged with empirical research, the picture would not be as clear-cut.

Habermas sets up mutual obligations for the interaction between the secular and religious in a postsecular society. The awareness by those who are secular that they are living in a postsecular society is due to their recognition of the persistence of religious communities. At the same time, those who are religious must adapt to a society that is becoming increasingly secular (Habermas 2008). In this context, the secular and the religious have something to learn from each other, which assumes that they both have an understanding of secularization 'as a *complementary* learning process' (Habermas 2008: 111; see also 140, 2017: 223). This learning process ensures that religious voices are given a legitimate place in postsecular society (Habermas 2017). So, what do they have to learn from each other? Those who are secular will find that their own 'normative truth contents' are rooted in religious traditions (Habermas 2008: 131). However, while they should be tolerant of religion, they should not fall prey to religious dogma.

The postsecular is a situation in which those who employ 'secular reason' have a relationship with those who have a reflexive 'form of religious consciousness' (Habermas 2017: 64), the latter seeming to refer to critical theologians or their equivalent. Excluded from this view are those who are religious but not reflexive, which could be a significant portion of the religious population. Regardless, the postsecular aims at civil social relations between the religious and the secular in a multicultural and pluralistic society, but it presents its own paradox. Bernstein (2013: 175) asks 'if secular reason is intrinsically a critique of faith, then how could there be a postsecular society that was not a repudiation of reason?' Indeed, this refrain from external critique is a refrain from exercising one's reason (or at least the voicing of it), granted that both theology and religion can be rational within certain parameters.

Conclusion

In his embrace of the postmetaphysical and postsecular, Habermas follows redemptive criticism rather than the critique of religion; however, they are not mutually exclusive. The process of critique involves rescuing the rational content of religion but at the same time discarding the mystical shell. Engaging in only one while refraining from the other is really not possible since, in order to rescue or redeem, one also has to discard or negate. The reformation of religion has never come solely from within; it has often been the result of external worldly pressure. Criticism must therefore not only come from within; it is also aided by criticism from without. In what comes across as an attempt to be polite and not to offend, while Habermas encourages dialogue, he avoids debate. For the religious and the secular to live together in a democratic postsecular society, they need to engage with each other even on the most contentious of issues—their beliefs and the epistemological basis upon which they rest. While secular reason can critique faith, faith traditions can critique the anomie of secular society, and it is in this latter aspect that the secular and religious would find they have much in common. Religious traditions, like secular ideologies, have both repressive and emancipatory potentials. However, as long as religious traditions, like secular powers, become mired in the pursuit of wealth and power, engage in dishonesty and corruption, resort to violence and bloodshed, and hold on to untruths to achieve these ends, there is still a crucial place for the critique of religion.

Further reading

Boer, R. (2007–2014) *On Marxism and Theology*, 5 vols, Leiden/Boston, MA: Brill (hardcover); Chicago, IL: Haymarket Books (paperback).

This encyclopaedic compendium explores the intersection between Marxism and theology in major Marxist and critical theorists.

Hewitt, M. A. (1995) *Critical Theory of Religion: a feminist analysis*, Minneapolis, MN: Fortress Press.

This book brings feminist theology into conversation with the Frankfurt School and derives from it the basis for a feminist critical theory of religion.

Siebert, R. J. (2001 [1985]) *The Critical Theory of Religion: the Frankfurt School*, Lanham, MD/London: Scarecrow Press.

This groundbreaking book derives a critical theory of religion from the writings of members of the Frankfurt School (including Walter Benjamin, Theodor Adorno, Erich Fromm, and Jürgen Habermas, among others).

References

Bernstein, J. M. (2013) 'Forgetting Issac'. In Calhoun, C., Mendieta, E. and J. VanAntwerpen (eds.) *Habermas and Religion*, Cambridge, MA: Polity Press, pp. 154–75.

Best, S. and D. Kellner (1997) *The Postmodern Turn*, New York, NY: The Guilford Press.

Calhoun, C., Mendieta, E. and J. VanAntwerpen (2013) 'Introduction'. In Calhoun, C., Mendieta, E. and J. VanAntwerpen (eds.) *Habermas and Religion*, Cambridge, MA: Polity Press, pp. 1–23.

Casanova, J. (2013) 'Exploring the postsecular: three meanings of the "secular" and their possible transcendence'. In Calhoun, C., Mendieta, E. and J. VanAntwerpen (eds.) *Habermas and Religion*, Cambridge, MA: Polity Press, pp. 27–48.

——— (1994) *Public Religions in the Modern World*, Chicago, IL: University of Chicago Press.

Dillon, M. (2012) 'Jürgen Habermas and the post-secular appropriation of religion: a sociological critique'. In Gorski, P. S., Kyuman Kim, D., Torpey, J. and J. VanAntwerpen (eds.) *The Post-Secular in Question: religion in contemporary society*, New York: New York University Press and Social Science Research Council, pp. 249–78.

Goldstein, W. S. (2016) 'The dialectical pattern of secularization: a comparative-historical approach'. In Wendel, S. and J. Könemann (eds.) *Religion, Öffentlichkeit, Moderne: transdisziplinäre perspektiven*, Bielefeld: Transcript Verlag, pp. 19–39.

Goldstein, W. S., Boer, R. and J. Boyarin (2013) 'Editorial', *Critical Research on Religion*, 1(1): 3–8.

Gorski, P. S., Kim, D. K., Torpey, J. and J. VanAntwerpen (2012) 'The post-secular in question'. In Gorski, P. S., Kyuman Kim, D., Torpey, J. and J. VanAntwerpen (eds.) *The Post-Secular in Question: religion in contemporary society*, New York, NY: New York University Press, pp. 1–22.

Habermas, J. (1979) 'Consciousness-raising or redemptive criticism: the contemporaneity of Walter Benjamin', *New German Critique*, 17: 30–59.
——— (1987a) *The Philosophical Discourse of Modernity: twelve lectures*, Cambridge, MA: The MIT Press.
——— (1987b) *The Theory of Communicative Action, Volume Two, Lifeworld and System: a critique of functionalist reason*, Boston, MA: Beacon Press.
——— (1992) *Postmetaphysical Thinking: philosophical essays*, Cambridge, MA: The MIT Press.
——— (2002) *Religion and Rationality: essays on reason, God, and modernity*, Cambridge, MA: The MIT Press.
——— (2008) *Between Naturalism and Religion*, Cambridge/Malden, MA: Polity Press.
——— (2013) 'Reply to my critics'. In Calhoun, C., Mendieta, E. and J. VanAntwerpen (eds.) *Habermas and Religion*, Cambridge, MA: Polity Press, pp. 347–90.
——— (2017) *Postmetaphysical Thinking II: essays and replies*, Cambridge, MA: Polity Press.
Habermas, J. and J. Ratzinger (2005) *The Dialectics of Secularization: on reason and religion*, San Francisco, CA: Ignatius Press.
Mendieta, E. (2013) 'Appendix: religion in Habermas' work'. In Mendieta, E. and J. VanAntwerpen (eds.) *The Power of Religion in the Public Sphere*, New York, NY: Columbia University Press, pp. 391–407.
Mendieta, E and J. VanAntwerpen (2011) 'Introduction'. In Mendieta, E. and J. VanAntwerpen (eds.) *The Power of Religion in the Public Sphere*, New York, NY: Columbia University Press, pp. 1–14.
Moberg, M., Granholm, K. and P. Nynäs (2012) 'Trajectories of post-secular complexity: an introduction'. In Moberg, M., Granholm, K. and P. Nynäs (eds.) *Post-Secular Society*, New Brunswick, NJ: Transaction Publishers, pp. 1–25.
Roof, W. C. and W. McKinney (1987) *American Mainline Religion: its changing shape and future*, New Brunswick, NJ: Rutgers University Press.
Taylor, C. (2007) *A Secular Age*, Cambridge, MA: Belknap Press of Harvard University Press.
Wickström, L. and R. Illman (2012) 'Environmentalism as a trend in post-secular society'. In Moberg, M., Granholm, K. and P. Nynäs (eds.) *Post-Secular Society*, New Brunswick, NJ: Transaction Publishers, pp. 217–38.

6

Postatheism and the phenomenon of minimal religion in Russia[1]

Mikhail Epstein

The end of atheism and the varieties of post-atheistic experience

The seven decades of Soviet atheism (1917–1987), whether one calls it 'mass atheism', 'scientific atheism', or 'state atheism', was unquestionably a new phenomenon in world history. Mass heterodox movements and heresies were known before, but these developments did not change the core of a religious perception of the world; they did not suspend the belief in God, the Holy Scriptures, or the possibility of the soul's salvation. The German Anabaptists are a case in point (Bowman 1995; Durnbaugh 2003; Meier 2008). In the past there have been periods of libertine thought, but these touched only the intellectual tip of the social iceberg without altering the religious mood of the masses. The French Enlightenment is a relevant example here (Chartier 1991; Gomez 2003; Bronner 2004).

Only in the Soviet Union did militant atheism penetrate into the masses and create three generations of non-believers. Even if they themselves did not destroy churches and burn icons, neither did they pray or invoke the name of God; indeed, they forgot about His very existence. Even if the majority were not antagonistic to religion, they became profoundly indifferent to it.

Could religion, which had been subjected to such a long period of persecutions and negations, be simply reborn in its earlier traditional forms? Or, on the contrary, could it be inferred that, since atheism was a historically new phenomenon, postatheist spirituality, which superseded it, would have to be interpreted as even more of a novelty?

What is taking place in postatheist Russian society can be divided into three major tendencies. One of these tendencies constitutes a 'religious revival' proper. It is a return of Russian society to its *preatheist traditional beliefs*. The traditional religions—Orthodoxy, Catholicism, Islam, Buddhism, Judaism—are regaining their status and influence in the spiritual, cultural, and even political life of Russia. Not surprisingly, these newly converted believers bring an emotional ardour and dogmatic ignorance to the life of their new church. These believers are also bearers of a protective, romantic nationalism and messianism. The Orthodox hierarchy accumulates huge property, including land and buildings, and in exchange obediently cooperates with the State bureaucracy, wilfully fulfilling its ideological assignments, and supports patriotism as an allegedly authentic expression of the Christian faith. In the end, however, scant little goes beyond the confines of tradition: it merely fine-tunes the tradition to meet current needs.

Another tendency seeks to restore not the prerevolutionary but the archaic, pre-Christian layers of religious traditions and can be characterized as *neopaganism* (Pilkington and Popov 2009). According to its adherents, Russia's salvation is not to be sought in a religion of the spirit, but in the ancient cult of nature. What is required is an immediate restoration of the pre-Christian Russian and Arian pantheon. More often than not, neopaganism is mixed with elegiac ecological sentiments, in which the pagan cult of nature is presented as the defender of the environment against the encroachments of civilization. It is even more common to see Orthodox Christianity interpreted in the pagan spirit, as a special branch of Christianity, intimately connected to Russia's state and army and its God-bearing people. The advantage of Orthodoxy vis-à-vis other Christian confessions lies in its doubling as both a religion of the Heavenly Father and an ancient cult of Mother Earth. Orthodoxy in this context appears as a militant form of patriotism, destined, from time immemorial, to defend Holy Russia from the 'heresies' of Judaism, Catholicism, Freemasonry, and other 'foreign contaminations'. The new paganism also features the cultivation of magic, extra-sensory perception, para-psychology, spiritism. and other similar beliefs, which hark back to the earliest animistic and fetishistic practices.

Minimal religion

Together with the return to traditional religion and the parallel immersion in pagan and Orthodox archaism, a third tendency—*minimal religion*—can be observed in contemporary Russian religious life (Epstein 1999a, 1999b; cf. Taylor 2007). To date, it has attracted the least attention because it tends to escape all forms of objectification. Its 'minimality' almost precludes the formation of dogma or ritual and can be identified only as an internal impulse, a state of spirit, or a disposition of mind. This is what I call 'poor' or 'minimal' religion. For a Western observer, a more 'recognizable' name for it would be religious modernism, universalism, or ecumenicism, even if these terms do not exactly correspond to the Russian phenomenon. The spiritual vacuum, created by Soviet atheism, gave rise in the 1970s and 1980s to a new type of religiosity.

Imagine a young man from a typical Soviet family which, for a few generations, was resolutely cut off from all religious traditions. Suddenly hearing a spiritual call in his soul, he cannot decide where to go in search of truth and salvation. He tries the Orthodox Church, the Catholic Church, the Jewish Synagogue, Baptist, and also Lutheran services. He finds historically shaped traditions of faith everywhere. Yet he is eager to experience spirit as whole and indivisible. Looking for faith he finds only different religions. It is in this disparity between faith and denominations that 'minimal' religion emerges. It is a religion without an order of service, holy books, or specific rituals. It is notable that many more people in post-Soviet Russia were abandoning atheism than joining specific denominations. These people can be characterized as 'poor believers' who do not subscribe to any specific set of conventional religious practices. They belong to 'religion' as such, without further definitions or qualifications. Their relationship to God is holistic, mirroring the wholeness and indivisibility of God Himself.

In an essay written even before perestroika, I called this postatheist spirituality 'poor', or 'minimal' religion (Epstein 1982; 1999a, 1999b). It took the form of 'faith pure and simple', without clarifications or addenda, without any clear denominational characteristics. It manifested itself as an indivisible sense of God, outside historical, national, and confessional traditions. The atheistic negation of all religions gave rise to a 'minimal' religiosity negating all positive distinctions among historical religions. Paradoxically, this 'faith as such', 'faith in general', was prepared by the atheist denial of all faiths.

A typical expression of such 'minimal religiosity' can be found in the simple 'creed' of a well-known contemporary Russian artist, Garif Basyrov (1944–2004). Asked if he could be considered a faithful Muslim, Basyrov replied:

> That's ridiculous. As any normal person I feel I am approaching something... To use high-flown words, you can say I am on my way. But I am neither an expert in Islam, Buddhism, nor Christianity. All these rituals are not for me. One thing I do know for sure: God exists.
>
> *(Basyrov 1991: 31)*

For a minimal believer, God exists above and beyond all religions, thus nullifying their historical divisions. What is at issue here is the possibility of establishing a unitary religious consciousness through the experience of the negative void of the atheist world. This postatheist spirituality is as historically unprecedented as the phenomenon of mass atheism that preceded and conditioned it.

Thus, the religious revival in postcommunist Russia is not only a renaissance of traditional beliefs, which were widespread before the atheistic revolution, but is also the *naissance* of a qualitatively new, postatheist kind of spirituality. In the soul of a 'poor believer' there are no dogmatic or ritual preferences created by either a continuous historical tradition or long-standing family religious commitments. Just as the divisions among farmers' holdings were destroyed during collectivization, turning fertile land into wasteland, so the confessional divisions were also erased. This prepared the post-Soviet wasteland not only for a revival of old traditions, but also for a renewal of the religious consciousness as such, capable of transcending historically established boundaries.

In the prophet Isaiah we read:

> 3 The voice of him that crieth in the wilderness, Prepare ye the way of the Lord, make straight in the desert a highway for our God.
>
> 4 Every valley shall be exalted, and every mountain and hill shall be made low: and the crooked shall be made straight, and the rough places plain:
>
> 5 And the glory of the Lord shall be revealed, and all flesh shall see it together: for the mouth of the Lord hath spoken it.
>
> *(Isaiah, 40: 3–5, King James version)*

A few questions arise: was not this prophecy implemented, with uncanny precision, by the ideology of Soviet atheism? Did not atheism actually prepare a road for 'the glory of the Lord' by persecuting all beliefs, levelling all mountains and valleys, smoothing out different beliefs so that a trans-confessional spirituality could arise? (see Berdyaev 2009).

In the 1980s, there was no way to substantiate those observations with sociological data. Later, with the development of democratic procedures in Russia, there appeared instruments to test statistically the validity of these theoretical assumptions. A poll conducted in December 1995, by the Center for Sociological Research of Moscow State University, under the direction of S. V. Tumanov, shows (from a sampling of 3,710 respondents) that 37.7% of the Russian population characterized itself as believers who do not observe religious rituals; only 12.8% of the respondents characterized themselves as observant believers.

With regard to confessional self-determination, 12.8% of the religious population identified themselves as Christians in general (non-denominational), as compared to 71.0% Orthodox, 0.2% Catholics, and 0.7% Protestants. In addition, 2.7% of the believers did not perceive any essential difference among denominations, and 2.5% had their own perception of God. Based on these statistics we can conclude that approximately 18% of the Russian religious population is non-denominational (Dobrynina et al. 2000).

Lyudmila Vorontsova and Sergei Filatov (1994: 401–2 quoted in Epstein et al. 1999a: 469) reveal even more striking statistical data concerning what they call 'just Christians' in contemporary Russia.

> The growth of religiosity and the increase in those who believe in God has not been accompanied by a growth in the popularity of Orthodoxy. Indeed, the years 1990–1992 saw a sharp fall in its popularity. /.../ Orthodoxy's main competitor is not other religions, but the swiftly growing category of people with no denominational adherence: 'just Christians.' They grew two and a half times over the three years 1989–1992 and made up 52 percent of the population, while the number of Orthodox (of all jurisdictions) decreased. /.../ 'Just Christians' are the neophytes who believe in God and have come to faith but are not prepared to enter the church unconditionally and to accept church disciplines... The fact that the number of 'just Christians' is growing at the expense of Orthodoxy testifies to the rebirth of Christianity not in the form of Orthodoxy as it was 70 years ago, but on a more modern, universal level.

The phenomenon of 'minimal religiosity' is illuminated by numerous works of Russian literature of the 1970s and 1980s. The literary protagonists of Andrei Bitov, Yuz Aleshkovsky, Yuri Trifonov, Vasily Aksyonov, Bella Akhmadulina, and Venedikt Erofeev all crave for a higher spirituality. To satisfy this craving, the heroes of this late Soviet literature cannot turn to traditional forms of religiosity, because, for several decades, such forms had no common currency in Russian life. The new religiosity, which these fictional characters come to embody, is alienated from all objectified historical traditions.

The autobiographical protagonist of Venedikt Erofeev's (1969, 1995) novella *Moscow to the End of the Line* ('Moskva-Petushki') is a typical 'poor believer'. He occasionally turns to God with the plea: 'Oh, Lord, you see what I possess. But do I need all *this*? Is *this* what my soul pines for?' (*Op cit.*: 25). Venia's 'Lord' exists outside all traditions and confessions. He has no temple in this world other than the littered train from which the hero's soul addresses itself to Him. The hero also has no church, no preconceived notion of religion, and no other method of proof of God's existence than the 'hiccups' that overpower and release him with equal suddenness.

> The Law is higher than all of us. Hiccups are higher than the Law... We are trembling creatures while hiccupping is almighty. It is God's Right Hand, which is raised over us all... *He* cannot be conceived by the mind, therefore *He* exists. So be perfect as your Heavenly Father is perfect.
>
> (op cit. *p. 55)*

The 'sixth' proof of God's existence—proof by hiccups—may appear blasphemous or at the very least parodic in relation to canonical theological discourse. Remember that in *Summa Theologiae*, Thomas Aquinas elaborated five logical proofs of God's existence (Aquinas 1989). The main function of Venia's hero, however, is not comedy or parody. It acts, instead, as a revelation of the apophatic spirit of 'poor' religiosity, which is on a par with the sacrilegious and eccentric behaviour of the Russian fools-in-Christ, or 'holy fools'. The same logic of the negative knowledge of God applies here. Man knows God through things that man cannot control by his will or reason. Hiccups are an elementary example of such a thing: they are a sequence of unwilled bodily movements at unregulated temporal intervals. Using this negative logic, Erofeev can make the assertion, which appears to have come straight out of a treatise on apophatic theology: '*He* [God] cannot be conceived by the mind, therefore *He* exists'. (Erofeev 1969; 1995: 55).

The transitional stage between (Marxist) atheism and (Christian) faith is admirably illustrated in Yuz Aleshkovsky's (1993) narrative *The Ring in the Case* ('Persten v futliare') with the characteristic subtitle 'A Christmas Novel'. This didactic and grotesque narrative portrays the transformation into a 'poor believer' of one Helio Revolverovich Serious, a die-hard atheist and 'third-generation fighter against bourgeois prejudices', whose efforts to stamp out the belief in God have earned him a position in the upper ranks of the party hierarchy. Shaken by the disintegration of his relationship with his beloved woman, and reduced to breaking point by the misery of his physical and spiritual existence, Helio suddenly experiences the need for prayer, not knowing to whom to address himself or about what to pray:

> Perhaps I should have another try at praying to... but what can one pray about? That's the question... About love? ... It's too late... About salvation?
>
> (Op cit.: *44)*

The numerous marks of omission are significant. They are a graphic embodiment of his impoverished religious feeling or, to be more specific, the apophatic nature of this religiosity, which cannot use images or words to describe the One, Who...

The hero then resorts to poetry and addresses Him with a line from Pasternak: 'O Lord, how perfect are Thy works!'[2] However, he never specifies—neither for himself nor for the reader—whom he actually means when he says 'Oh, Lord!'. The phrase is pronounced almost like an interjection, a sigh—'Oh, Lord!'. But in this case the universal character of the interjection does not signify indifference or automatism. On the contrary, it bears the stamp of a meaning attained through suffering and deeply felt experience. What would seem more natural for this seasoned atheist, erudite in the dialectic of relativism ('The Moslems picture him as Allah, while for the Christians he is simultaneously Father, Son and Holy Ghost, which shows that the assumptions of the different religions contradict one another, revealing their complete implausibility...'), than to ask himself: whom am I addressing, *which* God?

And that is precisely the point: twentieth-century atheism used the diversity and historicity of the world's religions and spiritual traditions as its most powerful argument against faith; that is, since there are so many religions, each with its own god, then there is no god. It was atheism that asked all those official, 'passport-like' questions, trying to specify to which tradition or confession 'god' might appertain: 'date of birth', 'nationality', 'place of residence', and so on. It was a natural reaction on the part of postatheist religiosity to erase this entire panoply of historical and national 'essences' and to make a fresh start by setting up a pure, universal, 'poor', and singular name, analogous to the interjection 'Oh, God!' or 'Oh, Lord!', devoid of specificity and without any determinants whatsoever. Atheism had used the diversity of religions to argue for the relativity of religion. Consequently, the demise of atheism signalled the return to the simplest, virtually empty, and infinite form of monotheism and monodeism.[3] If God is one, then faith must be one.

In most places around the world, people are raised within a specific religious tradition and are brought to a church, a mosque, or a synagogue from childhood. In postatheist Russia, however, many people first experience the spirit in their hearts and then come to houses of worship. There are two different ways of conversion. One is a conversion to God through the church. This is the normal way of 'conversion' in the world with established religious traditions. The other is a conversion to the church through God. This was the way of conversion in the times of Moses, Christ, and Luther. It is also the way of conversion in late atheist and postatheist Russia.

One might speculate that this thrust towards religious reformation will dominate the spirit of twenty-first-century Russia. The restoration of preatheist traditions is the focus of the current religious revival, but the atheistic past, the experience of the wilderness, cannot pass without a trace, and this trace of 'the void' will manifest itself in a striving for a fullness of spirit, transcending the boundaries of historical denominations. Those people who have found God in the wilderness feel that the walls of the existing temples are too narrow for them and should be expanded. Thus, we have seen that the three tendencies can be discerned in the misty dawn of Earth's first postatheist society. One is traditionalism, which is housed in existing churches and subscribes to the existing religious subdivisions. A second is neopaganism, which is focused on archaic objects of worship such as the soil, blood, and national identity. The third is 'poor' or 'minimal' religion, which is free from historical divisions and seeks the unification of all religions in the gap between existing churches and the fullness of a future epiphany.

It is significant that Russian religious thought of the early twentieth century, which now serves as a general reference point on all questions of Russian spirituality, furnishes models for all three tendencies. Traditionalism is connected with the figure of the priest Pavel Florensky. Its firm basis is in the philosophically interpreted church canon and the heritage of the church fathers. The second, archaist tendency is connected with the name of Vasily Rozanov. It is close to paganism and the primal cults of the Sun and the Earth, consecrating the universe of sex drives and fecundity. The third, modernist tendency, inspired by Nikolai Berdiaev, issues from the apophatic conception of pure freedom, which posits itself as anterior to God and the act of creation. It presupposes an ultimate unification of all religions in anticipation of the eschatological end of history.

Theses on 'poor faith'[4]

In the following text I resort to the form of theses for the very reason that they provide the most succinct means of setting out ideas which belong to a worldview still in the process of formation. I will divide 32 theses into several thematic sections to prompt further discussion.

Statistics

1 According to the results of the largest sociological survey of opinions involving 56,900 respondents and carried out by the *Sreda* research centre in Russia in December 2012, one in four people fall into the category of 'poor religion'—that is to say, a simple belief in God without any affiliation to a belief system or denomination. This proportion (25%) comes second only to Orthodox Christian believers (41%) and exceeds Muslims (6.5%).[5]

2 The growth of supra-confessional awareness is a worldwide tendency. According to data from the Pew Research Center, the 'extra-confessional' (religiously non-aligned)

population of the world makes up 1.1 billion people (or 16% of the total world population). In China that is 52% of its population, or 700 million, and in the USA 20%, or 46 million. These figures include atheists and agnostics. However, the proportion of believers among these 'religious nones' is considerable: 7% in China, 30% in France, and 68% in the USA affirm that they do believe in God. And 10% of those religiously unaffiliated even pray every day.[6]

Foundations of poor faith

3 In the religious context the word 'poor' has a strong positive connotation, as in 'Blessed are the poor in spirit'. The adherents of poor faith do not possess any symbolic capital in the form of generally recognized and specific denominational traditions, church buildings, and socially recognized prestige or image. Poverty is a Christian virtue, which one could use as a measure for judging Christianity itself and its dogmatic wealth.

4 When applied to the realm of faith, the word 'poor' has connotations for synonyms like *free, direct, living, open, creative*.

5 Poor faith corresponds to apophatic theology which denies the possibility of knowing God or conceiving of Him in positive forms, symbols, or definitions. This faith lives beyond the boundary of all faith systems.

6 In Russia, poor faith was the consequence of seven decades of atheism. By rejecting all religious confessions at a single stroke, militant atheism created a favourable setting for the emergence of 'religion in general'. It was precisely the faithlessness of the Soviet years which formed the kind of person who could only be defined as a 'believer'.

7 An important feature of faith is the dialectical relationship between induction into the life of the church and disengagement from the church, between entry into the historical and denominational body of religion and distancing from it.

8 The 'trans-religious' is not only a 'going out' beyond the boundaries of historically formed religions, but also beyond *religion as such*. The trans-religious is Christ prior to Christianity, that is, the stage of faith before religion, or Christ after Christianity, Christ in apocalyptic times, the stage of faith after religion. Thus, faith constantly brings forth a crisis in religion and then supersedes it, in the form of an instant impoverishment, an 'emptying out' of the dogmatic and ritualistic wealth accrued in the long course of history.

9 If we consider just Russian instances of the phenomenon, then Lev Tolstoy, Daniil Andreyev, and Grigory Pomerants all represent various versions and stages of this movement towards the trans-religious. Tolstoy represents the discovery of trans-religious space in its pure form, as critique of the church. Andreyev sees the trans-religious as a means of gathering together all the historical forms of religion in a supra-historical unity, 'the Rose of the World'. For Pomerants, the trans-religious amounts to the individual's existential experience of contact with various religions.

10 There is another, fourth, path, which is that of poor faith. It is a minimalist, not maximalist, extension of religious openness. Poor faith is not only a postatheistic but also trans-religious consciousness.

11 Poor faith does not criticize specific beliefs and belief systems, but rather, from the position of a complete toleration of faiths, it signifies openness towards their revelations, their spiritual beauty, their historical meaning and searching, not indifference towards these aspects.

12 In the contemporary world, two tendencies can be observed: (a) strengthening of militant forms of confessionalism, of clerical and fundamentalist movements; and (b) growth

in poor faith and other forms of trans-religious and supra-confessional awareness, forms that are able to unite people above the level of all ethnic and religious barriers. A distinguishing feature of this postsecular age is the polarization of these two tendencies. Religion is acquiring a new significance, both as an instrument for political struggle and also as the path of a new spiritual unification of the world.

13 The struggle between faith and religion constitutes the spiritual tension and the inner conflict of our time experienced by many believers. Disillusionment with the symbiosis of church and corrupt power structures ('religarchy') in Russia may push society yet further and closer towards poor faith.

14 A putative postsecular epoch is not confronted with the question 'To believe or not to believe?' Rather, the question is framed this way: 'Is your faith church-based or not-church-based?', or 'Is your faith based on rites or not based on rites?' There is a time for gathering stones and a time for taking stones apart, including the stones of the temple within your own soul. The history of humankind, as well as the history of each soul, is not only a struggle of faith with faithlessness, but also a struggle of faith with religion.

The dynamics of poor faith

15 Henri Bergson distinguishes between static religions and dynamic religions, identifying those which reproduce stable stereotypes for behaviour and support stability in society and those which are based on man's experience of direct communication with God (Bergson 1932; 1977).

16 Poor faith is not a constant and self-identical spiritual condition. It has many levels and many stages. It moves from a state of naivety to a state of reflection, from a preconfessional to a supra-confessional stage. Preconfessional faith lacks any knowledge and experience of specific religious confessions or denominations, whereas supra-confessional faith attempts to reach beyond their boundaries.

17 Five stages of poor faith can be identified. The initial stage is a non-reflective 'I believe in Something'/'I believe in Someone', which constitutes the first step in distinguishing oneself from non-believers.

18 The second stage comes in the form of an attraction to mystical yearnings, esotericism as popularly understood: spiritualism, theosophy, yoga, and folk superstitions.

19 The third stage consists of a parallel or subsequent affiliation to one or several confessions, an experience of interaction between them, participation in their mysteries, and a full or partial induction into church life.

20 The fourth stage is that of disillusionment in organized and ritual forms of religion and their dogmatic nature, commercialization, politicization, and their confluence with state power. This is the stage of *dis*-engagement from church, and of the individual person's aspiration to find support in a direct standing before God.

21 The fifth stage is that of a conscious faith outside the confines of religious confession, a faith enriched by varied spiritual experience and encounters with various belief systems, Holy Scriptures, teachings, and traditions.

22 Poor faith is a path of spiritual becoming which can pass by and through other stages and in a different order from that outlined earlier. Poor faith can arise within specific religious confessions as an aspiration to feel the living spirit beyond the outward forms. The fire of faith is common to all religions; national and historical forms vary in the ways that this fire of the spirit dissipates.

23 Poor faith does not adhere to old dogmas nor creates new ones. Rather, it takes the traditions of various religions as raw material for constructing its own personal experience. Just as an artist creates her or his own canvas, using a plethora of lines and colours, so the poor believer creates her/his own faith, employing those forms and meanings which are accessible through the whole legacy received from previous testimonies and revelations of faith.

24 Poor faith can go through crises and through 'dark nights of the soul', and lack all forms of support other than its own efforts to further set down its path towards God.

25 Just as various paths lead to poor faith, so also various paths lead away from it. These include a weakening of, or a break in, communication with God, together with a drying up or withering of the inner life; a move into atheism, a loss of faith; joining one of the non-traditional religious cults which claim to possess a 'non-confessional' nature; mono-confessionalism, that is, returning to a former confession or attaching to a new one; multi-confessionalism, the experience of belonging to two or several confessions; and a synthesis of various confessions and the construction of a universal religion.

26 Poor faith can be a primary, naive impulse of faith, but it can also absorb a multifaceted spiritual experience. It is able to assimilate the values of various mysteries and sacraments, but it remains the person's inner work, a communication of the unique individual with the One, which takes place completely outside the framework of any religious cult. However well-endowed poor faith might be, it remains poor to the extent that it is not identical at all to any confession, and it does not permit confessions to take a hold and to form within itself.

27 Poor faith remains poor precisely because it is not organized. As soon as anyone begins to form something like a 'community of adherents to poor faith' with its own customs and core teaching, it is no longer poor faith, but yet another church, albeit a radically protestant and non-conformist one.

Poor messianism

28 The beginnings of all religious traditions are various, but the end can only be a commonly shared end. Monotheism turns into *theomonism*, a coming together of all forms of monotheism in the unity of faith itself. The historical significance of the postsecular age may lie precisely in this transition.

29 The messianism which is an inherent feature of all faith systems in the Abrahamic tradition is the expectation of a Messiah and a constant state of readiness for His coming. 'Watch therefore, for ye know neither the day nor the hour wherein the Son of man cometh' (Matt, 25: 13, King James). Poor messianism, or 'messianicity' (a term coined by Jacques Derrida 1996; 2002) is a broader *not-knowing*: of the very possibility of the coming of the Messiah.

30 Such messianicity, as a projection of poor faith into the absolute future, does not guarantee the appearance of a real Messiah. Rather, it consciously attributes to any 'claimant' the possibility of personifying false messiahship and remains a vanishing horizon for our looking and striving.

31 At the same time, the open structure of expectation is preserved, akin to the structure of hospitality, of 'inviting-in'. Absolute hospitality does not know in advance who the guest is who will appear on one's threshold. The same holds true for poor messianism: it is elevated into the absolute and at the same time reduced to a minimum. It is an expectation that is open to any unexpected happenings or events, including the *non*-appearance of the Messiah. It is faith in its initial source, a precondition for any other faith, a prefaith for all faiths.

32 That same Godot whom the other characters in Beckett's play of the absurd are waiting for, does not arrive either. It is he who is the true God (God O!), as he is defined in the relationship between the maximum and minimum of faith. This is the general condition for all other forms of expectation: the expectation of good, of justice, of perfection. This 'supra-expectation' is transforming the world by the power of its openness, although it does not promise fulfilment but only allows for such a possibility.

Minimal religion and Western postsecularism

In the previous sections I emphasized the specificity of minimal religion as a uniquely postatheist phenomenon. There are, however, many lines of convergence between Russian postatheism and postsecularist tendencies in the West. The Canadian philosopher Charles Taylor highlights this confluence in his acclaimed book *The Secular Age* (2007). Close to the end of the book, Taylor attempts to envisage 'a time in which the hegemony of the mainstream master narrative of secularization will be more and more challenged....[I]ts overcoming would open new possibilities' (p. 534).

What are these possibilities of postsecularism that Taylor is anticipating? The first, and in fact, the most articulated phenomenon that he mentions in this context is the *minimal religion* to which he devotes several pages of his book (pp. 533–5; 849). Paradoxically, this 'irreligious' spirituality has originated from the country that was the first in the twentieth century to affirm and then to survive atheism, the tragic experience of 'super–secularization' that turned out to be the devastation of all religions—to allow the new form of spiritual poverty to be born from this void.

Taylor writes in reference to my work:

> Epstein introduces the concept of "minimal religion". He also speaks of an overlapping category, the people who declare themselves "just Christians" in surveys of religious allegiance, as against those who adhere to one or other Christian confession, like Orthodox, or Catholic. This kind of religious (p. 533) position Epstein sees as "post-atheist"; and this in two senses. The people concerned were brought up under a militantly atheist regime, which denied and repressed all religious forms, so that they are equidistant from, and equally ignorant of, all the confessional options. But the position is also post-atheist in the stronger sense that those concerned have reacted against their training: they have acquired in some fashion a sense of God, which however ill-defined places them outside the space of their upbringing.
>
> "Minimal religion" is spirituality lived in one's immediate circle, with family and friends, rather than in churches, one especially aware of the particular, both in individual human beings, and in the places and things that surround us. In response to the universalist concern for the 'distant one' stressed in Marxist communism, it seeks to honor the 'image and likeness of God' in the particular people who share our lives.
>
> But because this religion was born outside of any confessional structures, it has its own kind of universalism, a sort of spontaneous and unreflective ecumenism, in which the coexistence of plural forms of spirituality and worship is taken for granted. Even when people who start with this kind of spirituality end up joining a church, as many of them do, they retain something of their original outlook. <...>
>
> Perhaps something analogous can be said about the situation in "post–secular" Europe. I use this term not as designating an age in which the declines in belief and practice of the last century would have been reversed, because this doesn't seem likely,

> at least for the moment; I rather mean a time in which the hegemony of the mainstream master narrative of secularization will be more and more challenged.
>
> *(p. 534)* <…>
>
> Being "spiritual but not religious" is one of the western phenomena which has some affinity with Epstein's "minimal religion" in Russia; it usually designates a spiritual life which retains some distance from the disciplines and authority of religious confessions. ….In both cases, a certain diffuse ecumenical sense is widespread, and even those who subsequently take on some confessional life, and thus become "religious", retain something of this original freedom from sectarianism.
>
> *(p. 535)*

Charles Taylor indicates that the phenomenon known in Russia as poor faith or minimal religion in the West, is often called 'spirituality' in opposition to organized religion. I would argue, however, that spirituality is not quite a relevant term to characterize postsecular trends. Spirituality as such includes belief in high ideals, adherence to supreme moral values, or the aesthetic experience of the sublime and the beautiful—all this can belong to a purely secular worldview. Spirituality as such can be both secular and postsecular, and by itself it draws an insufficiently clear line of demarcation between the two. What makes the distinction between the secular and the postsecular more articulate is that the latter still retains the direct relationship with *faith* and *religion*. But postsecularism is *poor* faith; *minimal* religion. The degree of minimality may differ in each individual case, for each personal experience of faith. But without at least minimal faith, the boundary between the secular and postsecular disappears. In this sense minimality is also liminality.

Minimal religion is transcendental in two senses. As a *religion*, in distinction from a purely secular position, it transcends the 'earthly' world of immanence. As a *minimal* religion, in distinction from all organized religions, it transcends the world of all existing religious traditions, dogmas, rituals, and affiliations. It is only in the liminal 'between' that this uniquely postsecular phenomenon can be defined. Religion, yet minimal; poor, yet faith.

Notes

1 This article draws upon and integrates several pieces published earlier in Russian (see Epstein 1996; 2014), with new material added in English.
2 From Boris Pasternak's poem 'V bol'nitse'. See https://books.google.com/books?id=puymuobJIoC&pg=PA111&lpg=PA111&dq accessed online 28-05-2018.
3 I refer here to monodeism as the default standard concept of deism, distinct from polydeism, pandeism, and spiritual deism.
4 See Epstein (2014) translated from Russian by Dr. Jonathan Sutton, Leeds, UK.
5 See ARENA: *Atlas of Religions and Nationalities of Russia*, http://sreda.org/arena, accessed online 28-05-2018. Full survey results can be downloaded from the ARENA website.
6 See www.pewforum.org/2012/12/18/global-religious-landscape-unaffiliated/accessed online 28-05-2018; see also www.pewforum.org/2012/10/09/nones-on-the-rise/accessed online 28-05-2018.

Further reading

Daniel, W. L. (2016) *Russia's Uncommon Prophet: Father Aleksandr Men and his times*, DeKalb: Northern Illinois University Press.

This biography is of Aleksandr Men (1930–1990), a most influential Russian Orthodox priest, prolific writer, and theologian who attempted to open Orthodoxy to other religious traditions and to universal human values. Growing up during the darkest, most oppressive years in the history of the former Soviet Union, he attracted large, diverse groups of people in Russian society and made an enormous impact of the postatheistic consciousness of late Soviet intelligentsia.

Epstein, M. (1994) *Vera i obraz. Religioznoe bessoznatel'noe v russkoi kul'ture XX veka* ["Faith and Image: the religious unconscious in twentieth century Russian culture"], Tenafly, NJ: Hermitage Publishers.

The book treats the phenomenon of unconscious religiosity as it is revealed in the philosophy, literature, and art of the Soviet epoch, and compares the Russian-Soviet tradition of apophatism-atheism with Eastern non-theistic religions, such as Buddhism and with the Western processes of secularization.

Epstein, M. (2013) *Religia posle ateizma: Novye vozmozhnosti teologii* ["Religion after Atheism: New Possibilities for Theology"]. Moscow: AST-Press.

According to sociological surveys, 25% of the Russian population believe in God but do not associate themselves with any particular confession. This book analyzes the new phenomenon of a 'minimal religion' (non-denominational faith) and many other trends of religious thought following the demise of Soviet atheism and the crisis of Western secularism. The book explores the following questions. What new opportunities for theology become open in the advent of a post-secular age? How dangerous is a historically precipitous transition from militant atheism to a dominant role of the church? How does a religious worldview incorporate recent scientific discoveries?

Erdozain, D. (ed.) (2017) *The Dangerous God: Christianity and the Soviet experiment*, DeKalb: Northern Illinois University Press.

The book explores the contribution of Christian thought and belief to the Soviet culture of dissent and survival, showing how religious and secular streams of resistance joined in the self-awareness of Soviet intelligentsia. Against the Marxist notion of the 'ideological' function of religion, the authors set the example of people for whom faith was more than an opiate, and propose the centrality of religious faith in the intellectual, political, and cultural life of the late modern era.

Fagan, G. (2014) *Believing in Russia: Religious Policy after Communism*, London: Routledge.

This book presents an overview of religious policy in Russia since the end of the communist regime, exposing many of the ambiguities and uncertainties about the position of religion in Russian life. It reveals how religious freedom in Russia has, contrary to the widely held view, a long tradition, and discusses the pursuit of privilege for the Russian Orthodox Church and other 'traditional' beliefs.

Furman, D. E. and K. Kaariainen (2000) *Starye tserkvi, novye veruiushchie. Religiia v massovom soznanii postsovetskoi Rossii.* ["Old Churches, New Believers: Religion in the Mass Consciousness of Post-Soviet Russia"]. St. Petersburg/Moscow: Summer Garden.

The book is based on unique polls, on a strictly sociological approach to analysis. What do modern believers and unbelievers think? Is their faith deep or is their faith pure formality, a tribute to a new tradition? How do different elites of Russian society treat religion? The book sums up the research project "Religion and values after the fall of communism" (1991–1999).

Johnson, J. Stepaniants, M. and B. Forest (eds.) (2005) *Religion and Identity in Modern Russia: the revival of Orthodoxy and Islam*, London: Routledge.

This book discusses and compares the role of Russia's two major religions, Orthodoxy and Islam, in forging identity in the modern era and brings a blend of sociological, historical, linguistic, and geographic scholarship to the problem of post-Soviet Russian identity.

Jones, M. V. (2005) *Dostoevsky and the Dynamics of Religious Experience*, London: Anthem Press.

This book, relying in particular on Mikhail Epstein's concept of poor faith, shows how Fyodor Dostoevsky, despite his passionately proclaimed devotion to Orthodox Christianity, presents a much deeper and broader dedication to the theology of minimal religion, as presented in the characters and spiritual revelations of Elder Zosima and Alyosha Karamazov.

Louth, A. (2015) *Modern Orthodox Thinkers: from the philokalia to the present*, Downers Grove, IL: IVP Academic.

The book offers historical and biographical sketches of Orthodox religious thinkers and includes a chapter on theology in Russia under communism (Alexei Losev, Sergei Averintsev, Aleksandr Men).

Sutton, J. (2006) "'Minimal religion' and Mikhail Epstein's interpretation of religion in late-Soviet and post-Soviet Russia", *Studies in East European Thought*, 58(2): 107–35.

The entire special issue 'Two readings of the dynamic between religion and culture: Sergej Averintsev and Mikhail Epstein', edited by Jonathan Sutton and especially his article on minimal religion, deals with two different interpretations of relations between religion and culture in the context of decades of mass atheism.

References

Aleshkovsky, Y. (1993) *Persten' v Futliare: rozhdestvenskii roman* ["The Ring in the Case: a christmas novel"], published in the journal Zvezda (St. Petersburg), No. 7: 7–89.
Aquinas, T. (1989) *Summa Theologiae: a concise translation*, trans. Timothy McDermott, London: Eyre and Spottiswoode.
Basyrov, G. (1991) *Grafika: katalog vystavki*, Moscow: Sovetskii Khudozhnik.
Berdyaev, N. (2009) *The End of Our Time*, English edition, trans. Donald Attwater with a foreword by Boris Jakim, New York, NY: Semantron Press.
Bergson, H. (1932; 1977) *The Two Sources of Morality and Religion*, Notre Dame, IN: University of Notre Dame Press.
Bowman, C. F. (1995) *Brethren Society: the cultural transformation of a peculiar people*, Baltimore, MD: Johns Hopkins University Press.
Bronner, S. (2004) *Reclaiming the Enlightenment: toward a politics of radical engagement*, New York, NY: Columbia University Press.
Chartier, R. (1991) *The Cultural Origins of the French Revolution*, trans. Lydia G. Cochrane, Durham, NC: Duke University Press.
Derrida, J. (1996; 2002) 'Faith and knowledge: the two sources of "religion" at the limits of reason alone', trans. S. Weber. In Anidjar, G. (ed.) *Acts of Religion*, London: Routledge, pp. 40–101.
Dobrynina, V. I., Kukhtevich, T. N. and S. V. Tumanov (eds.) (2000) *Kul'turnie miry molodykh Rossiyan: Tri zhiznennye situatsii* ["Cultural worlds of young Russians: three life situations"], Moscow: Moscow State University, pp. 167–91.
Durnbaugh, D. F. (1968; 2003) *The Believers' Church: the history and character of radical protestantism*, Eugene, OR: Wipf & Stock Publishers.
Epstein, M. (1996) 'Post-ateizm, ili bednaia religiia' ("Post-atheism, or poor religion"), *Oktiabr (Moscow)*, 9: 158–65.
——— (1999a) 'Minimal religion'. In Epstein, M., Genis, A. and S. Vladiv-Glover (eds.) *Russian Postmodernism: new perspectives on post-Soviet culture*, New York, Oxford: Berghahn Books, pp. 163–71.
——— (1999b) 'Post-atheism: from apophatic theology to "minimal religion"'. In Epstein, M., Genis, A. and S. Vladiv-Glover (eds.) *Russian Postmodernism: new perspectives on post-Soviet culture*, New York, Oxford: Berghahn Books, pp. 345–93.
——— (2014) 'Tezisy bednoi very' ("Theses of poor faith"), *Zvezda*, 8: 221–34.
Erofeev, V. (1969; 1995) *Moskva-Petushki* ["Moscow to the End of the Line"], trans. H. W. Tjalsma, Evanston, IL: Northwestern University Press.
Gomez, O. (ed.) (2003) *The Enlightenment: a sourcebook and reader*, London/New York, NY: Routledge.
Meier, M. (2008) *The Origin of the Schwarzenau Brethren*, Issue 7 of Brethren encyclopedia monograph series, Ambler, PA: Brethren Encyclopedia, Inc.
Pilkington, H. and A. Popov (2009) 'Understanding neo-paganism in Russia: religion? ideology? philosophy? fantasy?'. In McKay, G., Williams, C., Goddard, M., Foxlee, N. and E. Ramanauskaite (eds.) *Subcultures and New Religious Movements in Russia and East-Central Europe*, Cultural Identity Studies, Volume 15, Oxford/New York, NY: Verlag Peter Lang, pp. 253–304.
Taylor, C. (2007) *A Secular Age*, Boston, MA: The Belknap Press of Harvard University Press.
Vorontsova. L. and S. Filatov (1994) 'Religiosity and political consciousness in Postsoviet Russia', *Religion, State and Society*, 22(4): 397–402.

7

The performative force of the postsecular

Herbert De Vriese and Guido Vanheeswijck

Introduction

The term postsecular is a fluid and multidimensional concept. Not surprisingly, it includes a range of meanings and connotations that are dependent on, and derivative from, a variety of views and interpretations of the secular within scholarly research on secularism and secularization. Even if adopted by various academic disciplines as sociology, political science, philosophy, and history, the actual meaning of the term postsecular remains vague and ambiguous. Yet, within this intricate web of meanings and connotations, the term postsecular also expresses the idea of putting an abrupt end to conventional ways of thinking about secularism, secularization, and the transformation of religion in modern societies. The term itself seems to suggest the emergence of a new historical epoch in which a new type of global world order or pluralist society has arisen. Moreover, in various fields of academic scholarship, the increasingly common usage of the term postsecular has the effect of imposing a paradigm shift in thinking about the role, place, and significance of religion in the contemporary world. This dimension we call the performative force of the term postsecular. And it is the analysis of this performative dimension of the usage of the term that will be the main subject of this contribution.

Of special interest to us are the decisive motives and arguments underlying this rhetorical gesture to take leave of an established research tradition on religion and secularism and to launch a completely different approach. What fundamental change in our theoretical understanding of religious transformation has motivated this gesture? What has affected the fate of religion and secularism in the modern world so profoundly that it calls for a sudden change of perspective and legitimizes the decision to use the term postsecular? These questions will guide us in an attempt to identify the deeper causes and justifications behind the seemingly self-confident speech act of heralding a postsecular condition.

At first sight, the term postsecular seems to refer to a major historical transition, one that has been aptly captured by the idea of the return or resurgence of religion in the modern world. At the same time, however, the question arises whether it denotes a factual and empirically verifiable change in the historical development of religion, or rather a change in the way we understand and formulate religious transformations in the modern world.

To put it more sharply, is the postsecular turn founded on real historical change of factual data, or rather on a profound shift in our theoretical frameworks and narratives? This question may be seen as the central problem related to the performative use of the term postsecular. For the increasingly common usage of the term tends to obscure the crucial distinction between a genuine historical shift of the role, place, and significance of religion in the modern world on the one hand, and a profound theoretical shift in our thinking and discourse about these issues on the other hand. We will argue that disregard for this distinction threatens to reinforce exactly those questionable ideas and presuppositions of classical secularization theory which are credulously assumed to have been overcome. We thus believe that the term postsecular should be used more thoughtfully, rather as an appeal to critically examine our indebtedness to classical secularization theory and our reasons to move beyond it, than as a manifest reference to a condition completely outside its scope.

In the first part of this paper, we focus on the crisis of classical secularization theory and on the issue of the so-called return of religion in the modern world. We ask whether this transition has gained its prominence from a real return of religion in contemporary society, or from an ostensible return of religion in the minds of scholars. As a consequence, we develop two different arguments to explain where the postsecular turn comes from. In the second part, we deepen our two-way analysis by distinguishing two entirely different research approaches to secularization. While the dominant approach highlights the way modernity has affected the fate of religion, we address another noteworthy approach, which is about religion's contribution to the emergence of modernity. Because secularization theories of the second type of approach have developed a view on the relationship between modernity and religion, very different from the first type but still from a Eurocentric perspective, a third motive for making the postsecular turn arises. We conclude with an overall assessment of both the potentially beneficial effects and adverse consequences of the performative force of the term postsecular.

After the dominance of classical secularization theory

During the past two decades, classical secularization theory has come under increasing pressure. Cultural theorists and scholars of comparative religion have accumulated evidence against a too hasty, cross-cultural generalization of the theory. Historians have challenged the notion of a 'golden age of faith' from where the transition from a deeply religious society to a more worldly one was supposed to start, and have documented the intensification of religious vitality in later periods. Social scientists have questioned the predictive power of secularization theory and now content themselves with a strictly descriptive approach. As a result, the long-standing theoretical model to examine the secularization of the world has been forced to accept methodological, historical, and geographical limits. What was once a common and relatively simple assumption is now fragmented into different, and often mutually contradictory, theorems, methods, claims, and conceptions (Bruce 2006).

In the wake of these successive assaults and readjustments, a change in intellectual climate has taken place, with philosophers and social scientists increasingly sympathetic to the 'secularization is dead' thesis. The idea that modernization, with a sort of inner necessity, leads to the progressive weakening of religion in all of its dimensions, with no other outcome than the eventual disappearance of religion, is hardly a defensible one today. Some have pronounced the verdict of 'the end of secularization theory' (Joas 2007: 13–4), whereas others have jumped to the conclusion that it was merely a product of wishful thinking (Stark 1999).

This conclusion, however, requires some qualification. First, in spite of numerous historical data that challenge confidence in the secularization thesis and despite the lack of objective, empirical evidence in support of its hegemony, the general thrust of the classical thesis still remains implicitly accepted within broad sectors of the population in the Western world. One way or another, it remains a popular notion that modernity in all its ramifications has, if not extinguished, at any rate diminished the significance of religion, and will continue to do so. Second, there are voices within the academic community who insist on the explanatory power of the 'orthodox model' in interpreting modern processes of religious change (Wallis and Bruce 1992: 27). For some, it remains a 'highly plausible' hypothesis, supported by many indicators, over long time periods, of the decline in the social significance of religion (Wilson 1998). Others argue that classical secularization theory actually consists of three separate hypotheses—the 'differentiation thesis', the 'decline-of-religion thesis', and the 'privatization thesis'—of which only the first is consistent with the findings and standards of empirical scientific research. Not everyone is equally prepared to dispense with the theory altogether. Some merely call for its revision and redefinition.

Nevertheless, the atmosphere in the intellectual community has changed. The number of adherents to classical secularization theory is gradually being outweighed by an increasing amount of sociological and philosophical scholars who explicitly reject its hegemony. In 'After Secularization?', Gorski and Altınordu (2008) note how much the terms of the debate have shifted since the late 1960s:

> Today, secularism qua political project and secularization qua sociological theory both find themselves in an increasingly defensive and even beleaguered posture. Once hegemonic, liberal secularist philosophies and sociological theories of secularization are violently rejected by many outside the West, very much on the defensive in North America, and under fire even in Western Europe.
>
> *(Gorski and Altınordu 2008: 57)*

Their conclusion is that the study of secularization is entering a new phase, where the classical and 'scientifically unanswerable question' concerning the future of religion—will it survive or will it die?—will evaporate (*ibid.*: 75). Instead, the focus will be on analytically specific types, not on the general direction of religious change.

This change of focus confirms José Casanova's observation, as early as 1994, that 'a typical Kuhnian revolution in scientific paradigms' was taking place. He saw the majority of sociologists of religion abandoning the classical paradigm 'with the same uncritical haste with which they previously embraced it' (1994: 11). This notion of a 'paradigm shift' became applicable to the entire field of social sciences and humanities, where a theoretical watershed was drawn between old and new approaches to the fate of religion in the modern world. Within a growing community of scholars who were dissatisfied with, and strongly critical of, the classical paradigm, the need to make a clean break with the past and to coin a new term, that is, the term postsecular, was felt. In the next subsection, we try to give more insight into the deeper sources of this need.

The revenge of the facts

It has often been remarked that classical secularization theory, while being one of the few theories to reach a truly paradigmatic status within the modern social sciences, had never been rigorously tested. Apparently, the consensus in the intellectual community was so overwhelming that the need to ground the theory on objective, empirical evidence seemed patently absurd. It was not until the 1960s that the first serious methodological and conceptual questions were raised. Even then, casting doubt upon the scientific validity of the theory was destined to fall on deaf ears: survey data clearly refuting the secularization thesis were likely to be reinterpreted as *incidentally* religion and *essentially* something else (Martin 1991).

As a theoretical framework, secularization theory seemed unshakable. It took far-reaching social and political developments all across the globe to drain confidence in the theory. The Iranian Revolution of 1979, the fall of the Berlin wall in 1989, the upsurge of the Moral Majority, the New Christian Right and Evangelicalism in the USA, political upheaval due to fundamentalist radicalization in several countries: all around the world, the face of religion began to change. From liberation theology in Latin America to Islamist movements in Southeast Asia, religion broke out of the private sphere to reclaim its role on the social and political levels. Religion returned to the public sphere, and most conspicuous was the upsurge of outspoken conservative, orthodox, and fundamentalist religious movements. These and analogous developments have undermined the paradigmatic status of the classical secularization thesis and have given rise to a huge controversy within modern social sciences and humanities as to its objective basis.

Finally, the tide began to turn in the intellectual community. A telling 'sign of the times' came from Peter Berger, who, in his 1999 edited volume *The Desecularization of the World*, offered a global overview of the worldwide resurgence of religion. One particular statement could not pass unnoticed:

> My point is that the assumption that we live in a secularized world is false. The world today, with some exceptions to which I will come presently, is as furiously religious as it ever was, and in some places more so than ever. This means that a whole body of literature by historians and social scientists loosely labeled 'secularization theory' is essentially mistaken.
>
> (*Berger 1999: 2*)

The idea of a gradually secularizing world lost its credit. From a global perspective, a totally different picture emerged. Religious traditions in most parts of the world seemed to have either experienced some growth or maintained their vitality, notwithstanding rapid increases in industrialization, urbanization, education, and so forth (Casanova 1994: 26). The real problem was not why religion had not declined in most parts of the world, but why it had declined in those particular milieus—Western Europe and the campuses of American universities according to Berger—that had given birth to secularization theory. 'Strongly felt religion has always been around; what needs explanation is its absence rather than its presence' (Berger 1999: 11–2). All of a sudden, not modern religiosity, but modern secularity, became the most puzzling phenomenon.

Berger's personal case helps to understand the abruptness of this switch. In the 1960s, Berger himself had been one of the founding fathers of classical secularization theory. In *The Sacred Canopy* (1967), he had argued for a theory of secularization that considered modernization and secularization as causally intertwined. The general thrust of his argument was that modernity inevitably undermined the social significance of religion. And despite a divergence of opinion as to the causes and the results of this diminishing significance, most sociologists of religion were eager to embrace this argument during the late 60s and early 70s. In view of that, the completely different approach to religion that arose at the turn of the twenty-first century boiled down to a deeply felt experience of having to abandon deep-rooted and seemingly self-evident presuppositions that had been instrumental during several decades in shaping a generally accepted worldview. This new approach was not so much about progressive insight into amendable shortcomings of an accepted theory, but about theoretical embarrassment with an uncomfortably uncritical conjecture. Hence, the performative act of heralding a postsecular age, a postsecular world, or a postsecular modernity can be regarded as a decision in the intellectual community to emancipate itself from a one-dimensional evolutionary account of religious change in the modern world.

Different frames for the return of religion

Over the course of the past decade, the term postsecularity has gained acknowledged currency. It seems to imply that we are experiencing a new era where modernity and religion are no longer seen as inherently antagonistic and mutually exclusive forces. Entering this new era means opening our eyes to the vitality and self-transformative capacity of religion in the modern world and simultaneously rejecting, on the levels of theory and discourse, the once-dominant classical secularization theory. However, when Jürgen Habermas coined the term postsecular in his influential paper ‘Notes on a post-secular society’ (2008), it seemed to convey a different understanding of the return of religion, one that did not completely brush aside classical secularization theory, but mainly its geographical and historical reach. In his descriptive account of a ‘post-secular society’, Habermas’ focus is not on a religiously vital world society, but on the return of religion in the largely secularized societies of Europe. A ‘post-secular society’ is one in which religion has regained influence and relevance within national public spheres, so as to challenge and undermine the once-dominant secularist view that religion will gradually disappear in the course of modernization. So, far from endorsing Berger’s statement that classical secularization theory is essentially mistaken, Habermas’ concept of postsecularity appears to presuppose a historical development in the past that can be accurately described in terms of classical secularization theory (see also Mendieta 2012). His only proviso to the theory is that this historical development has been interrupted by new developments in the present age that no longer fit this theory and therefore need to be described in a postsecular framework.

In other words, Habermas’ use of the term postsecular assumes not only a chronological perspective that moves from a premodern, religious society, over a secular society, to a postsecular society. It also implies exclusive applicability to those societies which have gone through the very process of secularization. Put differently, the background of a secular society is the necessary condition of possibility for the development of a postsecular society:

> A postsecular society must at some point have been in a ‘secular’ state. The controversial term can therefore only be applied to the affluent societies of Europe or countries such as Canada, Australia and New Zealand, where people’s religious ties have steadily or rather quite dramatically lapsed into the post-War period. These regions have witnessed a spreading awareness that their citizens are living in a secularized society.
>
> *(Habermas 2008:17)*

As already mentioned, a significant number of sociologists and philosophers currently dispute that classical presupposition. Not only do they refute it with respect to our present age, as Habermas does, but unlike Habermas, they also reject it with respect to the entire history of religion and modernity. The central reason why, for instance, Charles Taylor refuses to use the term postsecular in *A Secular Age* (only once does he make use of the term, when referring to the situation in postsecular Europe) must be seen in this respect:

> I use this term not as designating an age in which the decline in belief and practice of the last century would have been reversed, because this doesn't seem likely, at least for the moment; I rather mean a time in which the hegemony of the mainstream master narrative of secularization will be more and more challenged. This I think is now happening. But because, as I believe, this hegemony has helped to effect the decline, its overcoming would open new possibilities.
>
> *(Taylor 2007: 534–5)*

Obviously, the word postsecular in this quotation does not refer to a new period in our culture, following that of a secular age, but to an *alternative story* about the genesis and evolution of the secular age.

To conclude, we can articulate two different arguments for the contemporary use of the term postsecular. The first argument refers to the actual return of religion 'on the ground' of what were assumed to be secular, Western societies. In Habermas' reasoning, the need for the new term expresses both the novelty of an empirically verifiable phenomenon and the extent to which it is perceived as fully unexpected and unanticipated by the established theoretical models. Hence, the return of religion, related to the emergence of a religiously pluralist society in the West, requires a completely new approach on social and political fronts alike, beyond the confines of former secularization theory. The second argument has a different tone. Here, the term postsecular expresses the idea that classical secularization theory, and its core view that the modernization of society leads to a weakening of the role and significance of religion, are essentially mistaken. The rhetorical forcefulness of the term to take leave of the former paradigm is not so much derived from the unexpected return of religion in the public sphere, but rather from the shocking awareness of a deep-rooted prejudice in Western scholarship.

The common usage of the term postsecular, however, runs the risk of covering up a crucial difference between the main arguments and motivations that have helped establish a new theoretical paradigm. In the next section, we will deepen this analysis by relating it to a distinction between two different research areas within the broader domain of secularization theory. We will argue that the distinction between these two research areas is illuminating to gain a better understanding of where these two entirely different motivations underlying the performative action of the postsecular turn come from.

Genealogical versus diagnostic secularization theories

During the second half of the twentieth century, scholarly interest in the problem of secularization split up into two fields of research, each with its own distinct methodology and specific set of research questions. The first was mainly the field of philosophy, intellectual history, and interpretive sociology, with a hermeneutic methodology focused on a deeper understanding of modern culture. Here, scholars of different intellectual backgrounds wished to determine whether and to what extent key features of the modern world had their roots in premodern, religious culture. Accordingly, this field of research tended to establish an intrinsic relationship between Christianity and Western modernity, contrary to the latter's self-comprehension, in particular to the grand narratives it had developed and cultivated from the Enlightenment onwards.

The second was a social scientific field of study, mainly occupied by the disciplines of sociology of religion and history of religion. Here, the interest in the topic of secularization derived from the question of religious change in contemporary, modern societies. Supported by an analysis of empirically measurable indicators of religious adherence, like church attendance or denominational membership, much of the research in this area was centred around the hypothesis that processes of modernization have a negative impact on the vitality, stability, and social significance of religion. Hence, secularization theories of all sorts pretended to offer an objective, scientifically rigorous and value-free description of the changing role of religion in modern societies.

Earlier in the twentieth century, these two different approaches to the question of secularization had not been so clearly distinguished. In Max Weber's writings, for instance, one recognizes both types of investigation. A clear parting of the ways took place soon after the middle of the twentieth century (Monod 2002: 112). The then predominant social scientific discourse on secularization, firmly grounded on empirical examination of changes in both individual religiosity and religious authority over the past decades or even centuries, became more and more alienated from the erudite and far less popular secularization theories in the fields of philosophy, interpretive sociology, and intellectual history, which, by virtue of their unified and comprehensive accounts of the emergence and development of modernity, can be considered as the last significant master narratives of Western culture (*ibid.*: 113).

By the 1960s, scholarly study of secularization had definitely split into two separate approaches, each using the same term but applying it in a very different way. An important attempt to draw the demarcation line between these two branches came from Hans Blumenberg (1985):

> There is after all a difference between, on the one hand, saying that in a particular state the "secularization of the countryside" is very advanced, and that this is indicated by the empirical decline of obligations owed by village communities to the church, and, on the other hand, formulating the thesis that the capitalist valuation of success in business is the secularization of 'certainty of salvation' in the context of the Reformation doctrine of predestination.
>
> *(Blumenberg 1985: 10)*

So, in Blumenberg's view, the watershed between two areas of research was defined by the difference between the intransitive sense of the term secularization ('secularization is progressing rapidly') and the transitive sense of the word ('one thing is secularized into another').

Even if this criterion may be too strict, it is helpful to articulate the increasing estrangement between two entirely different approaches to secularization. Two crucial differences must be underlined here. First, there is a methodological difference: hermeneutic inquiry versus empirically based research. Second, the research questions and objectives are poles apart. Whereas the first field of inquiry concentrates on the transition from the Middle Ages to the Modern Age, and is set up as a *genealogical* inquiry into the religious roots of modernity, the second is a *diagnostic* (and therefore frequently also a prognostic) inquiry into the vitality of religion in the modern world, examining quantitative indicators of various types of religious change, inducing general patterns and relationships from them, in order to systematize these data into basic and lawful patterns of linear development. In what follows, we will define these two areas of research as, respectively, the genealogical and the diagnostic type of secularization theory. Whereas the former sees an inner alliance between religion and modernity, the latter has always tended to see religion as an irrational force and modernity as the product of rational progress, and therefore, considered both as strongly opposed.

In the first part, we already analyzed the diagnostic research field, showing how classical secularization theory came to an end and why the need for a new theoretical approach was felt so forcefully. Now, we turn our attention to the genealogical research field and see how it informs a different source of motivation for making the postsecular turn.

The genealogical research field: the religious roots of modernity

Even if Auguste Comte's positivistic view of the historical relation between religion and modernity initially functioned as a model for later views about the irrationality or superfluity of religion in light of scientific progress, it—paradoxically enough—paved the way for studying the genealogical relationship between modernity and religion. It was, however, Max Weber (1958, 1978) who inaugurated this genealogical approach by retracing the dynamic of rationalization in modern culture to its deeper sources in religion (see also Weidner 2004). Monotheistic religious traditions, by vehemently criticizing magical practices in the so-called 'natural' religions, Weber argues, unwittingly and unwillingly created the conditions of possibility for the breakthrough of modern science, eventually leading to what he famously called the disenchantment or *Entzauberung* of the world. Weber's connection between capitalism and protestant ethics can be regarded as well as an example of this internal capacity of religion for self-transformation, and a warning against simplifying secularization theories which state that 'science has substituted religion'—or will eventually do so.

In line with Weber, the Jewish philosopher Karl Löwith questioned modernity and its self-acclaimed originality. In *Meaning in History* (1949), Löwith coined a new (philosophical) secularization paradigm, that of modernity as an illegitimate transfer of previously religious elements to a worldly level. The transfer is illegitimate, because one cannot untie secular concepts or practices from their religious framework, without perverting the original purpose of the religious concepts and practices. For example, the Christian concept of 'hope' was deeply intertwined with a transcendent, eschatological perspective on life after death, by no means directed towards an inner-worldly total solution for suffering. When modernity illegitimately transformed Christian hope into the modern idea of progress, totalitarian ideologies saw their chance to promote inner-worldly goals leading to ultimate salvation *in this life*.

Hans Blumenberg opposed Löwith's paradigm of secularization as an illegitimate transfer of religious concepts to modern, secular concepts by defending modernity's original claims in his classic book *The Legitimacy of the Modern Age* (1985). Secularization, he claimed, did not come about by an illegitimate transition from Christian hope to the modern idea of progress. Medieval eschatology was not primarily based on the hope for salvation, but on the fear of condemnation. This fear was theologically deepened by the nominalist account of a voluntarist and almighty God, who could at any time alter the 'ratio' of his own creation arbitrarily and totally whimsically. Hence, man could no longer be considered as the image of God, nor could nature be seen as the reflection of a divine plan of creation. As a result, man was left on his own, squeezed between an almighty but capricious *Deus absconditus* on the one hand, and a whimsical nature on the other hand. According to Blumenberg, this predicament gave rise to the modern idea of self-assertion: man had to enhance his unenviable position by becoming master and commander of nature by his own rationality, and not by understanding God's rationality. Modern science developed in this context as a perfectly legitimate endeavour.

In more recent decades, philosophers like Marcel Gauchet and Charles Taylor have widened the scope of the Löwith-Blumenberg debate. While Löwith's and Blumenberg's perspectives on secularization remain true for their part, for Taylor and Gauchet, they only tell a relatively small part of the story. In other words, Gauchet and Taylor continue to recognize the pivotal importance of Christianity for modernity, but place that importance in a larger story about the evolution of religion in general.

This widening of the scope is carried out from a twofold angle. First, the whole history of religion is taken into account. In *The Disenchantment of the World*, Gauchet (1997) considers secularization as the very last stage in the irreversible transition from heteronomy to autonomy. Originally, in the era of primeval religion, the 'gods' were everywhere; there was no spatial gap between the 'invisible' and the 'visible'. Gauchet's somewhat paradoxical but central claim is that the 'reign of the invisible' or 'transcendence' first needed to be articulated more deeply, in order to eventually drift away from the 'visible' or 'immanent' world. The axial age, and the Judeo-Christian tradition, precisely constituted historical stages that articulated ontological dualism between the realm of the transcendent/invisible and that of the immanent/visible, and thus made it possible to ultimately separate the transcendent from the immanent. Gauchet sees Christianity as the penultimate stage of disenchantment (prior to the Enlightenment, with the rise of modern science), famously depicting Christianity as 'the religion for the departure from religion'—*la religion de la sortie de la religion*.

In his *opus magnum* on secularization, *A Secular Age*, Taylor (2007) raises exactly the same question as Gauchet: which conditions made it possible for Western culture to evolve from a clearly religious age around 1500 AD to a secular age in our times? But he wishes to enlarge the scope of his predecessors by emphasizing the partial role of sheer intellectual explanations and motivations. Like Gauchet, he takes into account the whole history of religion as well. Unlike Gauchet, he widens the scope of the Löwith-Blumenberg debate by stressing that a shift in ideas can never entirely account for complex processes such as secularization. Consequently, Blumenberg's account is explicitly criticized, while Löwith's account implicitly. Different conceptions of God or eschatology play their part but can never account for the large-scale reforms that changed the *social imaginaries* of people.

While Taylor's and Gauchet's genealogies of modernity and secularization broadly converge on empirical and transcendental levels, their ways part when it comes to their appreciation of secularization and modernity. Gauchet's story about the 'dynamics of transcendence' logically results in his thesis about the irreversibility of secularization in the public sphere. Taylor, on the other hand, believes that 'nothing is ever lost'. The historic sediments of foregone ages may always be dug up again and may provoke us to criticize our contemporary ways of living. It is by no means predictable which road the future will take: will our society eventually ban all reference to religion, or will it rediscover the richness of its religious past?

Taylor's dictum (borrowed from Bellah) that 'nothing is ever lost' also echoes, be it inadvertently, in much of contemporary deconstructionist philosophy. Jean-Luc Nancy, for example, develops in *La Déclosion* (2005) a deconstruction of Christianity that draws attention to secular elements inherent to Christianity. Since the secular is essentially a product of monotheistic religions, it can never be fully understood without reference to its religious predecessors. Hence, it is precisely religion that withstands any full definition of 'the secular'. Paraphrasing Taylor, religion is the 'un-thought' of the secular. From different perspectives, this very tension within Christianity is elaborated by such divergent philosophers as Gianni Vattimo and René Girard.

It may have become clear by now in what respect the genealogical approach to secularization, undertaken on the research fields of philosophy, interpretive sociology, and intellectual history, differs from the diagnostic research area and its social-scientific study of secularization. While the latter, until very recently, has supported the view that the relationship between modernity and religion is mutually exclusive, the former has established an alternative story of secularization depicting an inner relationship between modernity and religion, with Christianity taking a predominant role. Yet, the genealogical research area has its limitations too, and they help us to understand the performative force of the term postsecular in a fresh way. Since the genealogical approach to secularization underlines a Eurocentric, Christianity-centred view on global processes of modernization, the performative use of the term postsecular can be seen as derivative from the decision of scholars no longer wishing to operate within a paradigm in which modernity and global processes of modernization are traced back to their roots in one specific historical context, due to the inner workings of one specific religion.

Postsecular: the West versus the rest

The growing sensitivity for the complexity of the concept of secularization in the Western world and the alleged need for another term, like, for instance, postsecular, indeed is intensified when it is transferred to non-Christian religions and to other areas in the world, where the sociological strains and tensions between religion and the state, secularism, and democracy, or the philosophical ones between transcendence and immanence have taken a completely different shape. The case of India is especially instructive in this regard. Since its independence from Britain, secularism has been a defining feature of its political system, whereas at the same time, it has become a deeply contested concept. A whole gamut of scholars claim that Western secularism, imposed on Indian society as a hegemonic ideology indebted to Protestant Reformation, is not viable for Indian society. They argue that minority rights for the traditional believers and religious toleration offer a more reliable guarantee for ethnic and religious tolerance than secularism does. In a similar vein, analogous questions and controversies apply to the world-wide problem of migration in general, and of Muslim migration in Europe in particular. The term postsecular seems not only applicable to Western Europe, but first and foremost to the world as a whole (Gorski and Altimordu).

In order to cope with the difficulty to find a concept covering the worldwide relations between religion and contemporary societies, the classic category of secularization hardly seems to be applicable. Since religions as Confucianism or Taoism, for instance, have, in a sense, always been 'worldly' and 'lay', the Western concept of secularization does not make much sense as a descriptive account of their recent evolution (Casanova 2006). A number of scholars (Eisenstadt 2003; Casanova 2006; Delanty 2007; Taylor 2007) have therefore developed the concept of 'multiple modernities' as a more adequate tool to designate global trends in this respect than either secular cosmopolitanism or a clash of civilizations between religious and secular regimes (Casanova 2006). Those who support this 'multiple modernities' position reject both the idea of a modern radical break with religious traditions (classical secularization paradigm) and the position that defends an essential continuity with tradition. In a postsecular world society, traditions are forced to adjust to modern conditions, in the process of which they also help to shape the particular forms of modernity (Casanova 2006). The lively discussion about topics like 'multiple modernities' and the relation between the secular and postsecular shows that the original self-evident belief in the inevitable decline and eventual extinction of religion has made room for a versatile debate not only around a whole range of aspects of the relation between modernity and religion, but even for a renewed conceptual analysis of the term 'secularization' itself:

> Which is to say that the definition of the secular and its relationship to the religious are as hotly disputed now as ever, and that the scope of the debate is probably wider than ever. Whether this period of contestation marks the beginning of a post-secular age, or merely a period of secular ebb and religious flow, only time will tell.
>
> *(Gorski and Altimordu 2008: 77)*

Conclusion

In this contribution, we attempted to articulate several motives and arguments behind the rhetorical gesture of heralding a postsecular age. Because these motives and arguments appear to be remarkably different, their distinctive thrust tends to be covered up by the performative use of the term postsecular, and finally risks being erased by its increasingly widespread and thoughtless use. We argue, therefore, for a more attentive use of the word postsecular, reflecting the exact motivation to move beyond classical secularization theory.

In the first part, we concentrated on what we defined as 'diagnostic' secularization theories. From two different angles, diagnostic theorists have arrived at the conclusion that our contemporary society is postsecular. While some, like Berger, refer to a profound shift in our theoretical discourse about the role and significance of religion in the modern world, others, like Habermas, focus on a factual historical shift. These two angles have resulted into two different understandings of postsecularity which, if used uncritically, tend to reinforce the very presuppositions of classical secularization theory, such as the idea of an evolutionary scheme of religious change for which the heartlands of Western Europe are supposed to serve as a model.

In the second part, we studied 'genealogical' secularization theories. Although genealogical theorists have moved beyond a Eurocentric and predominantly Christian perspective, in their theories, the West, specifically the relation between (Christian) religion and secularity, is often singled out as a distinctive trajectory of modernity, so implicitly serving as a role model for religious modernization. In this context, the performative use of the term postsecular may challenge this theoretical hierarchy of the West versus the rest, opening up fresh avenues for the study of religion in the contemporary world. However, it remains important to take into account questionable presuppositions of classical secularization theory, such as the idea of Christianity's privileged relation to global processes of modernization, and to investigate in what respect they continue to inform and shape current academic scholarship.

Against that double backdrop, we conclude that the term postsecular should be used with more precision and caution. An enduring attempt, therefore, to examine deep-rooted beliefs and prejudices of classical secularization theory may be more important than simply considering them as innocent relics of a surpassed historical period or paradigm.

Further reading

Betz, J. R. (2009) *After Enlightenment: the post-secular vision of J.G. Hamann*, Oxford: Wiley-Blackwell Publishers.

An introduction to the life and work of eighteenth-century German philosopher, J. G. Hamann, often considered as the founding father of what is known as 'Radical Orthodoxy'.

Gorski, P. S., Kyuman Kim, D., Torpey, J. and J. VanAntwerpen (eds.) (2012) *The Post-Secular in Question: religion in contemporary society*, New York: New York University Press.

This collection of essays by leading philosophers assembles original and interdisciplinary interventions in the ever-transforming discussion of the role of religion and secularism in today's world.

Milbank, J. (1992) 'Problematizing the secular: the post-postmodern agenda'. In Berry, P. and A. Wernick (eds.) *The Shadow of Spirit: postmodernism and religion*, London: Routledge, pp. 30–44.

In this article, Milbank raises the question whether postmodern thought does not somehow have to redefine the modern critique of religion so as to make it no longer conducted on behalf of the 'human', in opposition to the illusory intrusions of the sacred.

Parmaksız, U. (2016) 'Making sense of the postsecular', *European Journal of Social Theory*, 15: 98–116.

This article critically examines the postsecular literature so as to get rid of the scepticism about the concept's theoretical import and critical power. In the last section, the focus shifts to the analytical utility of the concept, examining 'postsecular society' and 'postsecularization' in the light of the previous discussion.

References

Berger, P. L. (1967) *The Sacred Canopy: elements of a sociological theory of religion*, New York, NY: Doubleday.

——— (ed.) (1999) *The Desecularization of the World Resurgent Religion and World Politics*, Grand Rapids, MI: William B. *Eerdmans*.

Blumenberg, H. (1985) *The Legitimacy of the Modern Age*, Cambridge, MA: The MIT Press.

Bruce, S. (2006) *God is Dead: secularization in the west*, Oxford: Blackwell.

Casanova, J. (1994) *Public Religions in the Modern World*, Chicago, IL: University of Chicago Press.

——— (2006) 'Rethinking secularization: a global comparative perspective', *Hedgehog Review*, 8: 7–22.

Delanty, G. (2007) 'European citizenship: a critical assessment', *Citizenship Studies*, 11: 63–72.

Eisenstadt, S. N. (2003) *Comparative Civilizations and Multiple Modernities*, Vol. I, Leiden: Brill.

Gauchet, M. (1997) *The Disenchantment of the World: a political history of religion*, Princeton, NJ: Princeton University Press.

Gorski, P. S. and A. Altınordu (2008) 'After secularization?', *Annual Review of Sociology*, 34: 55–58.

Habermas, J. (2008) 'Notes on a post-secular society', *New Perspectives Quarterly*, 25: 17–29.

Joas, H. (2007) 'Gesellschaft, staat und religion: ihr verhältnis in der sicht der weltreligionen'. In Joas, H. and K. Wiegandt (eds.) *Säkularisierung und die Weltreligionen*, Frankfurt a.M.: Fischer.

Löwith, K. (1949) *Meaning in History: the theological implications of the philosophy of history*, Chicago, IL: University of Chicago Press.

Martin, D. (1991) 'The secularization issue: prospect and retrospect', *The British Journal of Sociology*, 42(3): 465–74.

Mendieta, E. (2012) 'Spiritual politics and post-secular authenticity: Foucault and Habermas on post-metaphysical religion'. In Gorski, P. S., Kyuman Kim, D., Torpey, J. and J. VanAntwerpen (eds.) *The Post-Secular in Question: religion in contemporary society*, New York: New York University Press, pp. 307–44.

Monod, J.-C. (2002) *La Querelle de la Sécularisation: théologie politique et philosophies de l'histoire de Hegel à Blumenberg*, Paris: J. Vrin.

Nancy, J. L. (2005) *La Déclosion: déconstruction du Christianisme 1*, Paris: Galilée.

Stark, R. (1999) 'Secularization, R.I.P.', *Sociology of Religion*, 60(3): 249–73.

Taylor, C. (2007) *A Secular Age*, Cambridge, MA: The Belknap Press of Harvard University Press.

Wallis, R. and S. Bruce (1992) 'Secularization: the orthodox model'. In Bruce, S. (ed.) *Religion and Modernization: sociologists and historians debate the Secularization thesis*. Oxford: Clarendon Press.

Weber, M. (1958) *The Protestant Ethic and the Spirit of Capitalism*, New York, NY: Charles Scribners' Sons.

——— (1978) *Economy and Society: an outline of interpretive sociology*, edited by G. Roth and C. Wittich, Berkeley: University of California Press.

Weidner, D. (2004) 'Zur rhetorik der säkularisierung', *Deutsche Vierteljahrsschrift für Literaturwissenschaft und Geistesgeschichte*, 78: 95–132.

Wilson, B. (1998) 'The secularization thesis: criticisms and rebuttals'. In Laermans, R., Wilson, B. and J. Billiet (eds.) *Secularization and Social Integration: papers in honor of Karel Dobbelaere*, Leuven: Leuven University Press, pp. 45–65.

8

Postsecularism, reason, and violence

Michiel Leezenberg

Introduction

The term 'postsecularism' appears to have been coined in the 1960s by Andrew Greeley, but it was undoubtedly Jürgen Habermas who did most to popularize it, around the turn of the twenty-first century.[1] As developed by Habermas, the postsecular involves the idea that religious convictions and claims may not only persist or gain new strength in secular environments, but may also be considered reasonable by secular actors if they accept the rules and procedures of the secular nation-state. Increasingly, however, one witnesses religious positions and movements that do not simply demand recognition within and from a secular environment, but openly reject the very concepts or values of (gendered) morality, life, law, and reason on which modern secular Western societies rest. Arguably, it has been the increasingly visible assertive presence of Muslim population groups in Europe and America, and the dramatic appearance of both state and non-state Islamic actors on the world political stage, rather than the continuing or renewed self-confidence of Catholic and evangelical Christian demands (let alone Hindu nationalism in India, neo-Confucianism in China, or the Orthodox Christian revival in Eastern Europe), which have most visibly posed a challenge to theoretical debates on the role of religion in the public sphere—that is, on questions of secularism and postsecularism. Hence, as illustrations, I will briefly discuss two forms of contemporary Islam, namely, quietist Salafism and potentially violent Salafi-jihadism.

The experience of partly violent forms of contemporary Islam indeed has been central to the writings on postsecularism by some of the most prominent contemporary thinkers. Here, I will focus on the liberal philosopher Jürgen Habermas and on the anti-liberal anthropologist Talal Asad. In particular, I will discuss the views of both these authors on the explanation and justification of violent politicized religion. Although Habermas and Asad seem at odds with each other on many points, they appear to share several important assumptions: first, a positive appreciation of rational debate and a sweeping condemnation of violence as not only illegitimate but irrational; second, a distinct and problematic notion of terrorism; third, and finally, a somewhat reductionist view of the contemporary world as having arisen out of a confrontation of a generic 'Islamic tradition' or 'traditional Islam' with an equally generic Western 'imperialism', 'colonialism', or 'modernity'.

Violent politicized religion: a short history

The rapid decline of communism after the 1989 fall of the Berlin Wall left a gap that was quickly filled by newly public and politicized religious ideologies. These could be peaceful and involve civil political action; but increasingly, they could also motivate acts of political violence, most visibly, but by no means exclusively, in non-liberal, conflict-ridden settings like the Middle East.[2] Such acts are usually labelled—and condemned—as 'terrorism'. In earlier decades, conceptions of terrorism had included state terrorism and state-sponsored terrorism; however, after the collapse of the communist East Bloc, and especially in the wake of the 11 September 2001 assaults, terrorism has virtually come to be identified with political violence by non-state actors against existing states. As such, it is illegitimate virtually by definition; to the extent that it is practiced by religiously motivated groups or individuals, it has also increasingly come to be seen as irrational and premodern.

It should be emphasized, however, that this 'Islamic terrorism' is a qualitatively novel phenomenon that resists reductionist explanations in terms of either a timeless Islamic essence or a generic imperialist influence. Both analytically and normatively, one cannot and should not isolate it from the complex whole of interlocking factors out of which it has emerged. These factors include a decline or collapse in state power, a heritage of violent authoritarian regimes, Cold War polarization, and interference by neighbouring countries or foreign superpowers. Thus, Osama bin Laden's al-Qa'ida network, which resorted to spectacular transnational actions, and the so-called Islamic State, which briefly held large swathes of territory in Syria and Iraq between 2014 and 2017, are but the most visible, and most widely publicized, examples of a rather broader pattern. The very novelty of these forms of (political) violence by non-state actors, and indeed their unprecedented scale and character, precludes any attempt to seek their causes in any premodern, let alone timeless, Islamic religious tradition.

The same holds for suicide assaults by jihadist actors, perhaps the most spectacular form of violence in the name of Islam carried out in recent years, in the Middle East as well as in the USA and Western Europe. Three aspects of these assaults stand out. First, their perpetrators actively sought to die in their action; second, they were generally transnational, many of them having been committed by second-generation immigrant male youths; third, they have been called 'depoliticized' and 'deterritorialized', as it is increasingly difficult to link them to any clear or coherent political or territorial cause, like, earlier, the Israeli occupation of Palestinian lands or the presence of American troops in Saudi Arabia.

It should also be emphasized that traditional Islamic jurisprudence does not sanction suicide for any reason. The first theological justifications of suicide bombings did not appear until the 1980s, that is, *after* the first such bombings had been carried out (cf. Leezenberg forthcoming). Suicide bombings remain controversial in the Islamic world: the (vast) majority of Muslim scholars unambiguously disapprove of them. Clearly, present-day terrorism cannot be explained from religious tradition, and openly defies traditional religious authority.

Habermas on postsecularism and Islam

The persistence, strengthening, and/or revival of religious claims in secular environments have forced liberal thinkers like John Rawls and Jürgen Habermas to reconsider their secularist assumptions. In developing his ideas on the legitimacy of religious claims, Habermas originally focused on topics like abortion, euthanasia, and genetic manipulation in liberal and secular Western societies; however, the topic of violent political Islam almost forced itself on him. Thus, 'Faith and Knowledge', his acceptance speech for the 2001 Peace Prize of German booksellers, was originally meant to address the rational acceptability of religious objections against, for example, euthanasia or genetic engineering in liberal Western societies. It was presented, however, a mere month after the 11 September 2001 assaults against the World Trade Center in New York and the Pentagon in Washington. Hence, its introduction—clearly written after the main body of the text—addresses the question of religiously motivated or legitimated violence (Habermas 2003: 102). Habermas here sees these attacks as 'motivated by religious convictions' rather than political grievances, claims, or calculations. In later statements, he would shift his position regarding the motivations of these and other suicide attacks; thus, Habermas (2015: online) argues that: 'Jihadi fundamentalism expresses itself in religious codes but it is no religion... It could use any other religious language'.

To put it differently, according to Habermas' later writings, violent politicized forms of religion are not part of that religion itself, as a distinct 'sphere' of social life or as a tradition; as a result, it would seem, they cannot embody or express legitimate *religious* claims almost as a matter of definition.

Apart from this particular change, however, Habermas' views on the character and explanation of such violence have been remarkably consistent over the years. In 2001, he argues that the assaults are 'fundamentalist', and as such, a thoroughly modern phenomenon: for him, fundamentalism marks an 'uprooted' form of modernization and a 'blocked spiritual development'; likewise, in 2015, he states that 'Jihadism is a thoroughly modern form of reaction to uprooted ways of life' (2003: 102; 2015). Thus, Habermas appears to explain violent religious activism in cultural and civilizational terms of modernization, at the expense of attention for possible political motives, while tacitly presuming the Western European process of secularization as the normal course of development. This explanation is not only tilted or biased, in ignoring any political motive or grievance (whether or not legitimate), but also unabashedly Eurocentric, in assuming the Western European secularization process as the normal form of modernization (see Leezenberg 2010). This liberal ethnocentrism is also expressed in Habermas' call for a greater self-reflection 'necessary if we want to present a different image of the West to other cultures' (2003: 103) rather than a self-criticism of what, if any, recent Western actions might have triggered such reactions.

In other words, for Habermas, religiously motivated suicide terrorism falls outside of religion, and requires sociological or civilizational explanation rather than political legitimation. This implies that such acts of violence can *never* be legitimate, let alone 'translated' into rationally justifiable political or societal demands, almost as a matter of definition.

Undoubtedly, rejecting the legitimacy of religiously motivated terrorist violence sounds eminently reasonable—not to say inevitable—from a liberal and secular normative perspective; however, analytically, such an a priori rejection is less than satisfactory. In fact, an ambivalence in Habermas' concept of postsecularism comes to the fore here. Initially, he seems to use the term 'postsecularism' to describe a purely *factual* state of affairs in which religious claims persist or reappear in a secularized environment; later, he also uses it as a *normative* term to denote a 'mutual learning process' between religious and non-religious citizens. Accordingly, he shifts from his earlier view that religious claims are legitimate only to the extent that they can be translated into secular political claims and arguments, to the position that 'religious reasons' as such are not only legitimate but indeed essential parts of a liberal deliberative democracy. Problems like euthanasia, abortion, and genetic manipulation, he argues, are so complex that one cannot be certain that only the non-religious perspective has the correct moral intuitions. In line with this shift, Habermas reinterprets the separation of church and state as a separation between the public sphere and the formal institutions of the state: in the former, he argues, religious claims are permitted; but they must be translated into secular contributions if they are to be accepted in the latter (2008b).

On closer inspection, however, the concept of the postsecular presupposes both a descriptive historical component, namely, the secularization thesis, and a normative component, namely, secularism, already from the start. This normative dimension appears when Habermas qualifies the Islamic world as marked by 'thwarted spiritual change' (*blockierte Geisteswandel*) and 'derailing secularization' (*entgleisende Säkularisierung*) (2003: 102; 2015). Comparably to Rawls's vision of political liberalism as a neutral framework rather than a 'comprehensive doctrine', Habermas calls his own position 'postmetaphysical', in that it no longer requires the rejection of religiously formulated metaphysical claims. Unlike anti-religious naturalism, which he calls 'pure ideology', postmetaphysical thinking 'is prepared to learn from religion while at the same time remaining agnostic'; he continues by saying that both religious and secular consciousness can, and indeed must, become more reflective over time (2008a: 143). Despite such proclamations of neutrality and acknowledgements of historical change, however, Habermas tacitly assumes religion as, by definition, non- or anti-modern, and as founded on faith rather than rationally justifiable belief; likewise, he characterizes religious as opposed to secular subjectivity as essentially heteronomous rather than autonomous, and as shaped by social power rather than individual liberty. This very opposition, however, presupposes a distinctly modern definition of religion as ultimately irrational and authoritarian, and as primarily, if not exclusively, involving inner faith rather than public practice.

Habermas' position has been criticized for making religious claims appear more reasonable and less authoritarian than they are (see e.g. Flores d'Arcais n.d.). Yet, Habermas, too, would undoubtedly want to distinguish between legitimate and illegitimate religious claims and reasons. Initially, he accepts religious claims as legitimate to the extent that they accept the constitutional state and deliberative democracy (2008a: 143). This position presupposes the state's monopoly on legitimate violence, and thus would seem to exclude a priori the legitimacy of any resort to violence by non-state actors. In international politics, by contrast, and especially in constellations in which there is no well-functioning constitutional state or deliberative democracy, acts of violence or civil disobedience may be legitimate (*ibid.*). Yet, an ambiguity remains: Habermas' comment that jihadi terrorism involves the 'abuse of religion for political purposes', suggests on the one hand that it is political rather than religious or civilizational, and on the other hand that in itself, religion is by definition, non-modern, irrational, and/or separated from rational political or other worldly purposes; yet, elsewhere, Habermas' characterization of religiously legitimated terrorism in terms of a failed or derailed modernization process boils down to a *rejection* of its political (that is, rational and potentially legitimate) character.

It is indeed difficult to see how Habermas' account could be extended or modified in such a way as to allow *any* religiously motivated terrorist violence to have a political motivation—let alone justification. Yet, the 11 September assaults at least may be called 'rational' in the restricted sense of 'goal-rational', or having concrete and realistic (political) motivations and aims. As Osama bin Laden's speeches suggest, these assaults aimed at ending US military presence in Saudi Arabia, where American troops had been stationed since the 1990 Gulf Crisis (Lawrence 2005). It is unclear how, if at all, Habermas' account could accommodate such aspects, let alone address questions of the legitimacy of political claims and grievances underlying terrorist acts, even when one may for good reasons reject the legitimacy of those acts themselves. In the following section, I will argue that there is a deeper reason for this inability.

Talal Asad and Islamic communitarianism

Despite allowing for historical changes, Habermas appears to assume the categories of the religious and the secular, and the spheres of religion and politics, as substantially given and essentially unproblematic. Perhaps no author has been as influential in calling attention to the temporally and geographically specific, constantly renegotiated, and power-saturated character of these categories as the anthropologist Talal Asad. Normatively, too, Asad's anti-liberal position may seem very different from Habermas'; on closer inspection, however, both authors turn out to share a positive appraisal of argument as rational (even if they mean different things by 'rationality'), and a negative valuation of violence as both illegitimate and irrational; that is, as belonging neither to 'reasonable' politics nor to religious traditions.

Proceeding from a genealogically inspired critique of the anthropological category of 'religion', Asad (1993: chap. 1) argues that seemingly universally applicable concepts like 'religion' and 'the secular' have a specifically Christian history, which renders problematic their application to other traditions, like Islam. To the extent, moreover, that such concepts *have* been applied, they reflect a history of Western (imperialist) power. Because of its implication in a post-Enlightenment history of power, modern liberal secularism therefore does not form a neutral framework, and its main concepts do not form a universally valid vocabulary; rather, liberalism is just one moral tradition like others. Subsequently, Asad (*ibid.*) even argues that liberalism is not just a tradition, but a secular political *myth*, and, as such, no more rational or universally valid than 'religious' myths. And indeed, he dismissively characterizes the idea of liberal democracy as 'our contemporary holy cow' (Iqbal 2017: 210).

Paired to this genealogy of the concepts of religion and the secular is Asad's equally influential communitarian view of Islam as a 'discursive tradition', that is, as an essentially reasonable and continuous tradition of Muslim discourse and debate that authorizes Islamic practices of the present by referring them to an Islamic past (1986: 14–5). Asad has repeatedly emphasized that this view involves not a substantial claim about what 'real Islam' is, but a methodological suggestion to construe Islam as an object of anthropological inquiry in a particular way. Despite its emphatically methodological character, however, it does have substantive implications; or at least, it has been read that way by some of his followers, like Saba Mahmood, Charles Hirschkind, and Wael Hallaq. This normative position may be characterized as *Islamic communitarianism*; indeed, Asad quite explicitly acknowledges his indebtedness to MacIntyre's brand of communitarian moral philosophy (e.g., Asad 1986: 21n, 2015b: 167, n8, n12). Against both Marxist and genealogical modernists, Asad emphasizes that religious traditions are given rather than invented (2015b: 168); against both Foucault and MacIntyre, he argues that genealogical critique involves not the rejection but the purification of tradition. This purification, he adds, puzzlingly:

> uses violence (or the threat of violence) to restore an obscured origin that can then accommodate itself more smoothly to the real, progressive world. This process of critical purification (modernization) is a process of what must be transgressed if tradition is to become civilized.
>
> *(2015b: 168)*

In other words, Asad seems to preclude the very possibility that the Islamic discursive tradition is based on power or violence, and as such can be anything other than 'rational'. Instead, he sees not tradition but the *change* of traditions as violent: discontinuities and new traditions, he writes, originate in 'discursive rupture, which means through a kind of violence' (2009: 33).

Here, a remarkable asymmetry appears in Asad's treatment of Western and Islamic concepts, practices, and traditions. The notion of an Islamic discursive tradition, he has argued, is meant to foreground questions of embodiment and power (2015b: 166–7); in fact, however, neither notion functions very prominently in his analyses of things Islamic. Indeed, he turns out as keen to emphasize the reasonableness of the Islamic tradition as he is to expose the power implied in Western secular reason. Moreover, despite his seeming genealogical approach, Asad flat out contradicts Foucault's insistence that genealogy sets out to destroy the semblance or illusion of continuity conjured up by the notion of tradition (Foucault 2000: 373, 385). Normatively, this position is problematic because it places the rejection of tradition out of bounds, and amounts to an emphatic, if largely implicit, denial that violence, oppression, or domination may be located within the tradition itself, witness his suggestion that only the genealogical critique of, or rupture with, tradition, and not tradition itself, involves violence or the threat of violence (2015b: 168).

Most problematically, Asad systematically downplays the power processes involved in creating and reproducing this Islamic tradition. He does so, among others, by strictly distinguishing the religious authority of Islam or its revealed law, the sharî'a (which, significantly, he discusses rather than the legal scholars, or 'ulama), from the political power of amîrs or other rulers. Further, he consistently characterizes this tradition as shaped by disagreement, that is, by '*arguments* about what it means to be a Muslim' (2017: 200; emph. in original). That is to say, he sees this discursive tradition as defined, maintained, and developed by rational debate rather than the exercise of power or the imposition of authority. Following MacIntyre, Asad adds that critique is central to it, albeit not in abstract theories but in 'embodied (and yet criticizable) ways of life' (2015b: 167).

After this discussion of some of the conceptual and normative issues raised by Asad's appeal to discursive traditions, let us now look at some of the empirical dimensions of his writings. Unlike Habermas, Asad sees the secular liberal nation-state, and post-Enlightenment (Western) Europe more generally, as matters of identities rather than rights or obligations. In an article originally published a year before the 11 September assaults, Asad (2003: chap. 5) argues that Muslim immigrants have a problematic position in Europe, not because their religious and cultural backgrounds clash with Enlightenment values and principles, like religious tolerance, secularism, and autonomy, but because these universalist principles themselves exclude Muslims from an identity construed as 'European'. For example, he argues, even when liberals and Far-Right politicians disagree on the question of whether or not Europe can 'tolerate' the presence of Muslims, they proceed from a tacit shared assumption that Muslims may *live in*, but do not *belong to*, Europe—precisely in so far as Muslims are represented as external to the essence of Europe, he concludes, the question of whether 'real' Europeans can coexist with them is problematic (2003: 164–5).

Given his rejection of the universal validity and neutrality of liberal secular claims and framings, Asad's own position may perhaps be qualified as 'postsecular'. And indeed, to the extent that it invites new ways of thinking about how religion can be accepted in the public sphere, Asad welcomes Habermas' concept. He criticizes it, however, for uncritically relying on the concept of the secular, and thus reproducing the assumptions and affects it indicates (Asad n.d.). Whereas Habermas presupposes the language of rational argument, rights and obligations, and of argument as giving and asking for reasons, Asad describes how appeals to such seemingly universalist principles and values in a modern European setting amount to the a priori exclusion of religious convictions as irrational in themselves, and of Muslim identities as essentially non-European.[3] Thus, affect and identity are crucial to Asad's analyses.

This is certainly an interesting position; now let us see how Asad views political violence, and in particular terrorist violence, carried out in the name of Islam. Asad (2010, 2015a) does not discuss possible political motivations or aims of such terrorist violence in detail. In fact, he does not even seem to countenance the possibility that (political) violence may *ever* be rational, let alone justified, in the sense of serving concrete and realizable goals, witness his rejection of utilitarian or functionalist views that focus on the aims and effects of violence (cf. Anidjar 2015). Instead, he shifts attention to the question of why we find particular forms of violence more abhorrent than others.

This line of argument leads to a systematic blurring of the opposition between legitimate and illegitimate violence, and between warfare and terror. Thus, Asad rejects the opposition between just or legitimate wars of states and the illegitimate if not evil terrorism of non-state actors, arguing that 'the liberal democratic state shares a space of violence with terrorism' (2010: 18). Likewise, he appears to reject the idea of a state monopoly on violence, identifying the 'terror of modern warfare' and 'terror created by militants', characterizing the state's torture of prisoners as 'individualized terror' (2010: 7), and arguing that the notion of 'humanitarian intervention' dissolves the distinction between war and peace (2010: 11). All this leads him to a blanket condemnation of all violence, which he generically qualifies as 'terror'. Unlike, and indeed against, Habermas and other liberal thinkers, he reduces even the international legal order to the sovereign power of the liberal democracies of the West, first and foremost the USA, arguing that the language of law is a mere fig leaf for the self-legitimations of victorious states: 'in international relations, what we call [...] war is ultimately brute force and not authority that is decisive [*sic*], even if that force is argued for in a legal language' (2010: 5). Asad undoubtedly has a point in calling attention to the violence of this international order; but my point here concerns his conceptualization of a generic 'terror' and 'terrorist violence' as not only illegitimate but also irrational.

Further, Asad tacitly identifies states with liberal secular democracies. This identification is most clearly visible, and most problematic, in his discussion of post-2011 Egypt. In a 2015 article, Asad returns to the question of Islam as a (discursive) tradition, elaborating on the relation between the discursive and the embodied aspects of this tradition, and exploring how religion, authority, and tradition were linked in the 2011 Egyptian uprising against, and ousting of, President Mubarak. Significantly, Asad only addresses *political* authority here, tacitly avoiding all questions concerning *religious* authority, as involved in, for example, the social power of Shaykh Usama and the Fatwa Council. Apparently, he reduces such forms of power to a presumed sovereignty of the sharî'a, under which a Muslim 'care of the self' is allegedly practiced (2015: 179–81).

Here and elsewhere, Asad once again evades questions of power within the Islamic tradition, while simultaneously conflating liberalism, secularism, and Mubarak's and Sisi's Egypt—as if the latter could by any stretch of the imagination be called a liberal state. This is no mere slip of the pen: elsewhere, Asad writes about 'Egypt's acquisition of liberalism', conflating the economic liberalization of Egypt's 1980s *infitah* policies, or free-market reforms, with the alleged spread of political ideas of individual freedom—a conflation that even leads him to insinuate that Nasser was a liberal ruler rather than, as is commonly thought, a pioneer of Arab socialism (2015b: 199–200).

In positing this Manichean dichotomy or opposition between the intellectual and political tradition of Western liberal secular democracies and the embodied discursive tradition of Islam, Asad not only downplays the non-liberal authoritarianism of the secular Egyptian state, but also overlooks a crucial intellectual and political factor that has arguably shaped both post-independence Arab states and post-Cold War politicized Islamic movements: Marxism-Leninism. Limitations of space preclude a fuller discussion here; suffice it to say that self-proclaimed 'socialist' Arab states like Egypt, Syria, Iraq, and Yemen were characterized by a Leninist form of strictly hierarchical one-party rule backed by the pervasive presence of intelligence services. These technologies of ruling were profoundly inspired, and practically and organizationally supported, by East Bloc states like the German Democratic Republic and the Soviet Union. Likewise, the personality cult surrounding Arab leaders like Nasser, Sadat, Mubarak, Hafez al-Assad, and Saddam Hussein followed recognizably Stalinist models.

Asad, however, routinely denies that communism has ever been a major political or intellectual force in the Arab Middle East.[4] But, despite such denials, Marxist-Leninist ideas, strategies, and organizational forms have in fact had an enduring relevance and importance for Islamist political parties and social movements in the post-Cold War and postsecular Islamic world. Arguably, Leninist and Stalinist models have also shaped Islamist thought and practice from the 1960s, if not earlier. Thus, Sayyid Qutb's famous pamphlet, *Ma'âlim fi'l-tarîq* ('Milestones') (2007 [1964]), is emphatically Leninist in its call for the uncompromising ideological purity of a vanguard of genuine Muslims, and even more clearly in its call for revolutionary violence against *jâhilî* states that are based on human rather than divine sovereignty. Hence, legitimations for revolutionary and/or terrorist violence in present-day Islamic authors are thoroughly modern in character, often inspired by modern Eastern European ideologies rather than classical Islamic traditions. Arguably, the challenge—if not the governmental reality—of Leninism has decisively shaped not only the institutions and governmental techniques of modern Arab states but also the outlook and organizational forms of various present-day currents of political Islam. Given his overstating of the continuity of the Islamic discursive tradition, as well as his binary opposition between this tradition and the liberal secular West, Asad's overlooking or downplaying of this Eastern European and/or communist genealogy appears to be quite systematic.

Salafism and jihadism in Europe: a genealogical sketch

One reason for Asad's tendency to slip from a genealogical to a communitarian position may be his focus on sovereign state power: many of his analyses focus on the repressive powers of the sovereign secular nation-state. Foucault himself, however, warned against assimilating all forms of power to a juridical model of sovereignty, provocatively stating that 'in political thought and analysis, we have still not cut off the head of the king' (Foucault 1978: 88–9). Famously, Foucault has explored discipline, a non-sovereign and non-repressive modality of power emerging in spaces and institutions like schools, hospitals, and psychiatric wards. He saw this disciplinary power as typical of early nineteenth-century European states, and as having come to an effective end by the late 1970s. He analyzed neoliberalism (then very much in the process of being formulated and indeed politically realized) in his 1979 Collège de France lectures on biopolitics, specifically suggesting that it is a non-disciplinary technology of government. Simultaneously, he also formulated his famous—or notorious—thesis that the revolt against the secular Pahlavi regime in Iran reflected a novel 'political spirituality', which rejected secular modernity in the guise of both liberal capitalism and communism (see Foucault 2008: 248ff).[5]

Continuing along these lines, one may ask whether early twenty-first-century forms of politicized Islam amount to a rejection of a secular political rationality while at the same time embodying novel forms of governmentality themselves. Let me therefore conclude this overview with a preliminary genealogy of contemporary Salafism—a reform or revivalist movement in Sunni Islam—as a neoliberal, postsecular post-Cold War phenomenon.[6] Although encouraged, and in part sponsored, by states like Saudi Arabia and Qatar, Salafism is not beholden either to any existing state power or to any centralized religious authority. Salafists may reject the liberal secular nation-state and various liberal-secular concepts and values including, in some cases, the idea of democratic elections; but they are generally quietist rather than politically activist. They are less concerned with state power than with lifestyle: theirs is a religiosity or spirituality which is practiced in private but shown in public, and does not necessarily involve political claims, let alone forming specifically Salafî political parties. In Western Europe, Salafist groups generally do not make political demands beyond visibility in the public sphere, whether in the form of Islamic (and specifically Salafi) schools, prayer rooms in secular public institutions, or face-covering dress for women. Arguably, this depoliticized Salafist concern with religion as lifestyle reflects a novel (dare one say neoliberal?) concern with self-government.

Next to this apolitical Salafism, one also finds a politicized, and potentially violent, jihadism. In Western Europe, the latter became increasingly visible from around 2012, with the escalation of the civil war in Syria. In Sunni Muslim circles, reports of the Assad regime's atrocities against Syrian civilians, which represented the regime in theological terms as Shi'ite 'hypocrites' (*munafiqun*) and/or apostates, mobilized significant (even if statistically almost negligible) numbers of local Muslim youths, mostly of North African descent, to join in what in some circles was called the 'Syrian jihad'. This jihad was led by groups like the al-Qa'ida-affiliated Jabhat al-Nusra, and the so-called Islamic State that had emerged in the power vacuum then obtaining in the predominantly Sunni Arab regions of Syria and Iraq. Substantial recruitment for the jihad in Syria and Iraq lasted roughly from 2012 to 2016, but in 2015 and 2016, jihadists also sought targets in Western Europe, leading to a wave of assaults legitimated in Salafi-jihadi terms, most notably in Brussels, Paris,[7] Nice, and Berlin. Generally, indeed almost invariably, the perpetrators were young males of North African backgrounds, with a personal history of petty crime and drug use rather than religious zeal. These assaults were widely seen as having been orchestrated by the so-called Islamic State (IS); indeed, IS was quick to claim responsibility for whatever assault in Europe might be construed as jihadist. On the whole, however, no clear organizational links or chains of command between IS and the perpetrators have been found. In Western Europe, suicide missions hardly if ever targeted state power, but that need not imply they have no political significance. They not only escape sovereign power in radically defying the logic of crime, law, and punishment, but their perpetrators also assert, or arrogate, a power over their very lives—and those of others. As such, they may be seen as biopolitical gestures (cf. Leezenberg forthcoming).

Apolitical and peaceful Salafism and potentially or actually violent Salafi-jihadism should be kept analytically separate, even if there may be doctrinal and, occasionally, organizational links between them. It is an open question how their rejection of various liberal-secular concepts and values should be understood conceptually and valued normatively. In one currently influential interpretation of both, Olivier Roy (2017) argues that Salafism is a deculturalized and deterritorialized form of globalized Islam that appeals to second-generation Muslims in migration settings and to converts precisely because it is no longer culturally or geographically specific, and that the 2015 and 2016 jihadist suicide attacks in Western Europe were not driven by political claims or caused by social marginalization or European state policies; rather, he writes, they were nihilist, expressing a purely aesthetic cult of death.

Indisputably, there is much to be said for both explanations, which focus on deculturation and youth culture rather than on religious traditions or modernization processes. There are also problems with Roy's claims, however: despite their rhetorical appeals to the entire ummah or community of the faithful, Salafism and Salafi-jihadism are widely seen among Muslims as specifically Arab forms of Islam, that is, as culturally and geographically specific. Obviously, its militantly Sunni Islam has little to attract Muslims of Shi'ite backgrounds or beliefs; but it has also relatively few sympathizers among Turks and Kurds, among whom organized Islam has generally been shaped by Naqshbandi Sufism, which has an enduringly anti-Salafi character (Yavuz 2003). Likewise, Roy's claim that jihadist violence is only contingently linked to Salafi ideology and to concrete political goals seems an overstatement.

On the whole, Roy thus appears to share Habermas' and Asad's denial of genuinely political motivations for terrorist acts. These denials stand in a long tradition of seeing terrorism in moral (or, in Roy's case, aesthetic) rather than social or political terms, and treating with suspicion any attempt to find causal social explanations or rational political motivations for terrorist acts of violence. This refusal to see terrorist violence as 'reasonable' (i.e., as either goal-rational, motivated by identifiable political causes, or justified by reasonable political demands) is not simply an analytical or conceptual shortcoming in the writings of these authors. Rather, it reflects a widespread and persistent tendency to minimize or deny the political dimensions of terrorist violence. American anthropologist Lisa Stampnitzky has argued that from the late 1970s, the question of whether terrorists may have rational, objective motives has itself become highly politically charged (Stampnitzky 2013: 65). Instead, terrorism studies have witnessed the emergence of what she calls a 'politics of anti-knowledge', that is, a predominant *rejection* of the very attempt to understand terrorism in goal-rational terms or to seek causal explanations as a covert or explicit expression of sympathy for its perpetrators (*ibid.*: 194–5). In other words, if terrorism is—implicitly or explicitly—*defined* as irrational and illegitimate political violence, it becomes virtually impossible to coherently raise the question of whether it may in some circumstances be justified or motivated by legitimate grievances. Stampnitzky continues by arguing that terrorism studies are not an established and depoliticized academic discipline, but a field that has long remained, and probably will remain for the foreseeable future, less than fully institutionalized. By extension, the notion of terrorism itself has remained and will remain an unstable, 'undisciplined' object of knowledge. Even if nowadays terrorism is virtually defined as evil if not irrational, this depoliticized and moralizing characterization remains fiercely contested.

Conclusion

Despite their considerable differences, the liberal postsecularist Habermas and the Islamic communitarian Asad appear to converge on a number of points. First, they share a conceptual and normative assumption of debate and discussion as 'reasonable'; that is, even if in practice they mean rather different things by 'reason' both agree that in itself debate is free and fair and rational. In doing so, both downplay, ignore, or deny the presence of power relations or power effects in reasonable debate, if not in the notion of rationality itself.

Second, both Habermas and Asad appear to presume a highly problematic and highly contested notion of terrorism, or terror, as illegitimate and indeed unreasonable; both also explicitly state that Salafi-jihadist assaults stand outside the Islamic religious tradition—which they characterize, respectively, as in essence a distinct sphere of social life and as an essentially reasonable and non-violent tradition of debate under a legitimate religious authority. Hence, both see terrorist violence, even if articulated and legitimated in religious terms, as non-religious almost as a matter of definition. In doing so, however, they fail to provide an adequate account of *why* in the early twenty-first century, such assaults tended to be articulated and legitimated in religious (and more concretely Salafi-jihadist) terms, rather than in the terms of revolutionary Marxism or anarchism, as would have happened several decades earlier.

Third, and finally, Habermas and Asad share a rather reductionist view of the historical development of modern forms of Islam as arising out of a generic confrontation between an Islamic world assumed to be traditional and a modern West assumed to be liberal and secular. This historical oversimplification has non-trivial conceptual and normative consequences, which become most clearly visible in Habermas' claims that the Islamic world is not yet modernized (or secularized), and that it has witnessed a modernization-as-secularization gone astray, in disregard of concrete political processes or conflicts. Likewise, Asad's sweeping condemnation of all political violence as 'terror', and his equally sweeping reduction of modern Islamic history to a confrontation between an Islamic world or 'tradition' and an imperial, liberal, and secular West, precludes him from raising the question of what is historically and conceptually specific about present-day Salafi-jihadism, and of how it has at least in part been shaped by the experience of communism.

The conceptual and normative implications of the genealogical account outlined above are far from clear, but a reorientation towards the (partly) Marxist-Leninist origins of Salafism and Salafi-jihadism and towards the forms of governmentality they imply may be of use here. Likewise, Stampnitzky's argument may force us to rethink the permanently contested character of a concept of terrorism that authors like Habermas, Asad, and Roy appear to take for granted. Against these and other authors, one may—and indeed should—emphasize that religiously motivated or legitimated violence *can* be explained from social and other causes and *can* be driven by political or other motivations or aims that may be qualified as 'rational'. But the very attempt to do so is likely to remain controversial for the foreseeable future, in academic discussions as much as in public debates.

Notes

1 For an overview, see Beckford (2012); see also Molendijk (2015). Beckford, incidentally, expresses scepticism about the analytical usefulness of the term, in view of the various uses to which it has been put.

2 For reasons of space, I can only discuss a small number of examples, all linked to the Islamic world. No implication that terrorist violence is in any way specific or unique to Islam should be read into this restriction.

3 As of 2018, even the secular European identity as signalled and criticized by Asad is increasingly challenged by new xenophobic, anti-European, and anti-Muslim assertions of national identity, which appeal to national and/or religious traditions as much as to liberal secular values. Equally significantly, sexuality plays a central role in defining these identities. Lack of space precludes a fuller discussion of this phenomenon.

4 See e.g., Asad (1986: 20n22): 'Apart from the important Communist parties of Iraq and Sudan (neither of which commanded a massive following), Marxism has had no real roots among Muslim populations'.

5 See Leezenberg (2018) for a fuller discussion of the relations between Foucault's writings on neoliberalism and the Iranian Revolution.
6 On the—surprisingly brief—history of modern Salafism, see especially Lauzière (2016). For a convenient introduction to jihadism, see Schmidinger (2016), which focuses on paths of radicalization and deradicalization among youths in Europe.

Further reading

Asad, T. (2007) *On Suicide Bombing*, New York, NY: Columbia University Press.
A sustained critique of the liberal horror at this particular form of—often religiously motivated—violence.
Kepel, G. (2017) *Terror in France: the rise of jihad in the west*, Princeton, NJ: Princeton University Press.
An empirical study by Roy's great academic rival, setting out to explain the 2015–2016 wave of suicide assaults in terms of both Islamic theology and sociological conditions.
Roy, O. (2004) *Globalized Islam*, London: Hurst.
Roy's first characterization of Salafi Islam as deculturalized and deterritorialized, and hence as particularly well adapted to a globalized world.

References

Anidjar, G. (2015) 'The violence of violence: response to Talal Asad's "reflections on violence, law and humanitarianism"', *Critical Inquiry*, 41(2): 435–42.
Asad, T. (1986) *The Idea of an Anthropology of Islam* (Occasional Papers Series), Centre for Contemporary Arab Studies, Washington, DC: Georgetown University Press.
——— (1993) *Genealogies of Religion*, Baltimore, MD: Johns Hopkins University Press.
——— (2009) 'Free speech, blasphemy, and secular criticism.' In Asad, T., Brown, W., Butler, J. and S. Mahmood (eds.) *Is Critique Secular? Blasphemy, injury, and free speech*, Berkeley: University of California Press.
——— (2010) 'Thinking about terrorism and just war', *Cambridge Review of International Affairs*, 23(1): 3–24
——— (2015a) 'Reflections on violence, law and humanitarianism', *Critical Inquiry*, 41(2): 390–427.
——— (2015b) 'Thinking about tradition, religion and politics in Egypt today', *Critical Inquiry*, 42(1): 166–214.
——— (n.d.) 'Interview, University of Chester'. See www.chester.ac.uk/cfpp/research/staff/prof-t-asad, accessed online 04-06-2018.
Beckford, J. A. (2012) 'Public religions and the postsecular: critical reflections', *Journal for the Scientific Study of Religion*, 51(1): 1–19.
Flores D'Arcais, P. (n.d.) 'Eleven theses against Habermas', *The Utopian*. See www.the-utopian.org/d%27Arcais_1, accessed online 04-05-2018.
Foucault, M. (1978) *The History of Sexuality, Vol. 1: the will to knowledge*, New York, NY: Pantheon.
——— (2000) 'Nietzsche, genealogy, history'. In Faubion, J. D. (ed.) *The Essential Works of Foucault, 1954–1984, Vol. 2: aesthetics, method, and epistemology*, London: Penguin, pp. 369–92.
——— (2008) *The Birth of Biopolitics: Lectures at the Collège de France, 1978–1979*, Basingstoke: Palgrave Macmillan.
Habermas, J. (2003) *The Future of Human Nature*, Cambridge, MA: Polity Press.
——— (2008a) *Between Naturalism and Religion: philosophical essays*. Cambridge, MA: Polity Press.
——— (2008b) 'Notes on a post-secular society', *Signandsight*, 18 June (originally published in *Blätter für deutsche und internationale Politik*, April 2008). See www.signandsight.com/features/1714.html, accessed online 04-06-2018.
——— (2015) 'The Paris attack and its aftermath', *Social Europe*, 26 November (originally published in French in *Le monde*, November 22, 2015). See www.socialeurope.eu/habermas-paris-attack, accessed online 04-06-2018.
Iqbal, B. K. (2017) 'Thinking about method: a conversation with Talal Asad', *Qui Parle*, 26(1): 195–218.
Lauzière, H. (2016) *The Making of Salafism: Islamic reform in the twentieth century*, New York, NY: Columbia University Press.
Lawrence, B. (ed.) (2005) *Messages to the World: the statements of Osama Bin Laden*, London: Verso.

Leezenberg, M. (2010) 'How ethnocentric is the concept of the postsecular?' In Molendijk, A. L., Beaumont, J. and C. Jedan (eds.) *Exploring the Postsecular: the religious, the political and the urban*, Leiden: Brill, pp. 91–112.

——— (2018) 'Foucault and the Iranian Revolution reconsidered: revolt, religion, and neoliberalism', *Iran Namag*, 3(2): 4–28

——— 'Suicide bombing and the modern nation state: antiliberal protest or biopolitical performance?'. In Leezenberg, M., Van Bruinessen, M. and Korte, A. M. (eds.) *Gestures: religion as performance*, New York, NY: Fordham University Press, forthcoming.

Molendijk, A. (2015) 'In pursuit of the postsecular', *International Journal of Philosophy and Theology*, 76(2): 100–15.

Qutb, S. (2007 [1964]) *Milestones*, Birmingham, England: Maktabah Publishers.

Roy, O. (2017) *Jihad and Death: the global appeal of Islamic State*, London: Hurst.

Schmidinger, T. (2016) *Jihadismus*, Vienna: Mandelbaum Verlag.

Stampnitzky, L. (2013) *Disciplining Terror: how experts invented 'terrorism'*, Cambridge: Cambridge University Press.

Yavuz, M. H. (2003) *Islamic Political Identity in Turkey*, New York, NY: Oxford University Press, pp. 133–50.

9

Theoretical framings of the postsecular

Manav Ratti

Introduction

The concept of the postsecular has a wide range of formulations and applications across highly diverse disciplines and fields of inquiry, such as cinema studies (Bradatan and Ungureanu 2014; Kilbourn 2017; Caruana and Cauchi 2018), feminism (Bracke 2008; Braidotti 2008; Butler 2008; Vasilaki 2016; Deo 2018), geography (Cloke and Beaumont 2013; Williams 2014; Gökariksel and Secor 2015; Della Dora 2018), and religion (De Vries and Sullivan 2006; Gorski et al. 2012; Ni 2016; Areshdize 2017; Mapril et al. 2017; Dillon 2018). With such an expanse of postseculars, how can we understand some of this concept's features and forms? I offer in this chapter some theoretical framings of the post-secular. These framings are not exhaustive. They address primarily postsecularism as it has been inflected by postcolonial studies and literary studies, with these framings necessarily overlapping with one another. Given the fecundity of postsecularism as both theory and methodology, the framings that follow can resonate across the disciplines, allowing for further theorizations.

Theoretical framing 1: The postsecular is situated in relation to secularism and religion because secularism and religion are inextricably bound up with one another

To understand postsecularism, it helps to understand some of the meanings and history of secularism. Secularism has historically been conceptualized in opposition to religion, with this mutual definition of secularism and religion emerging within Christian thought. The term *secular* has both spatial and temporal dimensions. It is derived from the Latin noun *saeculum*, which, according to the *Oxford English Dictionary* (OED), can denote 'age' or 'generation', and in Christian Latin 'the world' as opposed to the church.[1] The *secular* can thus mean the age, generation, or world to which humanity belongs, in opposition to the timelessness and world of God. *Secular* also referred to clergymen who lived outside the monastery, and therefore in the 'world' or 'secular world', as opposed to the world, abode, or kingdom of God. The OED defines *secular* as, 'of or belonging to the present or visible world as distinguished from the eternal or spiritual world; temporal, worldly'.[2] The following is a similar definition of the *secular* from the OED: 'belonging to the world and its affairs as distinguished from the church and religion; civil, lay, temporal. Chiefly used as a negative term, with the meaning non-ecclesiastical, non-religious, or non-sacred'.[3] *Secular* can thus designate the observable here-and-now, the visible and human world, a category distinct from religion, given the latter's doctrines about the transcendental. *Secularization* means the gradual separation of religion and the state (more on this in the next theoretical framing), a process that reinforces *secularism*. The OED offers the following philosophical definition of *secularism*: 'The doctrine that morality should be based solely on regard to the well-being of mankind in the present life, to the exclusion of all considerations drawn from belief in God or in a future state'.[4] *Secularism* can also mark the values, beliefs, and attitudes that one is modern, progressive, and rational, including the belief that state and religion should be separate. According to José Casanova, this secularism can signify, from a Christian perspective, that one is modern and rational in contrast to premodern Christianity and non-Christian religions (Casanova 2011: 69). We can thus appreciate how the concept of the secular emanates within and is inextricable from religion.

Given that postsecularism has significance within postcolonial contexts, what are some of the implications—political, cultural, social—of this definition of the secular against the religious? Here the work of Gil Anidjar, building on Edward Said's foundational work on orientalism and the constructions of the 'Orient' (Said, 1978, 1983, 1993), can shed light on how secularism has operated historically and into the present. In theorizing the inextricability of secularism from religion, Anidjar argues that secularism 'is a name Christianity gave itself when it invented religion, when it named its other or others as religions' (2006: 62). According to Anidjar, 'the religious and the secular are terms that... continue to inform each other and have persisted historically, institutionally in *masking* (to invoke [Talal] Asad's term) the one pertinent religion, the one and diverse Christianity and Western Christendom' (62; *emphasis original*). With Christianity thus producing the secular-religious distinction, it masks itself *as* a religion, with other religions—such as Hinduism, Islam, Judaism, and Sikhism—becoming seen as 'religion', allowing Christianity to attain a kind of normalcy, neutrality, and exceptionalism.

Within colonial and postcolonial contexts, to claim that the postsecular rethinks both secularism and religion means that the postsecular identifies and critiques the normalized interconnections among at least secularism, Christianity, and colonialism. According to Peter van der Veer, 'the very distinction between religious and secular is a product of the Enlightenment that was used in orientalism to draw a sharp opposition between irrational, religious behaviour of the Oriental and rational secularism, which enabled the westerner to rule the Oriental' (Breckenridge and van der Veer 1993: 39). Van der Veer's argument is one of many that show the politicization of the secular-religious divide across the colonizer and colonized. There is no historically straight line, as if to suggest 'progress' or a form of organizing people, from religion, to secularism, to postsecularism. Rather, postsecularism can be the 'post' to an ideological and political construction of secularism that in particular historical, cultural, political, and regional contexts masks itself and makes itself invisible; that supports, enhances, and perpetuates institutions that exercise domination and hegemony; that 'others' religions, which include constructing them *as* 'religion'; that functions as the ideological enabler of international force and violence, whether as colonization, annexation, or war. Even while as a face of Christianity, secularism operates on and marks Christianity, splitting this 'subject' that creates other religions as 'objects', so that scholars of Christianity seek to understand, and perhaps recover, Christianity from the weight of secularism's representations, history, and pressures, leading to formations such as a postsecular Christianity, a postsecular Catholicism. This is analogous to colonization, which indelibly transforms both the colonized and the colonizer (see Nandy 1983).

Theoretical framing 2: The postsecular does not mean a rejection of secularism, but a rethinking of it, both politically and philosophically

In the preceding section I argued that the postsecular can emerge in relation to a particular formation of secularism (and thus of religion) that is ideologically constructed and historically, culturally, politically, and regionally specific. In this section, I frame the specific ways in which the postsecular rethinks both political and philosophical secularism. The hard-won political gains of secularism can be productive and inclusive, and here we can think of the values promulgated by the French Revolution (inspired by the Enlightenment and the American Revolution), consisting of equality, liberty, legal protections of individual rights, fraternity, and constitutionalism, all based on the sovereignty of the people. But that same political secularism can have limitations and face crises (more about which in the next theoretical framing), which raises the question of how that secularism can be reworked.

As indicated by the postsecular's root words, the postsecular is necessarily in relation with the secular. This begs the question of what is meant by 'post'. According to the OED, 'post' can refer to both time and space. The more common usage of 'post' is 'referring to time or order',[5] specifically meaning 'subsequent to, later than, following, since'[6] or 'occurring or existing afterwards, subsequent, later'.[7] In this sense of time, 'post' often combines with nouns, forming words such as postsecularism, postmodernism, poststructuralism, and postcolonialism. A second, less common use of 'post' is with reference to space. Within anatomy and zoology, 'post' is 'prefixed to adjectives (rarely nouns) to form adjectives, with the sense situated, produced, or occurring behind, posterior to, (occasionally) in the posterior part of, or distal to (the thing denoted by the second element)',[8] as in postcerebellar (behind the cerebellum) and postoral (behind the mouth), or, outside anatomy, postscenium (behind the stage). Robert Young has analyzed this spatial sense of 'post' with respect to poststructuralism, given its suggestive visual metaphor of structure, to argue that 'structuralism itself can only exist as always already inhabited by post-structuralism, which comes both behind and after' (Young 1982: 4). Analyzing postsecularism in this spatial sense, if the concept of religion always informs and constitutes the secular, then the 'space' which the secular occupies will always already produce questions about its very constitution, provoking the possibility of other relations with and configurations of secularism, which can be imagined as 'always already inhabited' by postsecularism.

Many concepts could always already be inhabited behind and after by their alternatives, with Derrida's theory of deconstruction understanding this constitutive fragility in many ways (and extending it to language as a whole), including as aporia, as a play of signification, as *différance*, as instability, and even as self-contradiction. A consideration of regional and historical specificity helps us understand the particular fragility of the secular-postsecular interconnectedness because of the pressures exerted on secularism by the demands of politics, ones which are perhaps more visible and urgent in postcolonial societies. *Secularization* in Western societies designates the various processes that separate institutionalized religion (for example, 'the church') from the state, relegating religion to the private sphere. The term's oldest meaning is the state's expropriation of church property (Pecora 2006: 13). As a political term, secularization denotes the relation between the political and the religious as a model for the organization of the state, and its relation with the nation. Secularization can be viewed as synonymous with modernization, with many institutions viewing science as a corrective to superstition (Pecora 2006: 17). The *secularization thesis* or *secularization theory* asserts that as nations become more modern and industrialized, they become more secular, limiting religion to the private sphere. An exception to the secularization thesis is the USA, because, while it is a highly industrialized and modernized state, it also has a large concentration of religious believers and public religious institutions. The secularization thesis has been rethought and challenged by scholars such as Peter Berger (2014) and José Casanova (1994, 2006, 2011, 2017). In addition, Talal Asad (2003) and Saba Mahmood (2005, 2006) have critiqued the secular-religion opposition from a postcolonial and Global South perspective, with Asad arguing that 'a straightforward narrative of progress from the religious to the secular is no longer acceptable' (Asad 2003: 1). The posting of the political secular can involve precisely a challenge such as this to the secularization thesis.

To examine now the 'post' to the philosophical sense of the secular, we can begin by recalling the OED's definition of the secular as 'belonging to the world and its affairs as distinguished from the church and religion; civil, lay, temporal. Chiefly used as a negative term, with the meaning non-ecclesiastical, non-religious, or non-sacred'.[9] In this definition, the secular is seen as distanced from religion and the sacred, suggesting scepticism of belief towards ideas and doctrines about the transcendent and spiritual. In this framework of the secular, we can consider Max Weber's diagnosis that the combined effect of industrialization, secularization, and an emphasis on a scientific-technological-rationalist worldview has resulted in disenchantment, translated from the German word *Entzauberung*, which has 'magic' (*Zauber*) as its root (Weber 1946: 155). Charles Taylor figures prominently in debates about the postsecular, although he does not use 'postsecular' as such in his volume, *A Secular Age* (2007). In his concept of the 'immanent frame', Taylor argues that modern Western society is marked by a 'self-sufficient immanent order' (Taylor 2007: 543), in contrast to a supernatural order and transcendent world. The immanent frame is an impersonal and socially constructed order in which 'instrumental rationality is a key value, and time is pervasively secular' (Taylor 2007: 542). Taylor diagnoses a lack of 'fullness' or richness in secularized societies, where life could be 'fuller, richer, deeper, more worthwhile, more admirable, more what it should be. This [fullness] is perhaps a place of power: we often experience this as deeply moving, as inspiring' (Taylor 2007: 5). Taylor's work and postsecularism are discussed extensively at the interdisciplinary digital forum *The Immanent Frame: secularism, religion, and the public sphere* (https://tif.ssrc.org/). In this context, the postsecular can signify a search for some kind of belief or values in the wake of the dissatisfactions and disenchantments of secularism (and, in Taylor's formulation, immanence) as an existential outlook.

Postsecularism is thus a shifting term, moving among religion, state, secularism, and across cultures, such as the political secularism of India and the disenchanted secularism of Western Europe (for more about this combined posting of political and philosophical secularism, see the next theoretical framing's discussion of India). Here I view literature as amenable to exploring such shiftings in postsecularism, because the creativity and flexibility of literary language and works allow for multiple, ambiguous, contradictory, and heterogeneous representations that can reproduce the process of an experimental, highly individual search for postsecular values. It might be worthwhile to examine how postsecularism unfolds in specific regions, where I now turn.

Theoretical framing 3: The postsecular is region-specific

I have thus far examined how postsecularism is situated in relation to secularism and religion, both politically and philosophically, and how that postsecularism rethinks and reworks certain aspects of secularism rather than functioning as a total rejection of secularism (again, both politically and philosophically). Building on this varying relation of the postsecular with secularism and religion, I pursue in this section the formations of the postsecular across different world regions. Political secularism—as a relation between the state and the people—varies according to national context, so that secularism will be specific to each nation-state, informed by the diversity of people living within its borders. This diversity can be constituted in terms of religion, caste, race, ethnicity, gender, and language, among others, informing the degree of political secularism in nation-states as diverse as Canada, France, India, Israel, Sri Lanka, Turkey, the UK, and the USA, to name only a few. Because populations are dynamic and in flux, political secularism has the potential to change and shift—in all political directions, such as across liberalism and conservatism in the USA, with varying degrees of success—making it historically specific. Secularization theory originated in Europe, primarily as a response to the various trajectories of secularism (as linked with modernization) in European nation-states. For Jürgen Habermas, 'postsecular' designates the challenge faced by secularized societies in contemporary nation-states in which 'religion maintains a public influence and relevance', while 'the secularistic certainty that religion will disappear worldwide in the course of modernisation is losing ground' (Habermas 2008: 21). He uses 'postsecular' to describe a 'change in [public] consciousness' (Habermas 2008: 20) of citizens of modern Europe, based on three factors. First, the visibility of global conflict informed by religious strife; second, the increasing influence of religion within national public spheres, especially on contentious moral issues; and third, the transformation of European societies to now having greater religious diversity, especially including religions from the Global South as a result of postcolonial immigration from outside Europe (Habermas 2008: 20). Postsecularism has been pursued as a concept in numerous other world regions, including Africa (Jackson and Surh-Sytsma 2017), Australia (Sunderland 2007; Paranjape 2010), India (Ganguly 2005; Mandair 2009; Ratti 2013), Latin America (Maldonado-Torres 2008; Junge et al. 2014), the Middle East (Alijani 2011), and the USA (McClure 2007; Coviello and Hickman 2014; Fessenden 2014). Due to constraints of space, I will focus in this section on the USA and India.

The USA and postsecularism

The fact that the USA stands as an exception to the secularization thesis (given its combination of technological modernity and its large number of religious believers) is perhaps among the factors that has led some US-based scholars to examine postsecularism, as reflected in special issues of *boundary 2* and *American Literature.* In a 2013 issue of *boundary 2* on 'Antinomies of the postsecular', scholars with a range of interests—postcolonialism, queer theory, feminist theory, religion, philosophy—question, critique, and also resist the presuppositions of postsecularism. These scholars' criticisms of postsecularism include challenging the kinds of secularism that it assumes (historical, philosophical, Christian, transcendental, political, Western, among others; see Gourgouris 2013; Lambropoulos 2013; Robbins 2013); its majoritarianism, lack of representativeness of the people, and misleading views of Islam (see Abbas 2013; Mufti 2013); and its misunderstandings of the transnational affinities (across the West and non-West) of political religion and cultural nationalism (see Cooper 2013).

In a 2014 issue of *American Literature* on 'After the postsecular', Peter Coviello and Jared Hickman propose what they term *postsecular 1*, *postsecular 2*, and *postsecular 3.* According to Coviello and Hickman, *postsecular 1* is 'the attempt to examine the historical past unburdened by a particular fantasy of the inevitable or necessary supersession of something called "religion"' (Coviello and Hickman 2014: 646). *Postsecular 1* becomes a criticism of the secularization thesis, in that the former is a 'project of dislodging a particular style of progress narrative' (Coviello and Hickman 2014: 646). That this secularization narrative assumes 'disenchantment, a swing from superstition to rationality, credulity to skepticism, eschatological fanaticism to liberal tolerance' (*ibid.*: 645) no longer makes it an 'adequate conceptual framework for the post-Enlightenment movement of bodies and belief, of thought and authority' (*ibid.*: 645). The epistemological and methodological formation that results from Coviello's and Hickman's conception of *postsecular 1* is what they term *postsecularism 2*, which 'unwrites secularist presumption' (Coviello and Hickman 2014: 647). With these secularist presumptions and framework dislodged, Coviello and Hickman explore 'what habituated forms of thought, what orthodoxies major and minor, might get reconfigured from the ground up' (Coviello and Hickman 2014: 647). This leads to *postsecularism 3*, the ways in which modern life and the categories of critical thought—including sexuality, race, and the literary itself—can be reconsidered at a remove from the secular. It is here that Coviello and Hickman propose globality. By globality, they mean the 'difference within the emergent singularity of the globe that fragilized belief in an unprecedented way' (Coviello and Hickman 2014: 649). A framework of globality allows everyone to be a 'full subject of history', and also has the advantage of being more neutral (than secularism) to religion (Coviello and Hickman 2014: 649). The articles in this special issue focus on American literature from the nineteenth to the twenty-first centuries, exploring how such literature complicates and disentangles secularizing narratives, such as by showing, as in the case of Lydia Maria Child's novel *Hobomok* (1832), 'belief's fragilization by the proliferation rather than attenuation of belief under the early-national conditions of state disestablishment' (Coviello and Hickman 2014: 650).

India and postsecularism

Secularism in India has a specific meaning as a state policy, which carries with it special challenges given the religious diversity, large population, and long history of religiosity in the region (for an analysis of postsecularism in another postcolonial South Asian context, specifically Sri Lanka, see Abeysekara 2008). India gained independence from British colonization in 1947, the same year that witnessed Partition, the division of India into India and Pakistan on the grounds of religion, with the two countries having majority Hindu and Muslim populations, respectively. As the largest forced migration of people in history, Partition displaced about 12 million people, and its interreligious violence (largely between Hindus and Muslims) resulted in about one million deaths (Butalia 2017: 3). In the aftermath of Partition, independent India was guided by the principle of secularism, that all religions would enjoy equality and liberty. Secularism in India thus has also had an important nation-building function in order to ensure democracy and minority rights.

While state secularism has had its successes, it has also experienced numerous crises, and to understand these potentials and limits, we can consider some of the distinctive features of secularism in India. The Indian constitution was modelled in part on the French and American constitutions, which formally separate religion and state, called the *non-establishment* (of religion) model or *wall of separation* model. Guided by the values of democracy, liberty, and equality for all citizens and religions, state secularism in India also has several distinctive features, including what Rajeev Bhargava (2010) terms its 'principled distance' from all religions. This means that the state can intervene in the affairs of religious communities, such as to reform some of their practices, an example of which is prohibiting child marriage (Bhargava 2010: 28). Other distinctive features include allowing group-differentiated citizenship—citizenship based on identification with a particular religious group (see Bhargava 2010: 27, 92)—and recognizing group rights for religious communities, which means having a system of 'personal laws' for those religious communities (Bhargava 2010: 27) concerning matters like marriage, divorce, adoption, and inheritance.

The crises of secularism in India include continuing violence among religious groups, such as the massacre of Sikhs in 1984, the destruction of the Babri mosque in 1992, and the massacre of Muslims in Gujarat in 2002. Among other examples, 1985 saw an elderly Muslim woman, Shah Bano, successfully petition the Supreme Court for alimony. When some Muslim leaders subsequently argued that the state should not interfere in the divorce laws of religious communities, the Indian parliament, with Rajeev Gandhi as prime minister, reversed the Supreme Court decision, a move perceived as appeasing Muslim voters and as an affront to women's rights. The year 2014 saw the election of Narendra Modi as prime minister, representing a Hindu nationalist party, the BJP (Bharatiya Janata Party). Since his election, there has been an increase in hate-crimes, including lynchings, against Muslims and Dalits in India. Starting in 2014, the VHP (Vishwa Hindu Parishad), the RSS (Rashtriya Swayamsevak Sangh), and other organizations based on ideologies of Hindu nationalism and Hindu supremacy have converted Muslims, Christians, tribals, and Dalits to Hinduism, under the guise of *ghar wapsi* or 'home coming'.

In response to these crises, scholars of Indian secularism have offered several revisions of secularism. Rajeev Bhargava has argued for a 'spiritualized, humanist' secularism (Bhargava 1995: 341). Rajeswari Sunder Rajan and Anuradha Needham have considered an indigenous or non-state secularism based on indigenous traditions of tolerance, rationalism, and humanism, such as Buddhism, Kabir, Akbar's Din Ilahi, Dara Sikoh, Ram Mohan Roy, Brahmo Samaj, Ambedkar, and Periyar (Needham and Sunder Rajan, 2007: 21, 22). In my work, a postcolonial postsecularism explored by South Asian writers, including diasporic ones, is 'post' to both political secularism and philosophical secularism (Ratti 2013: xvii–xxvi, 1–31), thus informed by multiple genealogies of the secular, national, and cultural. This means that writers will seek values that retain some of the faith, awe, wonder, transcendence, and enchantment afforded by religious belief and practice, but devoid of the violence to which religious ideologies can attach themselves, especially in colonial and postcolonial contexts (including between religious communities). At the same time, they wish to retain the positive values of political or state secularism, which includes the legal protection of rights (and minority rights), especially for the sake of ensuring democracy and equality within a religiously, linguistically, racially, and ethnically plural nation—but 'post' to the crises that that political secularism has faced. The postcolonialism of the postsecular is thus marked by the edge of the political, which includes the history and present of colonial and postcolonial violence, the necessary protections of a democratic secular nation-state, and the search for everyday ethical values.

Theoretical framing 4: The postsecular addresses questions of ethics, as within policy and literature, with ethics as a marker of social relations among people

Secularism and religion of course shape private and public life, by informing people's sense of values and their commitments to one another and to the civic good—what might broadly be called a system of ethics. In this section, I frame another dimension of the postsecular's interconnectedness with secularism and religion by examining how ethics shape the postsecular. The postsecular is concerned with values, ones that guide choices about the ethical life and ethical conduct. These ethical values and questions are influential at both the private, individual level and at the public, institutional level. In the public domain, they are relevant at the level of society, nation, and relation between nations, especially among religiously, racially, ethnically, and linguistically diverse groups. For instance, the challenges of ethics are evident in Habermas' question, 'How should we see ourselves as members of a post-secular society and what must we reciprocally expect from one another in order to ensure that in firmly entrenched nation states, social relations remain civil despite the growth of a plurality of cultures and religious world views?' (Habermas 2008: 21). With postsecularism pursued in professional contexts such as policies in development (Horner 2017), education (Sutton 2018), and social work (Shaw 2018), it is in this 'applied' sense that the role of ethics is particularly visible.

The search for ethical values is also present in literature. With literature as an expression of the imagination, exploring postsecular possibilities is precisely that—an exploration. It is an individualistic and experimental search for values in the wake of the crises of political secularism and religion, with the individuality of the creative process mirroring the individuality of the ethical search. When published literature then circulates in public space, the postsecular can emerge as a public discourse. The literary explorations of the postsecular can influence, and in turn be shaped by, academic discourses across disciplines such as sociology, religious studies, history, and anthropology, and perhaps even influence policy, such as through civil society organizations (Ratti 2013: xxiii–xxv). Among South Asian diasporic writers, including Michael Ondaatje and Salman Rushdie, whose writings—including Rushdie's *The Satanic Verses* (1988)—I analyze in my work on postcolonial postsecularism, their search is for affirmative values that can tenuously emerge in the wake of postcolonial violence. These values include love, friendship, literature, art, music, the migrant's eye-view, hybridity, and newness (Ratti 2013: xxiii). These values in and of themselves might not be 'new', but my interest within a secular/postsecular/postcolonial framework is in how writers arrive at these values and the ways in which these values are constitutively marked (or not) by the edge of the political, which can include the postcolonial, the national, the diasporic, and the minority position. I have analyzed how writers depict religion by invoking its signifiers and ethics, and then translate and secularize them within the contingency—and urgency—of material and historical circumstance (Ratti 2013: xxiii, 164). The postcolonial literature I examine presents decisions made out of human choices and human risks, in fraught circumstances such as civil war and Partition. Such moments will not result in immediate juridico-legal change (which can include policy), but can gesture to an epistemic change, which is unpredictable and whose tenuous trajectories can produce an ethics we might consider as postsecular.

Theoretical framing 5: The postsecular can inform methodology

In this final framing, I focus on how the postsecular can function as a methodology. The methodological potential of the postsecular presents another facet of how the postsecular is informed by secularism and religion, and how the postsecular can show the limits of the religion/secularism dichotomy. The postsecular can emerge as a method of analysis where the language, distinctions, and binary oppositions of 'religion' and 'secularism' are no longer adequate for describing concepts, phenomena, possibilities, and circumstances that traverse the religion-secularism boundaries. In other words, the postsecular can emerge as a deconstruction of the 'religion' and 'secularism' opposition, a methodological mechanism that can—in the spirit of deconstructive and poststructural theory—identify and pursue the instabilities, faultlines, blindspots, exclusions, silences, contradictions, and fissures of a discourse, discipline, institution, and text. As one among many instances of the methodological reach of the secular/

postsecular analytical framework, my work on postsecularism has been invoked in diverse contexts. Within literary studies—such as African literature (Smith 2014; Dudley 2017; Jackson and Suhr-Sytsma 2017), American literature (Haque 2014; Jones 2018), British literature (Conway and Harol 2015; Nelson 2016; Müller et al. 2017; Zheng 2018), Canadian literature (Dickert 2017), Iranian literature (Sobhani 2015), and South Asian literature (Grosu-Rădulescu 2016; McNamara 2015, 2016, 2018)—my work's theorization of the postsecular as a challenge to the secular-religious opposition has been used as a methodology for close readings as well as a theoretical framework for understanding formations of belief within multicultural and postcolonial contexts. Similarly, within areas such as education (Sutton 2018), melodrama and film studies (Majithia 2015), musical arts (Koperniak 2015), philosophy (Goh 2015), political science (Mohamadi 2016), religious studies (Ni 2016), and trauma theory (Visser 2015), my work has been invoked in analyses of how the religion-secularism binary is inadequate in naming and organizing emerging modes of belief and practice, and how power structures can be decolonized using the spirit of a postsecular consciousness of the global, colonial histories of secularism's and religion's interconnectedness with domination and power.

Conclusion

The theoretical framings I have offered in this chapter capture some of the potentials, methodologies, limits, and criticisms of postsecularism. I believe the first three theoretical framings are among the most consequential because the postsecular inherits the intertwined and multifaceted history of the secular and the religious. Analyzing the postsecular's regional specificity, including outside Euro-America, brings to attention the nuances, representativeness, responsiveness, and historical contexts of the concept. The postcolonial rethinking of secularism that I have invoked here (Said 1978, 1983; Breckenridge and van der Veer 1993; Asad 2003; Mahmood 2005, 2006; Anidjar 2006; see also the third theoretical framing) demonstrates the regional distinctiveness and limits of secularism, especially in the ways that secularism has been imbricated with the creation and exercise of power, both positive and negative. In my work on postcolonial postsecularism, I have asserted that the postsecular in a South Asian context cannot represent a return to certain aspects of religion because of the historical and ongoing association of elements of religion with violence (Ratti 2013: 21). At the same time, this does not mean that the postsecular is simply a reincarnation of liberal, multicultural 'good' religion. Religion, including its more orthodox beliefs, is of course a source of belief, at the same time that the dimensions of political secularism ensuring democracy, liberty, equality, and minority rights must be maintained for civic conduct and formation. With a span that stretches from the theoretical dimensions of deconstruction (in response to discourse, binaries, and blindspots) to the realpolitik of ethical conduct and political formation (in response to majoritarianism, supremacy, and disenchantment), the postsecular can be a provocation. But the same deconstruction that allows it to look outward also enables it to look inward, so that the fragility from which it emerges can be the same fragility which produces its limits. The future of postsecularism is the inseparable future of secularism and religion themselves, between belief and politics.

Notes

1 *Oxford English Dictionary* [online]. Oxford University Press, March 2018. Available from: www.oed.com/view/Entry/174620 [Accessed 2 May 2018].

2 *Ibid.* [Accessed 2 May 2018].

3 *Ibid.* [Accessed 2 May 2018].
4 *Oxford English Dictionary* [online]. Oxford University Press, March 2018. Available from: www.oed.com/view/Entry/174621 [Accessed 2 May 2018].
5 *Oxford English Dictionary* [online]. Oxford University Press, March 2018. Available from: www.oed.com/view/Entry/148402 [Accessed 2 May 2018].
6 *Oxford English Dictionary* [online]. Oxford University Press, March 2018. Available from: www.oed.com/view/Entry/148401 [Accessed 2 May 2018].
7 *Oxford English Dictionary* [online]. Oxford University Press, March 2018. Available from: www.oed.com/view/Entry/148402 [Accessed 2 May 2018].
8 *Ibid.* [Accessed 2 May 2018].
9 *Oxford English Dictionary* [online]. Oxford University Press, March 2018. Available from: www.oed.com/view/Entry/174620 [Accessed 2 May 2018].

Further reading

Asad, T. (1993) *Genealogies of Religion: discipline and reasons of power in Christianity and Islam*. Baltimore, MD/London: The Johns Hopkins University Press.

In examining how knowledge, power, and history have shaped religion, Asad moves the analysis of religion away from anthropology. He critiques the Christian view of religion as a universal, trans-historical essence separate from endeavours such as science and politics.

Chakrabarty, D. (2002) *Provincializing Europe: postcolonial thought and historical difference*, Princeton, NJ: Princeton University Press.

Chakrabarty argues for the Europeanness, and thus limits, of concepts such as modernity, history, capitalism, time, and secularism. See especially Chapter 1, 'Postcoloniality and the Artifice of History' and Chapter 2, 'Translating Life-Worlds into Labor and History'.

Connolly, W. (1999) *Why I am Not a Secularist*. Minneapolis, MN: University of Minnesota Press.

Connolly marshals insights from political theory, philosophy, and psychology to critique the limits of secularism and liberalism as both a philosophy and politics.

Derrida, J. (2002) *Acts of Religion*, edited by Gil Anidjar, New York, NY: Routledge.

Derrida theorizes the opposition between religion and reason, religion and secularization, and religion and translation, and how these oppositions inform the workings of the state with respect to justice, violence, and law.

Masuzawa, T. (2005) *The Invention of World Religions: or, how European universalism was preserved in the language of pluralism*. Chicago, IL: University of Chicago Press.

Masuzawa analyzes how the construction and classification of world religions helped create the hegemony and universalism of Christianity.

Mufti, A. (1998) 'Auerbach in Istanbul: Edward said, secular criticism, and the question of minority culture', *Critical Inquiry* 25: 95–125.

Mufti argues that secular criticism, and its attendant ideas of culture, canon, and community, are fundamentally constituted by a concern with minority culture and existence.

Nandy, A. (1998) 'The politics of secularism and the recovery of religious tolerance'. In: Bhargava R (ed.) *Secularism and Its Critics*. Oxford: Oxford University Press.

Nandy presents an argument for the history and subjectivity of peaceful religious coexistence independent of state structures such as political secularism.

References

Abbas, S. (2013) 'The echo chamber of freedom: The Muslim woman and the pretext of agency', *boundary 2*, 40(1): 155–89.

Abeysekara, A. (2008) *The Politics of Postsecular Religion: mourning secular futures*, New York, NY: Columbia University Press.

Alijani, R. (2011) 'Pre-secular Iranians in a post-secular age: the death of god, the resurrection of god', *Comparative Studies of South Asia, Africa and the Middle East*, 31(1): 27–33.

Anidjar, G. (2006) 'Secularism', *Critical Inquiry*, 33(1): 52–77.

Areshdize, G. (2017) 'Taking religion seriously? Habermas on religious translation and cooperative learning in post-secular society'. *American Political Science Review*, 111(4): 724–37.

Asad, T. (2003) *Formations of the Secular: Christianity, Islam and modernity*, Stanford, CA: Stanford University Press.
Berger, P. (2014) *The Many Altars of Modernity: toward a paradigm for religion in a pluralist age*, Berlin/Boston, MA: De Gruyter.
Bhargava, R. (1995) 'Religious and secular identities'. In Baxi, U. and B. Parekh (eds.) *Crisis and Change in Contemporary India*, New Delhi: Sage.
Bhargava, R. (2010) *The Promise of India's Secular Democracy*, New Delhi: Oxford University Press.
Bracke, S. (2008) 'Conjugating the modern/religious, conceptualizing female religious agency: contours of a "post-secular" conjuncture', *Theory, Culture & Society*, 25(6): 51–67.
Bradatan, C. and C. Ungureanu (2014) *Religion in Contemporary European Cinema: the postsecular constellation*, London: Routledge.
Braidotti, R. (2008) 'In spite of the times: the postsecular turn in feminism', *Theory, Culture & Society*, 25(6): 1–24.
Breckenridge, C. and P. van der Veer (eds.) (1993) *Orientalism and the Postcolonial Predicament: perspectives on South Asia*, Philadelphia, PA: University of Pennsylvania Press.
Butalia, U. (2017) *The Other Side of Silence: voices from the partition of India*, New Delhi: Penguin.
Butler, J. (2008) 'Sexual politics, torture and secular time', *The British Journal of Sociology*, 59(1): 1–23.
Caruana, J. and M. Cauchi (eds.) (2018) *Immanent Frames: postsecular cinema between Malick and von Trier*, Albany, NY: SUNY Press.
Casanova, J. (1994) *Public Religions in the Modern World*, Chicago, IL: University of Chicago Press.
Casanova, J. (2006) 'Rethinking secularization: a global comparative perspective', *The Hedgehog Review*, 8(1–2): 7–22.
Casanova J (2011) 'The secular, secularizations, secularisms'. In Calhoun C, Juergensmeyer, M., and J. VanAntwerpen (eds.) *Rethinking Secularism*, New York, NY: Oxford University Press.
Casanova, J. (2017) 'Exploring the postsecular: three meanings of the "secular" and their possible transcendence'. In Calhoun, C., Mendieta, E. and J. VanAntwerpen (eds.) *Habermas and Religion*, Malden, MA/Cambridge, UK: Polity Press, pp. 27–48.
Cloke, P. and J. Beaumont (2013) 'Geographies of postsecular rapprochement in the city', *Progress in Human Geography*, 37(1): 27–51.
Conway, A. and C. Harol (2015) 'Towards a postsecular eighteenth century', *Literature Compass*, 12(11): 565–74.
Cooper, M. (2013) 'Why I am not a postsecularist', *boundary 2*, 40(1): 21–39.
Coviello, P. and J. Hickman (2014) 'Introduction: after the postsecular', *American Literature*, 86(4): 645–54.
Deo, N. (2018) *Postsecular Feminisms: religion and gender in transnational context*, London: Bloomsbury.
Dickert, M. (2017) *Canadian Muslim Writing: an introduction*, Norderstedt, Germany: Grin Publishing.
Dillon, M. (2018) *Postsecular Catholicism: relevance and renewal*, New York, NY: Oxford University Press.
Della Dora, V. (2018) 'Infrasecular geographies', *Progress in Human Geography*, 42(1): 44–71.
De Vries, H. and L. Sullivan (eds.) (2006) *Political Theologies: public religions in a post-secular world*, New York, NY: Fordham University Press.
Dudley, J. (2017) '"Along a road that may lead nowhere": J. M. Coetzee's *Disgrace* and the postsecular novel', *Studies in the Novel*, 49(1): 109–30.
Fessenden, T. (2014) 'The problem of the postsecular', *American Literary History*, 26(1): 154–67.
Ganguly, D. (2005) *Caste, Colonialism, and Counter-Modernity: notes on a postcolonial hermeneutics of caste*, London/New York, NY: Routledge.
Goh, I. (2015) *The Reject: community, politics, and religion after the subject*, New York, NY: Fordham University Press.
Gökarıksel, B. and A. Secor (2015) 'Post-secular geographies and the problem of pluralism: religion and everyday life in Istanbul', *Political Geography*, 46(3): 21–30.
Gorski, P. S., Kyuman Kim, D., Torpey, J. and J. VanAntwerpen (eds.) (2012) *The Post-Secular in Question: religion in contemporary society*, New York, NY: Fordham University Press.
Gourgouris, S. (2013) 'Why I am not a postsecularist', *boundary 2*, 40(1): 41–54.
Grosu-Rădulescu, L. (2016) 'Objectification of women and violence in what the body remembers', *Rupkatha Journal on Interdisciplinary Studies in Humanities*, 8(1): 86–96.
Habermas. J. (2008) 'Notes on a post-secular society', *New Perspectives Quarterly*, 25(4): 17–29.
Haque, D. (2014) 'The postsecular turn and Muslim American literature', *American Literature*, 86(4): 799–829.

Horner, L. (2017) 'Rethinking development and peacebuilding in non-secular contexts: a postsecular alternative in Mindanao', *Third World Quarterly*, 38(9): 2009–26.
Jackson, J. and N. Surh-Sytsma (2017) 'Introduction: religion, secularity, and African writing', *Research in African Literature*, 48(2): vii–xvi.
Jones, N. (2018) *Provincializing the Bible: Faulkner and postsecular American literature*, New York, NY: Routledge.
Junge, B., Rubin, J. and D. Smilde (2014) 'Lived religion and lived citizenship in Latin America's zones of crisis', *Latin American Research Review*, 49(special issue): 7–26.
Kilbourn, R. (2017) 'Affect/ face/close-up: beyond the affection-image in postsecular cinema'. In Daigle, C. and T. MacDonald (eds.) *Posthumanism Through Deleuze*, Bloomington: Indiana University Press.
Koperniak, M. R. (2015) *Christmas Music in American Public Schools: a genealogical inquiry*, DMA thesis, Boston University, viewed February 1, 2018. https://open.bu.edu/handle/2144/13291
Lambropoulos, V. (2013) 'Why I am not a postsecularist', *boundary 2*, 40(1): 77–80.
Mahmood, S. (2005) *The Politics of Piety: the Islamic revival and the feminist subject*, Princeton, NJ: Princeton University Press.
——— (2006) 'Secularism, hermeneutics, and empire: the politics of Islamic reformation', *Public Culture*, 18(2): 323–47.
Majithia, S. (2015) 'Rethinking postcolonial melodrama and affect', *Modern Drama*, 58(1): 1–24.
Maldonado-Torres, N. (2008) 'Secularism and religion in the modern/colonial world-system: from secular postcoloniality to postsecular transmodernity'. In Dussel, E., Jáuregui, C. and M. Moraña (eds.) *Coloniality at Large: Latin America and the postcolonial debate*, Durham, NC/London: Duke University Press, pp. 360–84.
Mandair, A. (2009) *Religion and the Specter of the West: Sikhism, India, postcoloniality, and the politics of Translation*, New York, NY: Columbia University Press.
Mapril, J., Blanes, R., Giumbelli, E. and E. K. Wilson (eds.) (2017) *Secularisms in a Postsecular Age? religiosities and subjectivities in comparative perspective*, London: Palgrave Macmillan.
McClure, J. (2007) *Partial Faiths: postsecular fictions in the age of Pynchon and Morrison*, Athens: University of Georgia Press.
McNamara, R. (2015) 'The uneven aesthetics of I. Allan Sealy's *The Trotter-Nama*: secularization, nationalism, and the marginalization of the Anglo-Indian community', *Postcolonial Text*, 10(2): 1–21.
——— (2016) 'Developing "a fine balance": secularism, religion, and minority politics in Rohinton Mistry's *Family Matters*', *South Asia: Journal of South Asian Studies*, 40(1): 54–70.
——— (2018) *Secularism and the Crisis of Minority Identity in Postcolonial Literature*, Lanham, MD: Lexington Books.
Mohamadi, O. (2016) *Modernity, Secularism, and the Political in Iran*, PhD thesis, University of California, Santa Cruz, viewed February 1, 2018. https://escholarship.org/uc/item/4bm1d2b3
Mufti, A. (2013) 'Why I am not a postsecularist', *boundary 2*, 40(1): 7–19.
Müller, A., Herrmann, S. and F. Burstyn (2017) 'Canon formation and social imaginaries in British children's literature'. In Kümmerling-Meibauer, B. and A. Müller (eds.) *Canon Constitution and Canon Change in Children's Literature*, New York, NY/London: Routledge, pp. 39–56.
Nandy, A. (1983) *The Intimate Enemy: loss and recovery of self under colonialism*, New Delhi: Oxford University Press.
Needham, A. D. and R. S. Rajan (2007) *The Crisis of Secularism in India*, Durham, NC: Duke University Press.
Nelson, J. (2016) *Clothed with Salvation: pastoral power and eighteenth-century Anglican satire*, PhD thesis, Rice University, viewed February 1, 2018. https://scholarship.rice.edu/handle/1911/95588
Ni, Z. (2016) 'Postsecular reading'. In Felch, S. M. (ed.) *The Cambridge Companion to Literature and Religion*, New York, NY: Cambridge University Press, pp. 51–70.
Paranjape, M. (ed.) (2010) *Sacred Australia: post-secular considerations*, New Delhi: Aditya Prakashan.
Pecora, V. (2006) *Secularization and Cultural Criticism: religion, nation, and modernity*, Chicago, IL/London: University of Chicago Press.
Ratti, M (2013) *The Postsecular Imagination: postcolonialism, religion, and literature*, New York, NY: Routledge.
Robbins, B. (2013) 'Is the postcolonial also postsecular?', *boundary 2*, 40(1): 245–62.
Rushdie, S. (1988) *The Satanic Verses*, London: Vintage.
Said, E. (1978) *Orientalism*, New York, NY: Pantheon.
——— (1983) *The World, The Text, and the Critic*, Cambridge, MA: Harvard University Press.
——— (1993) *Culture and Imperialism*, London: Vintage.

Shaw, J. (2018) 'Introducing post-secular social work: towards a post-liberal ethics of care', *The British Journal of Social Work*, 48(2): 412–29.
Smith, C. M. (2014) 'On not yet being Christian: J. M. Coetzee's *Slow Man* and the ethics of being (un) interesting', *Postcolonial Text*, 9(1): 1–18.
Sobhani, M. (2015) 'A postsecular look at the reading motif in Bahiyyih Nakhjavani's *The Woman Who Read Too Much*', *The Journal of Baha'i Studies*, 25(1/2): 73–99.
Sunderland, S. (2007) 'Post-secular nation: or how "Australian spirituality" privileges a secular, white, Judaeo-Christian culture', *Transforming Cultures eJournal*, 2(1): 57–77.
Sutton, T. J. (2018) 'Orthodoxy revisited: the postsecular classroom, *Journal of Beliefs and Value*, 39(1): 3–16.
Taylor, C. (2007) *A Secular Age*, Cambridge, MA: Harvard University Press.
Vasilaki, R. (2016) 'The politics of postsecular feminism', *Theory, Culture & Society*, 33(2): 103–23.
Visser, I. (2015) 'Decolonizing trauma theory: retrospect and prospects', *Humanities*, 4(2): 250–65.
Weber, M. (1946) *From Max Weber: essays in sociology*, trans. and eds. Gerth H. H. and C. Wright Mills. New York, NY: Oxford University Press.
Williams, A. (2014) 'Postsecular geographies: theo-ethics, rapprochement and neoliberal governance in a faith-based drug programme', *Transactions of the Institute of British Geographers*, 40(2): 192–208.
Young, R. (1982) 'Post-structuralism: the end of theory', *Oxford Literary Review*, 5(1): 3–20.
Zheng, S. (2018) 'Postsecular return of religion: Jewish and Zen elements in Zadie Smith's *The Autograph Man*', *Neohelicon*, doi:10.1007/s11059-018-0439-8.

10

Formations of the postsecular in education

David Lewin

Introduction

The place of religion in education has long been controversial. There is no simple answer to the question of how religion and education are related because the religious history of each nation-state gives manifold form to those relations (see Jackson 2007). This is far more than an issue of whether Religious Education ought to be taught in schools, not least because educational formation takes place across the whole of life. From a philosophical perspective, the relations between religion and education have been usefully arranged into themes: religious upbringing, faith schools, religious education in the curriculum, religious philosophies of education, and issues of religious identity (see Strhan 2014). In what follows these themes will provide a broad horizon for thinking about the postsecular in education.

Arguably more than any other domain, the spaces of education (from parenting, to schooling, and beyond) are where the interactions between public and private are most complex and unavoidable. Educational formation must affirm some kinds of change as being 'good' and therefore complicate one key principle of secularism, namely, remaining neutral about which values, beliefs, and activities are good or right. I argue, therefore, that education is a key concern for the postsecular, since it shows that some ultimate commitments are unavoidable in education. In other words, to speak of neutral educational formation is oxymoronic. Some key questions must be addressed: What do religion, education, and 'religious education' look like from the perspective of the postsecular? What philosophical and pedagogical issues are raised by the new context of the postsecular? The terms of this debate are by no means settled, and so I explore the varied conceptions of secularism and the postsecular, arguing that the postsecular complicates rather than refutes the secularization thesis. The argument challenges the view of religion as basically reducible to doctrines, creeds, or truth claims, showing how that conception of religion skews the discussion of the place of religion in education towards one that considers only issues of indoctrination, and the rights of parents or religious groups to reproduce themselves. I suggest that the postcolonial concern to reveal and challenge assumptions around Western liberalism provides a fresh context to articulate the postsecular and its influence upon education.

The rise and fall of the secular

Detecting the waning influence of secular narratives of culture around the turn of the millennium, Steve Bruce presented a powerful case for secularization in Western societies. He argued that many of the strongest accounts of secularization across the social sciences (e.g., Marx, Weber, Durkheim and Freud) were being displaced by weaker, less thoroughgoing counter-secularization analyses that emerged towards the close of the twentieth century. Bruce's basic argument, affirming 'a long-term decline in the power, popularity and prestige of religious beliefs and rituals' (Bruce 2002: 44), though richly descriptive of trends in nineteenth-century Britain, seems inconsistent with more recent developments and events across the globe. One year after the publication of Bruce's book, the anthropologist Talal Asad announced the end of a certain picture of secularism as a progressive, liberal project in his book *Formations of the Secular*, stating that '[i]f anything is agreed upon, it is that a straightforward narrative of progress from the religious to the secular is no longer acceptable' (Asad 2003: 1). Asad, and those have developed his lines of inquiry such as Saba Mahmood, José Casanova, William Connolly, Wendy Brown, and many others (see Scott and Hirschkind 2006), have enriched the terms of the debate around secularization, but the sense in which secularization has faltered remains an ongoing controversy (Scott and Hirschkind 2006; Bruce 2013). What follows elaborates these insights, by arguing that the formations of postsecularism influence, and are influenced by, the ways educational thinking interrupts some of the more simplistic accounts of secularism.

The turbulence across many parts of the world since the turn of the millennium (wars, revolutions, insurgencies, and the like) has revealed a world of dizzying complexity where perpetual conflict has become the norm, with tensions cutting across political, cultural, religious, and ethnic identities. Social, economic, and environmental problems magnify—and perhaps may even constitute a decisive influence upon—these issues and differences. A widespread *fear of secularism* in parts of the Middle East and elsewhere is hard to disentangle from increased fundamentalism leading to the now common observation that 'secularism and fundamentalism feed off each other' (Williams 2012: 16). A general fear of the 'other' is surely bound up with increasing religious fanaticism and fundamentalism that seems built into global tensions. While difficult to explain, these fears cannot be dismissed as a temporary regression that swims against the inevitable tide of progressive enlightenment rationalism. Rowan Williams sees the instrumentalization of social relations as the characteristic and problematic constitution of a programmatic secularism that excludes religion entirely from public life (Williams 2012). Society cannot be reduced to the administration of more or less successful methods of maintaining public order and upholding private freedoms, but inevitably relies upon moral and social orders grounded in ultimate principles. This is especially true in the context of education. The idea that these principles can be eliminated from public life does, to some extent, feed the fears of the secular and the reactions to the programmatic secularism that disavows principles and commitments in the public domain. Thus, religious revivalism and fundamentalism across the world cannot simply be explained away as conservative reactions against aspects of modernity and postmodernity, but may themselves be products of certain formations of the secular.

How can the 'postsecular' help us address these tensions? Does education have a particular role to play in forming a more inclusive society, somewhere between the secular and the confessional? Postsecularism offers us an opportunity to engage with the contributions of our religious traditions, implying neither blind obedience nor a denial of the achievements of modernity. Asad's insights concerning the limitations of the secular have encouraged the development of the concept of the *postsecular*, a term which Asad himself does not use in *Formations of the Secular* (Asad 2003), but one that has come to identify, among other developments, the so-called return of religion, and though a complex and contested term, it denotes a state of affairs of particular significance for education.

The postsecular in education

These interactions between religion and culture are perhaps most keenly felt in the domain of education. Understanding the postsecular as a complication rather than repudiation of secularism (Lewin 2016: chap. 2) encourages us to move away from the normative question of whether religion should (or should not) have a role in public education, to the more fundamental question of the ways in which religion already has (and probably inevitably will have) such a role. I make no claim that religion is a force for good (or bad) in education, but it should be recognized that religion is a fundamental force that shapes people, communities, and education, a force that is not going away. In that sense, the *postsecular* offers a descriptive term for exploring the continued significance of religion within education. Education here is broadly construed as any intentional formative activities. The strong claim here is that there is no neutral educational formation despite the fact that many so-called progressive educators might wish to argue for a neutral education (see Sommers 2002).

I would like to proceed by providing an unambiguous definition of the *postsecular* that goes beyond the notion of a return to religion, but as this volume shows, such a definition remains elusive. In a general sense, the postsecular refers to the idea that modernity no longer entails an inevitable march towards secularism and the loss of faith. But already we should be alert to problematic conflations: of *secularism*, a term used to define a worldview in which religion is largely absent from *secularization*, a historical process in which social 'progress' and modernity are associated with religion losing cultural and social significance (Casanova 2009). The latter process is itself easily related to the growth of atheism or humanism but ought not to be identified with it. This failure to distinguish between a secular worldview and secularization process makes it all too easy to overlook an arguably more fundamental distinction, namely, the secular as the public domain which is free of the private interests of particular individuals and groups (which does not necessarily imply a loss of faith, but rather a privatization of it), and the secular as a broader process in which society is generally less religious. Although secularism, a general loss of faith, and a rise in atheism are by no means identical, there are many complex intersections rendering a neat division between a secular public domain and a private religious sphere untenable.

Public life, especially public education, requires reference to evaluative discourse, often informed by, or related to, religious commitments. As Rowan Williams puts it:

> evaluative discourse leaks out into the public sphere, sometimes in the moralizing rhetoric of political leaders, sometimes in the improvised rituals (of celebration or mourning or solidarity) that sporadically take over some part of the public territory.
>
> *(Williams 2012: 13)*

Once conceived as an inclusive principle of non-discrimination, the privatization of religion might, in fact, stand in opposition to many central religious perspectives since 'the very idea of deriving law from sacred texts is a repudiation of the public/private distinction' (Fox 2002: 22). There are good reasons why we have been so keen to embrace the idea that religion should be contained to the private sphere. It respects the commitments and values of citizens and releases shared dimensions of social life from the weight of those commitments. But, as Asad suggests, there are equally good reasons why religion cannot be contained in this way, not least because religion is, by definition, a communal enterprise which could be seen to push against containment within the private sphere. William Connolly has neatly summarized Asad's critical perspective on the secular, a perspective which presents the division between private and public as reflecting a partisan view of social and religious life:

1 Secularism is not merely the division between public and private realms that allows religious diversity to flourish in the latter. It can itself be a carrier of harsh exclusions. And it secretes a new definition of 'religion' that conceals some of its most problematic practices from itself.
2 In creating its characteristic division between secular public space and religious private space, European secularism sought to shuffle ritual and discipline into the private realm. In doing so, however, it loses touch with the ways in which embodied practices of conduct help to constitute culture, including European culture (Connolly 2006: 75).

Asad's analysis is broadly consistent with Williams's view who adds a further distinction in the development of a secular public space, between procedural and programmatic secularism. *Procedural secularism* takes a more pluralist attitude which disavows favour to any religious grouping, while trying to maintain a broad representation for all (as, for example, in India), while *programmatic secularism* seeks to iron out any and every public manifestation of religious allegiance, with France often cited as the paradigmatic case (Williams 2012: 2).

If defining secularism as the neat division between private and public cannot be upheld, then nor can we straightforwardly define postsecularism as the reintegration of religion and public life because something recognizably religious has always formed part of our cultural identities. Thus, echoing Latour, the postsecular suggests that we have never been secular. Clearly, the postsecular idea of a 'return of religion' can only take us so far, since the postsecular is anything but a simple return. The analogy for educational theory here might be a simple return to confessional religious education in schools, where a single religious perspective is taught as true, an approach that is out of fashion in most Western education systems (Gearon 2013; Lewin 2017). For many, pluralism and multiculturalism signal cosmopolitanism, tolerance, and progress, values that underpin certain republican principles which reflect a particular progressivist view of history. Some forms of progressive education have also been identified with this cosmopolitanism, looking towards an education for global citizenship (Bamber et al. 2018). This broad alignment of progressivism, education, and secularism must be shown to be at best simplistic, if not outright wrong.

Complicating progressivism

Whether empirical or normative, our narratives of enlightened progress seem to align with, and reinforce, conceptions of human development that also structure ideas about education. In this sense, education is intrinsically teleological since it is concerned with the intended improvement of a particular skill, knowledge, capacity, or disposition. For many progressive educators, the liberal project is oriented by ideals of the formation of a rational and free subject (Gutmann 1982). While defying the aspiration to become rational and free might not, in itself, seem reasonable, such teleological thinking is in danger of making us blind to the variety and complexity of human identity and development because we become wedded to a linear conception of time and progress in which modern Western culture has inaugurated the end of history (Fukuyama 1992). Opposition to this rather limited view of human development and progress may be reflected in the concomitant suspicion, even fear, of secularism as a project of Western progress. No doubt the strident tone within 'aggressive secularism' (Glendinning 2009) and the increasingly passé 'new atheism' (Cimino and Smith 2011) reflects anxieties of resurgent religious bigotry, which itself seems more reactionary than substantive.

If we are to escape this *reactionary circle*, then we will need a more nuanced and circumspect analysis that first of all would challenge a naïve, rather Eurocentric, progressivism that places a narrow form of rational subjectivity as the singular anthropology. So, as many postsecular theorists have argued, history has turned out to be more complicated than any neo-Hegelian narrative of *Geistesgeschichte* (roughly understood as cultural history in progressive development).[1] This complication should expand the debate within educational philosophy away from rather narrow considerations of curriculum issues around how to reconcile competing truth claims within religious education, or how to express the rational civic core that all citizens must adhere to (in the UK expressed as the inculcation of 'British Values'), to a broader discussion of how our fundamental commitments and ultimate concerns figure in education today (Lewin 2014).

This raises questions of what religion is for people today, since the voluntaristic conception that religious life entails choices and decisions, to commit to a system of beliefs, truth claims, or worldviews, is itself a very particular, and rather unhelpful, one. As Wendy Brown puts it:

> The conceit of religion as a matter of individual choice…is already a distinct (and distinctly Protestant) way of conceiving religion, one that is woefully inapt for Islam and, I might add, Judaism, which is why neither comports easily with the privatized individual religious subject presumed by the formulations of religion freedom and tolerance governing Euro-Atlantic modernity.
>
> *(Brown 2013: 17)*

The conception of choice is related to a view of religion as a worldview, or set of truth claims, about which one makes that choice. This voluntaristic conception of religion tends to frame debates about the place of religion in education, contributing to a wide misunderstanding of the nature of (religious) commitment: as though only religious positions rely on commitments, while secular worldviews are based on firmer foundations of rationality alone. Secular perspectives embody a range of commitments which are not necessarily visible at level of propositions or worldviews (Lewin 2016: chap. 3). The postsecular encourages us to recognize that the kinds of binaries that structure our thinking about religion—religious/secular, faith/reason—are deceptive, since the terms and identities are more porous than these accounts assume.

In the field of education, we must work with the conceptual fluidity of the postsecular in developing the different senses in which religion plays a part in the social life of public educational institutions. In other words, religion—in one sense or another—will inform public education, whether we like it or not. This does not mean that we shouldn't consider creating spaces within public education that are, as far as possible, free of religious influence. Nor should we turn a blind eye to over-zealous religious inculcation in educational institutions (public or private) in the name of liberalism or tolerance. But the idea that we can leave our religious attitudes at the school gates presupposes a particular conception of what it means to be religious that, I argue, reinforces a parochialism, perhaps even imperialism, since it reflects, echoing Wendy Brown, what could be called a rather protestant view of what it means to be religious.[2] We cannot deny the formative dimensions of education which means that seeing religious/worldview 'neutrality' as an educational ideal is itself by no means neutral

(Cooling 2010). It entails a narrow and inadequate understanding of religion as the cognitive assent to truth claims, doctrines, or worldviews.

I argue against any neat separation of religion and education for both practical and theoretical reasons. From a practical point of view, I believe it to be vital that children and young people become religiously literate. From a theoretical perspective, a value-laden orientation to life is inevitable and essential whether or not we engage explicitly in religious or humanist formations of those values. In addition, there is an important pedagogical point here. It is not obvious that the job of educators is to encourage cognitive assent to truth claims, even if those claims are taken to be straightforwardly true. This is because learning is better served by a *pedagogy of inquiry* (or non-directive pedagogy) than a *pedagogy of assent* (or directive pedagogy) (see Sommers 2002).

A pedagogy of assent supposes that the substance of learning is constituted by relatively stable knowledge, skills, and dispositions, all of which can and should be directly inculcated into the student. Education cannot be entirely free of the process of absorbing or assenting to such forms of knowledge, but the craft of teaching is better understood to entail indirect forms, which I call a pedagogy of inquiry, that ultimately gives the student the freedom and responsibility to take what they can from it. Trevor Cooling uses the example of a student asking about creationism in a science lesson arguing that it might not be appropriate to simply 'dismiss it outright as wrong-headed' (Cooling 2010: 11). The fact that creationism is false is not enough since the educator must also employ a measure of what Van Manen calls 'pedagogical tact' in handling the situation (Van Manen 2016). Cooling's point is that 'grappling successfully with questions of meaning and significance contributes to developing into a healthy, balanced person and is a fundamentally important component of education' (Cooling 2010: 14). The shift from directive 'Religious Instruction' (often referred to as *catechesis* within Christian communities) to non-directive 'Religious Education' in the 1988 Education Reform Act in England and Wales, was in part intended to address this issue: that any perception of indoctrinatory intent within confessional religious instruction was mitigated by an exposure to more than one religious tradition. The postsecular complication of linear narratives of culture continues to be felt in the fact that the curriculum of Religious Education is the one place in which the issues around taking religion seriously while simultaneously being inclusive have seemed most intractable (Barnes 2009; Aldridge 2015). Can there be a meaningful religious education for all?

Education for all?

Given the vast range of concepts and contexts that the postsecular evokes, is there a core notion that can be identified and applied to any kind of general education? The political, sociological, philosophical, and theological registers of the term hint at a concern that might be understood as a postcolonial appreciation of other ways of knowing, being, and educating which can be related to Williams' definition of the non-secular as 'a willingness to see things or other persons as the objects of another sensibility than my own' (Williams 2012: 13). That the Western liberal project, which today is manifest in a highly urbanized and consumerist culture, is the 'only game in town' scarcely needs to be stated, so self-evident has it become. That this Western liberalism is bound up with conceptions of universal reason and progress that have supplanted local indigenous communities appears equally, and problematically, self-evident. Mass education in nineteenth-century Europe seems to have gone hand in hand with a colonialist reach of the civilizing power of Western European education, evident, for instance, in John Gast's painting from 1872, entitled 'American Progress' (see Figure 10.1).

Figure 10.1 American Progress by John Gast (1872)
Source: Wikipedia Commons: United States Library of Congress.

And 'we' (UNESCO, the World Bank, and the so-called International Community) now call for an 'Education for All' whose noble intention is to bring the benefits of an education 'to every citizen in every society' (World Bank 2014). Have we become inured to the one-dimensional nature of the cultural and educational imperialism at the heart of this project? In a series of articles and a powerful film, Carol Black strongly argues that:

> [i]n "developed" societies, we are so accustomed to centralized control over learning that it has become functionally invisible to us, and most people accept it as natural, inevitable, and consistent with the principles of freedom and democracy. We assume that this central authority, because it is associated with something that seems like an unequivocal good – "education"—must itself be fundamentally good, a sort of benevolent dictatorship of the intellect.
>
> *(Black 2012)*

Just as Jacques Rancière has pointed out the contradiction inherent to installing democracy (Rancière 1991), so Black argues that the installation of a particular conception of education for all negates the varieties of indigenous education and culture that have existed across all societies for hundreds of years. Black may be overstating her case, romanticizing indigenous cultures, or failing to recognize the complexity by placing mass education in opposition to indigenous education in this way. But it is hard to refute the totalizing intent of 'Education for all' and the near universal assumption that it represents an unequivocal good.

For the purposes of my argument, the postsecular offers one way to open up a space for alternatives to a disenchanted, industrialized, and urbanized future. It is a political, social, and spiritual concern that we find alternatives to the neo-liberal narratives of the future. But, to interpret Heidegger's enigmatic claim, only a god can interrupt the totalizing power of that narrative. In an interview for *Der Spiegel*, withheld from publication until after his death, Heidegger (1991: 107) stated enigmatically that:

> Philosophy will not be able to effect an immediate transformation of the present condition of the world. This is not only true of philosophy, but of all merely human thought and endeavor. Only a god can save us. The sole possibility that is left for us is to prepare a sort of readiness, through thinking and poeticizing, for the appearance of the god or for the absence of the god in the time of foundering [*Untergang*] for in the face of the god who is absent, we founder.

My intention with this reference is to indicate that an interruption to the prevailing liberal democratic order entails some radical reorientation that seems to require something 'transcendent'. Either that, or we will be visited by a global environmental catastrophe which will be impossible to ignore—an interruption of a different kind.

One might object that the context and image of colonialist education cannot be identified with secularism. After all, colonialism was initially motivated not just by economic and political interests, but by an evangelizing imperative that moved in a different direction to secularization: seeking to bring others into the faith. This is important in view of education since the missionary spirit of those colonial powers took great interest in schooling: the legacy of British colonialism is evident in that many of the elite schools in India today, for example, are Christian. Surely, then, secularization should be seen in opposition to colonialism. There are many complex points here that can only be briefly touched upon.

The influential theory of Carl Schmitt shows how the founding principles of secular society (concepts of science and progress) arise out of the theological traditions of Western Europe (Schmitt 2007). But this brief overview can only hope to indicate the geopolitical dimensions of this debate sufficiently to suggest that the missionary zeal of early colonialists seems largely reflected in the assumptions within liberal democratic societies: that modern industrialized societies are to be exported to every corner of the globe. It is the manner of being-in-the-world that admits of no alternatives, and that therefore fails to recognize itself as just one mode of being. It is, to use Heideggerian language, the oblivion of being.

I am not for a moment suggesting that the scientific revolution, the Enlightenment, and modernity in general are not momentous shifts resulting in extraordinary benefits to millions, even billions, of lives around the world. One might observe that modernity is a victim of its own success in the sense that it has become so all-encompassing, so totalizing or *enframing*, again to reference Heidegger, that all other modes of being are all but invisible (Heidegger 1977; Lewin 2015). Other ways of addressing global problems, environmental, social, political, and spiritual, are silenced. This is not, then, an attack on, or negation of, Western consumerist culture, technology, urbanization, or secularism more broadly, but only to acknowledge that this is not the only game in town. It may be that other ways of being, including religious, spiritual, poetic, aesthetic, and so on, draw more explicitly on the *postsecular*, inhabiting a world which is more than the projection of the power of subjectivity. We need to bring the discussion back to a more direct engagement with the question of religion, to which I return.

The return of religion and education

We have seen that the taking seriously the postsecular involves acknowledging that religion has an ongoing influence on culture and on education particularly. Straightforward theories of secularization have to be re-examined in light of the ongoing influence of religion. This means disabusing ourselves of the assumption that 'to be secular means to be modern, and therefore by implication, to be religious means not yet fully modern' (Casanova 2010: 59). This statement could be recast for educators in the following way: 'to be secular means to be critical, and therefore by implication, to be religious means not yet fully critical'. I would argue that the existence and development of critical thinking must be uncoupled from assumptions around secularization, and that the fear of secularization in parts of the Islamic world is associated with fears around certain kinds of education because of this association. Attacks on education in Pakistan, for example, are abhorrent. But understanding these fears as arising out of a misunderstanding of the relationship between criticality and credulity is an important step. In other words, I want to call for a softening of the polarization between criticality and credulity, because we all believe and affirm something before we critically engage. Criticality is not, then, in opposition to credulity or belief; rather, it is (or ought to be) a correlate dimension of belief. This idea has long been understood by philosophers as being positioned within a hermeneutical circle in which faith and reason work together (Lewin 2014). This is our hermeneutical condition, perhaps even our human condition.

Softening the opposition between those who are religious and non-religious might also involve giving voice to those who stand betwixt and between the secular and the religious. The site of the postsecular gives form to the spaces or cracks that many people in the present age would recognize, but do not often discuss: the spaces between the secular and the confessional. It is in these spaces that many people find themselves: unwilling to fully negate religious life, nor fully able to embrace the worldviews that seem untenable. Philosophers might see this disposition as heralding a *religion after metaphysics*. Taylor calls this space a third way between orthodoxy and unbelief, a view that Smith, quoting Taylor, neatly captures:

> All sorts of people feel themselves caught; 'in the face of the opposition between orthodoxy and unbelief, many, and among them the best and most sensitive minds, were [and are] cross-pressured, looking for a third way.
>
> *(Smith 2014: 64)*

After the putative death of God, the possibility of authentic religious life is an ongoing question, hence the rise in alternative forms of religion and interests in spirituality (King 2009). For those who consider themselves 'spiritual but not religious', there may emerge an appreciation of the opportunity to explore the postsecular defined in terms not reducible to either pole of secular/confessional binary. This contribution has attempted, in different ways, to interpret the postsecular as a kind of complication of the secular. This complexity is essential in the domain of education where intentional formative activities require some reflection on the ultimate concerns of education.

Conclusion

I began by asking what education might look like in a postsecular age. This entailed a discussion of the nature of secularism. I argued that the postsecular is a complication rather than repudiation of secularism. The neat division between private and public that certain formations of the secular espouse cannot be maintained within education, because education is a formation exercise which necessarily entails certain affirmations of the good. A certain reductive notion of criticality within educational thinking fails to recognize the affirmation present within any kind of educational formation. In other words, educators have sought to make education into a neutral or value-free enterprise. I have attempted to show both, that this is not possible and also how this view of education reflects a Western liberal bias. Recently, philosophers of religion have pointed out some of the Eurocentric biases within this notion of secularism and the propositional conception of religion it assumes, an analysis that I have suggested should be extended to education. One view of the postsecular, then, would be to develop the opportunity to 'return to religion' in ways that go beyond reductive propositional or voluntaristic views of religion—going beyond an understanding of religions as competing worldviews or belief-systems. I have touched on this here, but much more needs to be done to show the complex interactions between religion and education. This means that we should explore how different symbolic systems offer rich narratives to make meaning that cannot be reduced to worldviews, and what this means for education. This would encourage a respect for other ways of being-in-the-world that are too often disregarded as primitive, premodern, and uneducated. This kind of respect amounts to a postsecular education that lies between the secular and the confessional.

Notes

1 This raises the question of whether Hegel's philosophy of history, and European Enlightenment philosophy more broadly, are irredeemably Eurocentric and colonialist (Tibebu 2011).
2 The use the term *Protestant* here is meant to be broadly inclusive of Protestant traditions that, in the wake of Luther, regard *sola fide*, by faith alone, to be the sole ground for salvation (see Dupré 1993).

Further reading

Bowie, B., Peterson, A. and L. Revell (2012) 'Post-secular trends: issues in education and faith', *Journal of Beliefs & Values: Studies in Religion & Education*, 33(2): 139–41.
This journal special issue is the first extended discussion of postsecularism in relation to education. Topics include: RE in the curriculum (Lieven Boeve); human rights (Bob Bowie); religious knowledge (David Carr); faith dialogue as a pedagogy (Mike Castelli); and others.
Hotam, Y. and P. Wexler (2014) 'Education in post-secular society', *Critical Studies in Education*, 55(1): 1–7.
This journal special issue is broader in approach than that by Bowie et al., including discussion of George Grant (William Pinar), Palestine and Israel (Ayman Agbaria & Muhanad Mustafa), Mindfulness (Oren Ergas), and others.
Lewin, D. (2016) *Educational Philosophy for a Post-secular Age*, London: Routledge.
I offer an expanded argument for the relevance of postsecularism for educational theory. I challenge the ways educational theory frames debates about indoctrination, competing truth claims and the rights of parents to engage in religious upbringing, arguing that postsecularism offers a way beyond these framings.

References

Aldridge, D. (2015) *A Hermeneutics of Religious Education*, London: Bloomsbury.
Asad, T. (2003) *Formations of the Secular: Christianity, Islam, modernity*, Stanford, CA: Stanford University Press
Bamber, P., Lewin, D. and M. White (2018) '(Dis-)locating the transformative dimension of global citizenship education', *Journal of Curriculum Studies*, 50(2): 204–30.
Barnes, P. (2009) 'Religious education: taking religious difference seriously', *Impact*, 17: 9–56.
Black, C. (2012) *Occupy Your Brain: on power, knowledge, and the re-occupation of Common Sense*, Films for Action, www.filmsforaction.org/news/occupy_your_brain_on_power_knowledge_and_the_reoccupation_of_common_sense/
Brown, W. (2013) 'Introduction'. In Asad, T., Brown, W., Butler, J. and S. Mahmood (eds.) (2013) *Is Critique Secular? Blasphemy, Injury, and free Speech*, 2nd revised edition New York, NY: Fordham University Press.

Bruce, S. (2002) *God is Dead: secularisation and the west*, Oxford: Wiley-Blackwell.
——— (2013) 'The other secular modern: an empirical critique of Asad', *Religion and Society: Advances in Research*, 4(1): 79–92.
Casanova, J. (2009) 'The secular and secularism', *Social Research*, 76(4): 1049–66.
——— (2010) 'A secular age: dawn or twilight?'. In Warner, M., VanAntwerpen, J. and C. Calhoun (eds.) (2013) *Varieties of Secularism in a Secular Age*, Cambridge, MA: Harvard University Press, pp. 265–81.
Cimino, R. and S. Smith (2011) 'The New Atheism and the formation of the imagined secularist community', *Journal of Media and Religion*, 10(1): 24–38.
Cooling, T. (2010) *Doing God in Education*, London: Theos.
Connolly, W. (2006) 'Europe: a minor tradition'. In Scott, D. and C. Hirschkind (eds.) *Op cit.*, Stanford, CA: Stanford University Press, pp. 75–92.
Dupré, L. (1993) *Passage to Modernity: an essay in the hermeneutics of culture*, New Haven, CT: Yale University Press.
Fox, J. (2002) *Ethnoreligious Conflict in the Late Twentieth Century: a general theory*, Oxford: Lexington Books.
Fukuyama, F. (1992) *The End of History and the Last Man*, New York, NY: Free Press.
Gast, J. (1872) *American Progress* [Oil on canvas], Autry Museum of the American West, Los Angeles: California.
Gearon, L. (2013) *Masterclass in Religious Education*, London: Bloomsbury.
Glendinning, S. (2009) 'Japheth's world: the rise of secularism and the revival of religion today', *The European Legacy*, 14(4): 409–26.
Gutmann, A. (1982) 'What's the use of going to school?'. In Sen, A. and B. Williams (eds.) *Utilitarianism and Beyond*, Cambridge: Cambridge University Press.
Heidegger, M. (1977) *The Question Concerning Technology and Other Essays*, trans. W. Lovitt, New York, NY: Harper and Row.
——— (1991) 'Only a God can save us: Der Spiegel's interview with Martin Heidegger'. In Wolin, R. (ed.) *The Heidegger Controversy: a critical reader*, London: The MIT Press, pp. 91–118.
Jackson, R. (2007) 'European institutions and the contributions of studies of religious diversity to education for democratic citizenship'. In Jackson, R., Miedema, S., Weisse, W. and J.-F. Willaime (eds.) *Religion and Education in Europe: developments, contexts and debates*, Münster: Waxmann, pp. 27–55.
King, M. (2009) *Postsecularism: the hidden challenge to extremism*, Cambridge, MA: James Clarke and Co.
Lewin, D. (2014) 'The leap of learning', *Ethics and Education*, 9(1): 113–26.
——— (2015) 'Heidegger east and west: philosophy as educative contemplation', *Journal of Philosophy of Education*, 49(2): 221–39.
——— (2017) 'Who's afraid of secularisation? Reframing the debate between Gearon and Jackson', *British Journal of Educational Studies*, 65(4): 445–61.
Rancière, J. (1991) *The Ignorant Schoolmaster: five lessons in intellectual emancipation*, Stanford, CA: Stanford University Press.
Schmitt, C. (2007) *The Concept of the Political*, Expanded Edition, trans. G. Schwab, Chicago, IL: University of Chicago Press.
Scott, D. and C. Hirschkind (eds.) (2006) *Powers of the Secular Modern: Talal Asad and his interlocutors*, Stanford, CA: Stanford University Press.
Smith, J. (2014) *How (Not) to Be Secular: reading Charles Taylor*, Grand Rapids, MI: Eerdmans.
Sommers, C. (2002) 'How moral education is finding its way back into America's schools'. In Damon, W. (ed.) *Bringing in a New Era in Character Education*, Stanford, CA: Hoover Institution Press, pp. 23–41.
Strhan, A. (2014) 'Editorial: education and the "problem" of religion: a special virtual issue', *Journal of Philosophy of Education*, https://onlinelibrary.wiley.com/page/journal/14679752/homepage/education_and_the__problem__of_religion.htm.
Tibebu, T. (2011) *Hegel and the Third World: the making of Eurocentrism in world history*, Syracuse, NY: Syracuse University Press.
Van Manen, M. (2016) *The Tact of Teaching: the meaning of pedagogical thoughtfulness*, London: Routledge.
Williams, R. (2012) *Faith in the Public Square*, London: Bloomsbury.
World Bank (2014) *Education for All*, www.worldbank.org/en/topic/education/brief/education-for-all

Part II
Theological perspectives

11

Redeeming the secular

Matt Bullimore

Introduction

This chapter explores radical orthodoxy's (RO) genealogy of the secular, its description of the contemporary postsecular situation, and the task it has set itself in renewing Christian socialism in a postliberal form. RO narrates the origins of the modern notion of the secular as something imagined and created, and dependent on deviations in the Christian tradition. If the postsecular names the moment of awareness that the idea of the secular as a tradition-free space is imploding, RO makes use of this opportunity to reiterate a theological construal of the secular as time under grace. It questions the idea of the postsecular as 'postmetaphysical' because all thinking and practice implies an ontology—an account of the way the world is. RO attempts to resituate the world within a renewed Christian ontology in order to out-narrate rival mythologies and so allows for the emergence of a more genuine Christian practice today.

What is RO?

RO is a theological mood or sensibility (Ward and Hoelzl 2008: 152, 158) and a 'hermeneutical disposition and a style of metaphysical vision' (Pickstock 2001: 406). It is not a movement with definite parameters but a set of broadly similar projects and shared interests. The eponymous volume of essays that introduced RO was edited by the Cambridge philosophical theologians John Milbank, Catherine Pickstock, and Graham Ward, and includes contributions from their students and self-described 'fellow-travellers' like William Cavanaugh (Milbank et al. 1999).[1]

RO is less accommodating of secular reason than some modern theology. It holds that there is a difference to Christian reason and that there is no neutral, autonomous rationality disembodied from tradition. Yet this does not mean it is fideist or uncomplicatedly conservative. It is not merely traditionalist or exegetical in scope but is a mediating theology that has a proper interest in all spheres of culture as vehicles of divine revelation (Milbank 2000: 33). It thus calls for the renewal of a deeper Christian humanism that can show that the Christian story is finally the most persuasive narrative because it is the most beautiful and desirable

(Pickstock 2001: 411). It is the most fitting story because of its humane resonance and because it understands that reason has bodily, emotive, aesthetic, linguistic, social, historical, and natural registers (Bengston 2016: 141).

RO incorporates criticisms of Christian thought and practice that reveal weaknesses in the development of the tradition and expose deviations from genuine practice. However, RO does not merely amend the tradition in order to accommodate secular philosophies but is more radical in two senses. As a *ressourcement* project, it reaches deeper into the roots of the tradition to find resources for recovering a truer Christian theo-logic and practice. Yet, to be truly orthodox, it also develops and renews the tradition in fresh ways that re-form it and also thereby offers constructive ripostes to critics.

What is the secular?

Like Charles Taylor, John Milbank refuses a 'subtraction theory' of secularization (Milbank 2009a). The secular as an apparently neutral and desacralized realm gradually freed from tradition, superstition, and faith is a construct that 'had to be instituted or *imagined*, both in theory and practice' (Milbank 2006: 9). The secular is a historical contingency with its own metaphysical assumptions about the way the world is. At the beginning of his seminal *Theology and Social Theory: Beyond Secular Reason*, Milbank makes the bold (and often misinterpreted) claim that:

> Once, there was no 'secular'. And the secular was not latent, waiting to fill more space with the steam of the 'purely human', when the pressure of the sacred was relaxed.
>
> *(Milbank 2006: 9)*

Milbank's fond recounting of a time in which there was no secular sphere is often taken as a form of theocratic rhetoric that aims to reclaim the hegemony of theology and, by so doing, proclaims the redundancy of all other forms of discourse. Milbank does argue that presumptively atheist social science is itself a heterodox theology, the result of changes in the theological imagination, and that it rests upon a founding *mythos* just as much (he concedes) as any theology. If Christian theology does not position and qualify other discourses, he continues, then they will qualify it and so theology must rid itself of false humility and recover and proclaim its own form of social thought (Milbank 2006: 1). Milbank's claims are often seen as fideistic and sectarian, deliberately cutting off theology and the church from a secular world.

Yet this is to misunderstand Milbank's argument. Today, the secular has come to name a neutral autonomous sphere freed of the divine. However, as Milbank argues, in the medieval period the '*saeculum*...was not a space, a domain, but a time – the interval between fall and *eschaton*' (Milbank 2006: 9). Things that were described as secular were thus the activities pertinent to this age—things which are temporal, penultimate, and passing—such as the building of roads, the care of the sick, or marriage. All of these pursuits, however, are undergone following Christ's inauguration of the Kingdom and so simply cannot be understood as operating apart from a theological account of human personhood or sociality. So, as Milbank notes, in the early medieval period, monasteries were also farms, the church-maintained bridges were also shrines, and 'the laity often exercised economic, charitable and festive functions in confraternities that were themselves units of the church as much as parishes, and therefore occupied no unambiguously "secular" space' (2009b: 342). Graham Ward argues similarly that today the boundary between church and world cannot be rigidly demarcated without doing violence to both:

> Theologically, the church simply cannot renounce this world (the doctrines of incarnation and creation forbid it) and neither can it sociologically for this is the time in which the church speaks. It is through this world that Christ is mediated. Any retreats into fundamentalism, confessionalism, or sectarian communitarianism do not redeem the secular, they abandon it.
>
> *(Ward 2000: 69)*

Both Milbank and Ward resist the modern rigid demarcation of the sacred and the secular because the secular age is the time in which all things can be transformed under grace.

When the secular is understood as this time of redemption then it takes on a moral density. It is not simply neutral, chronological time but one that is crossed either by *desire* for that which is transcendent, ultimate, and eternal, or desire for that which is immanent, penultimate, and provisional. That is to say, the use one makes of this secular time is determinative. Either it is a time in which people are transfigured by their love of God (worship) and so participate in the ongoing divine interruption of the world's violence—the establishment of peace—or it is marked by love of self and love of domination (idolatry), and so by participation in cycles of violence and coercion. The secular, then, for members of the body of Christ, is 'just the time *we* have been given to redeem' (Ward 2010: 348).

Milbank concedes that a failure to maintain this theology of the secular meant that the medieval church did not 'adequately incarnate Christianity in the lay and material orders' (2009b: 344). The laity were removed from influence over sacramental and sacred matters, and daily work was no longer seen as also contemplative and holy. The church became the site of a clericalist investment in the sacred and spiritual on the one hand, and so opened up a relatively autonomous 'secular' space evacuated of spiritual import on the other. (The narrative is taken up by Charles Taylor (2007) with his account of the reform movements which arose in reaction to this increasingly differentiated hierarchy of the religious/clerical and the lay.)

Sources of RO's theology of the secular

There are three key influences upon RO's theology of the secular: Augustine's theology of history and the two cities; the voluntarist and nominalist revolution of the thirteenth century; and the twentieth-century *ressourcement* movement in Catholic theology (especially the work of Henri de Lubac). For RO, the narration of the genesis of the modern construction of the secular as tradition-free space and the elucidation of a theological account of the secular act as rhetorical tools for transforming the cultural imaginary in the contemporary postsecular situation.

Augustine

In *The City of God against the Pagans,* Augustine argues that what we love forms our character and our society:

> Two cities have been created by two loves: that is, the earthly by love of self extending even to contempt of God, and the heavenly by love of God extending to contempt of self.
>
> *(14.28. Dyson 1998: 632)*

This age comprises two forms of social practice which arise from what their citizens worship. The two cities should not be glibly identified with the church or political institutions. Nevertheless, the city of God is most visible as the continuing story of Christ active in the church, whilst a prideful love of dominion is often most manifest in political formations. Milbank argues that Augustine's contrasting narratives of the two cities ground a contrast between ontological antagonism and ontological peace:

> The *Civitas Terrena* is marked by sin, which means, for Augustine, the denial of God and others in favour of self-love and self-assertion; an enjoyment of arbitrary, and therefore violent power over others – the *libido dominandi*.
>
> *(Milbank 2006: 392)*

For Augustine, even the relative virtues of the earthly city always presume defeat of something (be it other nations, classes, or bodily passions), which makes violent conquest a necessary part of virtue. By contrast, the city of God embodies 'a real peace, which is a state of harmonious agreement, based upon a common love, and a realization of justice for all' (Milbank 2006: 393). The two cities, like wheat and tares, are intermingled in this age. They share the same world, enjoy the same benefits, and are afflicted by the same troubles, but 'do so with a different faith, a different hope, a different love' (18.54, Dyson 1998: 907–8).

Crucial for understanding RO's account of secular time is seeing that there is for Augustine no strict separation of church and state, of sacred and secular, of private and public. Both cities form a public whole and are not to be conceived as two elements taking their respective places within one overarching public space. What divides these publics, these modes of social existence, is how they use the temporal goods and to what end (Cavanaugh 2011: 57). The city of God is thus a fully political reality that promotes harmony, consensus, forgiveness, reconciliation, and love. It is the ongoing performance of a true society, whereas what is normally understood as political society is for Augustine only imperfectly social because it contains elements of constraint, compromise, and compulsion.

It is with later forms of more settled political Augustinianism that the idea of the secular as one *time* marked by contrasting devotions, desires, and practices is transformed into the idea of there being two *spheres* of existence and respective institutional government—the sacred and the secular, the church and the state.

Voluntarism and nominalism

The advent of nominalism and voluntarism in the thirteenth century continued to open the way for the privatization and hyper-spiritualization of the sacred while it also imagined nature as a sphere of autonomous power—a realm that could be seen, as Grotius postulated, *etsi Deus non daretur*—as if God did not exist (Pickstock 1998, 2005).

Voluntarism and nominalism began to rearticulate the relationship between God and Being. When God is no longer seen as the donating source of all Being, as a Creator ontologically distinct from creatures, and Being is univocally predicated of God and creation (both sharing in Being on the same plane of existence—though as infinitely different), then a metaphysical science of Being is made possible in which, for the first time, there could be an investigation into Being prior to theological speculation. Theology as a discourse about God is made secondary to metaphysics, and it becomes possible to speak of the existence of worldly things apart from their divine origination and end. With voluntarism, God is more properly defined as unconstrained will and absolute power, now construed apart from a consideration of his co-transcendent goodness and wisdom. As God's will is now inscrutable (as arbitrarily wilful), the ordering of nature no longer inchoately mediates knowledge of God's goodness and wisdom. And with nominalism's disavowal of universals (humanity, goodness, beauty) as only conventional names, it was no longer the case that individual created things were seen to participate in a common transcendent reality. Thus, theology no longer spoke of *all things* under God, but became the preserve of that which was spiritual as *opposed* to the material, the worldly, and the bodily. It thus became possible to posit a sphere that could be understood without recourse to contemplation of God.

Christian integral humanism

One of the corollaries of this loss of an integral relationship of the material and the spiritual, of nature and grace, is that it posits a pure nature with its own autonomous integrity, and so grace can only be seen as a supernatural addition arriving as something wholly extrinsic to nature. In his controversial work *Surnatural*, Henri de Lubac (1896–1991) set out to disrupt any dualism of 'pure nature' and 'supernatural grace' by arguing that a more theological account would reaffirm their integral relationship.

In the wake of the ripples caused by the changes in the thirteenth century, De Lubac was fighting a battle on two fronts. On one side was the radical Augustinianism of figures like Michael Baius (1513–1589) and Cornelius Jansen (1585–1638) who denied that there was any residual integrity to a corrupt and depraved nature at all. Grace, for them, arrived extrinsically as merely a corrective to sin and was no longer seen as God's work of bringing human nature into a deeper participation in his life.

On the other side were the interpreters of St Thomas Aquinas (such as Thomas Cajetan (1469–1534) and Francisco Suárez (1548–1617)) who sought to preserve both the integrity of nature and the gratuity of grace. They argued that human endeavour (in political, economic, and social life) was intelligible without grace because we can achieve the purely natural ends we desire by virtue of our own natural capacities. Grace, they argued, was thus preserved in its sheer gratuity as the gift that enables us to obtain the supernatural end of the beatific vision. De Lubac argued, however, that the consequence was that an autonomous nature was now evacuated of grace, and grace was styled as an *extrinsic* and somewhat alien gift super-added to nature.

On both sides, de Lubac discerned the loss of an *integral* relationship between created human nature, seen as a gift, and the intensifying gift of grace that raised and deified human nature. In neither case was there a Christian humanism that saw the secular as the time in which the world was being gracefully redeemed. Grace and nature were put in a contrastive relationship that would grant 'space' to the secular, either as a self-sufficient realm with its own order indifferent to its supernatural end and evacuated of grace (the neo-scholastic Thomists), or as deformed and without integrity (the radical Augustinians). In both cases they separated every aspect of human living from the mediations of grace. Society was secularized as the church was privatized and sacralized.

De Lubac argued that a truer Augustinian and Thomist account of grace taught that humans have a *natural desire for the supernatural*. Human nature is created as already called to a supernatural end that it can nevertheless neither anticipate nor achieve by its own power. Grace is the ever-new transformation of a nature that God created precisely to be open to that work—grace is both remedy for sin and the 'supernaturalizing' of the natural. Aquinas' oft-quoted words that 'grace presupposes nature' (*Summa Theologiae*, Ia, q. 2, a. 2, ad 1) and 'grace perfects nature and does not destroy it' (*Summa Theologiae*, Ia, q.1, a. 8, ad 2) suggest that there is some deep affinity between our nature and its supernatural destiny. That is, human nature does not receive its integrity by what it can offer by contrast or in competition with grace but has integrity by virtue of being a created gift that is being redeemed and perfected in this age.

RO follows de Lubac in arguing that the relationship between grace and nature can be imagined in a way that does not entail the creation of the secular as an immanent totality devoid of grace (Milbank 2005a). They argue that modern 'secular' beliefs and practices arose from the voluntarist and nominalist shifts in late medieval theological understandings (Pabst 2012; Milbank 2013). The secular can be reimagined in an Augustinian mode as being the time in which all things, seen in the light of their created origin and true end, can participate in God's life of shared charity. If RO has a postsecular task, it is to help transform the cultural imaginary by exposing the fragility and contingency of the notion of a secular space and by pointing to opportunities for the renewal of Christian thought and practice.

What is the postsecular?

In his work on the new visibility of religion, Graham Ward notes that the terms postsecularism and postsecularity are asked to bear a lot of weight. 'Postsecularism' seems to embrace a variety of widely differing positions, while 'postsecularity' is asked to describe a whole new cultural era. What both terms do point to is a realization—which chimes with the intuitions of RO—that secularization was neither wholly successful nor inevitable, and furthermore that it is not necessarily irreversible.

Key indicators of 'secularization' must be seen alongside changes in other social and cultural spheres. Decline in participation in organized religion happened alongside similar decline in participation in other social and political bodies (political parties, trades unions). Moreover, the 'religious' refuses to fade into oblivion but returns in new guises in response to cultural changes. Ward therefore prefers to speak of a 'new visibility of religion'—religion's resurgence in the public sphere rather than its return (Ward and Hoelzl 2008; Ward 2009: 117–58). This can take many forms: the rise of political Islamism, the renewed role of religious institutions in civil society, interest in bowdlerized forms of Eastern religions, the recourse to theological language in contemporary philosophy, and the simulacra of religious ideas and images that pervade the cultural sphere—religion as special effect (Ward 2003: 133).

The 'postsecular' attests to the fact that we have never really been purely secular at all. Ola Sigurdson, whose work shares a family resemblance with that of RO, argues that the postsecular does not just connate the non-secular but refers to the discussions that come after belief in an exclusive and hegemonic secularity. That is, the postsecular is what we call it when secularism has become reflexive (Sigurdson 2010: 191). Ward argues that there is a tendency to lump processes together—modernization, secularization, disenchantment, atheism, detraditionalization—and glibly to presume that each necessarily entails the others (Ward 2016). He contends that to speak of the postsecular is to acknowledge that the stories of Western modernity have been made more complex in the light of multiple modernities and the decolonization of hegemonic discourses. It is no longer possible to speak of a secular modernity that embraces all social situations univocally and universally. Whilst modernities often share similar themes and agendas—emancipation, the place of reason, the break with reliance on tradition and authority—they emerge in different cultures, out of many traditions, each with their own specific histories, traumas, and sets of cultural imaginaries. To say that they are secular is a simplification that abstracts people from the hybrid contingencies, processes, and practices of life that are witnessed around the world (Adams et al. 2013: 5).

Despite this new visibility of religion and the loosening of the whiggish association of secularization and modernization, the experience of the postsecular is often marked by a feeling of 'the malaise of immanence and the memory of transcendence' (Smith 2014: x, following Taylor 2007). Although people still express a desire for meaning and value, the transcendent is experienced as an absence that haunts the apparently secular immanent frame. Moreover, the malaise of immanence comes with very real political and social consequences.

For example, Ward argues that democracy has come under the sway of the myth-making of contemporary media. *Depoliticization* occurs when there is an aestheticization of political life whereby citizens are by turns entertained, dazzled, made to forget or ignore, and truth is obfuscated. Participation in intermediate levels of social and political life is neglected and so the sovereign centre is given more control. Allied to this is a process of *commodification* as common values are replaced by market-driven lifestyles. Citizens appear to be more like service users, and there are crises in the trust of increasingly remote representatives. A *dematerialization* occurs when virtual realities control our opinions as when the invisible costs of production in sweat shops or environmental devastation are hidden. These processes are dangerous because when the ability or desire to act politically is weakened then societies can succumb to dehumanizing forces (Ward 2009: 37–116).

Ward describes this as the process of the truly social dissolving into the merely cultural. This is not to deny that the social is always culturally embedded, imagined, and mediated. It is people who dream, hope, aspire, create, and relate—in ways that are real, material, and social—and those hopes and relationships are also culturally conditioned (Ward 2005: 160). But herein lies the chink of light for Ward because if the aim is to remake the social, then this is done by affecting the cultural imaginary. It is here that Ward discerns the role of religions as resources for new forms of humanism (Ward 2005: 163–4). The postsecular, it seems, is as susceptible to the transformative presence of communities of worship as was the antique pluralism of Augustine's day—precisely when both are understood as partaking of a theological construal of secular time.

Opportunities for seeing the world as graced are at the very least increased in a postsecular environment because religious options are not immediately shut down. If there is now no neutral field upon which rational consensus can be found, then it is impossible to de facto set aside religious commitments, hopes, desires, emotions, actions, and communities. It is harder to deny that they contain a genuine sense of in what the good life may consist or a good way to order society.

Ward also points to the persistence of Christian imaginaries with their rich history, often embedded in institutional structures, laws, and the arts. In all this Ward finds signs of hope that theology can begin to remould the cultural imaginary. Ward's sustained attention to what makes beliefs believable resonates with Charles Taylor's work on changes in the conditions of believing (Ward 2014, 2016: 255–85), and he dares to hope that there is once again the possibility of believing in belief, of trusting in the meaningfulness of the world as shot through with transcendence (Ward 2014: 212, 217. See also Blond 1998).

For RO, the church offers a critique of contemporary cultural imaginaries in hope and not scepticism, and in so doing opens up new social utopias for the *Zeitgeist* (Ward 2005: 168):

> Theological discourse relates...to the productive transformation of culture by directing such transformation towards a transcendent hope. It works not only to participate in but

> to perform the presence of Christ. In and through its working the cultural imaginary is changed, and alternative forms of sociality, community and relation are fashioned.
>
> *(Ward 2005: 172)*

Ideas become material cultures (Ward 2014: 177). Christianity has always known this. To return to Augustine, his own project of renarrating Roman history under the judgement of God and his narrating of a counter-narrative of grace, charity, and peace, was a work of reimagining a theological vision of the cosmos. It was a vision that had to be lived out in practices of charity and peace-making in a world in which other religious options were available. Likewise, Ward is arguing for the church to forge a new public apologetic discourse based on public practices, producing public truth and demonstrating its credibility through witness (Ward 2005: 173). Ward sees this as a lay vocation, the calling of all Christians, to reassemble the social through the micropractices of Christian living:

> We perform the Gospel in every social and cultural engagement, and in these performances we deepen a hopeful trend...and extend the public role that religion can play in the elaboration of new political, social, and cultural futures.
>
> *(Ward 2009: 166)*

He uses the term 'ecclesiality' to define this performance of the church

> whereby the church names not a political entity identifiable by spatially discrete boundaries enforced by a sacred hierarchy but is instead "what this body of Christians do" in an ever-expansive web of sociality spun by an array of intermediate associations and relations.
>
> *(quoted in Bell: 2015: 125)*

Postliberalism

Perhaps the most concrete example of radically orthodox thinkers engaging with the cultural imaginary in order to spur the reassembly of the social is through participation in postliberal political movements. Postliberal projects can themselves be described as postsecular because they question the hegemonic liberal discourses that have sought to privatize religion and silence religious voices in the public sphere. They also deem there to be goods that transcend merely procedural, utilitarian, or immanent ends.

The critique of liberalism

Milbank is critical of the advocacy of the open space of the formally liberal because of the ever-present danger of it being filled by extreme positivisms and fundamentalisms (Milbank 2006: xi). The liberal subject is styled as a pure will, a will to will, abstracted from the desires, commitments, and associations that real people have. Liberalism posits a vacuous space with no extra-natural norms, one that is inevitably filled with competitive positivisms. In response, Milbank discerns a need to reconnect ideas of freedom and the will to good ends. No form of human community can flourish in common without some shared account of the *good*. Merely formal appeals to emancipation (from what and for what?), social justice (which looks like what?), or equality (in what way equal?) lack substance and frequently resign us to 'the market regulation of competing desires' or the common good as 'the lowest common denominator on which we supposedly all agree' (Milbank 2013: 268).

In Augustinian fashion, John Milbank and Adrian Pabst argue that the pursuit of wealth, power, glory, and pleasure lead to economic instability, social disorder, atomization, and environmental catastrophe (Milbank and Pabst 2016: 1–2). The usual remedies of coercive state control or market competition have been found wanting. What is proposed is a vision that would abandon:

> the formalism of abstract representation and subjective rights in favour of a substantive account of the relational Good which orders relations within the *cosmopolis* in line with transcendent standards of justice and a fair share for all. In this way, we can imagine a radically communitarian and associative virtue politics and virtue economy.
>
> *(Pabst 2010: 597)*

This is not to deny that liberalism has provided some measure of increased freedoms and genuine access to opportunities for many. Milbank has said that he does 'affirm some continuing but ideally receding need for a merely "contractual" peace, as opposed to the real peace of consensus and gift-exchange' (Milbank 2006: xv). Nonetheless, *the* Christian virtue is charity, which is a reciprocal exchange that expresses concern for both the self and the other. It is something ecstatic, personal, joyful, and relational, and so necessarily interrupts the formal and contractual (Candler and Cunningham 2007: 523).

The postliberal alternative

Milbank and Pabst's *The Politics of Virtue: Post-Liberalism and the Human Future* (2016) arises out of their involvement with Blue Labour (see Geary and Pabst 2015) and, to a lesser extent, Phillip Blond's Red Toryism (Blond 2010). Blue Labour arose as a remedial riposte to current hegemonic forms of liberalism that have eroded the social bonds of trust and cooperation within civil society and which resulted from the collusion of the politico-economic liberalism of the 1980s and the social liberalism of the 1960s (Milbank and Pabst 2014: 7). In *The Politics of Virtue,* they analyze the 'metacrisis' of liberal thought and practice and argue that a secular/sacred divide has not led to human flourishing. The creation of a secular space without the mediations of transcendence led to the modern bifurcation of the political into the Right and the Left. It set up an oscillation of power between the state and the market and between the sovereign centre and the individual, and has gradually weakened the sphere of civil society.

As Luke Bretherton has argued, a flourishing civil society is constituted by those convivial relationships of integrity that are able to resist state or market monopolies of power and so prevent the instrumentalization of social relations for the sake of political order or the economy (Bretherton 2010: 92). Milbank and Pabst thus argue that churches have an important role in rejuvenating and strengthening civil society because they vividly exemplify a way of life that has 'ambitions towards reconciliation and harmony [that] exceed those of a law-governed state or a market dominated by commercial exchange' (Milbank and Pabst 2014: 9).

Following Ivan Illich, via Charles Taylor, this is an argument for a more charitable, personalist, and reciprocal form of political life (see Cayley 2005: 29–37; Taylor 2007: 737–9). As Milbank argues, 'Christian *agape* operates through a network of direct relations not through conceptual or legal imposition' (Milbank 2009: 99). This form of political life requires popular participation and virtuous elites. Virtue is democratic precisely as open to all and yet needs to be taught and exemplified by 'virtuous, skilled, generous and wise' elites. The hope is for a 'more purposive and culturally pluralist shaping of association around shared aspiration and ideal purpose, which seeks to integrate different human roles and traditions' (Milbank and Pabst 2016: 7). Milbank and Pabst argue that perhaps only a religious vision can genuinely move people towards shaping a politics of the common good because such a vision sees humanity teleologically directed in an ethical direction.

In the context of the UK, for example, Milbank and Pabst argue that 'the political role of the established church is, therefore, neither to sanctify the state nor to supplant the government as elected and representative, but, rather, to "inform" public institutions in the direction of both individual virtue and public honour' (Milbank and Pabst 2016: 232). As established, the church can remind the secular arm of government that it is answerable to transcendent concerns and can be informed by pastoral and reconciling social aims that exceed its strictly 'political' remit.

Establishment can actually defend the principle of secularity (as this provisional time) against the encroachments of aggressive secularism or fundamentalist theocracy. It is here that some of the political ramifications of de Lubac's thesis are apparent. Without the presence of the church bending the political in a more social and ultimate direction (grace perfecting nature), it is easy enough to either invest the political sphere with a quasi-sacral ultimacy (fascism) or adopt a political religion with a realized eschatology (theocracy). In both cases, the idea of secularity is undermined—either because it is absolutized (as pure and autonomous nature) or wholly usurped by the sacral (nature as depraved and subjugated by extrinsic grace).

Another example is seen in the work of the Ugandan theologian Emmanuel Katongole (2011) who puts some of the insights of RO to work in an African context. Katongole argues that concepts of 'tribe' and 'ethnicity' are inventions of the modern state that became naturalized in political discourse as unchangeable givens (Katongole 2014a: 320–1). Missionaries who arrived in the modern period to save souls similarly presumed such apparently natural identities, so making the blood of tribalism thicker than baptismal waters. Once these apparently prior natural identities are assumed as given, it is difficult to grant any new transformative political dimension to agents or communities. The modern African state, he argues, presents itself as saviour by offering a form of lowest common denominator coexistence between groups that it has constructed as irreconcilable. In looking for alternative ways of living that can be inscribed into Africa's modernity, Katongole uncovers the possibilities that Christian faith and practice can make in transforming this naturalized *mythos* of the ethnic. He tells the stories of people who have 'hopes the wrong size for this world' (Katongole 2014b, 422) and shows that as the church pursues its hopeful imagination for new forms of peaceful, charitable living, it finds echoes, resonances, and allies among other existing institutions and communities. The church is thus always missiological rather than purely structural or institutional— a set of practices that transforms the cultural imaginary. As Ward prefers to speak of ecclesiality, so Katongole advocates speaking of ecclesiological radiances (Katongole 2014b: 430) rather than ecclesiastical spaces, attesting to the improvizational nature of the church's servanthood of the world as it reperforms the body of Christ anew in this secular time.

In both these examples we see how postliberalism is 'ambitious for the convivial beyond mere coexistence' (Milbank and Pabst 2016: 235) and requires an account of the good which is unavailable to a purely procedural and secular account of political life.

Socialisms

Postsecular materialisms

Another form of postsecular intervention is seen in the conversations between the Slovenian philosopher Slavoj Žižek and John Milbank (Davis 2009; Milbank et al. 2010). It takes the form of a common desire to renew socialism in ways that are less wedded to progress, the state, a reductive materialism, or historical determinism.

Creston Davis, a pupil of both Žižek and Milbank, describes them as believing theology can serve as a 'wellspring capable of funding a materialist politics of subjective truth' (Milbank et al. 2010: 1). For both, following the incarnation, truth is based on fidelity to an event in space and time. It brings a freedom to choose new coordinates for thinking the truth beyond any present impasse. In the present, the options appear to be liberalism or fundamentalism. It is a choice between the rival solipsism of the rational atheist or irrational fideist who live within their own self-referential systems (Davis in Milbank et al. 2010: 7). It is a stale dialectic that allows no faith in the public square and no reason in dogmatism.

Both Žižek and Milbank, however, see the possibility for a truly materialist human flourishing and resistance by grounding truth in a theological event in time attested to by communities formed in response to that event. We must make a leap of faith in seeing the truth of the incarnation in order to open radical new ways of seeing the world. For Žižek, it is the advent of new terrifying freedom as we see the divine kenotically pouring itself out into new forms of community (transcendence is immanentized). By contrast, Milbank sees the Christ event as showing how transcendence is incarnated in the material and personal (the immanent is divinized).

Žižek's interest in religion is its potential for emancipation. He looks not to its kernel (subjective, private experience) but to its husk (institutional and doctrinal structures) because it is concrete practices and public instantiations of religion that provide its political impact (Sigurdson 2010: 181). This is how religion can repoliticize the public sphere. RO argues that Žižek's advocacy for the logic of the incarnational event but not its content (that Christ was both God and human) instrumentalizes religion and denies its true potential. Ultimately, Žižek takes the form but not the content of the Christ event and disincarnates theology. He uses religious language, but ultimately is not interested in the religious phenomena, communities, or experiences that RO sees as the true vehicles for transforming the cultural imaginary.

Christian socialism

The Christian socialism of RO would claim to be more truly materialist. Grace and transcendence really are mediated 'through the encounter with very public material objects, persons and practices' (Bell 2015: 121). This 'suspends the material' from a divine, donating source that provides an ontological and eschatological basis for socialist hope (Milbank 2006: xiv). For RO, the name for the constantly arriving revolution of the incarnation is socialism (Milbank 2005b: 405). The event of the incarnation to which scripture and doctrine bear witness is renarrated as an 'enacted reading, the non-identical repetition, of Christ's charitable practices, heeding the command to 'follow', to do as he does; in short upon the ecclesial tradition of discipleship' (Loughlin 1997: 55). It is, in other words, a 'politicised metaphysics or metaphysical politics' that has been enacted—for better and for worse—over two millennia as the society of the church (Milbank 2013: 15).

The Christian story retells the hope that the good continuously interrupts the world's stories of violence and domination and is performed through finite acts of goodness that reflect a perfect infinite peaceful power (Milbank 2006: xvi). The church is a lived project of reconciliation seeking a social peace based not on violent overthrow but upon the positive unleashing of charity. It finds true power in forgiveness, reciprocity, gift, and service. It does not seek ecclesiastical mastery of a secular realm, but to be the yeast, light, and salt that gracefully transform over time. So, peace is what 'rises "by grace" as a thousand different specific modes of social harmony, a thousand different gifts of specific social bonding, a thousand kinds of community' (Milbank 2003: 167).

Milbank sees the importance of modern socialism at its best in 'realizing latent tendencies in the medieval tradition' and reasserting the importance of the laity, the collective, labour, sexuality, arts, language, material realm, and history (Milbank 2006: xix). Modernity saw the release of human creativity and imagination and had the potential to work towards a greater equality, new freedoms, and less fear. If, as Milbank argues, there was a general tendency on the Right to a conservative advocacy of the One, and on the liberal Left to an advocacy of the Many, it was nevertheless socialism that rejected both absolutist monarchy of the *Ancien Régime* and technological industrial modernity and advocated the importance of the place of the Few (civil society) in a mixed constitution.

Christian socialism is open to alliances with other forms of socialism that seek to re-invigorate civil society but, in the end, RO claims that Christianity narrates the most plausible and desirable account of social peace because—by its insistence on the reality of a participation in divine charity—it can hold differences in harmony. Of course, it is a story that is desirable only if there is some inchoate and innate human preference for peace over violence and a human bias to reason rather than unreason (Milbank 2006: xvi). And this is not something that can be proved, but must be taken in faith and demonstrated in lived practice.

Prospects for the future

RO shows us that the task of the church is more about seeing how to use this time we call the secular well rather than defending a space (see Sigurdson 2010: 193). RO is not a church in and of itself. It is a variety of Christian voices that critically expose the idolatries that ultimately deform and diminish human life and who tell a story that does not inevitably end in the secular as desacralized and detraditioned space. Its constructive project is to show how a liturgical and ecclesial form of life can be a means of mediating redemption to the world today. As Milbank argues:

> If truth is social, it can only be through a claim to offer the ultimate 'social science' that theology can establish itself and give any content to the notion of 'God'. And in practice, providing such a content means making an historical difference in the world.
>
> *(Milbank 2006: 6)*

RO's success will rely on its persuasive ability to enliven ecclesial practice. This is a crucial task because a cultural imaginary will only be changed, and found credible, when it is felt that exemplary social practice is transfomative. There is still much constructive work to be done in mediating its ideas to a more general audience in a less academic mode. One successful example of finding a new tone is an edition of radically orthodox sermons (Milbank et al. 2017).

No doubt RO will continue to take a part in changing the conditions for belief today, as it seeks to show the reasonableness of faith. The urgent need to unthink the necessity of violence remains (Milbank 2006: 416) as does the need to relearn how to act in a violent world with a presumption of the ontological priority of non-violence. By telling the Christian story in such a way as to give pause, it can continue to help question the stories and presuppositions about belief, knowledge, and the good life that have been ascendant in modernity. It will seek to enable the church continually to make strange the Christian difference by repeating and re-enacting the mystery again in ever new ways. It will thereby help open up the possibility that it still makes sense to live within the Christian city in the Christian cosmos. This is both a romantic and penitential task—the continual reimagining of the Christian life and a continual penitence for the church's failure to live up to the radicalism of the incarnation. By continuing to emphasize charity, reciprocity, and personalism in alliance with others in civil society, it will keep alive the hope for renewed forms of humanism in the midst of rampant utilitarian and instrumentalizing pressures on the social (Hughes 2007; Bullimore 2016).

Given the stress on the intermediary and lay micropractices, RO could begin to help the churches break lingering clericalist biases and focus on what it might mean to revivify lay ecclesial action in the world. John Hughes' writing on work and labour could also be extended in the direction of encouraging an ecclesial focus on the importance of secular vocations as being wholly theological (Hughes 2007; Bullimore 2016). It is in focussing on lay modes of life that the church is seen more as an interpersonal mode of charitable living than as an institution.

Ward has begun to work more on the decolonization of theology and to promote local and regional forms of theological engagement (Ward 2017). It will be important to see local refractions, reformations, and extensions of RO in diverse voices and traditions from world Christianity. A significant engagement with Pentecostalism would be welcome. This work can continue with a more sustained self-examination of RO's present locutionary positions. J. Kameron Carter has, for example, asked RO for a more nuanced and self-critical attention to race and colonialist inheritance (Carter 2008: 388n5).

RO has already increasingly been diffused through the academy and across the world as students have gone on to teach and extend the project in new directions. It might now be possible also to hope for some further influence in more literary and artistic circles. Already Catherine Pickstock's poetry has been set to music by Jeremy Thurlow, and the composer James MacMillan is on the advisory board of the journal *Radical Orthodoxy: Theology, Philosophy, Politics*. John Milbank and Simone Kotva have published poetry, and Alison Milbank writes at the intersection of theology and literature (Milbank 2009). At a more popularly accessible level, Peter M. Candler's writing is influenced by the tradition of Christian integral humanism (www.candler.ink), and the TV series *Did Darwin Kill God?* was presented and written by Conor Cunningham.

A significant danger for RO in a postsecular situation is becoming embroiled in culture wars and radical politics of difference (Ward 2003: ix). RO has shown that it is not fideistic and sectarian, but that is still how it is perceived in some academic and ecclesiastical circles. Engagements like those with Žižek open RO to a wider public and demonstrate the desire to build even apparently unlikely alliances. RO has shown its desire to be ecumenical and interdisciplinary, and it will need to continue to break the siloes and risk new encounters.

Perhaps *the* postsecular question is whether nihilism is finally liveable. If not, then secularism will finally be 'a blip' (Milbank 2013: 115; Ward 2014: 180), and in the resulting postsecular situation, the Christian life of charity will continue to be a practice that is possible, desirable, and necessary. RO's main contribution may be to remind us—in the face of religious, secularist, and political fundamentalisms—that we continue to live in the time of the secular, this provisional time under grace in which redemption is possible even as we wait in hope for the fullness of peace.

Note

1 For further introduction to RO, see: Milbank (2000); Long (2003); Pickstock (2003); Smith (2004); Milbank and Oliver (2009); Bell (2015). Critical appraisals include: Hyman (2001); Insole (2004); Hankey and Hedley (2005); Ruether (2006); Shakespeare (2007); Isherwood and Zlomislic (2012). Although largely Anglican, it is an ecumenical collaboration, and there have been conferences exploring RO's relationship to the Reformed tradition (Smith 2005), Catholicism (Hemming 2000), and Eastern Orthodoxy (Pabst and Schneider 2009). Book series associated with RO include Illuminations (Blackwell), Veritas (SCM), and RO (Routledge).

Further reading

Bengston, J. (2016) *Explorations in Post-Secular Metaphysics*, Basingstoke/New York, NY: Palgrave MacMillan.

A critical exploration of the metaphysical presuppositions behind postsecular philosophies and politics, exploring in particular the way the relationships between transcendence and immanence, the religious and the secular, are figured in the work of John Milbank, William Connolly, and Charles Taylor.

Milbank, J. and S. Oliver (eds.) (2009) *The Radical Orthodoxy Reader,* London/New York, NY: Routledge.

A volume of exemplary essays spanning radical orthodoxy's range of interests, framed by a lucid introduction to radical orthodoxy and a retrospective of the first ten years, with commentary and ideas for further reading.

Milbank, J. and A. Pabst (2016) *The Politics of Virtue: post-liberalism and the human future*, London/New York, NY: Rowman & Littlefield International.

Eschewing overtly theological language and addressed to theorists, activists, practitioners, and all with an interest in renewing civil society, the authors offer an analysis of the crisis of liberalism and set out postliberal alternatives.

Smith, J. K. A. (2004) *Introducing Radical Orthodoxy: mapping a post-secular theology*, Grand Rapids, MI: Baker Academic.

A sustained appreciative analysis and synthesis of the range of radically orthodox writings from a Reformed perspective, which uncovers the sources upon which they rely and unpacks the debates in which radical orthodoxy is engaged.

Ward, G. (2009) *The Politics of Discipleship: becoming postmaterial* citizens, Grand Rapids, MI: Baker Academic.

A radically orthodox reading of the signs of the times, analyzing the state of democracy, globalization, and the postsecular situation. A theological account of the relationship between the church and the world shows how Christian theology can offer critical and constructive interventions in contemporary debates.

References

Adams, N., Pattison, G. and G. Ward (eds.) (2013) *Oxford Handbook of Theology and Modern European Thought,* Oxford: Oxford University Press.

Bell, D. M. Jr. (2015) 'Postliberalism and radical orthodoxy'. In Hovey, C. and E. Phillips (eds.) *The Cambridge Companion to Christian Political Theology*, Cambridge: Cambridge University Press, pp. 110–32.

Blond, P. (ed.) (1998) *Post-Secular Philosophy: between philosophy and theology*, London/New York, NY: Routledge.

——— (2010) *Red Tory: how left and right have broken Britain and how we can fix it*, London: Faber and Faber.

Bretherton, L. (2010) *Christianity and Contemporary Politics,* Oxford: Wiley-Blackwell.

Bullimore, M. (ed.) (2016) *Graced Life: the writings of John Hughes*, London: SCM.
Candler, P. M. and C. Cunningham (eds.) (2007) *Belief and Metaphysics*, London: SCM.
Carter, K. J. (2008) *Race: a theological account*, Oxford: Oxford University Press.
Cavanaugh, W. T. (2011) *Migrations of the Holy: God, state, and the political meaning of the church*, Grand Rapids, MI/Cambridge, MA: Eerdmans.
Cayley, D. (ed.) (2005) *The Rivers North of the Future: the testament of Ivan Illich*, Toronto: Anansi.
Davis, C. (ed.) (2009) *The Monstrosity of Christ: paradox or dialectic?*, Cambridge, MA/London: The MIT Press.
Dyson, R. W. (ed.) (1998) *Augustine: The City of God against the pagans*, Cambridge: Cambridge University Press.
Geary, I. and A. Pabst (eds.) (2015) *Blue Labour: forging a new politics*, 2nd edition, London/New York, NY: I. B. Tauris.
Hankey, W. J. and D. Hedley (eds.) (2005) *Deconstructing Radical Orthodoxy: postmodern theology, rhetoric and truth*, Aldershot: Ashgate.
Hemming, L. P. (ed.) (2000) *Radical Orthodoxy?: a catholic enquiry*, Aldershot: Ashgate.
Hughes, J. (2007) *The End of Work: theological critiques of capitalism*, Oxford: Blackwell.
——— (2016) 'Integralism and gift exchange in the Anglican social tradition, or avoiding Niebuhr in ecclesiastical drag'. In Bullimore, M. (ed.) *Graced Life: the writings of John Hughes*, London: SCM, pp. 148–62.
Hyman, G. (2001) *The Predicament of Postmodern Theology: radical orthodoxy or nihilist textualism?*, Louisville, KT: Westminster John Knox Press.
Insole, C. J. (2004) *The Politics of Human Frailty: a theological defence of political liberalism*, London: SCM.
Isherwood, L. and M. Zlomislic (eds.) (2012) *The Poverty of Radical Orthodoxy*, Eugene, OR: Pickwick Publications.
Katongole, E. (2011) *The Sacrifice of Africa: a political theology for Africa*, Grand Rapids, MI/London: Eerdmans.
——— (2014a) '"A blood thicker than the blood of tribalism": Eucharist and identity in African politics', *Modern Theology*, 30(2): 319–25.
——— (2014b) 'The sacrifice of Africa: ecclesial radiances of "a different world right here": a response to Anne Arabome, Elias Bongmba and John Kiess', *Modern Theology*, 30(2): 421–30.
Long, D. S. (2003) 'Radical orthodoxy'. In Vanhoozer, K. J. (ed.) *The Cambridge Companion to Postmodern Theology*, Cambridge: Cambridge University Press, pp. 126–45.
Loughlin, G. (1997) 'The basis and authority of doctrine'. In Gunton, C. E. (ed.) *The Cambridge Companion to Christian Doctrine*, Cambridge: Cambridge University Press.
Milbank, A. (2009) *Chesterton and Tolkien as Theologians*, London: T & T Clark International.
Milbank, A., Hughes, J. and A. Milbank (eds.) (2017) *Preaching Radical & Orthodox*, London: SCM Press.
Milbank, J. (2000) 'The programme of radical orthodoxy'. In Hemming, L. P. (ed.) *Radical Orthodoxy?: a catholic enquiry*, Aldershot: Ashgate.
——— (2003) *Being Reconciled: ontology and Pardon*, London/New York, NY: Routledge.
——— (2005a) *The Suspended Middle: Henri de Lubac and the debate concerning the supernatural*, Grand Rapids, MI/Cambridge, MA: Eerdmans.
——— (2005b) 'Materialism and transcendence'. In Davis, C., Milbank, J. and S. Žižek (eds.) *Theology and the Political: the new debate*, Durham, NC/London: Duke University Press.
——— (2006) *Theology and Social Theory: beyond secular reason*, 2nd edition, Malden, MA/Oxford: Blackwell.
——— (2009a) 'A closer walk on the wild side: some comments on Charles Taylor's *A Secular Age*', *Studies in Christian Ethics*, 22(1): 89–104.
——— (2009b) 'The gift of ruling'. In Milbank, J. and S. Oliver (eds.) (2009) *The Radical Orthodoxy Reader*, London/New York, NY: Routledge, pp. 338–62.
——— (2013) *Beyond Secular Order: the representation of being and the representation of the people*, Chichester/Oxford: Wiley Blackwell.
Milbank, J. and A. Pabst (2014) 'The Anglican polity and the politics of the common good', *Crucible: The Christian Journal of Social Ethics*, January–March: 7–15.
Milbank, J., Pickstock, C. and G. Ward (eds.) (1999) *Radical Orthodoxy: a new theology*, London/New York, NY: Routledge.
Milbank, J., Žižek, S. and C. Davis (eds.) (2010) *Paul's New Moment: continental philosophy and the future of Christian theology*, Grand Rapids, MI: Braxos Press.

Pabst, A. (2010) 'Modern sovereignty in question: theology, democracy and capitalism', *Modern Theology*, 26(4): 570–602.
Pabst, A. (2012) *Metaphysics: the creation of hierarchy*, Grand Rapids, MI/Cambridge, MA: Eerdmans.
Pabst, A. and C. Schneider (eds.) (2009) *Encounter between Eastern Orthodoxy and Radical Orthodoxy: transfiguring the world through the word*, Farnham/Burlington, VT: Ashgate.
Pickstock, C. (1998) *After Writing: on the liturgical consummation of philosophy*, Oxford/Malden, MA: Blackwell.
——— (2001) 'Reply to David Ford and Guy Collins', *Scottish Journal of Theology*, 54(3): 405–22.
——— (2003) 'Is orthodoxy radical?'. In Morris, J. (ed.) *Faith and Freedom: exploring radical orthodoxy*, London: Affirming Catholicism, pp. 5–16.
——— (2005) 'Duns Scotus: his historical and contemporary significance', *Modern Theology*, 21(4): 543–74.
Ruether, R. R. (ed.) (2006) *Interpreting the Postmodern: responses to radical orthodoxy*, London: T & T Clark International.
Shakespeare, S. (2007) *Radical Orthodoxy: a critical introduction*, London: SPCK.
Sigurdson, O. (2010) 'Beyond secularism? Towards a post-secular political theology', *Modern Theology*, 26(2): 177–98.
Smith, J. K. A. (2005) *Radical Orthodoxy and the Reformed Tradition: creation, covenant, and participation*, Grand Rapids, MI: Baker Academic.
——— (2014) *How (Not) to be Secular*, Grand Rapids, MI/London: Eerdmans.
Taylor, C. (2007) *A Secular Age*, Cambridge, MA/London: The Belknap Press.
Ward, G. (2000) *Cities of God*, London/New York, NY: Routledge.
——— (2003) *True Religion*, Malden, MA/Oxford: Blackwell.
——— (2005) *Cultural Transformation and Religious Practice*, Cambridge: Cambridge University Press.
——— (2010) 'History, belief and imagination in Charles Taylor's *A Secular Age*', *Modern Theology*, 26(3): 337–8.
——— (2014) *Unbelievable: why we believe and why we don't*, London/New York, NY: I. B. Tauris.
——— (2016) 'Secularism: the golden lie.' Unpublished paper.
——— (2017) 'Decolonizing theology'. Unpublished paper.
Ward, G. and M. Hoelzl (eds.) (2008) *The New Visibility of Religion: studies in religion and cultural hermeneutics*, London/New York, NY: Continuum.

12

Christianity or barbarism

Dritëro Demjaha

Introduction

Jürgen Habermas and Joseph Ratzinger's *Dialectics of Secularization* and Charles Taylor's *A Secular Age* established two arguably founding tropes of postsecular discourse: the critique of secularism and the productive intersection of continental philosophical thought and the Catholic theological tradition. This essay concerns John Milbank's *Theology and Social Theory* which predates these crucial contributions by over 15 years and its sequel *Beyond Secular Order* which in a sense reflects on them both. The essay expounds, furthermore, some of the perspectives of 'Blue Labour', the most immediate political expression of intellectuals of Milbank's persuasion and a crucial development in the world of the postsecular. My thesis is that Milbank's work, and that of Blue Labour more generally, renew the traditions of Catholic Christianity in an eclectic fashion and present us with an 'aberrant' (a term whose use I explicate later in the chapter) postsecularity in so far as it proceeds from presuppositions about the status of the secular and modernity typically uncommon in the works of other notable postsecular thinkers like Habermas and Taylor.

Milbank's work is often polemical. Consequently, any thematic exposition of his work ought to also expound the views whose rejection shapes his own. In what follows I will give an account of the intellectual context for Milbank's engagement in postsecular discourse before as well as a sense of the postsecular views Milbank is eager to reject; I will then expound in some more detail his account of the relationship between the discourses of theology and secular social theory and his rejection of postsecular projects which recognize the autonomy of the latter. Following the critique of secularism, I will outline Milbank's critique of the liberal privileging of dialogue between religions. Finally, I will explicate how Milbank envisions a postsecular socialism as a way out of the antagonisms of secular modernity. By and large, I find Milbank's views about the postsecular—such as they are stated here—convincing, and I argue for their plausibility throughout the essay. However, in the section on politics and *Blue Labour*, I question whether Milbank's postsecular vision, which is also postliberal and postmodern, similarly extends to being post-capitalist, and suggest a possible reply to this issue.

The prehistory of postsecularism

Milbank's project is partly founded on the rejection of various commonplaces of postsecular discourse. Such a rejection stems from disagreement with proponents of the latter about not merely the status of religion and the secular but also that of modernity and philosophy. For Habermas, the overcoming of metaphysics and the subsequent emergence of the so-called post-metaphysical philosophy designates a cognitive advance in a diachrony which includes that of the Axial age. The concept of the postsecular is a complement to that of the postmetaphysical since the former proceeds from the recognition of the shared Axial origin of philosophy and religion. Judeo-Christian *Heilswissen* is particularly bound up with this metaphysical prehistory of contemporary thought, and its co-origination with philosophy entails for Habermas that faith and knowledge both 'belong to the history of the origins of the secular reason'.

There is in this line of thought an arguably paradoxical attempt to align, on the one hand, the fates of metaphysics and the secular, and on the other, those of philosophy and religion. However, both are joined by an account of modernity as a conclusion, of sorts, of the history of the origins of secular reason. It is this disagreement over the status of modernity which cuts across any other converging positions found in the postsecular philosophies of Habermas and Milbank. For the former, there exists a close parallel between the phylogenetic (historical) advance to modernity's universalistic moral consciousness (conceived in Kantian terms as a form of rational self-legislation) and the ontogenetic (cultural and psychological) emergence of individual postconventional moral consciousness. As Albrecht Wellmer (1991: 145) has argued, this indicates that 'the emergence of a post-conventional moral consciousness is a response to the fact that previously unquestioned norms come to be seen as questionable and *in need of* justification'. Accompanying this transition to postconventional moral consciousness, as Wellmer emphasizes, is the transition to a new understanding of normative validity claims as achieved through rational argument and intersubjective recognition. The postsecular position consequently becomes primarily concerned with submitting religion to this process of postconventional justification.

This view arguably reproduces a number of arguments made by Hans Blumenberg in his canonical defence of modernity, *The Legitimacy of the Modern Age*. Here, Blumenberg conceived, in part as a response to Karl Löwith's major work *Meaning in History*, a contribution to the debates about modernity and the secular whose place in the prehistory of postsecular discourse is increasingly recognized (see Dosdad 2016). One can regard Milbank's *Theology and Social Theory* as belonging to the same tradition as *Meaning and History*. The former offers an answer to the puzzle of the resurgence of religion in an increasingly secularized world which has increasingly dominated social scientific and anthropological discourse since the 1990s. Milbank attempts the renewal of the Löwithian thesis of modernity's unoriginality vis-à-vis Christianity. For puzzlement regarding the revival of religion is the upshot of the presumed autonomy of the modern world from the logic of the premodern. If, however, modernity's defining features, as Milbank (2013) argues, reproduce (and often unwittingly so) theological categories and structures of thought, then the presumption of their relation to religious belief as antithetical becomes highly suspect. Secularization comes to refer to the process of how modernity reconfigures theological categories into sociological constructs devoid of vertical reference to the transcendent, a process which consequently retains in the secular world a space for the religious insofar as the latter is epistemically prior to, and necessary for, the former.

Habermas' position seems to include a built-in reply to the Milbankian argument insofar as debates such as those between Löwith and Blumenberg are repeated in those surrounding the postsecular. Notwithstanding this anticipatory defence, Milbank has articulated a powerful critique of the Habermasian account of the 'history of the origin of secular reason' by offering a counter-genealogy of both philosophical and theological reason vis-à-vis the secular. Milbank (2013) argues, following Deleuze and Guattari, that philosophy is ontogenetically theological, since the first philosophy was an 'anti-mythical and monistic thinking of immanence, incorporating both matter and mind, both energy and image' (p. 22; cf. Deleuze and Guattari 1994). The paradigm for this theological ontogeneticity is antique materialism and consequently designates both a pre-Christian theology and religiosity. Thus, according to Milbank, the view that Hellenic philosophy was a purely 'rational' enterprise was a projection of the later philosophical traditions of the Abrahamic faiths upon their rediscovery of Aristotelian and Neoplatonic material. This projection stemmed from the appropriation of such material which would have been problematized if it had been taken to be continuous with Pagan religious reflection, rather than as a product of universal human reason. 'In this way a category of "pure reason" started to come into being only as the shadow of the notion of "faith"' (Milbank 2013). In light of this critique, the claim that faith and knowledge both 'belong to the history of the origins of the secular reason' becomes increasingly doubtful.

The prehistory of the secular

Milbank's postsecular project may be described as 'neo-medieval'—a term he has increasingly adopted as a self-designation (Milbank and Pabst 2016: 84, 145, 213, 328–30). A part of what this entails is integrating into the genealogy of the secular the story of the Middle Ages in a way in which other postsecular thinkers like Habermas do not. Milbank has himself suggested that this is a way of keeping at bay the Habermasian position which emerges as a semi-Hegelian compromise between liberal (David Tracy) and conservative (Joseph Ratzinger) Catholic view (Milbank 2002, 272). The year 1300 is a recurring one in Milbankian historiography. While he is not alone in emphasizing the importance of the early fourteenth century in contradistinction to the later and allegedly more exciting periods of the Reformation, the Renaissance or the late Enlightenment (see Certeau 1992; Alliez 1996), he has articulated with greater clarity and force the resources of the premodern prior to 'the crisis of 1300' in accordance with the gestures of *ressourcement* theology but always in conjunction with 'an invocation of modern romantic expressivism and "postmodern" ultraconstructivism' (Milbank 2006: xxii). To the extent that Milbank maintains that modern thinking reproduces theological categories, it is because he assimilates the former's greatest mistakes, the collapsing of ontological difference between creature and Creator in a univocal being, its voluntarism, and, perhaps most crucially, its nominalism, to the history of Christian theology. In this way, it becomes possible to 'out-narrate' the secular truth claims of the grand narratives of modernity without, however, being susceptible to the postmodern reproach against such narration (which is fundamentally a reproach against modernity).

One such strategy with regard to postsecularity immediately suggests itself from the Milbankian views which have hitherto been glossed. Josef Bengston has suggested a threefold typology of postsecular visions corresponding, respectively, to Habermas, Milbank and Taylor, and Connolly (see Connolly 2000, 2002): Protestant postsecularism; French Catholic postsecularism; and Deleuzian postsecularism (Bengston 2015: 112–6). Bengston derives his typology from predictable differences among postsecular thinkers regarding the relationship between immanence and transcendence.

The Protestant postsecular

The Protestant postsecular involves a strict distinction between the transcendent and the immanent, a distinction itself constitutive of secular modernity. It is perhaps unsurprising, then, that the representative of this ideal type is the irreligious Habermas for whom any mediation of faith and reason which resists a strict boundary between the two compromises the latter's foothold (Habermas 2008: 242–3). Despite protestations to the contrary, Habermas propounds an account of religious and secular rationalities which subordinate the former to a semi-Hegelian *vorstellendes Denken* (Habermas 2010: 18; cf. Milbank 2002: 272). Milbank's reproach to this project is to point out that Habermasian secular reason does not proceed from the shared ontogeny of philosophy and religion. Instead, it grows out of theological developments chiefly associated with Ockham and Scotus, as Habermas himself inconsistently acknowledges (2010: 22). Consequently, the Habermasian call for the conversion of theological ideas into secular discursive moral arguments (as part of a postsecular integration of religion into public discourse) is bankrupt if secular discourse is always already involved in such an operation. If secularity:

> does not just borrow inherently inappropriate modes of expression from religion as the only discourse to hand (this is Hans Blumenberg's interpretation), but is actually constituted in its secularity by 'heresy' in relation to orthodox Christianity, or else a rejection of Christianity that is more 'neopagan' than simply anti-religious.
>
> *(Milbank 2006: 3; Chesterton 2009: 86–95)*

Thus, the postsecular debate for Milbank merely designates a new context for the outworking of the theological controversies of the late Middle Ages (and later those of the Reformation). In this case, the heresies which challenged the Christian orthodoxy of which he is a defender (albeit an eclectic interpreter), 'invented' and 'encouraged' secularization because their religious-theological 'unbelievability' encouraged agnostic and atheistic scepticism 'which eventually engenders nihilism as a kind of truncated theological *via moderna*' (Milbank 2013: 3).

The French Catholic postsecular

The French Catholic postsecular involves an asymmetric mediation of the immanent by the transcendent. Bengston aligns Milbank with Taylor here insofar as both, following French thinkers like Maurice Blondel and Henri de Lubac, deny the existence of a pure nature and stress the involvement of the supernatural in the natural. However, Milbank is critical of a tradition he would identify with Taylor and earlier with Troeltsch (1911)—which is convinced of both the fruitfulness of engaging with secular culture and of the ability of secular modernity to attain a self-understanding of its theological origins whilst still maintaining a critical distance (Harrington 2006: 44). Ultimately, though Milbank shares with thinkers like Taylor a 'Catholic' account of the relationship between transcendence and immanence, he is uniquely situated within Bengston's type insofar as he proposes a more radical thesis about the relationship between theological and secular social discourse and the way in which this relationship determines the role of religion in contemporary society. Milbank's claim is that, despite being correct with regard to a number of issues, the 'German' tradition of sociology which includes thinkers like Troeltsch and all those influenced by their work (and whose critique Milbank develops in *Theology and Social Theory*) is problematically attached to a project irreducibly defined by its attachment to liberal freedom and a positivist attitude to law and sovereignty (Milbank 2006: 75).

The Deleuzian postsecular

The Deleuzian postsecular is founded on an immanentist materialism which rejects the strict distinction between the transcendent and the immanent, but posits that this rejection entails not mediation but a collapsing of the transcendent into the immanent. The immediate reproach to this postsecular vision in Milbank's work is the identification of this immanentist philosophy with pagan religious reflection and the subsequent deployment of the Christian appropriation of pagan philosophy as the unlocking of the latter's truth as fulfilment of its typological anticipation[1]. Thus, Milbank's criticism of the Protestant postsecular is that it is committed to a post-Christian univocity in the wake of Scotistic developments in late medieval scholasticism. He also criticizes the Deleuzian postsecular that it is committed to a pre-Christian univocity of ancient materialism from Democritus to Spinoza in the wake of antiquity.

The end of dialogue

The Deleuzian postsecular is perhaps also characterized, such as in the case of Connolly, by a pluralist theology of religions. However, though Milbank has been explicitly critical of this enterprise (Milbank 1990: 174–92), there is no real distinction in his work between his theology of religions and his, as it were, theology of the secular. His former has variously been described as a form of exclusivism (Durward 2006: 39) or as going beyond the threefold typology (D'Costa 2005: 637). The 'charge' of exclusivism (and it is often a 'charge') is perhaps understandable, but inadequate. It is worth noting that the context of the claims cited by Durward is that of the relationship between theology and secular social discourse (which remains one of the *foci* of Milbank's work). For the repudiation of theological pluralism follows not directly from the refusal to grant other religious traditions salvific content, but from the necessity to affirm theology's autonomy from secular social discourse. One of the main theses of *Theology and Social Theory* is that when Christian theology becomes wedded to secular social theory, it subjects itself to a bifurcation of salvation to the extent that when the latter is proclaimed to be 'religious', it must also be 'formal, transcendental, and private', and when it is proclaimed to be 'social', it must also be secular (Milbank 2006: 250). 'What is occluded is the real practical and linguistic context for salvation, namely the particular society that is the Church' (*ibid.*: 250).

Milbank's alternative postsecular vision is that of a world where Christianity straightforwardly declares its vision through the linguistic and social practices of the church. This means that dialogue is denied as both a site of non-agonistic mediation and special epistemic access. A similar view is taken by Slavoj Žižek (2008: 49–62), Milbank's interlocutor, in a trilogy of books which mirror the postsecular debate between Habermas and Ratzinger. Žižek agrees with Milbank that the postsecular space opened up by postmodernity ultimately deprives religion of its content in order to affirm the continuity of its form—which Žižek associates with Proudhon's 'pseudo-Hegelian' move criticized by Marx in *The Poverty of Philosophy* (Žižek 2005). However, to the extent that Žižek may be described as a thinker interested in the perseverance of religion and in resisting secular modernity, it is as a peculiar inheritor of another 'left-Hegelian', Bruno Bauer.

Bauer had argued (and consequently also been reproached by Marx) that in order for Christians and Jews to live together they ought to abandon what separated them in their emancipatory endeavours, namely religion, which in the particular forms of Christianity and Judaism represent for Bauer merely different stages in the development of the human *Geist* (Marx 2000: 48) This is not dissimilar to the Habermasian view of confining religion to 'the history of the origins of the secular reason', though it distinguishes itself from the latter with its explicit recommendation that the religious abandon their religiosity. Žižek has renewed this view, arguing that:

> [o]ne cannot be religious in general. One can only believe in some god(s) to the detriment of others. The failure of all the efforts to unite religions proves that the only way to be religious in general is under the banner of the 'anonymous religion of atheism'.
> *(Žižek 2008: 112)*

However, Žižek does not go so far as to suggest that religious people ought to abandon their religiosity. In fact, he is opposed to the notion of a 'choice' between theism and atheism at all, because such a choice would already be 'within the field of belief'. A decidedly affirmed 'atheism' is thus a 'miserable, pathetic stance of those who long for God but cannot find him (or who "rebel against God")' (Žižek 2005: xxv). The truly atheistic subject instead finds the question of theism and atheist irrelevant.

Žižek's views reinforce Milbank's views about dialogue. Žižek provides a succinct formulation of the 'dialogue' as being 'like every true philosophical dialogue, [...] an interaction of two monologues' (Milbank and Žižek 2009: 235). This insight, and indeed Milbank's whole stance on dialogue, seems to have been missed by some commentators on the debate like Adam Kotsko, who writes that it lacked enough shared premises for 'genuinely productive dialogue' (Kotsko 2014: 197; see also Robbins 2016). However, there were clearly fewer disagreements between Milbank and Žižek than between Ratzinger and Habermas—and yet less 'dialogue'. Milbank's postsecular concerns chiefly involve *Christian* interecclesial dialogue. As he writes in *The Monstrosity of Christ*,

> at both the theoretical and the historical level the issue of Catholic versus Protestant is far more fundamental than the question of theism versus atheism—the latter is merely a subplot of the former conflict, which is today notably resurfacing.
> *(Milbank and Žižek 2009: 114)*

Varieties of postsecular thought may be thusly discerned, and Milbank's work particularly encourages such a construal insofar as the Protestant and Deleuzian postsecular visions may be taken, in his final analysis of contemporary philosophy, to be informed by shared ontological presuppositions influenced by Franciscan proto-modernity and connected to the invention of a 'pure nature' (Milbank and Žižek 2009: 113–4; Milbank 2013: 3, 28–9). It could be said that the differences between the Protestant and Deleuzian visions are also determined by their relation to Christianity to the extent that the former is humanist insofar as it is 'post-Christian' (and therefore upholds, following Kant, a universal communicative reason), and the latter anti-humanist insofar as it is 'pre-Christian' and is therefore informed much more by the non-cognitivist and anti-humanist work of thinkers like Foucault. The upshot of this is that the Protestant postsecular vision upholds a secularity congenial only to certain Protestant versions of Christianity, whereas the Deleuzian vision opens up a postsecular space for religion, as Žižek points out, but demands an essentially pre-Christian 'spirituality'.

As Grigory Gutner has pointed out, most postsecular discourse only distinguishes between the 'pre-secular', 'secular', and 'postsecular' phases (Gutner 2016: 39). However, the category of 'pre-secular' fails to make the distinction, so far crucial to this exposition of Milbank's postsecular vision, between the Christian and pagan worlds. For the former alone consisted of the uniquely 'integral' universe which secularity dissolves, and the postsecular, in Milbank's case, seeks to re-establish. As we have seen, 'Catholic' postsecular projects have been generalized in terms of this integral continuity of the natural and the supernatural, and the Second Vatican Council crucially embraced the 'integralist revolution'.

Milbank has made much of the moment in Christian theology after the Second Vatican Council following this development and has described his critical genealogical work as tracing the collapse of 'integral humanism' (Milbank 2013: 261). In *Theology and Social Theory*, he identifies two sources for this integralism: a French source, deriving philosophically from Maurice Blondel and theologically from the *nouvelle théologie*; and a German source, deriving philosophically from Martin Heidegger and theologically from Karl Rahner. The distinction between the two may be roughly summarized by saying that the French integralism 'supernaturalises the natural', while the German 'naturalises the supernatural' (Milbank 2006: 207). It is, in other words, a difference in the direction of mediation[2], and here we can notice the Milbankian influence on Bengston's typology which specifies the Catholic postsecular position as 'French'. In both cases, the claim that the whole of humanity is imbued with grace is taken to imply the inseparability of political and social concerns from 'spiritual' salvific concerns.

However:

> [t]he thrust of the [German] version is in the direction of a mediating theology, a universal humanism, a rapprochement with the Enlightenment and an autonomous secular order. While these themes are not entirely absent from the French version, its main tendencies are in entirely different directions: for the nouvelle theologie, towards a recovery of a pre-modern sense of the Christianized person as the fully real person; for Blondel, towards a similar reinstatement, but in terms which stress action, not contemplation, as the mode of ingress for the concrete, supernatural life.
>
> *(Milbank 2006: 207)*

This impetus to universalizing humanism (as well as the appeasing of the Enlightenment) is a familiar trait of what I have been referring to as a Protestant postsecular vision. However, because Milbank's 'historical target' is ultimately not Protestantism but late scholasticism (Milbank 2005: 26), he assimilates the problematic tendencies of the latter to a host of thinkers and traditions within the Catholic churches[3] including Charles Taylor. Taylor is someone who has recognized the influence of *nouvelle théologie* thinkers like Henri de Lubac but has perhaps been primarily influenced by the thought of Heidegger and the American pragmatists. Hence, Harrington is right to note the disparity between Milbank and Taylor, particularly on the issue of the relationship between theological and secular rationalities (his clear preference for the latter notwithstanding) (Harrington 2006: 44). Furthermore, what he identifies as a distinction between 'sociological theology'—namely, 'sociologically self-conscious theology'—and 'theological sociology'—namely, 'unconsciously or consciously theological social theory' (Harrington 2006: 44) captures the distinctions between the German and French integralisms. Unsurprisingly, Harrington's key representatives for 'sociological theology' are the liberation theologians (Milbank's key representatives) and the representative for the latter is Löwith.

Due to the castigation of liberation theologies and other discourses deriving from this German source which Milbank's position entails, it is worth quoting his qualification to this critique at length:

> Not without distress do I realize that some of my conclusions here coincide with those of reactionaries in the Vatican. But in no sense is it left-wing politics to which I wish to object; on the contrary, my fear is that, as the Marxist belief in the inevitability of

> socialism, or else a socialism arising merely from the lifting of restrictions on human freedom declines, so also does socialism itself atrophy. It should, in fact, be peculiarly the responsibility of Christian socialists at present to demonstrate how socialism is grounded in Christianity, because it is impossible for anyone to accept any longer that socialism is simply the inevitable creed of all sane, rational human beings. But this is not the main direction that has been pursued by the proponents of political and liberation theology; on the contrary, theirs has been simply another effort to reinterpret Christianity in terms of a dominant secular discourse of our day.
>
> *(Milbank 2006: 208)*

Another recent attempt to express a similar distinction in the postsecular literature comes from Pawel Rojek who explores the difference between 'philosophical theology' and 'theological philosophy' (Rojek 2016: 97–108). Drawing on the typology in John Paul II's encyclical *Fides et Ratio*, Rojek distils with great clarity the relationship between theological and secular rationalities. There are theological projects which call upon philosophy and use its concepts to analyze divine revelation. Such 'philosophical theology' *tout court* predates secularity (such as in the case of Aquinas)[4] though it has become a programme, in the wake of the Scotistic elaborations of Aristotelianism, closely associated with the secularization of religious discourse. Conversely, there are what may be primarily called 'philosophical' projects which involve speculation 'in dynamic union with faith' (Rojek 2016: 100).[5]

The paradigmatic theological or Christian philosopher is undoubtedly Blondel. What is key for the distinction, as Rojek points out, is that Blondel remains firmly in the language of theology (Rojek 2016: 102). The same may be said for Harrington's typology and may be generalized for Milbank's distinction whose contention is that political theologies founded on the Rahnerian integralism remain trapped within 'secular reason' (Milbank 2006: 207). This is in part because such 'Rahnerian' political theologies allow for the espousal of secular social theories and philosophies whose autonomies they respect. And yet, as Alan Thomson has observed, Roman Catholic thought has moved considerably ahead of Protestant engagements with culture and the secular world (Thomson 2014: 46). The transcendentalism of the Rahnerian theory of grace allows for the treatment of socio-political concerns as essentially extra-ecclesial matters because it takes the context for salvation as an 'unthematised' apprehension of the divine, rather than the particular society of the church.

Blue Labour: aberrant Leftism

The concrete proposals of Milbank's postsecular project and his political programme more generally are self-admittedly paradoxical and eclectic. Indeed, Milbank's most succinct definition for 'Blue Labour', the political faction with which he has been most associated with and has contributed most to, is the paradoxical formulation 'One Nation Labour' (Milbank 2015: 27). Blue Labour has moreover been associated with 'post-liberal' politics in the UK, and it is worth noting that for Milbank, who links the hegemony of secular reason with the hegemony in modernity of liberalism, we can only overcome the former by also overcoming the latter. Hence, any postsecular vision entails for Milbank a postliberal vision. Furthermore, and in accordance with the neo-medieval stress of his theology, Milbank has been part of a revival in interest in guild socialists like G. D. H. Cole. The account of the early economic roots of modernity in Milbank and Pabst's *The Politics of Virtue* traces the undermining and eventual abolition of guilds, often in the context of the historical victories of liberalism's centralization of state of power and the Reformation's and Janesenism's transformations of the European religious horizons (Milbank and Pabst 2016: 42–3).

Though Blue Labour is a relatively recent development in UK politics (having been launched in 2009), postliberalism is present in some of Milbank's earliest work. He has been critical of the concession in some pluralist circles that commitment to interfaith dialogue necessitates commitment to liberal values (Milbank 1990: 175). To the extent that such liberal values are 'embedded in a wider Western discourse become globally dominant', Milbank argues, they usher a narcissistic 'xenophilia' through which other cultures are obliterated by Western norms (Milbank 1990: 175). Milbank's contention, which he has maintained throughout the years and brought to Blue Labour, is that the 'good causes of socialism, feminism, anti-racism, and ecologism' are confined not only by various 'pluralisms' but more crucially by secular and liberal positions. However, these radical impulses cannot come into their own simply by being relocated beyond secularism and liberalism (i.e., 'socialism arising merely from the lifting of restrictions on human freedom'), but must be introduced into the context of the Western religious traditions (Milbank 1990: 175). Thus, the choice is not 'socialism or barbarism' (Rosa Luxemburg), but 'Christianity or barbarism'.[6]

More recently, Milbank has expressed interest in the work of the Cambridge historian Robert Tombs, particularly in relation to the historical thesis that post-Reformation controversies remain unresolved, albeit transformed in different terms (Milbank and Pabst 2016, 2017). This thesis has particular application in the UK where these controversies relate essentially to the context of the 'Glorious Revolution', a period whose historical importance for England's self-understanding Milbank has repeatedly emphasized (Milbank and Pabst 2017: 133–5). Such a view furthermore runs parallel to the thesis that all of modernity's most defining features are a product of theological debate in the late Middle Ages, that apparently secular understandings—such as those of contemporary party politics—continue to be coloured by a presecular past whose presentness is increasingly unavoidable. These old divisions, between Anglicans and dissenters, and even Cavaliers and Roundheads, have become particularly pertinent over issues like national sovereignty and membership of the European Union where Milbank has been a staunch critic of the nation-state, arguing that it stands in opposition to Christianity on the basis of its emergence out of the failure and collapse of Christendom[7] (Milbank 1990; Milbank and Pabst 2017). He has called for a neo-medieval dispersal of sovereignties which overlap and mutually qualify each other, and has stressed, in connection, the neo-medieval origins of the European Union (Milbank and Pabst 2016, 2017).

Notwithstanding the support for a united Europe, and for such bodies as Royal Commissions and constitutional conventions in centralized countries like the UK, France, and Spain, Milbank has highlighted, following Tombs, the unique history of religion in England:

> Britain has always had a religious division between two groups and a semi-official sanctioning of both. In nearly all other European countries the main division has turned out to be between religion and secularity, with the latter supposedly taking the form of the left and the former that of the right. One could argue that in the case of Europe, it's the divisions over religion that are really primary. But in Britain the divisions which take priority are internal to religion which is why we've never had secular and religious parties or anything like that. Our left has also tended to define itself in religious terms in so far as it is in some sense the heir of the Puritans and the non-conformists.
>
> *(Milbank and Pabst 2017: 133–4)*

This uniqueness of the British example is no doubt what partly explains the irreducibly 'British' character of both Milbank's work and the specifically postsecular character of Blue Labour's socialism.

Such a socialism is firstly post-liberal insofar as it resists the tendencies of modern liberalism from both of its sources; the Hobbesian variety (derived also from Grotius in the seventeenth century) which sees all humans as fearful and selfish, and the Rousseauean variety, which inverts the Hobbesian position and argues that while the individual (and isolated) human is 'good', association and corporate bodies corrupt persons who become egotistic due to rivalry (Milbank 2015). These positions roughly designate in the UK the conservative liberals of the Tory party and the progressive liberals of the postwar Labour Party. It is secondly also postsecular as these views are broadly secular and materialistic. The Blue Labour position is to move beyond these views by 'combining greater economic justice with individual virtue and public honour' (Milbank 2015: 27). Furthermore, it calls for a new recognition of the role of tradition as the contract between generations and subsequently for the 'non-democratic' defence of democracy through a mixed constitution (Milbank and Pabst 2016: 239–40):[8]

> It follows that one can only save democracy through paradox: by trying to restore, albeit in a more democratic way (which Christianity has always innately demanded), a genuine ancient constitution, pivoted about shared *telos* and virtue. The state must, indeed, become once again a true (and not ideological) *universitas*, or community of common purpose, because any mere Oakeshottian *societas*, or bond of only formal agreement, now manifestly secures a community confined to the economic participation and banned from real political participation.
>
> *(Milbank and Pabst 2016: 183)*

Despite affirmations of the postliberal and postmodern character of his work, there remains an open question for Milbank about the possibly 'post-capitalist' character of his proposed civil economy. Such an economy, particularly in light of Milbank's criticisms of Marx, tends to imply merely postliberal capitalist economics. However, for the former, a postliberal alternative itself entails an abandonment of Marx whose work is excoriated as liberal in *Theology and Social Theory* (Milbank 2006). The plausibility of such a critique notwithstanding, Milbank is right to insist in his more recent work that Marx's understanding of capitalism as an historically inevitable product of the contradictions of Feudalism made his account of primitive accumulation insufficient—even as he correctly realized that colonization and globalization would inevitably follow in response to falling profit rates in capitalism, leading to 'ever-further "primitive accumulation"' (Milbank and Pabst 2016: 99). Milbank's history of capitalism consequently relies much more on the works of Karl Polanyi and the dissident semi-Marxist Robert Brenner. Moreover, insofar as Milbank is right to resist Marx's insistence on the historical inevitability of secularization, he is perhaps right to affirm not merely the contingency of the emergence of capitalism, but the counter-factual thesis that capitalism could have developed under different historical circumstances and in continuity with the essentially precapitalist conjunction of contract and gift, which those endorsing a civil economy today condone. Thus, to the extent that Milbank's socialism is postcapitalist is so because it draws on precapitalist sources through which an alternative capitalism is envisioned.

Milbank's paradoxical political disposition often derives from a staunch rejection of 'left' and 'right' politics as equally wedded to the problematic tendencies of modernity. Hence, the appeal to the premodern (as invoked by socialists like Cobbett, Carlyle, Ruskin, and

Hugo (Milbank and Pabst 2016: 182)[9] is neither traditionalist nor reactionary, but part of the attempt to traverse the modern and secular framework of the agonistic shuttle between conservatives and progressives and thus of the specifically modern emphasis on favouring either the one or the many (Milbank 2013: 261). Milbank argues that the emergence of the Left-Right distinction (which supervenes on the many-one oscillation) 'is directly the result of the nominalist-voluntarist theological revolution in the late Middle Ages' because if the ontology of universals and real relations is denuded, then only the many individuals or the one central power is real (Milbank 2013: 262). Socialism, defined as the theorization of society as the work of human personhood made manifest as free labour, is 'the joker in the pack of left-wing options' and the only way out of this deadlock—though it is only 'left-wing' in the specified 'aberrant sense' of eschewing the sources which traditionally overdetermine Left-wing politics in opposition to the Right (Milbank 2013: 262).

Conclusion

Josef Bengtson has argued convincingly that Milbank should be seen as part of a new tendency in postsecular thought which attempts to re-establish the link between metaphysics and political ethics, following the postmetaphysical emphasis given to it by Habermas (Bengston 2002: 155). While this is undoubtedly true, I have tried to show some of the serious ways in which Milbank differs from other thinkers of this ilk (like Taylor). Despite a mixed reception in the literature (see Hedges 2014), Milbank's project remains a serious and increasingly attractive alternative to the mainstream postsecular discourse because it couples the overcoming of secularism with the overcoming of modern liberalism and critically dismantles those positions whose often unstated reliance on the latter compromises their viability in world where the limitations of modern liberalism are increasingly evident. His work is furthermore marked by an urgent need for postsecular answers to ageing secular questions to which he responds in an unfailingly eclectic and illuminating way.

Notes

1 The key figure in the early Christian appropriation of Hellenic philosophy is Justin Martyr (see *Beyond Secular Order*, p. 22); however, Milbank thinks this dimension in terms of the trans-naturality of natural reason as always already constituted by supperaddition of grace (*ibid.*: 213).
2 On the issue of the directionality of the mediation of the transcendent and the immanent, see Desmond (2011: 118–20).
3 Milbank subscribes to a branch theory which upholds the Catholicity of the Roman Catholic, Eastern Orthodox, and Anglican churches; see *Beyond Secular Order*, pp. 198–206.
4 Rojek takes the Thomistic synthesis as paradigmatic of philosophical theology; however, Milbank's reading of Aquinas on this matter is reliably idiosyncratic; see Milbank and Pickstock (2001: 1–52). Rojek, however, concedes that 'the clearest example of a [theologically philosophical] theory might be the Thomistic doctrine of *esse*', pp. 105–6.
5 Rojek is here quoting John Paul II.
6 Milbank's most characteristic argumentative strategy is arguably a kind of *reductio ad nihilism* in which a realist and Augustinian Christianity is offered as the only alternative to nihilism; see, for example, *Beyond Secular Order*, p. 89.
7 Curiously but perhaps unsurprisingly, and in accordance with Milbank's syncretism, Robert Tombs is a staunch Brexiteer.
8 This involves the mediation of the few, 'both in the expanded Christian sense of mediating free associations and in the antique Greek and Roman sense of the guidance of the wise'; Milbank, J., *Beyond Secular Order*, p. 263.
9 Milbank and Pabst argue that Marx's insights apply to capitalism, but that it is *only* capitalism that produces necessary class antagonism.

Further reading

Betz, J. R. (2009) *After Enlightenment: the post-secular vision of J.G. Hamann*, Oxford: Wiley-Blackwell.

Betz's text is a compelling example of a Christian postsecular vision retrieved from one of Radical Orthodoxy's main historical sources, the German proto-Romantic thinker J.G. Hamann.

Milbank, J. and S. Oliver (ed.) (2009) *The Radical Orthodoxy Reader*, London: Routledge.

This reader contains a collection of essays, old and new, from figures associated with Radical Orthodoxy and is often preferable to the now classic 1998 compilation *Radical Orthodoxy: a new theology*. Of particular interest may be Catherine Pickstock's 'Theology and the secular'.

Pabst, A. and C. Schneider (ed.) (2009) *Encounter between Eastern Orthodoxy and Radical Orthodoxy*, Farnham: Ashgate.

An engaging collection debates between thinkers associated with Radical Orthodoxy and Eastern Orthodox theologians. Of particular interest may be Michael Northcott and Adrian Pabst's papers on politics and ecclesiology.

Smith, A. P. and S. Whistler (ed.) (2010) *After the Postsecular and the Postmodern*, Newcastle upon Tyne: Cambridge Scholars Press.

A contemporary collection of works by writers doing continental philosophy of religion in a specifically postsecular and postmodern mode.

Smith, J. K. A. (2004) *Introducing Radical Orthodoxy: mapping a postsecular theology*, Grand Rapids, MI: Baker Academic.

A useful and sometimes critical engagement with many aspects of Radical Orthodoxy and John Milbank's thought which is always cognisant of the latter's specific relevance to the postsecular context. The engagement with Caputo's views on the postsecular is especially worthwhile.

Smith, J. K. A. and J. H Olthius (ed.) (2005) *Radical Orthodoxy and the Reformed Tradition*, Grand Rapids, MI: Ashgate.

A collection of essays from both Radically Orthodox and Reformed thinkers engaging with the issue of the Reformed tradition's place in Radical Orthodoxy's critique of secularism. John Milbank's 'Alternative Protestantism' clarifies many false but oft-held beliefs.

References

Alliez, E. (1996) *Capital Times*, trans. G. van dem Abbeele, Minneapolis, MN: Minnesota University Press.

Bengston, J. (2015) *Explorations in Post-Secular Metaphysics*, New York, NY: Palgrave Maximillian.

Certeau, M. (1992) *The Mystic Fable*, trans. M. B. Smith, Chicago, IL: Chicago University Press.

Chesterton, G. K. (2009) *Heretics in The Wit, Whimsy, and Wisdom of G.K. Chesterton, Vol. 4*, Landisville, PA: Coachwhip Publications.

Connolly, W. (2000) *Why I am not a Secularist*, Minneapolis, MN: University of Minnesota Press.

——— (2002) *Identify/Difference: democratic negotiations of political paradox* Minneapolis, MN: University of Minnesota Press.

D'Costa, G. (2005) 'Theology of Religions'. In Ford, D. and R. Muers (eds.) *The Modern Theologies*, 3rd edition, Oxford: Blackwell.

Deleuze, G. and F. Guattari (1994) *What Is Philosophy?* trans. H. Tomlinson and G. Burchell, New York, NY: Columbia University Press.

Desmond, W. (2011) 'Between finitude and infinity'. In Žižek, S., Crockett, C. and C. Davis (eds.) *Hegel and the Infinite: religion, politics, and dialectic*, New York, NY: Columbia University Press, pp. 118–20.

Dosdad, Á. I. (2016) 'From the secular to the Habermasian post-secular and the forgotten dimension of time in rethinking religion and politics', *Contexto International*, 38(3): 887–98.

Durward, R. (2006) 'Christian claims of uniqueness, the Problem of Violence and interfaith dialogue'. In Marsden, L. (ed.) *The Ashgate Research Companion to Religion and Conflict Resolution*, London: Routledge.

Gutner, G. (2016) 'Post-secularity vs all-unity'. In Mrowczynski-Van Allen, A., Obolevitch, T. and P. Rojek (eds.) *Beyond Modernity: Russian religious philosophy and post-secularism*, Eugene, OR: Wipf and Stock, pp. 39–47.

Habermas, J. (2008) *Between Naturalism and Religion: philosophical essays*, Cambridge, MA: Polity Press.

——— (2010) *An Awareness of What Is Missing,* Cambridge, MA: Polity Press.
Harrington, A. (2006) 'Social theory and theology'. In Delanty, G. (ed.) *Handbook of Contemporary European Social Theory,* London: Routledge.
Hedges, P. (2014) 'The rhetoric and reception of John Milbank's radical orthodoxy: privileging prejudice in theology?', *Open Theology,* 1(1): 24–44.
Kotsko, A. (2014) 'John Milbank'. In Butler, R. (ed.) *The Žižek Dictionary,* London: Routledge.
Marx, K., (2000) 'On the Jewish question'. In McLellan, D. (ed.) *Karl Marx: selected writings,* Oxford: Oxford University Press.
Milbank, J. (1990) 'The end of dialogue'. In D'Costa, G. (ed.) *Christian Uniqueness Reconsidered: the myth of a pluralistic theology of religions,* Maryknoll, NY: Orbis Books.
——— (2002) 'The last of the last: theology, authority and democracy', *Revista Portuegesa de Filosofia,* 58(2): 271–98.
——— (2005) 'Alternative Protestantism'. In Smith, J. K. A. and J. H. Olthuis (eds.) *Radical Orthodoxy and the Reformed Tradition,* Grand Rapids, MI: Baker Academic.
——— (2006) *Theology and Social Theory: beyond secular reason,* Oxford: Wiley Blackwell.
——— (2013) *Beyond Secular Order: the representation of being and the representation of the people,* Oxford: Wiley-Blackwell.
——— (2015) 'The Blue Labour dream'. In Geary, I. and A. Pabst (eds.) *Blue Labour: forging a new politics,* London: I.B Tauris.
Milbank, J. and A. Pabst (2016) *The Politics of Virtue: post-liberalism and the human future,* London: Rowman & Littlefield.
——— (2017) 'Theology and international relations beyond liberalism: the question of Europe', *Radical Orthodoxy Journal,* 4(1): 8–28.
Milbank, J. and C. Pickstock (2001) *Truth in Aquinas,* London: Routledge.
Milbank, J. and S. Žižek (2009) *The Monstrosity of Christ,* edited by C. Davis, Cambridge, MA: MIT University Press.
Robbins, J. W. (2016) *Radical Theology: a vision for change,* Indianapolis: Indiana University Press.
Rojek, P. (2016) 'Post-secular metaphysics: Georges Florovsky's project of theological philosophy'. In Mrowczynski-Van Allen, A., Obolevitch, T. and P. Rojek (eds.) *Beyond Modernity: Russian religious philosophy and post-secularism,* Eugene, OR: Wipf and Stock.
Thomson, A. (2014) *Culture in a Post-Secular Context: theological possibilities in Milbank, Barth, and Bediako,* Cambridge, MA: James Clare & Co.
Troeltsch, E. (1911) *The Social Teaching of the Christian Churches vol. 1,* trans. O. Wydon, London: Macmillan.
Wellmer, A. (1991) *The Persistence of Modernity: essays on aesthetics, ethics, and postmodernism,* trans. D. Midgley, Cambridge, MA: Polity Press.
Žižek, S. (2005) *Interrogating the Real,* London: Bloomsbury.
——— (2008) *Violence,* London: Profile Books.

13
Postsecular theology[1]

Hagar Lahav

Introduction

The postsecular 'breakdown of the [secular] philosophical prohibition of religion' (Vattimo 1998: 91) has been conducive to the emergence of exciting hybrid combinations of theology and secular humanities that reflect (and engender) renewed interest in spiritual matters. At one time, (post)secular theology would have been interpreted as a contradiction in terms. At present, however, it appears that even theology—which may well be considered the heart of religion, especially in Christianity—could not escape the secular shift, nor could the secular escape theology.

The preface *post* in postsecularism is not taken here in its temporal meaning (as if the postsecular presents the time after the secular), but rather as a critical awareness of the hybrid relationships between the secular and the religious. Postsecular theology is understood as a set of theologies that demonstrates such awareness. I thus address twentieth-century theologians and philosophers, some of whom wrote long before the booming of the postsecular paradigm in the last decades of the previous millennium. I pay special attention to post-Christian and Jewish theologies, aiming to expand the predominant Christian discourse concerning theology and the postsecular.

The chapter thus offers a new conceptualization of the relations between different postsecular theologies from a unique, semi-Western, non-Christian point of view. It concentrates on these theologies' reactions to the secular critique on the traditional monotheistic God, the metaphors they use to denote the divine, and key categories within them, such as transcendence vs. immanence or revelation. The chapter proposes that postsecular theologies are united by their rejection of the atheistic impulse of (Christian-oriented) secularism but are deeply divided concerning the ways to 're-enchant the world', going backwards, forward, and sideways to and from the traditional monotheistic understanding of God.

Theology within *the secular*

Under Weber's (1905/2012) understanding of secularism as 'the disenchantment of the world', the secular was mainly understood as a sphere dominated by naturalism and science that rejects 'supernatural' explanations of reality. It demands what Berger (1969: 100) calls 'methodological atheism'—the assumption that God, miracle, and sanctity cannot serve as explanations for phenomena in reality. It appears, then, that in a deep sense, 'secularism' in its purer form must include some sort of atheistic perception.

The project of unravelling the seemingly Gordian knot between secularism and atheism stands at the heart of the works of many of the postsecular paradigm's founding fathers (e.g., Milbank 1990; Bellah 1970/1991; Bellah et al. 1992; Asad 1993, 2003; Schmitt 1929/1996; Benjamin 1940/2005). All these commentators try, each in their own way and borrowing Walter Benjamin's (1940/2005) metaphor, to show how theology has always pulled the strings of secular history. The secular, they claim, is not atheological because (Christian) theological concepts are unavoidably intertwined within (Western) secular sociology and philosophy. As opposed to atheistic secularism, which strives to 'clear' culture from theological 'residue', postsecularism celebrates such hybridity and claims that the secular may benefit from theology.

From a theological point of view, a preliminary critical question here is who or what the rejected God of atheism is. Berger (1969: 25) claims that methodological atheism demands the denial of 'the sacred', which he defines as 'a quality of mysterious and awesome power, other than man [sic] and yet related to him [sic], which is believed to reside in certain objects of experience'. A closer look at writings of atheist flag-bearers, however, reveals that even they would not necessarily reject such an amorphous conception of sanctity. Dawkins (2006: 33), for example, declares that his work carries 'transcendent wonder'. He insists, however, that this wonder exists not only in traditional religion but also in what he calls *Einsteinian religion*, whose God differs significantly from the *supernatural religion*'s theistic understanding.

For Dawkins, then, atheism means rejection of the monotheistic understanding of God, which usually characterizes the divine as one transcendent Supreme Being without a body, who is eternal, infinite, omnipotent, all-knowing, the creator of all things, that should be worshiped and obeyed (Anderson 1998). Hereinafter I use the term *God* with a capital *G* to refer to this monotheistic, hegemonic understanding of the divine, but contrast this concept from *god* with a small *g*, or *the more*, which can be defined as 'whatever may be considered as the divine' (James 1902/2002: 38). *God* with a capital G is thus only one (very dominant) suggestion concerning the understanding of *god*. Hence, theology is conceptualized here as meditation on the reciprocal relations among *the more*, humankind, and the universe.

The realization that secular theology points its accusations to the monotheistic God (and thus preserves—directly or by an implication—a space for other understandings of god) is also pronounced in *Death of God theology* (e.g., Vahanian 1961; Altizer and Hamilton 1966; Caputo and Vattimo 2009). Here the dead god is a 'God to whom adoration, praise and trust were appropriate, possible, and even necessary.... God as problem solver, absolute power, necessary being, the object of ultimate concern' (Altizer and Hamilton 1966: 3). Post-Holocaust theologians (e.g., Jonas 1987; Rubenstein, 1992) articulated a Jewish version of this theology, according to which God died in Auschwitz. Jonas (1987) writes:

> "Auschwitz" calls, even for the believer, the whole traditional concept of God into question... Accordingly, one who will not thereupon just give up the concept of God [god] altogether – and even the philosopher has a right to such an unwillingness – must rethink it so that it still remains thinkable.... The Lord of History [God], we suspect, will have to go by the board in this quest.
>
> *(p. 3)*

The secular, hence, cannot be understood as lacking any references to god. Indeed, secular theology may be described as a two-step move: First, it adopts the institutionalized monotheistic understanding of god as God. Then it rejects the claim of this God's existence. Thus, to avoid desacralization, one can either preserve God, by dismantling the secular claims against Him, or introduce a different conceptualization of god. I will start with the first option, and then proceed to the second.

Postsecular theology as returning to God

In anglophone theology, the most prominent articulation of postsecular theologies that return to God is Radical Orthodoxy (hereinafter RO), which appeared in Britain in the 1990s. Crystallized in a series of books by Milbank (1990), Pickstock (1998), and Ward (1995) (see also Milbank et al. 1999; Oliver 2007; Oliver and Milbank 2009), RO uses a postmodern critique on secular modernity to claim that secular theology is a bad 'pseudo-theology'. Such theology replaces the Christian 'ontological basis' of peace between differences (as manifested in the notion of the Trinity) with an understanding that violence and conflict govern the world. This change in the perception of the world is a product of the dichotomist impulse of modernity that cannot contain differences harmonically and peacefully, interpreting 'difference' as a source of violent conflict. In contrast, Christian theology sees creation as a 'gift of grace' and God as a transcendent, self-subsisting being, who constantly and abundantly fills the world with grace (Oliver 2009).

The problem, then, is not with the transcendent Being known as God, but with the secular understanding of God. Instead of seeing God as the giver of the gift of creation and grace, as all loving and all containing Being, God is understood as an agent of power, control, and violence. Modernity substituted 'infinite interpersonal harmonious order' with 'impersonal chaos' (Milbank et al. 1999: 2). Thus, RO turns back to premodern Christian theology. Leaning on Neoplatonism and Middle-Ages Christian theologians such as Augustine and Thomas Aquinas, it seeks to revive traditional doctrine in which, to put it simply, God is a living good, and not an (maybe dead, maybe invented) evil.

As RO is quite a familiar approach and is discussed elsewhere in this *Handbook* (see Chapters 11 and 12), I will not analyze it here nor enter into the details of its intra-Christian quarrel with other theologies, such as liberation, reform, feminist, ecological, process, and even new-orthodoxy (e.g., Smith and Olthuis 2005; Radford Ruether and Grau 2006; Shakespeare 2007; Isherwood and Zlomislic 2012). Beyond the details of these disagreements, however, stands the claim that even if we adopt the (questionable) description of pre-modern Christianity as a worldview of peace and grace—and of (Western) secularism as worldview of conflict and violence—secularism cannot be the only source of violence. Such violence often took place even before modernity in the name of this supposedly peaceful God. From this perspective, (secular) modernity's 'obsession with power' exposes the simple truth that power was indeed involved in the implementation of 'participation in creation'. Can we, then, return to a peaceful and holistic religious world, without paying deep attention to what we learned along the way: that the danger of mistaking harmony with oppression always lieth at the door?

Postsecular theology as going beyond God

RO would interpret the question posted earlier as a preservation of the same conflictual modern worldview and its 'pseudo-theology'. Other theologies, however, reject the understanding of secularism as a mere 'fall', suggesting that Enlightenment humanistic ideas also have a positive (although not complete or sufficient) contribution to the quality of life on this planet. Instead, they adopt the secular critique of God and try to go beyond it, by suggesting new conceptualizations of *the more*. Here I will concentrate on two outer-Christian theological streams: post-Christian feminist and post-Halachic Jewish theologies.

Neither of these theological branches can be considered postmodern in the same sense that RO is (indeed, they may be characterized as highly modern in the sense that they consider power as a prominent factor), but this does not make them secular, as the notion of god and the assumption of its existence stand at their centre. For such theologies, the 'secular murder' of the monotheistic God does not mean the loss of sanctity, spirituality, or the grace of god. On the contrary, it is good news, for it opens the door to a postsecular step—presenting new understandings of god(s) and 'participation in creation' that are not cauterized by institutionalized religion and at the same time resist spiritual void.

Post-Christian feminist theology

With its intellectual mother—modernity—the feminist movement was quick to describe monotheistic religions as oppressive and excluding systems, in which women's oppression is a part of the official ideology and praxis. Thus, many feminist theologians understand God as a hierarchal and dichotomous patriarchal construction, see transcendence as a fault, and/or emphasize His androcentric image (e.g., Daly 1973; McFague 1982; Christ and Plasko 1992b; Isherwood and McEwan 1993; Radford Ruether 1993; Johnson 1994; Anderson 1998). Such a harsh critique raises three main reactions: departing from all notions of god and desacralization of the world ('a secular response'), preserving God while trying to change religion from within ('a religious response'), or looking for a new conceptualization of *the more*, while asserting that traditional religion and its God are infected beyond repair with patriarchy ('a postsecular response'). While ignoring the 'secular response', Christ and Plasko (1992a: 9–11) also distinguish between 'reformist' (which offers 'religious responses') and 'revolutionary' (which offers what I call 'postsecular responses') attitudes in feminist theology. To present the postsecular response in feminist theology, I use the works of post-Christian feminist theologians Mary Daly (1928–2010) and Daphne Hampson (born 1944) (see Lahav 2016a).

Mary Daly was born in the USA, was raised Roman Catholic, and left the church in 1972. Perceiving theology as a metaphor that enforces social structures, she is concerned with transforming God-language in a way that will 'encourage human becoming toward psychological and social fulfilment' (Daly 1973: 21). In her most prominent book, *Beyond God the Father* (1973), she claims that 'To exist humanly is to name the self, the world, and God' (p. 8). She insists that the power of Naming was stolen from women throughout history, leaving them in a state of non-being. Re-Naming god is thus 'women's active process of self and social construction [that] relocates women on the map of being, from objects to living, world-making subjects' (Hoagland and Frye 2000: 3).

For Daly, god is no longer God, a noun, an object of masculine projections, but rather an intransitive *Verb—Be-ing*. 'Be-ing is the verb that says the dimensions of depth in all verbs... that are always there when one is really living' (Daly 1978: 40). Furthermore, 'Be-ing, the Verb, cannot without gross falsification be reified into a noun, whether that noun be identified as "Super Being," or "God," or "Goddess" (singular or plural)' (Daly 1984: 26). The process of Be-ing, of Verb-ing, of Naming, is a process of transcendence. It is not 'transcendent in the old sense of "outside" or "beyond" the world of our lives, but transcendent in a new sense-a way of be-ing that we participate in and that is constantly transformative'

(Hoagland and Frye 2000: 3). Daly thus offers a complex, innovative, and consciously political feminist theology that aims at empowering women 'by the help of god', wherein god is no longer understood as a (bad or good) agent, but as a verb, a continuous, active process of/in human life.

Almost 30 years after Daly's departure from the Church, Anglican British feminist theologian Daphne Hampson announced her shift from a Christian to a post-Christian position. In her most prominent book, *After Christianity* (2002), she suggests that Christianity, and indeed all Abrahamic religions, is untrue and unethical. Like Daly, Hampson offers an anthropocentric theology, considering people, rather than god/God, to be her starting point. Unlike Daly, however, Hampson does not focus on women but rather on human beings who possess an awareness of the sacred.

For Hampson (2009), who uses the term God with capital G to denote both the monotheistic God that she rejects and her own understanding of *the more,* God is no longer an agent, but a dimension of reality beyond that which meets the eye, an 'ultimate sense of things' (p. 172), a 'dimension of the totality, everywhere and at all times available' (p. 182). Regardless of how little we may understand this dimension, awareness of it denotes a fuller sense of reality and connectedness. It is crucial because 'without... that ultimate sense of things that humanity has variously called "God"... we cannot hope to be free' (p. 172). When we are aware of god, we become 'integrated but, also and therefore, profoundly open' (p. 175) to 'that which moves between people, to which we can be present, as to that with which we are profoundly interconnected.... There is a process of osmosis between self and that which lies immediately beyond self, which is God' (p. 195). Hampson determines that this awareness exists before and beyond language, although 'how we speak of God is fundamental to what "God" is to us' (p. 173).

According to Hampson, because of their awareness of that which we call God, human beings have always tried to find ways to refer to this dimension by adopting different 'vehicles' (sets of myths, concepts, and manners of beliefs) that are by no means as neutral as they appear. Christianity (and indeed all Abrahamic religions) is a particular vehicle that must be rejected on both ethical and epistemological grounds. Ethically, Christianity is to be rejected because of the *scandal of patriarchy*, as it was 'that ideology that has served to keep woman in her place: "man's world, woman's place"' (p. 172). Epistemologically, Christianity is based on revelation, the claim of a particular, unique break in the causal nexus through which God's knowledge was revealed to humanity. According to Hampson (2002: p. XIII), taking the Enlightenment and modern scientific knowledge into account, claims of an 'interruption of history' that is the 'locus of knowledge as to what is the case' are no longer valid. Furthermore, in view of contemporary thinking about space and time, it makes little sense to speak of God as 'in some way "outside" or "before"' (Hampson 2009: 173).

Thus, the mission, as Hampson (2009: 171, 173) sees it, is 'to shift one's conception of God' and to 'find ways of speaking that honor the integrity of all of us'. Indeed,

> if all theological language is metaphorical (there may be wisdom in the theological adage that language and concepts will always be inadequate for God), let us at least exercise judgment as to what might be true metaphor and what is palpably false.
>
> *(p. 173)*

Both Hampson and Daly thus offer viewpoints imbued with spiritual feelings but reject what they see as the theocentric, anthropomorphic, androcentric, and chauvinist theology of traditional Christianity. Hampson's and Daly's self-definition as post-Christians denotes, however, not only their leaving traditional religion but also their awareness of their Christian origin. Christianity remains the primary source of reference against which they construct their theologies. Daly cites, for example, Aquinas and Tillich, while Hampson develops her theology in dialogue with Schleiermacher. Their work carries footprints of other liberal Christian theologies, such as those with liberation, reform, process, and ecologic impulses.

Post-Halachic Jewish theology

Bearing Hampson's and Daly's Christian orientation in mind, I turn backward in time and eastward in space to two prominent post-Halachic Jewish thinkers—Aharon David Gordon (Ukraine, 1856-Palestine, 1922) and Martin Buber (Austria, 1878-Israel, 1965). Buber was internationally recognized, and his writings were translated into dozens of languages, while Gordon, who influenced Buber considerably, is familiar only within Israeli cultural space (Shapira 1990).

To understand their positions, one should bear in mind that at the core of the Jewish religion stands the *Halacha*—a set of traditional binding laws and practices. Post-Halachic theology refers to the works of a few Jewish philosophers who reject the observance of Halacha and its conceptual God, while insisting that the Jewish bookshelf may, and should, be a source of spiritual ideas that take into account secular critiques of monotheism. Biale (2011) suggests that the roots of this trend, which he identifies as postsecular, can be found in Jewish thought since the Middle Ages.

As in the case of Daly and Hampson, at the heart of both Gordon's and Buber's writings is a deep spiritual attitude toward the world and a sense of divinity, accompanied by rejection of traditional (Jewish) religion, law (Halacha), and theology that are perceived as frozen, archaic, oppressive, and even idolatrous. They thus separate between *religion*, which they reject, and the embraced *religiosity* (Buber) or *religious sentiment* (Gordon). There is consensus among scholars that Gordon and Buber should be read as Jewish philosophers not only by virtue of their ethnic origin but also because their philosophies were conceived within the interpretive framework of Judaism (Lahav 2014). They are by no means post-Jewish, as they perceive themselves as operating within Judaism, which is not only a religion and culture, like Christianity, but also an ethnic and national identity with a particular history. They are willing to leave the traditional Jewish religious institute, but not their Jewish identity.

Gordon understands *religious sentiment* as an authentic sense of human unity with the universe and responsibility for all that thrives therein (Ramon 2007). He observes that a religious attitude, 'like all living sentiments, requires a living and self-renewing manifestation' (Gordon 1951: 330). His anthropocentric starting point is that the supreme objective of humanity is to experience divinity, because 'the human spirit... cannot be calmed without religion' (p. 89). A religious sentiment or attitude 'originates in human nature, at its point of intersection with worldly nature' (p. 120) in the human soul's aspiration 'to live within the very life of the world... not only its limited individuality but also the infinite generality of worldly life, to live infinitely' (p. 274).

Gordon's conception of divinity is pantheistic, identifying it with nature and addressing divine immanence at all levels of creation (Schweid, 1990). He employs various cognomens for godhood, including the Kabbalistic concepts of *the infinite* (*einsof*) and *Creation* (*bria*). Its immanent manifestation is called *nature* or *being* (*havayah* in Hebrew, derived from the Hebrew name of God—YHVH—and thus bearing a sense of the sacred). Gordon thus believes in a benevolent, comprehensive, practical, and subjective unity of divine natural processes that human beings attract to themselves and their environment through their actions—first and foremost through their work in nature. In this manner, they ascend, filled with vitality and coalesce with godhood (Schweid, 1990). 'Atheistic narcissism', by contrast, leaves human beings alone, in a deep spiritual void.

Buber (1973: 80) identifies *religiosity* as 'man's [sic] sense of wonder and adoration, an ever anew becoming, an ever anew articulation and formulation of his [sic] feeling'. In his renowned work, *I and Thou* (1923/1970), he proposed a conception of god as the *eternal Thou*, manifested in human life in the I-Thou dialogic relationship, to be differentiated from the instrumental I-It relationship. The concept of *eternal Thou* comes quite close to the traditional monotheistic transcendent God. This led some scholars to see Buber as a religious and not (post)secular philosopher (Stetman and Sagi, 1986; Schweid 1993; Ross 2010; Biale 2017), although Buber himself said that 'in no way am I a believing Jew in the representative sense' (cited in Werblowsky 1982). I, however, follow Moore (1996) and Margolin (2008), who identify Buber's philosophy as a *secular religiosity* that may be better understood in terms of its secular roots. The way Buber combines transcendent and immanent, his anthropocentricity, his understanding of central categories (such as creation and revelation), and his allocation of the scriptures as human (and not divine) creations suggest that Buber's philosophy may be read as postsecular.

For Buber, who uses the term with capital G, God is made present at rare, special, and sanctified moments—that are firmly implanted in everyday life—when one proceeds from I-It to I-Thou relationship. Only when we address the other as Thou, rather than as It, can we also address the eternal Thou. 'The lines extending from [the I-Thou] relationships intersect in the eternal Thou' (Buber 1970: 123). The eternal Thou is thus an interlocutor in all (of what Buber sees as) authentic dialogue with another (person or animal) and the full realization of the I-Thou relationship. The transcendent *eternal Thou* is immanent in the hyphen of the I-Thou relationships that take place here and now. Buber's God, then, is revealed through relationships between living entities, and with whom people conduct 'relationship of nearness' (Buber 1960: 195). At the same time, the *eternal Thou* himself [sic] is beyond the conception of human beings.

Like Hampson, Gordon and Buber are also concerned with revelation and reject its traditional understanding. To Gordon (1951: 120), idolatry is 'faith in revelation of eternal faiths and a living Torah from heaven, with all the commitment it demands and implies'. That commitment is what religion maintains towards what it ostensibly received from a transcendental source, to which a claim of divine truth is ascribed (Katz 2009). Buber (1973: 161), in turn, states that the mighty revelations on which religions are based essentially resemble the covert revelations that take place everywhere, at all times. Revelation is thus not the giving of the Torah, but realization, here and now, of the encounter of the individual with god through his/her Thou relationship with the other. This is clearly a non-verbal revelation: 'The eternal source of strength flows, the eternal touch is waiting, the eternal voice sounds, nothing more' (Buber 1970: 160).

Postsecular theology as going behind or sideways from God

If Daly, Hampson, Gordon, and Buber try to take their own respective religious tradition as a starting point and go beyond it to replace the traditional God with new concepts of *the more*, other attempts to offer non-monotheistic conceptions of god(s) face sideways or behind. Such trends look at non-Abrahamic spiritual traditions, such as Buddhism, Hinduism, and Shamanism, or at pre-Abrahamic traditions, such as The Goddess, for sources of theological ideas. Within this framework, one can find a spectrum of beliefs ranging from claims about the existence of (transcendence or immanent) entities, spirits, energies, or dimensions to claims about an absolute void. From a feminist perspective, a good example of such variety is *thealogy* (theology of the Goddess). Some thealogians conceptualize the Goddess as a woman-God, or God-the-Mother, seeing Her as a (mainly immanent) divine female entity, who may or may not have a male consort and who should be worshiped and celebrated. Others, on the other hand, stay away from such God-like perception, pantheistically identifying the Goddess with the universe or seeing the Goddess as a human construction that represents women's physical, emotional, and intellectual experiences (Griffin 2003).

Countless combinations and varieties of theological ideas, which are rooted in non-Abrahamic or pre-Abrahamic traditions, appear in the very liquid and eclectic sphere of New Age and its theologies. This does not mean, however, that (Western) New Age is clear of traditional Abrahamic influence. For example, several studies on Jewish New-Agers in Israel and the USA reveal the tendency to incorporate Jewish beliefs, practices, and texts with New Age praxis (e.g., Rafael 1998; Loss 2010; Ruah-Midbar and Klin Oron 2010). Furthermore, deep New-Age conceptions, such as the understanding of spiritual life as an individual and self-elected journey with therapeutic and pragmatic outcomes, can be connected on several levels to Protestant Christianity (Lahav 2017b). Yet, at least on the surface, the dominant theological move of the New Age is not to go beyond monotheism, as Daly, Hampson, Gordon, and Buber do, but to look for outside sources to introduce conceptions of *the more* as a substitute for God.

Conclusion

Expending the theological view beyond Christianity enables me to perceive postsecular theology as an umbrella term for different theological branches, inspired by different religious and non-religious spiritual traditions but united by their desire to prevent spiritual void amidst various secular critiques of God. Whether it is perceived as positive or negative, the secular voice is heard and taken into deep consideration within this theological contemplation, while its atheistic impulse is rejected. This theology thus reflects a postsecular hybridity of the secular and the religious.

Postsecular theology, however, is not unified and homogeneous but rather fluid and dynamic, torn by conflicting desires to proceed forward, backwards, and sideways, to return and to depart, to leave and to believe. It may be claimed that the concept exists only in plural form, as *postsecular theologies* rather than as a single theology. Western and semi-Western postsecular theologies are divided first and foremost by their responses to the secular and its seemingly non-existing or disappearing God. While one strand reacts by deconstructing the secular as a way to reject its critics and return to a monotheistic God, the other adopts a secular critique while resisting its desacredness by suggesting non-traditional understandings of god(s). The latter goes beyond, behind, or aside monotheism to propose such conceptions of *the more*.

The concept of the divine thus varies in these theologies, and god is understood as an agent, force, process, or dimension. There is also no agreement concerning transcendence. Not only is there disagreement whether the divine may be understood as transcendent, immanent, or a combination of both, there is also disagreement concerning the notion of 'transcendence' itself, where different writers ascribe different meanings to it. The attitude toward traditional sources (including the scriptures) also varies, as does the approach to categories such as creation or revelation.

The suggested understanding of postsecular theology should not remain closed within theological discourse. Postsecularism in general, and postsecular theology in particular, expose the social and the political meanings of faith, not only in the public sphere but even in its most private and individual (Protestant-oriented) sphere of inner beliefs. Spiritual beliefs not only fulfil the genuine needs of many, but also deeply shape the ways in which people sense reality and act within it. The ways in which we think about god, the metaphors we use, and the meanings they carry concerning relations with creation, as well as the methods we use to combine past perspectives with present experiences and future goals in our spirituality, are all extremely important components of our social world.

Thus, as postsecular theologians (re)introduce understandings of god, they should consciously use them as a political and social endeavour, aiming to raise a commitment and responsibility towards the improvement (or the repair, in Jewish terminology) of the world and everything in it. Two dangers, however, are to be faced here. The first is the danger that postsecular faith would become only one more path to personal pleasure and self-fulfilment offered in the capitalistic global supermarket (as it often the case in the New-Age sphere). The second is that postsecular theologies would be mobilized (sometimes against the will of their original writers) to the neo-revisionist backlash against humanistic values which has characterized the beginning of the twenty-first century.

Within this framework, my main concern as a Jewish feminist sociologist of faith is with the postsecular interlaces of Judaism and feminism as a political tool for women's empowerment (Lahav 2015). Hence, I recently have devoted some studies to secular-believer Jewish women in Israel (Lahav 2016b, 2017a, 2017b). These women identify themselves as secular in the Jewish-Israeli manner, which denotes detachment from Halacha as a binding religious law, combined with a selective observation of religious practice and rituals that are interpreted as cultural manifestation of Judaism. At the same time, they also believe in god. Most of my interviewees articulated theological understandings consistent with what I call here 'postsecular theology as going beyond God'. Almost none expressed a desire to return to a religious past; most perceived religion as an oppressive patriarchal mechanism. At the same time, most of them attributed an outmost importance to their Judaism, seeing it not only as a cultural, ethnical, and national identity but also as their main spiritual source. Even those who expressed theological ideas originating in non-Jewish traditions usually combined them with Jewish perceptions. Based on these findings, I am excitedly looking forward to postsecular feminist Jewish theologies, which would go beyond God, towards *the more*.

Note

1 I would like to thank Prof. Yehuda Shenhav and Dr. Avner Dinur for their important insights and contributions to this chapter.

Further reading

Kien, J. (2000) *Reinstating the Divine Woman in Judaism*, Parkland, FL: Universal Publishers.

Heavily influenced by the Western Goddess movement, Kien aims at revealing traces of the Goddess in monotheistic Judaism and retrieving Her 'proper place' in Jewish theology as a feminist act.

Plaskow, J. (1991) *Standing Again at Sinai: Judaism from a feminist perspective*, New York, NY: HarperOne.

Written from within Liberal Judaism, Plaskow's book presents the first serious theological dissection of feminism and Judaism, based on the three traditional categories of religious Jewish thinking: the Torah, Israel, and God.

Raphael, M. (2003) *The Female Face of God in Auschwitz: a Jewish feminist theology of the Holocaust*, London: Routledge.

Rafael offers a Jewish feminist theology of the Holocaust, arguing that the patriarchal bias of post-Holocaust theology, including the claims about God's death, becomes fully apparent only when women's experiences and priorities are brought into historical light.

Ross, R. (2004) *Expending the Palace of Torah: orthodoxy and feminism*, Waltham, MA: Brandeis University Press.

Based on the theology of the Orthodox Rabbi Abraham Isaac Kook, Ross tries to reconcile the patriarchal nature of Orthodox Judaism with Western feminism.

References

Altizer, T. J. J. and W. Hamilton (1966) *Radical Theology and the Death of God*, Indianapolis, IN: The Bobbs-Merrill Company.

Anderson, P. S. (1998) *A Feminist Philosophy of Religion,* Oxford: Blackwell Publishers.

Asad, T. (1993) *Genealogies of Religion: discipline and reasons of power in Christianity and Islam*, Baltimore, MD: Johns Hopkins University Press.

——— (2003) *Formation of the Secular: Christianity, Islam, modernity*, Stanford, CA: Stanford University Press.

Bellah, R. N. (1991) *Beyond Belief: essays on religion in a post-traditionalist world*, Berkeley: University of California Press (original work published in 1970).

Bellah, R. N., Madsen, R., Tipton, S. M., Sullivan, W. M., and A. Swidler (1992) *The Good Society*, New York, NY: Knopf Doubleday Publishing Group.

Benjamin, W. (2005) *On the Concept of History*, www.marxists.org/reference/archive/benjamin/1940/history.htm (original work published in 1940), accessed online 05-05-2017.

Berger, P. L. (1969) *The Sacred Canopy: elements of a sociological theory of religion*, New York, NY: Anchor Books.

Biale, D. (2011) *Not in the Heavens: the tradition of Jewish secular thought*, Princeton, NJ: Princeton University Press.

——— (2017) Personal correspondence.

Buber, M. (1960) *Attestation and Purpose, Vol. 1: articles on Judaism*, Jerusalem: Bialik Institution. [Hebrew].

——— (1970) *I and Thou*, Edinburgh: T & T Clark (original work published in 1923).

——— (1973) *On Judaism*, New York, NY: Schocken Books.

Caputo, J. D. and G. Vattimo (2009) *After the Death of God,* New York, NY: Columbia University Press.

Christ, C. and J. Plasko (1992a) 'Introduction: womenspirit rising'. In Christ, C. and J. Plasko (eds.) *Op cit.*, New York, NY: HarperOne, pp. 1–17 (original work published in 1978).

——— (eds.) (1992b) *Womenspirit Rising: a feminist reader in religion,* New York, NY: HarperOne.

Daly, M. (1973) *Beyond God the Father: toward a philosophy of women's liberation*, Boston, MA: Beacon Press.

——— (1978) *Gyn/Ecology: the metaethics of radical feminism*, Boston, MA: Beacon Press.

——— (1984) *Pure Last: elemental feminist philosophy*, Boston, MA: Beacon Press.

Dawkins, R. (2006) *The God Delusion*, London: Bantam Press.

Gordon, A. D. (1951) *The Writings of A. D. Gordon, Vol. 1: man and nature*, Tel Aviv: The Zionist Library. [Hebrew].Griffin, W. (2003) 'Goddess spirituality and wicca'. In Sharma, A. and K. K. Young (eds.) *Her Voice, Her Faith*, Boulder, CO: Westview Press, pp. 243–82.

Hampson, D. (2002) *After Christianity*, London: SCM Press.

——— (2009) 'That which is God'. In Howie, G. and J. Jobling, (eds.) *Women and the Divine: Touching Transcendence*, New York, NY: Palgrave Macmillan, pp. 171–86.

Hoagland, S. L. and M. Frye (2000) 'Introduction'. In Hoagland, S. L. and M. Frye (eds.) *Feminist Interpretations of Mary Daly*, University Park, PA: Pennsylvania State University Press, pp. 1–16.

Isherwood, L. and D. McEwan (1993) *Introducing Feminist Theology*, Sheffield: Sheffield Academic Press.

Isherwood, L. and M. Zlomislic (eds.) (2012) *The Poverty of Radical Orthodoxy*, Eugene, OR: Pickwick Publications.

James, W. (2002) *The Varieties of Religious Experience: a study in human nature*, www2.hn.psu.edu/faculty/jmanis/wjames/varieties-rel-exp.pdf (Original work published in 1902), accessed online 05-05-2017.

Johnson, E. A. (1994) *She Who Is: the mystery of God in feminist theological discourse*. New York, NY: Crossroad.

Jonas, H. (1987) 'The concept of God after Auschwitz: a Jewish voice', *The Journal of Religion*, 67(1): 1–13.

Katz, G. (2009) 'The secular basis of Aharon David Gordon's thinking', *Yiunim Be'tkumat Israel*, 11: 465–85. [Hebrew].

Lahav, H. (2014) 'Postsecular Jewish theology: reading Gordon and Buber', *Israel Studies* 19(1): 189–213.

——— (2015) 'Postsecular Jewish feminist theology? The view from Israel', *Journal of Modern Jewish Studies*, 14(3): 355–72.
——— (2016a) 'Postsecular theology: Daly, Hampson, Buber and Gordon', *Theology Today* 72(4): 415–30.
——— (2016b) 'What do secular-believer women in Israel believe in?', *Journal of Contemporary Religion*, 31(1): 17–34.
——— (2017a) 'Jewish secular-believer women in Israel: a complex and ambivalent identity', *Israel Studies Review*, 32(2): 66–88.
——— (2017b) 'William James in the Holy Land: religious experience and secular-believer Jewish women in Israel', *Israel Studies*, 22(2): 55–77.
Loss, J. (2010) 'Buddha-Dhamma in Israel: explicit non-religious and implicit non-secular localization in religion', *Nova Religio: The Journal of Alternative and Emergent Religions*, 13(4): 84–105.
Margolin, R. (2008) 'The implicit secularism of Martin Buber's thought', *Israel Studies*, 13: 64–89.
McFague, S. (1982) *Metaphorical Theology*, Philadelphia, PA: Fortress Press.
Milbank, J. (1990) *Theology and Social Theory: beyond secular reason*, Oxford: Blackwell Publishing.
Milbank, J., Pickstock, C. and G. Ward (eds.) (1999) *Radical Orthodoxy: a new theology*, London: Routledge.
Moore, D. (1996) *Martin Buber, Prophet of Religious Secularism*, New York, NY: Fordham University Press.
Oliver, S. (2007) *Radical Orthodoxy: an introduction*, London: Routledge.
——— (2009) 'Radical orthodoxy: from participation to later modernity'. In Oliver, S. and J. Milbank (eds.) *The Radical Orthodoxy Reader*, London: Routledge, pp. 2–27.
Oliver, S. and J. Milbank (eds.) (2009) *The Radical Orthodoxy Reader*, London: Routledge.
Pickstock, C. (1998) *After Writing: on the liturgical consummation of philosophy*, Oxford: Blackwell.
Radford Ruether, R. (1993) *Sexism and God-Talk: toward a feminist theology*, Boston, MA: Beacon Press.
Radford Ruether, R. and M. Grau (eds.) (2006) *Interpreting the Postmodern: responses to "radical orthodoxy"*, New York, NY: T & T Clark.
Rafael, M. (1998) 'Goddess religion, postmodern Jewish feminism, and the complexity of alternative religious identities', *Nova Religio*, 1(2): 198–215.
Ramon, E. (2007) *A New Life: religion, motherhood and supreme love in the works of Aharon David Gordon*, Jerusalem: Carmel. [Hebrew].
Ross, N. (2010) *A Beloved-Despised Tradition*. Beersheba: Beersheba University Press. [Hebrew].
Ruah-Midbar, M. and A. Klin Oron (2010) 'Jew age: Jewish praxis in Israeli new age discourse', *Journal of Alternative Spiritualities and New Age Studies*, 5: 33–63.
Rubenstein, R. (1992) *After Auschwitz: history, theology, and contemporary Judaism*, 2nd edition, Baltimore, MD: Johns Hopkins University Press.
Schmitt, C. (1996) *The Concept of the Political*, Chicago, IL: University of Chicago Press (original work published in 1929).
Schweid, E. (1990) *A History of Jewish Philosophy in the Twentieth Century*, Tel Aviv: Am Oved. [Hebrew].
——— (1993) 'Buber's conception of Judaism and its significance in our time'. In Kelman, Y. and P. Mendes-Flohr (eds.) *Mordechai Martin Buber in the Test of Time*, Jerusalem: Magnes Press [Hebrew].
Shakespeare, S. (2007) *Radical Orthodoxy: a critical introduction*, London: SPCK.
Shapira, A. (1990) 'Whole systems in twentieth-century Jewish thought: Buber and Gordon—between parallelism and affectism', *Research in Jewish Studies*, 16–17: 697–722.
Smith, J. K. A. and J. H. Olthuis (2005) *Radical Orthodoxy and the Reformed Tradition*, Ada, MI: Baker Academic & Brazos Press.
Stetman, D. and A. Sagi (1986) 'Studies of the question of the relationship between religion and morality in Buber's philosophy', *Daat*, 17: 97–118. [Hebrew].
Vahanian, G. (1961) *The Death of God: the culture of our post-Christian era*, New York, NY: George Braziller.
Vattimo, G. (1998) 'The trace of the trace'. In Derrida, J. and G. Vattimo (eds.) *Religion*, Trans. David Webb and others, Cambridge, MA: Polity Press, pp. 79–94.
Ward, G. (1995) *Barth, Derrida, and the Language of Theology*, Cambridge: Cambridge University Press.
Weber, M. (2012) *The Protestant Ethic and the Spirit of Capitalism and Other Writings*, New York, NY: Routledge (original work published in 1905).
Werblowsky, J. Z. (1982) 'Buber and religion studies'. In Kelman, Y. (ed.) *Here and Now: studies in the philosophy of Mordechai Martin Buber*, Jerusalem: The Interfaith Committee, pp. 144–8. [Hebrew].

14

Political theology and postsecularity

Francis Schüssler Fiorenza

Introduction

'Political theologies' have emerged in the twentieth century in the face of modern political and societal challenges. In dealing with these challenges, political theologians have often interpreted them as the result of secularization. Such an interpretation has often focused their endeavours, but it has also limited their view of contemporary society insofar as they have overlooked the possibilities for the diversity and vitality of religion in public life. This paper begins with a typology of three contrasting directions in political theology. One approach seeks to overcome secularization by restoring the past; another approach seeks to separate the church as an alternative mode of life from modern society; and the third approach in its critique of secularization is open to the postsecular and seeks to bring the religious into public political discourse.

The second section argues that unilineal and universal interpretations of secularity are inadequate to describe modern societies. These interpretations overemphasize the substructures of society; they underestimate the contingencies of historical, political developments, and they work with an ambiguous understanding of secularization. A third section proposes, in contrast, the notion of multiple modernities as an alternative category for understanding modernity that allows one to take much more seriously the vitality and diversity of the post-secular within societies.

The final section points to the implications of the notion of multiple modernities for understanding modern society. It places the task upon political theology to take into account not only the tendencies toward privatization and detraditionalization within religious traditions, but also the emergence of new spaces, different formations, and the fusion of diverse religious identities within religiously pluralistic societies in a postsecular age. Religion as a visible force and presence within our society takes place in the fusion of new identities and groupings within our society. The question then is does political theology change from its historical roots in view of alternative views of modernity that include elements of secularity and elements of what has made it customary to refer to our age as 'postsecular' (Habermas 2017).

Diverse twentieth-century conceptions of political theology

Political theology as a concept has a long history, extending back to antiquity (Fiorenza 1977, 2015–2016). Today we understand political theology with different tasks, challenges, and prospects. This first section analyzes three twentieth-century conceptions of political theology that have each made secularity central to their conceptions of political theology. After questioning the understanding of secularity, the question emerges whether these conceptions have adequately taken into account the notion of multiple modernities and the existence of postsecularity. These provide the background and context for more constructive reflections on political theology and the relation of religious discourse within a modern society that deals not only with secularity but also with postsecularity.

Schmitt's political theology: secularization of sovereignty

Carl Schmitt first published *Political Theology: Four Chapters on the Concept of Sovereignty* in 1922, and a revised edition in 1930. The historical context was the aftermath of Germany's defeat in World War I and its economic plight. The Weimar Republic had a liberal and democratic parliamentary system. Schmitt's *Political Theology* is directed against that system for undermining the state's sovereignty. Arguing that political life is defined by the friend-enemy distinction, he affirmed that a state requires a single centre of power to survive in the face of enemies (Schmitt 1996; 2007). A strong sovereign is necessary.

Schmitt uses his understanding of deism to evaluate the modern Enlightenment. The deistic natural law and theology reduced God's sovereignty to a naturalism without exceptions (miracles) in the natural world. The God of deism no longer has absolute sovereignty over the world. Its God can no longer make exceptions to the laws of nature through miracles. This secularized God lacks the sovereignty of the traditional Creator God. Secularism for Schmitt is the naturalism of the deistic Enlightenment that undercuts the sovereignty of God.

Modern constitutional democracies are secularized insofar as they transform political sovereignty in a way analogous to the transformation of the understanding of God. They limit the power of the sovereign like deism limited the sovereignty of God. Both the constitution and the legislative power of the parliament limit the sovereign's power. Just as the laws of nature limit a deistically conceived God from contravening the laws of nature, so too does a constitutional and a parliamentary system limit the power of the modern sovereign who cannot contravene the law. The parliamentary system secularizes sovereign power because it limits the sovereign's ability to make exceptions to the law.

Schmitt's understanding of secularization is complex (Schmitz and Lepper 2007). In *Political Theology,* he argued that 'all the quintessential concepts of the theory of the modern state are secularized theological conceptions' (Schmitt 2005: 57, 2008). Schmitt notes that a structural identity exists between theological and judicial concepts. A functional parallelism exists between the principles that underlie the modern state and modern theology. The significance of this functional parallelism and structural identity comes to the fore in his political theology. Carl Schmitt appealed to the political theology of the French restoration in his criticism of the secularization and weakening of the sovereignty of the modern state. In contrast to his conception of political theology, the much more recent conception of political theology is quite the opposite. William Cavanaugh, influenced by Stanley Hauerwas's sectarian interpretation of the Christian church's relation to the state, argues that the modern state is too powerful.

Schmitt as political philosopher has been viewed with both disdain and praise. His critique of the liberal democracy of the Weimar Republic was accompanied with a strong advocacy for Hitler and Nazi regime (Scheuerman 1984, 1999). Schmitt's critique 'paved the way for a movement that wanted to do away with German democracy as soon as possible' (Ullrich 2016: 239). He became an apologist for the Nazi abrogation of the basic rights of citizens. His virulent anti-Semitism dovetailed with the horrible and consequential anti-Semitism of the Nazi regime (Mehring 2014). His understanding of sovereignty led him to affirm that the Führer's actions as leader 'are not subordinate to justice, but rather it is itself, supreme justice' (Schmitt 1935, in Rabinbach and Gilman 2013: 64–67).

Nevertheless, Schmitt's insight into the role that power plays in politics has had a diverse reception. His political theology criticized the proceduralism of liberal democracy and the democratic attempt at reasoned consensus. This very critique has become key today to the agonistic political philosophy that argues that the contemporary politics has become too rationalistic. Liberal conceptions have minimized the role that power plays in the political process. Politics entails the gaining of power to exercise certain actions and to enact specific laws (Mouffe 1999, 2000).

Schmitt's political theology correlates with his understanding of secularism. The modern democratic state entails a secularization insofar as it inherits the naturalism of the Enlightenment. To counter the Enlightenment, he appeals to the French restoration (Joseph de Maistre), with its emphasis on original sin and the need for a strong authority. The centrality of naturalism to his interpretation of secularism has re-emerged in contemporary criticisms of the modern age, as we shall see in our discussion of Charles Taylor, who identifies secularism with a naturalism that negates transcendence.

Political theology and the soteriological ideology of the modern secular state

If Carl Schmitt interpreted secularization as the undermining of the power of the sovereign, and thereby, the sovereignty of the modern state, William J. Cavanaugh offers the very opposite interpretation of secularization: the modern state has assumed a monopoly, and it perpetuates itself through an ideology of the modern state (Cavanaugh 2002). He writes, 'it is not enough to see what is called "secularization" as the progressive stripping away of the sacred from the profane remainder. What we have instead is the substitution of one *mythos* of salvation for another' (Cavanaugh 1999: 190). The successor mythos has been successful to the extent that it mimics its predecessor. The state's offer of salvation is a false unity and peace at odds with the Christian story. Moreover, Cavanaugh maintains that 'essential to state soteriology is the unity and uniqueness of the sovereign; the sovereign is a jealous God' (Cavanaugh 1999: 191).

In this view, political theology's task entails criticizing the ideologies of the modern secular state and promoting an alternative political vision implicit in different Christian practices that give witness to the world. The state propagates the ideology that the state saves us from the violence of religion through a self-serving interpretation of the seventeenth-century wars of religion as the start of religious freedom. In contrast, Cavanaugh argues that nation-state has in effect become the greatest instigator of violence in the modern period. The wars that have ensued from the violence created by the modern state belie the very claim that the modern liberal modern state ensures pluralism, and thereby peace.

Political theology and the memory of suffering in modernity

Johann Baptist Metz and Jürgen Moltmann have developed parallel though distinct visions of political theology in the postwar period of the 1960s. Both Metz and Moltmann had been drafted as teenagers, and became prisoners of war (Metz in the USA, and Moltmann in England). On their return to Germany, they studied theology, rose to professorships, and argued that Germany had to face its history: the rise of Hitler and Nazism. Each in their own way developed the importance of the relation between religion and political society.

Originally, Metz developed a theory of secularization drawing on the dialectic of Karl Rahner's understanding of Christology. The Christian belief in the incarnation affirms the full freedom of the human nature of Christ, and so too must the relation between the church and the world be interpreted with a dialectic of dependency and freedom that understands the implied freedom and autonomy of the world. Metz's development of Rahner's theological understanding of creation and incarnation sees secularization as rooted in the Christian affirmation of God's transcendence in relation to the world. This dialectic becomes further explicated in Metz's later development of political theology, at first influenced by Moltmann's emphasis on hope and eschatology, and later developed with an emphasis on the memory of suffering.

In Moltmann's view, the modern society that has established itself with the emergence of the modern industrial state is not so much the state as it is 'that sphere of public life which is governed by the conduct of business, by production, consumption, and commerce' (Moltmann 1967: 305). This bourgeois industrial society, with its 'system' of needs, has liberated itself from the traditional understanding of the relation of religion to society. Thus, 'the Christian church can consequently no longer present itself to this society as the religion of society' (Moltmann 1967: 305). Against this view of modern society, Moltmann relates his understanding of Christianity. Christian faith consists of an eschatological hope; Christians understand themselves as a pilgrim people and their church as an 'Exodus Church'. In some ways, this understanding of Christian life both contradicts and parallels modernity. Though it criticizes the reduction of religion to the private sphere, its advocacy of a pilgrim people and Exodus Church relates to society as a separate society rather than as one seeking to become integral to society to change and transform it.

Metz takes a distinctive and changing path. When he had at first coined the phrase 'political theology' in the 1960s, he was unaware of Schmitt's earlier conception of political theology. Influenced by Jürgen Moltmann's *Theology of Hope*, Metz emphasized the need to interpret the biblical message of the Kingdom of God, not existentially as Rudolf Bultmann did, but rather politically as an eschatological proviso. The point of political theology was not to advocate for a specific political position or even party (such as the Christian parties in Europe) but rather to critique the injustice and failures of all programmes. 'Political theology is therefore not the attempt to burden the church with a quite distinct and concrete politics, but first of all the attempt to hinder such an endeavor' (Metz 1977: 65).

Much later, having become aware of Carl Schmitt's political theology, Metz started referring to his proposals as the 'new political theology'. He took issue with Schmitt's views on modernity as secularization, the relation between monotheism and absolute sovereignty, and the critique of the Enlightenment and democracy. He argued that political theology should not understand modernity primarily as a demise or a malaise. Political theology should not necessarily be hostile to the Enlightenment, nor should that hostility lead to a political decisionistic theory that eliminates any universal claim (Metz 1997a, 1997b).

There are two significant observations in Metz's view of secularization that point to the postsecular. The first is that

> the world under the secular metaphor of the death of God is not a world free of religion, and it is not even a world hostile to religion. Indeed, the opposite, for it appears as if on the grave of God innumerable new religions blossom.
>
> *(Metz 1977: 21)*

The second is his observation that the world is becoming a polycentric world. Metz makes biblical monotheism central to his political critique, but in contrast to the two other approaches (Schmitt and Cavanaugh), such a biblical monotheism should therefore be neither an unqualified affirmation of political sovereignty, nor is it a sectarian separation of the church from political life. Instead, a Christian biblical monotheism is a remembrance of the passion and suffering of Jesus. The remembrance of the suffering of the other becomes the focal point for the significance of a political theology for public consciousness and memory and as a corrective of any political programme.

Implications and goals resulting from these diverse political theologies

These conceptions of political theology have assumed the secularization of modern society. However, they each have interpreted secularization quite differently. For Carl Schmitt, the expression of secularity is modernity's democratic constitutional system of government that limits the power of the sovereign. The remedy for secularization is the elimination of these limits and the abrogation of basic rights. William T. Cavanaugh's secularism is the myth of the modern state as a saving institution. If Schmitt sought to strengthen the sovereignty of modern state because modern liberal democracy has secularized and weakened it, Cavanaugh thinks the modern state has glorified (deified) itself into an agent of salvation and peace. His political response underscores the church as a community, primarily as a Eucharistic and sacramental community. In some sense his vision is close to Rod Dreher's recent book, *The Benedict Option* (Dreher 2017), that emphasizes the Christian community's otherness from modern society, except Cavanaugh emphasizes the Roman Catholic vision of the community as a sacramental and Eucharistic community (Cavanaugh 1998, 2011).

Despite the changes in the political theologies of Moltmann and Metz, their basic impulse has been that the churches need to bring religion into the public sphere much more strongly than it had in the past, against the National Socialist regime. Moltmann links his political theology with human rights and its importance for society. Metz consistently developed the importance of religion to counter the forgetfulness of the suffering of the victims of injustice within society. His political theology has underscored the importance of the struggle against anti-Semitism and the memory of the passion of Jesus as crucial for modern society.

Inadequacies of unilineal and universal approaches to secularism

A decisive question is whether these political theologies or theories of secularization overlook the complexities of modern society and the modern world. Two questions emerge. The first: does the interpretation of the modern European or North American state under the category of secularization deal adequately with the phenomena of 'religion' and its diversity even within the modern West? Does the theory of secularization entertain certain naturalistic assumptions about the development of history and culture (Blakey 2016)? And, as a result, does it thereby overlook the diverse existence of multiple religiosities that have led to the emergence of 'postsecularity'? The second question: does the view of the universalization of secularization minimize or overlook the significant differences between the developments in Europe over the last centuries and those in North America (Hempton and Mcleod 2017)? Does not the claim that the secularity of Western Europe will become dominant in other areas of the world overlook the existence of plural modernities? Is it not an oversight that results from a too monolithic view of modernity, including the state, society, and even religion?

The notion of 'multiple modernities' is a more adequate framework for cultural change in a complex and diverse modern world. The latter challenges a unilineal view of secularity and modernity that links modernization with aspects of Western life and culture. Could a pluralistic conception of modernities undercut a sharp distinction between public reason and religious reason? A brief comparison of the secularization thesis with the multiple modernities thesis may help answer this question.

These differences have impacted the understanding of political theology and the role of religious discourse within modern public discourse. In taking these differences into account, one important consideration is the diversity of various strains or methodological approaches to the issue of secularization and modernity. These strains often are intermixed, yet they are distinct influences in the interpretation of modernity. One methodological approach emphasizes the impact of the substructure of societies on religion and culture. Another approach focuses on historical political contingences. The third focuses on the cultural superstructure of societies. These approaches are often intertwined, but often one strain is more decisive for the interpretation. The importance of taking into account the differences among these approaches has consequences for how one interprets modernity, views the limitations of the secularity thesis, and can view the significance of postsecularity for political theological discourse.

Secularity and the contingencies of history

The diversity of patterns of secularity and modernity speaks against any unilateral view of secularity. The socio-political differences between Europe and the USA have led to quite diverse political patterns of religion and secularization. In Europe, the Peace of Westphalia, signed in Münster and Osnabrück, brought an end to the 'Thirty Years' War', and the 'Eighty Years' War'. This Westphalian arrangement prohibited interference in another state's affairs. It did not affirm religious freedom but the arrangement that the ruler of each nation determined its religion: *cuius regio, eius religio*. Distinct cultural patterns developed in reaction to the dominant church of a particular sovereignty. The Westphalian peace resulted not so much in separating the religious and political, but rather in linking them within the nation-state. Religious freedom and pluralism originated with the sectarian protests protesting against the dominant religious institutions. Their migrations to the USA led to the beginning of the right of religious freedom (Jellinek 1901).

In Europe, one model of secularism developed in reaction to the monopolies that the churches had in the wake of the Westphalian model. Yet even here important differences exist, as between the secularity in Sweden and the more active anti-clerical secularism in France. In the USA, however, with diverse motives for early immigration, some of which entailed religious freedom and others economic reasons, a context was provided for elements of religious pluralism that differs from the European. The confessional identity of later immigrant communities initially had a strong religious identity. In the face of urban secularity, these communities represented an intensification of religious identity—an important difference between the USA and Europe in modernity. The problem of diversity becomes even more complex if one considers the uncertainty about when modernity began. One could claim, if different political and economic institutions are considered, that modernity is 'a phenomenon that can be found in some parts of Western Europe during some periods of the twentieth century' (Wittrock 2000: 31–60).

Secularity and modernity in relation to substructures

The methodological approach that takes into account the importance of substructures or the organizational structures of society often takes the view that secularization is one-directional and inevitable. It points to the increased urbanization, industrialization, communication, and mobility. At the same time, the bureaucratization of society has led to the increased distinction of different spheres of life, including education, business, law, and religion. It is often assumed that these trends, which have contributed to secularization in the modern West, will inevitably affect other parts of the globe. Consequently, the secularization of the West will extend to the rest of the world.

Such a claim for the universal secularization analysis overlooks that these same trends have often led to contrary results. Increases in urbanization have led to the growth of megacities and urban centres where pluralism and diverse religious identities have thrived: some traditional and some new. These have fused together in the growth of postsecular identities. The Internet with its social spaces has impacted the personality of people in diverse areas of the world (Inkeles and Smith 1974). It has led to the emergence of new groupings and allowed postsecular groups, ideas, and communities to thrive.

Ambiguities in the concept of secularization

A central problem in the affirmation of secularization is the diverse specifications of its meaning. As Henrich Lubbe has shown, the notion of secularization originally had a legal origin. It was to specify the transfer of property from the church to the state, or the transition of persons from a clerical or religious state to that of a layperson. It was a clearly delimited concept (Lubbe 1965). Today, the concept is applied to a much more complex process of cultural change: the role of the religion in modern society. When a concept is transferred from the legal realm to the realm of culture and society, its demarcation and meaning are much more ambiguous, if not murky.

José Casanova has proposed that the theories of secularization operate with three different frameworks (Casanova 2001: 137–86). One understands secularization as the decline and, possibly, eventual disappearance of religion in the wake of the Enlightenment's critique and the emergence of modern sciences. A second framework, following Max Weber and Talcott Parsons, interprets secularization as the differentiation of societal spheres: economic, scientific, educational, religious, and political. This framework does not entail so much the decline of religion as its separation from other institutional spheres within modern society. Secularization results in the privatization, but not elimination, of religion and entails a reconceptualization of religion's role in society. In contrast to unilateral views of secularization, Casanova's *Public Religions in the Modern World* argues that our contemporary situation is experiencing 'deprivatization' of religion in the modern world. Religious traditions are 'refusing to accept the marginal and privatized role which theories of modernity, as well as theories of secularization, had reserved them' (Casanova 1994: 6). Social movements emerge as religious or as challenges to the autonomy and primacy of the secular sphere, such as the economy and state.

Taking a different tack, Charles Taylor interprets secularization from the perspective of the history of ideas. His books dealing with Hegel, the modern self, the malaise of modernity, social imagination, and the secular age focus on the cultural transformations that influence modern identity (Taylor 1989, 1991, 2011, 2017). Secularism for Taylor involves the turn towards the human subject as the filtered (buffered) self. He notes that community is mediated through constructed images even of the nation. Nationalism mediates the bond that underlies citizenship. However, the secularism of the modern self leaves little room for a common good that is transcendent. Taylor approaches modernity and secularism with a certain ambiguity. On the one hand, he disagrees with monolithic understandings of secularization. He maintains, therefore, that secularization should not be understood simply as the decline of religion. An increased proliferation of religious identities has taken place in modernity. Nevertheless, he contends, the modern self is one that searches for individual fulfilment. It is not self-oriented to what transcends the self.

Taylor explicates secularism and religion in terms of the immanent/transcendent distinction. Though he concedes that this distinction is 'tailor-made for our culture' and that it 'may be seen as parochial, incestuous, navel-gazing' (Taylor 2017: 15), he, nevertheless, reaffirms his immanent/transcendent distinction as 'wise' and with broader significance. He appeals to the Buddhist doctrine of 'anatha' as implying a distinction between an immanent human flourishing and what lies beyond it, and as contrary to the naturalism of modern secularism.

In his discussion of political theology, Taylor sees the secularization of a previously cosmic religious order of society now as a 'bottom-up' view of society and the cosmos. 'Its normative view is the liberty of all members, equality among them, and rule is based on consent' (Taylor 2011: 318). Yet this derogation of the rule of consent as a bottom-up view that expresses the naturalism of modernity brings him very close to Carl Schmitt's attack on the secularization of modernity. A different view of increased liberty, equality, and consent shows in acknowledging transcendence in the increased pluralism of religion that exists today and is present in a postsecular age.

Multiple modernities as an alternative category for political theology

The category of 'multiple modernities' counters the prevalent view that modernization and the USA would eventually expand and prevail in all parts of the world. The great variety in the patterns of modernization has become clear in the last decades of the twentieth century. These were influenced, if not determined, by specific cultural premises, religious tradition, and historical experiences (Eisenstadt 2000: 1–29). The category of multiple modernities as an alternative understanding of modernization is not unambiguous and can be understood diversely. How many modernities are there? Are there as many modernities as there are modernized societies? Are there not only varieties of modernities outside of the West but also within it? Björn Wittrock argues that significant differences exist among the various states in Europe. There are French, British, German, and Scandinavian modernities (Wittrock 2000: 31–36). Yet, such differences are not the main point of the new paradigm, but rather that modernities exist outside of the concepts and developments that make sense of Western modernity, and these often stand in contrast to Western institutional forms and structures (Schmidt 2006: 77–97).

In this context, many reject the understanding of secularization as an inevitable process that has begun in the West and will spread over the globe. It is not simply that they reject the identification of modernization and Westernization. Much more importantly, they reject an interpretation of modernization that identifies it with certain structural changes: urbanization, industrialization, and technologies of travel and communication. Instead, they underscore the importance of the different cultures, different lived communities, and the possibility and facticity of diverse modes and types of modernization.

> The civilization of modernity, which developed first in Europe, spread to the entire world but has crystallized in numerous forms. The resulting condition of multiple modernities challenges the classical theories of modernization and signals a particular view of the contemporary world in terms of a multiplicity of cultural and political projects.
> *(Delanty 2004: 391–404)*

One can challenge this approach by pointing to the complexity of analyzing a historical phenomenon such as Pentecostalism. Is Pentecostalism a development of the modern West's individualism because of its emphasis on individual religious experience? Or is Pentecostalism a reaction to the rationalism of the modern Enlightenment? Is Pentecostalism merely a continuation of Methodism, Pietism, and various revival movements? How does its emphasis on female agency, voluntarism, personal experience, emotion, and affect, and assistance to one another fit in one's interpretation (Shantz 2013)? Thus, a religious movement could be seen both as an expression of modernity as well as a reaction to modernity.

Another example: Talal Asad, in the discussion of the diverse formations of the secular, takes issue with Charles Taylor on two basic points. The first is the association of secularism with the emergence of the modern state. The second is the endeavour to make political ethics independent of religion (Asad 2003, 2013). Saba Mahmood follows Talal Asad's direction on the inadequacies of the prevalent interpretation of secularity. She suggests that the traditional understanding of secularism involves a 'generative contradiction'. The claim that the state separates church and state and therefore relegates religion to the private sphere overlooks the fact that the state's regulation of socio-religious life dissolves the distinction between public and private spheres. Hence, the line between what is public and what is private, or what is religious and what is political, is constantly subject to contestation (Mahmood 2016). Commenting on the situation in Egypt, Mahmood seeks to show the inadequacy of traditional versions of secularism as the separation of church and state along with religion and law. In her view, much more different and fundamental changes are taking place. She concludes that

> the ideal of interfaith equality might require not the bracketing of religious differences but their *ethical* thematization as a necessary risk when the conceptual and political resources of the state have proved inadequate to the challenge this ideal sets before us.
> *(Mahmood 2016: 212)*

Asad and Mahmood understand and construct a relationship between religion/theology and politics in a way that is distinctive from the usual Western approach to secularization and modernity.

Challenges and consequences of multiple modernities for political theology and postsecularity

To view the contemporary world as consisting of a plurality of modernities presents challenges and consequences for political theology and the discourse about secularism and postsecularity. These concern historical change, the nature of religion, the interpretation of contemporary movements as religious or secular, and the nature of political and ethical discourse. The perspective I am sketching here challenges a unilineal view of the world from the perspective of the West in which one moves from Christianity to secular modernity.

Any language about the postsecular might appear as a period that follows secularism. The problem for political-theological discourse is that the sequence is not simply temporal.

Depending on the conception that one has of secularization, the audience, language, and goal of that discourse have changed. Secularism is not simply a matter of the number and frequency of church attendance, nor the degree of religiosity professed by individuals, nor how strictly they adhere to the individual religious traditions or are integrated into their religious institutions. Instead, as has been noted, 'Secularism, in its dominant expression, combines a distinctive organization of public space with a generic understanding of how discourse and ethical judgment proceed in that space' (Connolly 1999: 20).

In the academic study of religion, especially comparative religion, it has been noted that often specific characteristics of Western religion or Christianity were taken as the norm of what constitutes religion and, contrarily, what constitutes superstition or non-religious practices. If, however, one takes into account the various native religions in various continents, then the question of what is religious becomes a very complex question. In addition, how does one describe Confucianism—as a religion or as an ethics? Or how does one describe the practices of yoga?

What is religious and what is secular is not only a matter of comparative religion, but it is also a problem of interpreting contemporary phenomena about what is religious or not in a world caught between secular and postsecular, and between multiple modernities. For example, in the USA, the growth of the Christian Prosperity Gospel movement is obvious and yet open to diverse interpretation. On TV, some of its prominent ministers preach that if one donates, and donates generously, then God will reward them in this life and the next. Some would interpret this movement as a sign of an increased influence of evangelical or Pentecostal forms of religion. Others, because of the emphasis on prosperity, might interpret this phenomenon as a sign of a secularization of religion within a capitalistic culture focused on prosperity. Yet, another example, with quite a contrasting and distinct orientation, is the faith-based communities that have emerged in European cities, 'establishing shelters for the homeless, soup kitchens for the hungry and employment training for the unskilled'. These are not to be interpreted as a return to a bygone age, but rather 'a highly significant aspect of the latest phase of the shifting relations between religion, state, and society' (Beaumont and Cloke 2012: 266).

Another example is the phenomenon of the growth of interest in mindfulness and meditation. Yoga classes have become increasingly popular, with their emphasis on breathing and mindfulness. How does one interpret this movement? Is this an increase of religiosity, with borrowing from Buddhist and Hindu traditions making an impact upon the West, in contrast to more traditional forms? Or is it a secularization, because the meditation and mindfulness are often related to the overcoming of the stresses and vicissitudes of life rather than to a transcendent deity? Despite the differences of interpretation, the last example shows the influence of non-Western forms of religiosity within the West that need to be taken into account in understanding the present situation not simply as secular but also as postsecular.

The third point consists of the sphere of public theology or public discourse within the context of multiple modernities. Religions within diverse nations and areas of the world are entering into the public sphere and political debates. They do so not merely to reaffirm traditional positions and standpoints, but also to engage with the very problems that the secular and postsecular aspects of modernity engender. Charles Taylor, as we have noted, uses the transcendent/immanent distinction to criticize forms of ethical reasoning that appear to him as based on consensus or a universalizing of reason.

We have to be careful of a certain reading of multiculturalism that often presupposes the 'incommensurability' of worldviews as if their discourses were closed universes of meaning with isolated standards for making arguments and asserting truth claims (Habermas 2017).

Likewise, we must be equally careful of the sharp distinction between religion and society as a diagnostic tool or a means of identifying societies (Martinelli 2005). Today, thinkers as prominent and diverse as John Rawls and Jürgen Habermas have understood that public discourse, in referring to public discussion (in distinction to judicial decisions), often uses religious language to make political judgements. They have noted that everyone who heard Lincoln's Second Inaugural Address or who heard Martin Luther King's March on Washington Address understood and was moved by what was said (Fiorenza 2014, 2018). A more abstract philosophical language would not have been more understandable or efficacious in such a context. These philosophers have made a significant move beyond their initial arguments for justice and public discourse; this is not to deny that some inconsistency still exists. However, the issue poses a challenge to political-theological discourse. One has to be aware that we should not make appeals to abstract language as if the world were secular.

Instead, legitimacy exists for the use of religious language and religious traditions in the public sphere. Such a discourse has to take into account that religious traditions and their continued presence in contemporary society represent a resource for political and ethical discourse. One has to acknowledge that resource, and at the same time realize the importance of a process of learning on the side of the religious and non-religious sectors to bring that resource to bear (Dillion in Gorski et al. 2012; Habermas 2017). Such discourse has to take into account the important switch entailed in the move away from a virtue ethic geared to the classic 'Mirror of a Prince' to an ethics geared to the diverse structures of society, as well as the presence of secular and postsecular cultures. In this context, political discourse moves into public discourse about justice and charity (Fiorenza 2018). Yet, it is nevertheless open to new movements and subcultures as they emerge within contemporary societal structures.

Further reading

Beaumont, J. and C. Baker (2011) *Postsecular Cities: space, theory and practice*. London: Continuum.

This book explores the fusion of urban space and how the dynamics of religion change in regard to religious diversity, public life, justice, and multiculturalism.

Eisenstadt, S. N. (1973) *Tradition, Change, and Modernity*, New York, NY: John Wiley.

A significant collection of essays dealing with the issue of modernity and multiple modernities.

Fiorenza, F., Tanner, K. and M. Welker (eds.) (2013) *Prospects for Political Theology: contemporary challenges and future directions*, Louisville, KY: Westminster John Knox Press.

A collection of essays on political theology, including texts by Johann B. Metz, Jürgen Moltmann, and Elisabeth Schüssler Fiorenza.

Gómez, L. and W. van Herck (2012) *The Sacred in the City*, London: Continuum.

Collection of essays exploring the role of religious and sacred identity in the modern metropolis as a postsecular space.

Gorski, P. S., Kim, D. K., Torpey, J. and J. VanAntwerpen (eds.) (2012) *The Post-Secular in Question: religion in contemporary society*, New York, NY: New York University Press.

A collection of essays from diverse philosophical and sociological perspectives underscoring the limits of secularism and possibilities of postsecular religiosity.

References

Asad, T. (2003) *Formations of the Secular: Christianity, Islam, modernity*, Stanford, CA: Stanford University Press.

——— (ed.) (2013) *Is Critique Secular? blasphemy, injury, and free speech*, New York, NY: Fordham University.

Beaumont, J. and P. Cloke (eds.) (2012) *Faith-Based Organizations and Exclusion in European Cities*, Bristol: Policy Press.

Blakey, J. (2016) *Alasdair MacIntyre: Charles Taylor and the demise of naturalism*, Notre Dame, IN: University of Notre Dame Press.
Casanova, J. (1994) *Public Religions in the Modern World*, Chicago, IL: University of Chicago.
——— (2001) 'Secularization'. In Smelser, N. J. and B. Baltes (eds.) *International Encyclopedia of Social and Behavioral Sciences*, Oxford: Pergamon, pp. 13786–91.
Cavanaugh, J. T. (1999) 'The city: beyond secular parodies'. In Milbank, J., Pickstock, C. and G. Ward (eds.) *Radical Orthodoxy: a new theology*, London: Routledge, pp. 182–199.
——— (2002) *Theopolitical Imagination*, London: T & T Clark.
——— (2011) *Migrations of the Holy: God, state, and the political meaning of the church*, Grand Rapids, MI: W.B. Eerdmans.
Connolly, W. E. (1999) *Why I am not a Secularist*, Minneapolis, MN: University of Minnesota.
Delanty, G. (2004) 'An interview with S. N. Eisenstadt: pluralism and the multiple forms of modernity', *European Journal of Social Theory*, 7(3): 391–404.
Dreher, R. (2017) *The Benedict Option: a strategy for Christians in a post-Christian nation*, New York, NY: Sentinel.
Fiorenza, F. S. (1977) 'Political theology as foundational theology', *Proceedings of the Catholic Theological Society of America*, 32: 142–77.
——— (2014) 'Faith and political engagement in a pluralistic world: beyond the idols of public space'. In Min, A. (ed.) *The Task of Theology: leading theologians on the most compelling questions for today*, Maryknoll, NY: Orbis, pp. 93–104.
——— (2015–2016) 'Foundational theology as political and sacramental public theology', *Louvain Studies*, 39(2): 132–40.
——— (2018) 'Faith, hope, and love and the challenges of justice'. In Min, A. (ed.) *Faith, Hope, Love, and Justice: the theological virtues today*, Lanham, MD: Lexington Books, pp. 79–114.
Habermas, J. (2017) 'Religion in the public Sphere of post-secular society'. In Habermas, J. *Postmetaphysical Thinking II: essays and replies*, Malden, MA/Cambridge, MA: Polity Press, pp. 210–25.
Hempton, D. and H. McLeod (2017) *Secularization and Religious Innovation in the North Atlantic World*, Oxford: Oxford University Press.
Inkeles, A. and D. Smith (1974) *Becoming Modern: individual change in six developing countries*. Cambridge, MA: Harvard University Press.
Jellinek, G. (1901) *The Declaration of the Rights of Man and of Citizens: a contribution to modern constitutional history*, New York, NY: T. Holt.
Lubbe, H. (1965) *Säkularisierung: geschichte eines ideenpolitischen begriffs*. Freiburg: K. Alber.
Mahmood, S. (2016) *Religious Difference in a Secular Age: a minority report*, Princeton, NJ: Princeton University Press.
Martinelli, A. (2005) *Global Modernities*, London: Sage.
Mehring, R. (2014) *Carl Schmitt: a biography*, Malden, MA: Polity Press.
Metz, J. B. (1977; 2007) *Faith in History and Society: toward a practical fundamental theology*, originally published 1977, a new translation by J. Matthew Ashley, with study guide, New York, NY: Crossroad.
——— (1997a) 'Wie rede ich von Gott angesichts der säkularen Welt?'. In Henrich, D., Metz, J. B., Hilberath, B. J. and R. J. Werblowsky (eds.) *Die Gottrede von Juden und Christen unter der Herausforderung der Säkularen Welt*, Münster: Lit-Verlag, pp. 21–33.
——— (1997b) *Zum Begriff der neuen Politischen Theologie 1967–1997*, Mainz: Matthias-Grunewald-Verlag.
Moltmann, J. (1967) *Theology of Hope: on the ground and implications of Christian eschatology*, London: SCM Press.
Mouffe, C. (1999) *The Challenge of Carl Schmitt*, London/New York, NY: Verso.
——— (2000) *Deliberative Democracy or Agonistic Pluralism*, Vienna: Institute for Advanced Studies.
Scheuerman, W. E. (1984) *Between the Norm and the Exception: the Frankfurt School and the rule of law*, Cambridge, MA: The MIT Press.
——— (1999) *Carl Schmitt: the end of law*, Lanham, MD: Rowman & Littlefield.
Schmidt, V. H. (2006) 'Multiple modernities or varieties of modernity?', *Current Sociology*, 54(1): 77–97.
Schmitt, C. (1935; 2013) 'Der führer schützt das rechte: on Adolf Hitler's reichstag address, 13 July 1934'. In Rabinbach, A. and S. L. Gilman (eds.) *The Third Reich Source Book*, Berkeley: University of California, pp. 63–67.
——— (1996; 2007) *The Concept of the Political*, Chicago, IL: University of Chicago Press.
——— (2005) *Political Theology: four chapters on the concept of sovereignty*, trans. George Schwab, Chicago, IL: University of Chicago Press.

——— (2008) *Political Theology II: the myth of the closure of any political theology*, Cambridge, MA: Polity Press.
Schmitz, A. and M. Lepper (eds.) (2007) *Hans Blumenberg, Carl Schmitt: briefwechsel 1971–1978 and weiter materialmen*, Frankfurt: Suhrkamp.
Shantz, D. H. (2013) *An Introduction to German Pietism: Protestant renewal at the dawn of modern Europe*, Baltimore, MD: John Hopkins University Press.
Taylor, C. (1989) *Sources of the Self: the making of modern identity*, Cambridge: Cambridge University Press.
——— (1991) *The Malaise of Modernity* (CBC Massey Lectures), Toronto, ON: House of Anansi Press, reprinted in the U.S. as *The Ethics of Authenticity*, 1992, Cambridge, MA: Harvard University Press.
——— (2011) *Dilemmas and Connections: selected essays*, Cambridge, MA: Harvard University Press.
——— (2017) *A Secular Age*, Cambridge, MA: Harvard University Press.
Ullrich, V. (2016) *Hitler: ascent, 1889–1939*, New York, NY: Alfred A. Knopf.
Wittrock, B. (2000) 'Modernity: one, none, or many? European origins and modernity as a global condition', *Daedalus*, 129: 31–60.

15
Postsecular prophets

Robert Joustra

Introduction

The secular, to say nothing of the postsecular, is an academic industry with few rivals. It is a debate whose terms are at root essentially contested: that is, in offering clear definition of the secular or postsecular, we necessarily say something normative, something not purely empirical, something—some would ironically say—religious about the world. This underscores the point of Daniel Philpott's excellent 2009 article in *The Annual Review of Political Science*, that we have come to ask 'not why the political influence of religion has returned but why it ever went away. Or, better yet, why anyone ever thought it went away' (Philpott 2009). Perhaps this thing called secularism never really arrived, or if it did, it arrived rather unevenly, in lecture halls on the coasts of the North Atlantic world, but hardly at all in the villages of sub-Saharan Africa, the madrasas of the ancient Silk Road, or the temples of Southeast Asia. My point is not that there is no such thing as a postsecular, because the secular itself is so famously contested,[1] but rather that an especially *global* perspective is critical, lest we become lost in the fashionable ideologies of the North Atlantic world and forget, as Scott Thomas points out, that 'strong religions and weak states' still make up much of the world (Thomas 2005).

This, in the argument of the editor, with which I heartily agree, is the reason why the term postsecular is important: not because it is has one, clear, unambiguous meaning but because any definition must presume that 'relations between religious, secular and humanist forces, previously viewed in isolation, are now placed together at the forefront of analysis of empirical developments on the ground, political advances for social transformation and theoretical explanations of macrosocial developments at large' (see Chapter 1). By postsecular, therefore, I mean that it is now largely recognized that one, neutral, secular account of society, politics, and knowledge has passed from fashion in the North Atlantic world, and this passing marks an important transition. It is in that moment of transition I propose, somewhat counterintuitively, to study two 'presecular' figures, themselves students of major global transition: Abraham Kuyper (1837–1920) and the Great War, and Herbert Butterfield (1900–1979), and World War II and the Cold War.

I do not turn to history in an exercise in nostalgia, as though the antidote to the presentism of the secular was to repeat the world of Kuyper and Butterfield, which could justifiably be called a kind of Christendom. It is rather simply to ask the basic question: how did *these people* cope with the problems of the pluralism endemic to what we now call the postsecular? What were *their* models, understandings, stories, and assumptions? And can these 'presecular' commentators, in fact, serve as orienteers for us, as 'postsecular prophets' in their own day for our time, as we come face to face with a world that so much of our secular age has left us unprepared for?

The structure of this chapter answers this question in two basic ways. First, I provide a survey of Abraham Kuyper's social and political thought for the purposes of global politics. While I have done this at greater length in other places (Joustra 2017), for our purposes I describe three basic sign posts: the necessary pluralism of international law, the undeniably religious character of nations, and primacy of freedom of religion or belief in global politics. Second, I provide a contrast in the historian and scholar Herbert Butterfield. These include his Augustinian account of the cultural and historical relativity of rationality, the dilemmas of sovereignty and anarchy amid such relativity, and the problem of an ethics of conviction amid such wide scale pluralism.

What unites these two otherwise seemingly disconnected thinkers is not only what I call their postsecular Augustinianism (neither kept their 'religion' out of it), but at least three points that serve as postsecular guides: (1) the variability of religion and worldview, (2) the problem of law and of ethics amid both worldview pluralism and anarchy, and (3) practical advice on how these conditions shape both the problem and the opportunity for an empathetic, non-hegemonic, postsecular politics of pluralism.

Abraham Kuyper among the nations (1837–1920)

We might ask, why refer to Abraham Kuyper to focus a conversation on postsecular politics in global affairs in the present age? Why, indeed, a man whose own foreign affairs during his short-lived time as Dutch prime minster (1901–1905) been described simply as a 'failure', which 'in fact did not become a foreign policy at all' (Kossmann 1978: 429)?

Abraham Kuyper (1837–1920) was many things in life: journalist, editor, pastor and theologian, founder of a university, founder and leader of a political party, and even prime minister of the Netherlands. Much has been made of his intellectual and political legacy, and his writings, which were voluminous, are still being translated into English, Korean, and more. Less has been made of his foreign affairs.[2]

But it is not Kuyper's curriculum vitae, his success or failures, that justify our excursion into this thought and life. Rather, it is his unique kind of Augustinian realism, a very European (rather than American) approach to international law and society, culture and religion, that fits a picture of postsecular politics partly for denying that a purely 'secular kind' could ever really exist. His approach not only takes for granted the fact of pluralism, but he even treats it with a kind of theological seriousness that acknowledges certain basic pluralisms as *good*, as intrinsic to the work of any politics reasonably called just and right. Far from the hegemony of the secular, Kuyper laboured to deny the monolithic, homogenizing forces of modernity which sought to scalp the religious and cultural particularities of his time. A Dutchmen, Kuyper was uncommonly well-travelled, writing memoires about his encounters 'around the Old-World Sea' (the Mediterranean) with Muslims, his lectures in America, his foreign policies in Indonesia and South Africa.

Three points are of note. The first is Kuyper's conviction about the capacious definition of religion, orientations at the heart of nations, whether acknowledged or not; second, by way of implication, the wide-scale pluralism of international law as a model for postsecular politics (see Kaemingk 2018 for contemporary commentary on Dutch and American society via this Kuyperian lens); and, third, a primacy on freedom of religion or belief as fundamental to postsecular politics, in part to preclude the kind of hegemony, secular or religious, that Kuyper saw in his day (Christendom), that would come in Butterfields (totalitarian fascism and communism), or the soft-secularism of (post)positivist projects in later generations. What is at stake for Kuyper, fundamentally, is the religious root of people and their freedom of religion or belief.

The religious life of the nation

Abraham Kuyper's Calvinism, which carries with it much Augustine, had what to some postmodern minds may be a peculiar definition of religion. By religion, Kuyper meant not only gathered worship of a specific confession, but the foundational beliefs and practices of cultures and societies. The 'religious' life of a nation was therefore not just for religious people, but was about core beliefs and motivating mentalities that rooted things like the dignity of human persons, equality before the law, and so forth. Kuyper believed that these 'foundations' of a society required regular, concerted attention, and were the wellspring from which national policies took life and force. I use foundations in the plural even if Kuyper himself had his preference for Calvinistic Christianity, partly because that preference does not rule out the expectation that others would have their own convictions. Foundations may be plural, but they should not—Kuyper would say cannot—be simply ignored or ruled out. Religion, then, is not a by-product of national life, but it is one of those wellsprings from which national life emerges. To even *do* foreign policy is first, then, to ask *religious* kinds of questions: what is it my neighbour loves? And how have they arranged their social and national life around the objects of their worship?

Foundations, in the plural, also meant for Kuyper a qualified endorsement of state sovereignty. The challenges of the twentieth century—its rival passions, loves, and religions—demanded, for him, a balancing and coordinating set of powers and institutions. And with the growth of technological and economic power, he foresaw the need to grow the institutions of justice alongside them. Governance, in other words, must keep pace with the scale of problems to be governed, and not lapse into the tyrannical states Kuyper so feared, but recognizing that public justice in a century dominated by the pace of technology and commerce that was before unseen would need governance regimes capable of maintaining the *boundaries* of those spheres of life. State sovereignty, then, for Kuyper was not about the absolutist authority on the part of the nation-state, but rather about investing in it sufficient authority so that it may successfully accomplish its task of doing public justice. 'Sovereign in its own sphere', as Kuyper would say, means a different thing than Westphalian 'state sovereignty'. The former speaks of balanced, bounded powers, the latter of uncontested, anthropocentric authority. For Kuyper, sovereignty could never be absolute, for where states fail in their stewardship, there may well be a role for higher authorities, international laws one among them, to mitigate such failure.

The laws of the nations

International law was, for Kuyper, also therefore about the *necessarily plural* foundations of any global order, about the need for religious and secular alike to recognize the instability and ultimate danger in proposing any singular, hegemonic foundation for something as vast and plural as international law.

The laws of the nations, for Kuyper, may perhaps be thought of as 'secular' but only in the sense of being commonly agreed too, not in the sense of being without multiple, often religious, justifications. The analogy here might be the negotiations over the Universal

Declaration of Human Rights. Here, certainly, there were Christians of conviction arguing from those convictions in theologically serious ways about the rights and responsibilities of citizens in the world. But here, also, were co-religionists, Muslims, Hindus, and others, making their own arguments for those rights.

For Kuyper, there could be no *other* way of imagining international law in a plural world. The foundations of law were always religious in some respect, which is to say they always depended on values, virtues, perspectives on what is right, true, and good that could never be purely empirical. They depended on beliefs. Kuyper called these founding 'beliefs' religious ground motives, the motivating core or life force of a society or civilization. In a world where we can not only expect but even strive to *protect* such diversity, the work of relations among nations in postsecular times becomes as much *interreligious dialogue* as it does politics. All laws, per Kuyper's thought, have one foot in religion, broadly understood. The key, then, is finding a *common* cause, rather than abdicating particularistic visions. Here there is a lesson for both the Christian heirs of Kuyper, and the students of religion and foreign policy more generally: *both* the singular reliance on Christian foundations (Christendom) *and* the hope for other universalistic (North Atlantic Secular) foundations are unstable and ultimately dangerous because they neglect the particularistic grounds of other, especially non-Western, societies and cultures. Such an exercise would repeat the colonial failures of Europe's past, rather than recognize *the necessarily particular and plural roots* of international law and its common causes.

Freedom of religion or belief: worldview pluralism in political practice

Finally, there is no question that perhaps Kuyper's most driving concern was the free proclamation of the gospel. For Kuyper, a fundamental feature of foreign policy was making safe the world for the proclamation of the gospel, and so he would be a passionate advocate today of what is called freedom of religion or belief. For while Kuyper unquestionably had a special priority on Christian evangelism, he also recognized that fundamental biblical precept that there can be no coercion in religion, and therefore *all religions* must have the same freedom he so cherished for Christianity. If, indeed, the character of nations is shaped in part by their religious worldviews, freedom of religion or belief not only ensures the Great Commission, it also ensures the *organic* character of nations; the freedom to contest and debate what a nation *is*, its identity, its future, and its loves. Here, Kuyper's Calvinism dovetails with scholars of religion and foreign policy: for no nation that cannot debate its spiritual parentage can be truly free in any meaningful sense of the term (Toft et al. 2011). There is no 'self-determination', and no 'organic' character where the fundamental questions of life and God are denied to people. The incapacity to do political theology (see Joustra 2017), for Kuyper, would be a denial of the organic character of a nation; it would be an authoritarian formalism, frozen and fragile, destined not to bend and grow, but simply to break.

Herbert Butterfield (1900–1979) and the early English School

Herbert Butterfield was an academic historian, a Cambridge man, and famous foremost for his historiographical work, particularly on the 'Whiggish interpretation of history'. He was also, somewhat unexpectedly, one of the key founders, in fact the *original* founder, of the British Committee on the Theory of International Politics, and thereby the English School of International Relations (Dunne, 1998). Butterfield was not a journalist, theologian, or politician like Kuyper. He often had to be 'pushed' into areas he did not consider his own, and despite his devout Methodism, he was no theologian (McIntire 1979).

He had a traditional academic's disdain for fashionable causes of the day, as Adam Watson recalled him often saying 'I value every one of the fifty-two miles that separate Cambridge from Westminster' (Coll 1985: x).

He did, however, produce much more original and focussed work on the international system than Kuyper did. His first major foray into political theory, *The Statecraft of Machiavelli* (Butterfield 1940), worked out classic issues of statecraft and ethics. Butterfield's concern with major problems in international affairs took clearer focus after the war, with *Christianity, Diplomacy and War* (1953) and *International Conflict in the Twentieth Century: a Christian View* (1960). It was in 1958 that Kenneth W. Thompson, Butterfield's friend and admirer from across the Atlantic, invited Butterfield, through Rockefeller Foundation, to begin the British Committee on the Theory of International Politics. This group, including Martin Wight, Michael Howard, Desmond Williams, Donald MaKinnon, Hedley Bull, and Adam Watson was a mix of historians, philosophers, and political scientists. Their first series of essays was edited by Butterfield and Martin Wight, *Diplomatic Investigations* (1966).

Unlike Kuyper, then, we need not glean through Butterfield's policies from his time in office, party platforms, opinion editorials, or longer treatises on issues of common grace, the church, theology, and so forth, the pillars of his international thought. Butterfield's international theory is published for us in a more traditional manner, and the subject itself of a small secondary literature on Butterfield's international thought (for example Coll 1985; Dunne 1998; Jones 2003). What we can summarize are at least two broad points.

First, Butterfield's Augustinian temper informed his ferocious criticisms of what was called the Whiggish interpretation of history, the tendency of historians to write and analyze history from their own moral and cultural positions rather than try to hear voices from the past in their own voices, detached in some sense from the tyranny of their present sensibilities. This criticism has, in fact, been a subject of protracted historiographical debate, even earning it the title, the 'Herbert Butterfield problem' (Sewell 2003). I aim to show that this first pillar of Butterfield's thought, however, is not the nascent positivism he has been accused of, but just the reverse: the inevitability and contestability of rationality, and the impossibility of understanding other places, times, and cultures without a deep exercise in empathy and sympathy.

Second, Butterfield's Augustinian inspiration also strongly implied that precisely *because* of this variability of rationality, misunderstanding and fear are prominent features in the international system. This was, of course, not merely the liberal lament of the tragedy of diplomacy in an unequal world, but a very sobering Augustinianism that fear is basic, because real evil persists, human beings and systems do indeed often seek self-benefit, and diplomatic interventions are therefore extremely difficult. He saw an intimate connection between the methods of historical science and those of good statecraft (Coll 1985). But this makes the diplomats' job doubly difficult for the historians, because not only must understanding be somehow achieved across wide epistemic gaps (a historian's job) but that understanding must somehow be leveraged into productive diplomatic relations which ease tensions and make for a more just world. The stakes are therefore very high and the odds extremely poor, underlining the tragedy of ethics and anarchy in international politics.

These, says Charles A. Jones, are Butterfield's pillars:

> each embedded firmly in his Christian faith… a deontological ethics, recognition of the dilemmas provoked by the anarchical structure of international politics, and an emphasis on the radical individuality of the person, and by implication, the cultural and historical variability of rationality.
>
> *(Jones 2003: 372)*

The Whig interpretation of history and the variability of rationality

Butterfield made his reputation with his short book, *The Whig Interpretation of History* (1931), which challenged his day's liberal historians for letting their moral values 'intrude into their historiography' to construct a 'kind of drama of good versus evil, in which the good eventually and inexorably prevailed' (Hall 2013: 16). This, he argued, 'was the worst kind of history' (Hall 2013: 16), because it marshalled the lives and loves of other human beings as little more than fodder for the fashionable debates of the present. Butterfield, a reader and admirer of Leopold van Ranke, even felt that historians had a *moral* and *religious* obligation. Ranke wrote, 'nothing in the world... can ever be regarded as existing merely for the sake of something else' (Coll 1985: 49). He was fond of recalling Ranke's dictum that all generations of humankind are equidistant from God (*ibid.*: 49; Butterfield 1955). For Butterfield, it was not simply being a professional historian that demanded a better historiography, it was the command of God to love his neighbour, which implied *knowing* his 'historical neighbour' for who he or she *really was*, not merely who we'd like them to be, or how they're useful to us, for our own purposes. If, in fact, the historical subject really is our neighbour, then for Butterfield the same table of laws applies as those for our relations with our present neighbours, foremost among them the injunction to never bear false witness.

How could such a thing be achieved? It is possible, of course, that it *cannot*, in a perfect sense, be achieved, but Butterfield was convinced that there were *better* and *worse* ways of going about the work of history, and that the Whig interpretation was, indeed, the worst. What was demanded of the historian, then, was an enormous effort of empathy and sympathy, driven first by an encounter with the people, the age, and the culture. Judgement, in other words, should be suspended, not abandoned, but paused, to truly hear the other, hear the 'foreigner' (for the dead are arguably more foreign to the living than the living to one another, regardless of their present-day cultural diversity) on their *own terms*, understand their motivations carefully, thoughtfully, and without rushing to the sorts of present moral judgements that eradicate their agency and morality in service of powerful, contemporary purposes. Writes Alberto R. Coll in *The Wisdom of Statecraft* (1985: 8–9):

> This preoccupation with transcending the historian's cultural and ideological perspective – with putting oneself in the place of others through a mighty exercise of sympathetic imagination, understanding, a suspension of judgment, and ultimately Christian love – would bear fruit later in Butterfield's analyses of diplomacy and his prescriptions to statesmen.

Butterfield was criticized for what was called his 'neutrality postulate' or, for the later term he used, 'technical history' to describe this methodological concept. The debate about whether, for example, Butterfield was attempting a kind of proto-positivist reading of history (to get at *true* history, unencumbered by bias) or even relativist historiography (that each truth is equal in its own way) was fierce (Sewell 2003). But there is good reason to rescue Butterfield from his critics on this count, not least because either criticism seems badly at odds with his Augustinian temperament.

Butterfield expert Keith Sewell writes of 'The Herbert Butterfield Problem' that critics have failed to observe the distinction he made between the 'Whig political method' and the 'Whig historiographical method' (Sewell 2003). This difference becomes clear in reading the relationship between *The Whig Interpretation* (1931) and *The Englishman* (1944), and in the latter he plainly states that he has issued no attack against the Whig political tradition, but rather against its historians. Ian Hall writes, that Butterfield had argued 'that while the Whig historians had misrepresented the past, they had shaped good political thinking and practice which had served to build, advance, and sustain English liberal democracy' (Hall 2013: 17).

What Butterfield is, therefore, arguing is not dissimilar to what Abraham Kuyper might say about basic worldview: to understand these, we must orient ourselves towards them affectively and empathetically, to grapple with what they themselves believed and practised as *though it were really true*, and so enter imaginatively into this foreign country of history. This neither pretends towards perfect positivism, as though bias and prejudice can be shed in historical method, nor does it default on moral judgement, as though all sin and evil can be excused as a matter of the 'times'. Rather, it is a historiography, developed as genuinely as possible, to understand a people and a past through their own convictions on their own terms. Those terms, for Butterfield, as for Kuyper, were often fundamentally religious, and such rival existential realities were hardest of all for the historian, like the statesman, to penetrate, because they inevitably shape the content of what we call rationality. Dialogue, then, becomes extraordinarily difficult, both with the past and the present, a condition exacerbated by Butterfield's second pillar, the conditions of fear and anarchy.

Ethics and anarchy

Butterfield, like Kuyper, saw certain basic orientations, worldviews, religious or otherwise, as shaping rationality and ultimately the context and conduct of nations, both in history and in contemporary affairs. And, so too like Kuyper, Butterfield struggled with how such wide-scale worldview pluralism could produce a pluralist rather than hegemonic, anarchic, and ultimately violent international politics. For Kuyper, part of the answer lay in international law, rules we might all agree on, to paraphrase Jacques Maritain, provided nobody asks us why. Butterfield struggled at greater length with this basic problem of anarchy, not primarily because of the problem of pluralism in global affairs, but because of what he called the reality of sin and evil. The reality of sin and evil justified a certain degree of fear and anarchy, which often precluded clear communication and international cooperation.

Butterfield described man's fundamental political problem, both within the state and in international relations, 'as that of a building the glasslike, fragile edifice of civilization atop the inherently volcano of human passions, irrationality, and disorder' (Coll 1985: 81). The problem of international politics was 'inseparable from and exceedingly complicated by the problem of human nature' (Coll 1985). Writes Tim Dunne, 'mutual fear lies at the heart of Butterfield's conception of the tragic nature of international conflict' (1998). Butterfield himself writes in *History and Human Relations* (1951: 21):

> It is the peculiar characteristic of the situation I am describing – the situation of what I shall call Hobbesian fear—that you yourself may vividly feel the terrible fear that you have of the other party, but you cannot enter into the other man's counter-fear, or even understand why he should be particularly nervous. For you know that you yourself mean him no harm, and that you want nothing from him save guarantees for your own safety; and it is never possible to realise or remember properly that since he cannot see the inside of your mind, he can never have the same assurance of your intentions that you have.

Here again we see how Butterfield's historiographical method is basic to his theory of international politics, that knowing the *other*, past or present, is a primary moral and political duty of any diplomat. For Butterfield, writes Dunne, 'diplomacy was the filter through which "so much of the world's past experience" flowed' (Dunne 1998: 78). Self-righteous statecraft, of the kind that Butterfield was bitterly critical during the Cold War, was dangerous precisely because it took as given the ethical righteousness of the self, and did not often hazard to imagine how we, ourselves, could be the monsters of other's moral imaginations. Kenneth Thompson described Butterfield's warnings on such ethics as his 'most lasting and valuable contribution' (Dunne 1998: 77).

'We cannot penetrate to the roots of fear', wrote Butterfield, 'if we merely condemn the other party moralistically. It is necessary to attack rather the structure of that fundamental dilemma which is the prime cause of international deadlock' (Butterfield 1960: 90). And again:

> There is aggression; there is tyranny; there is revolutionary ferment; but if we wish to civilize international affairs we must do more than arrogantly hold our own against the barbarians, merely meeting them with their own weapons. Everything is going to depend in fact upon what we do over and above the work of self-defence. There can be no international system until somebody finds a way of relieving the pressure and begins the task of creating confidence
>
> *(Butterfield 1960: 90)*

The presecular prophecy which Butterfield has for our postsecular world is therefore a version of what Charles Taylor calls the St. Francis of Assisi test: since rationality is variable, ethical calculations shift with it, and we must never make the fatal error of presuming our calculations are the same as other calculations, that the ethical traps and boundaries that define our diplomatic imaginations are the same globally. Rather, provided the reality of evil *and* worldview pluralism, our first set of diplomatic questions is essentially Augustinian: what is it our neighbours love, how have they have arranged their common life around it, what do they fear, and how can we build confidence around that problem to support an international system of mutual concern?

'We seem unable to subdue the demon of frightfulness', said Butterfield,

> in a head-on fight. Let us take the devil by rear, and surprise him with a dose of those gentler virtues that will poison him. At least when the world is in extremities, the doctrine of love becomes the ultimate measure of our conduct.
>
> *(Butterfield 1960: 98)*

He took his own advice to what many of his colleagues considered a dangerous, logical extreme, arguing that because power demanded 'moral responsibility' and thus significant investments into a 'tolerably civilized international order', the West must 'not manufacture, store, or ever use strategic nuclear weapons' (Coll 1985: 119). Fear and anarchy may have been endemic features of politics, but they did not have to have the last word: 'imperial hegemony or unlimited international anarchy were not the only avenues open to international statesmanship' (Coll 1985: 121).

Butterfield was fond of the Old Testament maxim, 'Fret not thyself because of evildoers' for 'much presumptuousness and many disasters in foreign policy could be traced to excessive fear and anxiety on the part of statesmen toward future possible or imagined contingencies' (Coll 1985: 68–9). There was a demand that statesmen begin with their historiographical method of empathy not only for rival rationalities towards understanding, but for a deontological ethic which demanded to act in the first instance and pressed actors in international affairs to mitigate, as best they could, the realm of fear and anarchy and build a world of greater confidence and understanding. The diplomat's efforts are always, in a sense, a fool's errand. There is no complete work for the diplomat, any more than for the historian, who must tirelessly exercise their imagination, while holding steadfast to their own ethical rationale for doing so. Writes Albert M. Coll his conclusion (1985: 153):

> The elimination of the tragic character of international politics and the resolution of its perpetual moral and political dilemmas through some transcendent principle or ideology were beyond the statesman's realm of action. To recognize this and act accordingly was, indeed, the beginning of the wisdom of statecraft.

Conclusion

It is hard to imagine two more temperamentally different men: Abraham Kuyper, the renaissance man, a scholar-activist who never found a project whose enormity could daunt him, and whose work has become the inspiration and preoccupation of a wide range of interdisciplinary projects; or Herbert Butterfield, the quiet Cambridge historian, whose interdisciplinary excursions had to be nearly forced upon him, and whose life and legacy, despite significant contributions to international theory and theology, has been confined mainly to debates in historiography. And yet, as this brief excursion has hopefully demonstrated these two unlikely colleagues have some measure of wisdom to share with the postsecular politics of our present world. We can, perhaps, summarize these, by way of conclusion, in three points: (1) religion and worldview; (2) the problem of law and of ethics, amid both worldview pluralism and anarchy; and (3) how these shape both the problem and the opportunity for an empathetic, non-hegemonic postsecular politics of pluralism.

For both Kuyper and Butterfield there was a profound acknowledgement of the epistemic limits of politics and history, what Kuyper's heirs came to call worldview, and what Butterfield in his Whiggish criticisms called the variability of rationality. Echoing in the background we hear, perhaps, Alasdair MacIntyre's (1988) now much-quoted question: *Whose Justice? Which Rationality.* This flips the secular script outside of its often-hegemonic position in the North Atlantic world, and demands a normative, religious account of the non-empirical presuppositions that inevitably shape history and politics. Or, as Kuyperian philosopher Nicholas Wolterstorff puts it,

> It is often said… that everyone has a 'set of presuppositions' or a 'perspective on reality' to bring to a theoretical inquiry. That may be true. But saying such things cannot be the end of the matter. It must at best be the beginning.
>
> *(Wolterstorff 1976: 22)*

The second signpost that Kuyper and Butterfield offer us is how to address the problem of politics amid such wide-scale worldview pluralism. Kuyper and Butterfield both offer us interesting options here, and their perspectives hardly exhaust approaches to postsecular pluralism. Kuyper is perhaps more practical, addressing at length the issue of international law as a set of principles set down by plural nations, all for their own domestic rationale. He sees such cooperation not as a new secular code, but rather as a set of civilizing boundaries that mark out the limits of international society, agreed upon and sustained by as capacious a group as possible. Butterfield, of course, is a better international theorist, struggling with *how* to build confidence amid the *justified* conditions of fear and anarchy in a world in which he finds sin and evil very real.

For both Kuyper and Butterfield there is no ready-made 'resolution' to this Augustinian dilemma, but there are *better* and *worse* ways to address it, the better being a commitment in the third place to empathy in widely plural politics, what Butterfield called the origin of his historiographical method. Ultimately the demand for both Kuyper and Butterfield came from a deontological ethic, in their case the commands of God, but also the demand of Christian love and charity as not separate from, but woven into, the needed and present virtues of international affairs. We may not share their specific faith, but as biographies and—to share Butterfield's own historical method—colleagues who, like us, wrestle now with a postsecular politics, marked more often than not by serious religion, we could do worse than ponder these signposts, the lives of these sometimes peculiar men, and share, at least, a common cause with facing the fear and anarchy of the international system with honest self-awareness, clear-minded empathy, and the possibility of growing confidence. For men who talked at length about fear, it is ironic, but nonetheless appropriate, that the final word these prophets offer us in the postsecular age, echoing Bob Goudzwaard (2007), is hope in troubled times.

Notes

1 In Philpott's article at least nine definitions are offered. I present six in the context of the North Atlantic world (Joustra 2018); Jonathan Fox (2015) offers many more.

2 Much of this argument is borrowed, with permission, from a longer reading of Abraham Kuyper (Joustra 2017) and the newly available English translations of Abraham Kuyper's work made available by the Abraham Kuyper Translation Project. I am grateful for the earlier drafts of these works; published versions can be found at www.abrahamkuyper.com/, accessed online 20-11-2017.

Further reading

For further reading on the Augustinian tradition of Kuyper and Butterfield as a form of postsecular politics, I recommend four journals:

The Review of Faith & International Affairs

Published by the American think-tank the Institute for Global Engagement and Taylor & Francis (Routledge), edited by Dennis Hoover (2001–), this journal regularly features scholars of all faiths (and of none) publishing on postsecular issues in global politics and society.

Providence: A Journal of Christianity & American Foreign Policy

Published by the Institute on Religion & Democracy and the Philos Project, edited by Mark Tooley and Marc LiVecche, this journal features commentary from American-evangelicals part of a broader tradition of Christian Realists.

Comment magazine

Published by the Canadian think-tank Cardus, edited by James. K.A. Smith (2012–), a prominent Kuyperian and scholar, this journal features issues dedicated to the Anti-Revolutionary Party (Kuyper's party) and problems in postsecular society.

Public Justice Review

Published by the American think-tank The Center for Public Justice, edited by Robert Joustra (2018–), this journal connects public theology to public policy using the framework of Kuypern pluralism expounded in this contribution.

References

Butterfield, H. (1931) *The Whig Interpretation of History*, London: W.W. Norton.

——— (1940) *The Statecraft of Machiavelli*, Ann Arbor, MI: G. Bell and Sons Limited.

——— (1944) *The Englishman and His History*, Cambridge: Cambridge University Press.

——— (1951) *History and Human Relations*, London: Collins.

——— (1953) *Christianity, Diplomacy and War*, New York, NY: Abingdon-Cokesbury Press.

——— (1955) *Man on His Past: the study of the history of historical scholarship*, Cambridge, MA: Cambridge University Press.

——— (1960) *International Conflict in the Twentieth Century: a Christian view*, London: Routledge.

Butterfield, H. and M. Wight (eds.) (1966) *Diplomatic Investigations*, London: George Allen & Unwin Ltd.

Coll, A. R. (1985) *The Wisdom of Statecraft: Sir Herbert Butterfield and the philosophy of international politics*, Durham, NC: Duke University Press.

Dunne, T. (1998) *Inventing International Society: a history of the English school*, London: Macmillan Press.

Fox, J. (2015) *Political Secularism, Religion and the State: a time series analysis of worldwide data*, Cambridge: Cambridge University Press.

Goudzwaard, B. (2007) *Hope in Troubled Times: a new vision for confronting global crises*, Grand Rapids, MI: Baker Academic.

Hall, I. (2013) 'Martin Wight and the whig tradition of international thought and practice', paper given at the Centre for British Studies, University of California, Berkeley.

Jones, C. A. (2003) 'Christian realism and the foundations of the English school', *International Relations*, 17: 371–88.

Joustra, R. (2017) *The Religious Problem with Religious Freedom: why foreign policy needs political theology*, London: Routledge.

——— (2018) 'Abraham Kuyper among the nations', *Politics and Religion*, 11(1): 146–69.

Kaemingk, M. (2018) *Christian Hospitality and Muslim Immigration in an Age of Fear*, Grand Rapids, MI: Eerdmans.

Kossmann, E. H. (1978) *The Low Countries: 1780–1940*, Oxford: Oxford University Press.

MacIntyre. A. (1988) *Whose Justice? Which Rationality?* Notre Dame, IN: University of Notre Dame Press.

McIntire, C. T. (ed.) (1979) *Herbert Butterfield: writings on Christianity and history*, Oxford: Oxford University Press.

Philpott, D. (2009) 'Has the study of global politics found religion?', *Annual Review of Political Science*, 12: 183–202.

Sewell, K. (2003) 'The "Herbert Butterfield Problem" and its resolution', *Journal of the History of Ideas*, 64(4): 599–618.

Thomas, S. (2005) *The Global Resurgence of Religion and the Transformation of International Relations*, New York, NY: Palgrave Macmillan.

Toft, M., Philpott, D. and T. Shah (2011) *God's Century: resurgent religion and global politics*, New York, NY: W.W. Norton & Company.

Wolterstorff, N. (1976) *Reason within the Bounds of Religion*, Grand Rapids, MI: Eerdmans.

16

Anticipating postsecularity

Christopher Rowland

Introduction

The theme of this chapter is the recovery of the secular in the biblical tradition as a necessary vehicle for understanding the import of the theological content of the Bible, which has been a crucial, if neglected, part of Christian intellectual history. The chapter concerns two proponents of secular theology, Gerrard Winstanley and William Blake, which arguably anticipates elements of postsecularity today (Rowland 1988, 2010, 2014, 2017a). While some have referred to a 'postsecular sensibility' (Blanton et al. 2016; see Chapter 1), what Winstanley and Blake are doing in their work is better described as indicative of the necessarily secular character of Christian incarnational theology rather than 'crossover narratives' of postsecularity (Habermas 2010; Beaumont and Baker 2011; Cloke and Beaumont 2013; Cloke et al. 2019). Their work exemplifies the impact of religious language as the primary content of secular and human worlds.

The extant writings of Gerrard Winstanley (1609–1676) come from a short period in his life (c. 1646–1652), though there is some doubt about the date of the earliest writings. Most of them were written in the years preceding, or in the immediate aftermath of, the execution of Charles I in 1649. In that year, Winstanley and his colleagues began to dig the common land, which was an earnest of the kind of social equality of a new age which the historical moment offered. That experiment lasted little more than two years, but in Winstanley's writings produced some of the most remarkable political theology ever written. William Blake (1757–1827) also lived during a period of upheaval, and his writing career spans the years immediately following the French Revolution and the secession of the American colonies, both of which are subject of illuminated writings by Blake, *America a Prophecy* and *Europe a Prophecy* (Bindman 1978, 2000). Both Winstanley and Blake well exemplify an important aspect of the Christian tradition, the hope for heaven on earth and the anticipation in contemporary social life of that hope for the future.

Blake may seem a strange candidate, when so much of his later work resorted to a complex mythology which seemed to be remote from the world of flesh and blood. Decades before Ludwig Feuerbach's *Das Wesen des Christentums*, Blake saw theology as about human nature, and social arrangements and the often-complex interaction between his mythical beings expressed the ways in which men and women organized themselves, or not as the case may be, and how the imbalance in society and the individual needed to be set right and fraternal relations established, so that human relating and equality could flourish:

> And if God dieth not for Man & giveth not himself Eternally for Man Man could not exist. For Man is Love: As God is Love: every kindness to another is a little Death: In the Divine Image nor can Man exist but by Brotherhood.
>
> (Jerusalem *96: 25–8, E256)*

For Winstanley, the Christian story of the Fall and apocalyptic terminology offer a way of understanding the rise of private property and the political struggle between the rich and poor.

Even if neither was deliberately attempting any 'crossover' between religion and secularity, they do exemplify what Cloke and Beaumont (2013: 37) describe, namely, 'religious and secular utterances are mixed together in the pre-political broth, and may not be easily distinguishable within the discourses that become associated with political and ethical praxis'. Following Jürgen Habermas, they suggest that 'as reason reflects on its deepest foundations, it discovers other origins, other directions, and effectively undergoes an exercise of philosophical repentance in which, with no theological intention, reason becomes aware of its own limitations and thus transcends itself into an openness towards other frameworks of reason, including religious tradition'. Perhaps unknowingly, these words echo Blake's in *The Marriage of Heaven and Hell* about imagination, energy, and desire providing the resources for reason to build on. In their writings, Winstanley and Blake reconnect religious concepts with their roots (Rowland 2017a).

The secular character of the biblical message

One doesn't have to be a postsecular Christian to think that the foundation of whatever Christianity means is about the secular. If the fundamental doctrine of Christianity means anything at all, it is that the secular and the physical are key to the beginnings of the knowledge of God: 'The Word became flesh and dwelt among us' (John 1:14); and 'Those who say, "I love God," and hate their brothers or sisters, are liars; for those who do not love a brother or sister whom they have seen, cannot love God whom they have not seen' (1 John 4: 20). Centuries before the New Testament and millennia before it was picked by liberation theology, the prophet Jeremiah asserted that knowing God is doing justice:

> Woe to him who builds his house by unrighteousness, and his upper rooms by injustice; who makes his neighbours work for nothing, and does not give them their wages; who says, 'I will build myself a spacious house with large upper rooms,' and who cuts out windows for it, panelling it with cedar, and painting it with vermilion. Are you a king because you compete in cedar? Did not your father eat and drink and do justice and righteousness? Then it was well with him. He judged the cause of the poor and needy; then it was well. Is not this to know me? says the Lord,
>
> (Jeremiah *22: 13–6)*

Similarly, Gustavo Gutiérrez put it that:

> to be followers of Jesus requires that [we] walk with and be committed to the poor; when [we] do, [we] experience an encounter with the Lord who is simultaneously revealed and hidden in the faces of the poor.
>
> *(1984: 37–8)*

Gerrard Winstanley: rising up of Christ in sons and daughters

Winstanley's writings offer one of the best examples of theology in context (Rowland 2014; 2017a, 2017b). Most of Winstanley's extant writings come from just before, and shortly after, Charles I's execution, much of it connected with the setting up of the Digger colony first on St George's Hill in April 1649, weeks after the execution of Charles I. Winstanley was prompted by a revelation that he and his companions should dig the common land, thus claiming what they regarded as their rightful inheritance ('True Levellers Standard' or 'Declaration to the Powers of England', Corns et al. 2009 [CHL hereafter] ii.13–5). In most of his extant writings, the mix of the emphasis on social and political change, the emphasis of the experience of the divine within, and the suspicion of book learning, anticipating Blake's contrast between 'Inspiration' and 'Memory', buttressed by a robust socio-political biblical interpretation, testify to his conviction about the possibility for political change as Christ rose in men and women.

Key themes in Winstanley's theology

In two works written within a short time of each other in 1648–9, *The Saints Paradice* and *The New Law of Righteousnes* (CHL i.313–407 and i.472–600), we find enunciated some of Winstanley's most distinctive theological themes. *The Saints Paradice* is in large part a challenge to those 'professors' who may know the Bible well and its history, but 'who worshipped a God, but neither knew who he was, nor where he was'. Winstanley criticized those who buried their heads in study about what happened in '*Moses* time, in the Prophets time, in the Apostles, and in the Son of Mans time' without waiting 'to find light and power of righteousness to arise up within his heart' ('New Law', CHL i.547; Rowland 2017a).

The Bible and experience

It is not knowledge of the words of the Bible that count but experiential understanding of God.

> It is very possible, that a man may attain to the literal knowledge of the Scriptures .... and may speak largely of the History thereof, and draw conclusions, and raise many uses for the present support of a troubled soul ... and yet ... may be not only unacquainted with but enemies to that Spirit of truth, by which the Prophets and Apostles writ.
>
> For it is not the Apostles' writings, but the spirit that dwelt in them that did inspire their hearts, which gives life, and peace to us all ....
>
> (Saints Paradice *1, CHL i.322)*

He argued that the scriptures had been written 'by the experimentall hand of Shepherds, Husbandmen, Fishermen and such inferiour men of the world' (*Fire in the Bush*, CHL ii.200). Indeed, in language reminiscent of twentieth-century liberation theologians, Winstanley, echoing Matthew 11:25–6, stressed the interpretative ability of those who match experience and the Bible (Hill 1993: 223–4; see also Hill 1983). Thus, the 'plough man' is in as good a position as the university scholar to understand God.

> Nay let me tel you, That the poorest man, that sees his maker, and lives in the light, though he could never read a letter in the book, dares throw the glove too al the humane learning in the world, and declare the deceit of it, how it doth bewitch & delude man-kinde in spiritual things, yet it is that great Dragon, that hath deceived the world,

> for it draws men from knowing the Spirit, to own bare letters, words and histories for spirit: The light and life of Christ within the heart, discovers all darknesse and delivers mankind from bondage, And besides him there is no Saviour.
>
> *(The New Law of Righteousnes, CHL i.537)*

Anticipating indeed the hermeneutical privilege of the poor found in liberation theology, Winstanley suggested that it was the poor and outcast who would be the instruments of change:

> The Father now is rising up a people to himself out of the dust, that is, out of the lowest and despised sort of people, that are counted the dust of the earth, man-kind, that are trod under foot. In these, and from these shall the Law of Righteousnesse break forth first, for the poor they begin to receive the Gospel, and plentifull discoveries of the Fathers love flows from them, and the waters of the learned and great men of the world, begins to dry up like the brooks in Summer. Math. 11.25. 1 Cor. 1.27.
>
> *[text citations in the original] (The New Law of Righteousnes, CHL i.508)*

Like William Blake after him, Winstanley emphasized 'the sight of the King of glory within', which does not depend on:

> strength of memory, calling to mind what a man hath read and heard, being able by a humane capacity to joyn things together into a method; & through the power of free utterance, to hold it forth before others, as the fashion of Students is in their Sermon work.
>
> (New Law, *CHL i.557)*

It is the 'inward power of feeling experience' which counts, which even 'a plough man that was never bread in their Universities may have' (*The New Law of Righteousnes,* CHL i.557). What was required was 'a teacher within yourselves (which is Spirit)' who 'will teach you all things, and bring all things to your remembrance, so that you shall not need to run after men for instruction' (*The New Law of Righteousnes,* CHL i.502; *Truth Lifting up its Head,* CHL i.435).

In Chapter VI of *The Saints Paradice*, Winstanley makes an appeal to 'the Powers, Governours and Armies of the Land' to 'learn to worship the Lord in righteousness'. He writes that 'the Father is about the work .... That all enemies to Christ are made his footstool'. The reason is that 'Jesus Christ is upon his rising from the dead, and will rule King of Righteousnesse in flesh' for 'now is the time come that he will lift up himself, and tread the powers of the flesh under his feet' (CHL i.355).

The secular import of theological language

This kind of language about the eschatological vindication of God's ways at Christ's coming is unexceptional. But what is completely unexpected is the way in which it is glossed by Winstanley in what is an explicit critique of a transcendent theology and an explanation of how key features of Christian doctrine are to be understood. The following extract makes this point clear:

> And friends, do not mistake the resurrection of Christ. You expect he shall come in one single person, as he did when he came to suffer and die, and thereby to answer the types of *Moses* Law; let me tell you, if you look for him under the notion of one single man after the flesh, to be your Saviour, you shall never, never taste salvation by him.
>
> *(Saints Paradice, CHL i.356)*

Salvation does not come by believing there was a man, Jesus, that lived and died at Jerusalem but by 'the power of a meek spirit come into you'. There is also no need to 'look for a God .... to be in a place of glory beyond the Sun, Moon and Stars, nor imagine a divine being you know not where, but you see him ruling within you, and not onely in you, but you see him and know him to be the spirit and power that dwells in every man and woman; yea, in every creature, according to his orbe, within the globe of Creation'. Gone should be the time when one 'looks for a God without himself, and worships God at a distance, he worships he knows not what'. Rather, 'he that looks for a God within himself and submits himself in the spirit of righteousness that shines within; this man knows whom he worships, for he is made subject, and hath communitie with the spirit that made all flesh, that dwells in all flesh, and in every creature within the globe' (*The Saints Paradice*, CHL i.356–9). What is needed is for the power that dwelt in the humane flesh of Christ, which was spread abroad at the Ascension, to appear in many men and women. So, Christ is now rising and spreading himself abroad in his sons and daughters, and thus enlightening the whole creation 'in every branch of it, and cover the earth with knowledge, as the waters cover the seas' (CHL i.356). What is essential is not theological abstractions but transformed human lives.

In the slightly later *The New Law of Righteousnes,* he took up a similar theme he had expounded in *The Saints Paradice*. Here in Chapter 1, Winstanley wrote of the need for Christ to give way to the Holy Ghost, so that the distinctive power of Christ can extend far beyond one historical person to many 'sons and daughters'. Thus, this will take place: 'the spreading of the Divine power ...the one Law of Righteousnesse, being the teacher of all. So, that upon the rising up of Christ in sons and daughters, which is his second comming, the ministration of Christ in one single person is to be silent and draw back, and set the spreading power of Righteousnesse and wisdom in the chair, of whose Kingdom there shall be no end' ..... 'So, the Son delivers up the Kingdom unto the Father (cf. 1 Cor 15:24); And he that is the spreading power, not one single person, become all in all in every person, the one King of Righteousnesse in every one' (*New Law of Righteousnes*, CHL i.485). The final coming of Christ is the manifestation of Christ-like attitudes in all.

Later he argued that 'the fleshly man seeks content and peace from Sermons, Prayers, Studies, Books, Church-fellowship, and from outward Forms and Customs, in Divine Worship. But that peace that is built upon the foundation of gold, silver and precious stones, will fall and come to nothing'. He criticized those who seek for 'the New Jerusalem, the City of *Sion,* or Heaven, to be above the skies, in a local place, wherein is all glory, and the beholding of all excellent beauty, like the seeing of a show or a mask before a man. And this is not to be seen neither by the eyes of the body still the body be dead. A strange conceit'. When Christ, 'the second *Adam* rises up in the heart, he makes a person see Heaven within' ... 'This is heaven that will not fail us' .... 'This Christ is within you, your everlasting rest and glory' (*The New Law of Righteousnes,* CHL i.550).

Winstanley's emphasis on present experience continues an ancient tradition in Christianity in which the need to internalize the facets of doctrine is what counts. But there is more than that, a suspicion of transcendence and an emphasis on the God-within. The application of language to personal and social renewal also has its antecedents, not least in the writings of Paul (especially Colossians 3:1 and Romans 6:4–5). The difference is that Paul retains a strong interest in what happened in history in the foundational character of the first appearance of Christ to the disciples.

'The rising of Christ in sons and daughters which is Christ's second coming' may seem to be a rather fanciful interpretation of the Son of Man coming on the clouds of heaven drawing the elect from the four winds (e.g. Mark 13: 26). Nevertheless, as many modern biblical scholars have pointed out, the origin of this image in Daniel 7 clearly identifies the human figure who comes with the clouds (Daniel 7:13) with the saints of the most high who will inherit the kingdom (Daniel 7: 18; 27). So it looks as if Winstanley may have anticipated the modern interpreters in suggesting this communal interpretation by identifying the coming of the Son of Man with the rising of Christ in sons and daughters. It is consistent with the ways in which Winstanley took major theological themes and interpreted them in terms of human actions.

What is it to walk righteously

Winstanley demonstrated the priority he gave to the ethical in answering the question he poses, 'What is it to walk righteously'. It echoes some of the themes from the Gospel of Matthew, especially Matt 25:31–45:

> First, When a man lives in all acts of love to his fellow-creatures; feeding the hungry; clothing the naked; relieving the oppressed; seeking the preservation of others as well as himself; looking upon himselfe as a fellow-creature (though he be Lord of all creatures) to all other creatures of all kinds; and so doing to them, as he would have them doe to him; to this end, that the Creation may be upheld and kept together by the spirit of love, tendernes and one-nesse, and that no creature may complain of any act of unrighteousnesse and oppression from him.
>
> Secondly, when a man loves in the knowledge and love of the Father, seeing the Father in every creature, and so loves, delights, obeyes, & honours the Spirit which he sees in the creature, and so acts rightly towards that creature in whom he sees the spirit of the Father for to rest, according to its measure.
>
> *(Truth Lifting up his Head, CHL i.418–9)*

Winstanley's emphasis on the earth as 'common treasury' seems to maintain a humanocentric view of the reality of the world. His play on the word Adam, however, indicates a grasp of the closeness of the bonds between humanity and the earth. Politically he is as aware as any one of his generation of the 'devastation' caused by monarchy and the maintenance of the interests of the elite, ideologically and institutionally.

William Blake: challenging the contemporary views of the Bible

In 1790, towards the end of *The Marriage of Heaven and Hell,* the note Blake struck is entirely optimistic, as the clarion-call 'Empire is no more! And now the lion & wolf shall cease.... For everything that lives is Holy' (E45) indicates. Elsewhere in the same work Blake insisted that 'the worship of God is: Honouring his gifts in other men each according to his genius, and loving the greatest men best, those who envy or calumniate great men hate God, for there is no other God' (*Marriage of Heaven and Hell* 22–3, Erman and Blake 1988 [E hereafter]: 43).

There can be few more pungent challenges to orthodoxy from one who regarded himself a Christian than Blake's *The Marriage of Heaven and Hell.* It is a very deliberate, and unsurpassed, tilt at received theological wisdom, not least in its title. Throughout *The Marriage of Heaven and Hell,* the devils take the role of advocates of that which Blake deems to be true to what he considers the gospel—the Spirit, the freedom from subservience to rules, the elevation of desire over repressions and energy over restraint, which offer a sustained critique of dualistic thinking promulgated by the errors of 'Bibles or sacred codes' and the advocacy of 'energy' as 'the only life' channelled through the body, and reason, which are 'the bound or outward circumference of Energy' (*The Marriage of Heaven and Hell*, 5–6, E35–6)

One of Blake's major critiques of the Bible concerned the way in which he considered major themes had been co-opted by 'the primeval Priests assum'd power' in the service of a monarchical state. There is 'One King, one God, one Law' *First Book of Urizen* 2:2; E70; 4:40; E72). Blake saw religion and politics intertwined. The enthroned monarchical divinity endorsed the polity of the British state and others like it. It is the divine 'brazen Book, That Kings & Priests had copied on Earth Expanded from North to South' (*Europe* 11:5; E64). The hegemony of this type of religion served the interests of a monarchical state, and it was the elite that benefitted from it.

In several of the plates early in the *Illustrations of the Book of Job,* completed a few years before Blake's death, God is depicted as a divine monarch, transcendent, surrounded by the heavenly host, and with a book open on his lap; below, Job and his wife discuss, as they too consult their books surrounded as they are by their family. The opening plate of the *Illustrations of the Book of Job* suggests that Blake interpreted the *Book of Job* as the story of a conventionally upright man, a creature of habit (hence the quotation of the words 'Thus did Job continually'), and a devoted adherent to a holy book. But as the series goes on, as a result of his bitter experience, he comes to a different understanding of God, a God who dwells not far off but with, and in, humanity. Across the top of the first image are the opening words of The Lord's Prayer, 'Our Father which art in heaven'. That is Job's view at the outset of the story from which the experience of suffering and vision helped deliver him.

Blake interpreted the meaning of God's appearance to Job in the whirlwind as Job coming face to face with God in the incarnate Christ, thus picking up on an ancient theme in Christian theology. Words from Job 42:5 dominate the textual commentary on this page, and indeed are a key to Blake's interpretation of the *Book of Job* as a whole. In this image, Blake has, as the main caption, his version of the words from the King James Version of the Bible 'I have heard of thee with the Hearing of the Ear but now my Eye seeth thee'. Blake regarded Job as a type of person who went through a profound change in his theological understanding from a transcendent God to a God who is with, and in, humanity, which is the point of the climactic plate 17 of Blake's *Illustrations of the Book of Job.*

For the first time in the Job series, Blake prints the biblical quotations in the open books in the marginal illustrations. Their position, subordinate to the image, is another way in which Blake emphasized the priority of what had been seen rather than merely relying on what had been received by word of mouth. The textual commentary has a selection of texts from the Gospel of John in which the incarnate Christ dwells with and in those who see God in Jesus ('I am in my Father, and ye in me, and I in you'; 'I and the Father are One'; 'He who has seen me has seen the Father'). God is not far off but with and *in* Job and his wife. It is also worth noting here that Blake, perhaps inspired by his relationship with Catherine, his wife, includes Job's wife in the process of theological education throughout the series. The inclusion of both Job *and* his wife is important. Henry Crabb Robinson's reminiscence of Blake saying 'Jesus is the only God' but then immediately adding 'And so am I and so are you' is an anecdote which reflects the theology evident in the selection of the biblical texts (Bentley 2004: 696; see also Bentley 2001).

In the very next image Blake chose other words from Job 42: ('And the Lord turned the captivity of Job when he prayed for his friends') to indicate that it is at the moment that Job prays for his friends that his redemption is sealed. The vision of God by itself is insufficient, for, in addition, forgiveness of sins is essential, and in the next illustration in the sequence, the words are from Matthew 5: 44: 'Love your enemies, bless them that curse you, do good to them that hate you, and pray for them which despitefully use you, and persecute you' (Rowland 2010: 58–64).

Would to God that all the Lord's People were Prophets

The stanzas in Blake's 'Preface to *Milton*' (William Blake, *Milton A Poem* Preface, E95–6), commonly known as 'Jerusalem', need little introduction. These familiar words are found in the context of a manifesto, in which 'Inspiration' is commended as superior to 'Memory', inspired by some words of John Milton (*Reason of Church Government*, Preface to Book II), and the Bible is preferred to classical authors. The whole is summarized by a clarion-call to prophecy epitomized by the quotation of a verse from Numbers 11:29. These words echo various biblical themes, for example Elijah's chariot, John's vision of the New Jerusalem come down to earth from heaven, and the spiritual and intellectual warfare mentioned at several points in the New Testament. The poem is a stirring summons to emulate the 'Prophetic Character', in the spirit and power of Elijah. Indeed, Blake's fervent hope was that *all* the Lord's people should be prophets, a quotation from Numbers 11:29 being appended to the biblical words. For Blake, the Poetic or Prophetic Character is a human characteristic, whose development in people Blake sought to kindle ('the Poetic Genius which is every where call'd the Spirit of Prophecy', *All Religions Are One* 5; E1). Its stimulation expands the horizons of human imagination, to avoid a 'repeat' of 'the same dull round over again' (*There is No Natural Religion* b Conclusion; E3).

For Blake, prophecy didn't mean predicting what would happen in the future but understanding more deeply what was going on and telling the truth as one saw it, whether concerning the hostile reaction of Britain to the American colonies, or the resistance to change of the *ancien regime* in Europe. Blake believed that prophecy was the responsibility of all. 'Every honest man is a Prophet', *Annotations to Watson's Apology,* 14, E617) and 'the voice of honest indignation is the voice of God' (*Marriage of Heaven and Hell* 12, E38). It is not a charisma for a few. The Poetic or Prophetic Character is a human characteristic, whose development in people Blake sought to encourage.

The suspicion of abstract theology

Blake's and Winstanley's suspicion of abstract theology and their emphasis on the mutual forgiveness of sins and the community of goods, respectively, indicate the various ways in which they perceive that the secular is the necessary context of what constitutes the Christian response. This should not surprise us. After all, it is a repeated theme in the Bible. The Torah is about the demonstration of obedience in lives lived, and a persistent theme in the prophetic corpus is the priority given to just acts rather than the perfection of worship. It is a theme taken up in the New Testament. Confessing the name and being part of an ecclesial community is not what counts (cf. Matt. 7:21–3 cf Matt 25:31–45). How best to 'love a brother or sister whom [we] have seen' and thereby 'love God whom we have not seen' (1 John 4:20) is absolutely key.

Practical discipleship is the context within which theological understanding takes place, and understanding something of the identity of God comes through service to those who are the least of Christ's brothers and sisters (cf. Matt 25:31–45). The Beatitudes reveal the divine identification with the poor and marginalized. Indeed, Matt 25:31–45 suggests that the wretched 'are the latent presence of the coming Saviour and Judge in the world, the touchstone which determines salvation and damnation' (Moltmann 1977: 127). Notwithstanding 'Death of God' kinds of statements such as 'Thou art a Man God is no more Thine own humanity Learn to adore' which Blake puts on the lips of Jesus in *The Everlasting Gospel* (E520) and the consistent critique of the patriarchal law-giving God, Blake's personal piety was a curious mixture of the conventional and the radical, as his letters indicate (see also Merton 1981). The same was probably true of Winstanley, if the hints we have about his later years are anything to go by (Gurney 2013).

While the similarities in form and content between what Winstanley wrote and what we find in liberationist exegesis are striking, there are differences. Experience is the motor of Winstanley's exegesis, but how solidarity with the poor and marginal offers a crucial perspective on the Bible and tradition and that how one lives, and what one's commitments are, and how they determine one's ideas and political and theological preferences, are never explicitly stated. Winstanley saw as clearly as anyone the ways in which social institutions, indeed theology as the ideological justification of the status quo, served the interests of the rich and powerful in his day, no better seen in his critique of ideology inspired by apocalyptic texts like Daniel and Revelation.

Liberation theology led to a discovery of significant strands in the Bible where practice is regarded as fundamental for theological epistemology (Graham 1996). Biblical passages like Jeremiah 22: 16; Matthew 25: 31–45, and 1 John 4: 20–1 summarize this practice, but none does it better than Winstanley's 'Action is the life of all' (CHL ii.80 cf. CHL i.508, 516, Gurney 2013: 43). Winstanley's experiences led him to a way of reading the Bible, Genesis 2–3, for example, with a keen political edge, coloured by his appreciation of the divisions in society and church and the action necessary to initiate a change in society's structure. What drove his understanding is experience—of poverty, of oppression, and of actual injustice. Indeed, we may note John Gurney's judgement that 'Winstanley certainly learnt much from his short time in Cobham, for one of the most telling aspects of the Digger programme was its successful fusion of religious with social radicalism and its skilful appropriation of traditional languages of rural discontent' (Gurney 2013: 21).

Conclusion

The emphasis on this world, its contradictions, and its political shortcomings are evident in the writings of both Blake and Winstanley. But whereas Winstanley encourages a focus on this world, rather than what is beyond the skies, Blake probes the political character of hegemonic theology, and his stories of the deities in his own mythology explore psychological conflict and imbalance, as well as the ways in which an alliance of religion and politics works against human integration. There is a sense that Winstanley's career as a Digger or True Leveller exemplified his words 'Action is the life of all and if thou dost not act thou dost nothing' (CHL ii.80). Blake was never an activist like Winstanley. Even his dangerous brush with the law in 1803–1804 was probably the result of a moment of impetuosity rather than a deliberate political protest, however heartfelt the strong language he may have used when he spoke to the English soldier in his garden in Felpham.

The themes explored in different ways in the works of Gerrard Winstanley and William Blake have many affinities with what we find in contemporary liberation theology (Rowland 2017a: 131–52). As this example from popular education material from the northeast of Brazil shows, the liberation theology method expands the horizon of what constitutes divine revelation beyond Bible, doctrine, and religious ritual to include, and prioritize, everyday experience. The image comes from material prepared for use in grassroots communities in the northeast of Brazil (see Figure 16.1). A chance meeting in 1990 with a group of catechists, who were preparing material for use in local communities, gave me an opportunity to witness the genesis of material, which is typical of the way liberationist reading has evolved. The image expands the horizon of the understanding of divine revelation beyond the usual ecclesial channels of Bible, teaching, and sacraments, and embraces the many different experiences of creation and, crucially, includes the political as well as the religious. All these together point to the knowledge of God (Rowland 2017a: 135–42). The image includes *criação*' (creation), *fatos* (events/actions), *natureza* (nature), *vida* (life), *historía* (stories and history), *martírio* (acts of

Figure 16.1 Material prepared for use in grassroots communities in the north east of Brazil
Source: photo: author.

witness), and *luta* (struggle). Together these combine to be the raw material of an apocalyptic process, leading to *revelação de Deus*, enabling a different understanding and perspectives on life and opening up new ways of acting and relating in this world.

What the evidence surveyed in this essay suggests is that inspired by the biblical writings themselves, a strongly secular understanding of theology and the secular, too often neglected, has existed alongside the hegemonic abstract tradition of Christian theology.

Further reading

Altizer, T. (1967) *The New Apocalypse: the radical Christian vision of William Blake,* East Lansing: Michigan State University Press.

An exposition by an advocate of 'Death of God Theology' of the way in which William Blake's writing anticipated such ideas.

Bennett, Z. and C. Rowland (2016) *In a Glass Darkly: the Bible, reflection and everyday life*, London: SCM.An essay on the contextuality of theology.

Bradstock, A. (2011) *Radical Religion in Cromwell's England: a concise history from the English civil war to the end of the commonwealth*, London: I.B. Tauris.

A theologically sensitive and historical study of radical movements in the seventeenth century, which complements Christopher Hill's study.

Freedman, L. (2011) 'Tom Altizer and William Blake: the apocalypse of belief', *Literature and Theology,* 25: 20–31.

A recent study of the significance of Altizer's study of Blake's theology.

Hill, C. (1972) *The World Turned Upside Down,* London: Penguin.

The classic study of radical groups in seventeenth-century England in which Gerrard Winstanley's religious and political context is expounded and which offers a concise summary of the secular character of his theology.

References

Beaumont, J. and C. Baker (eds.) (2011) *Postsecular Cities: space, theory and practice*, London: Continuum.

Bentley, G. E. (2001) *The Stranger from Paradise: a biography of William Blake*, New Haven, CT: Yale University Press.

——— (2004) *Blake Records: documents (1714–1841) concerning the life of William Blake (1757–1827) and his family*, 2nd edition, New Haven, CT: Yale University Press for the Paul Mellon Centre for Studies in British Art.

Bindman, D. (1978) *William Blake's Illustrations of the Book of Job: the engravings with related material*, London: William Blake Trust.
——— (2000) *William Blake's Illuminated Books*, New Ed edition, London: Thames and Hudson Ltd.
Blanton, W., Crockett, C, Robbins, J., and N. Vahanian (2016) *An Insurrectionist Manifesto: four new gospels for a radical politics,* New York, NY: Columbia University Press.
Cloke, P. and J. Beaumont (2013) 'Geographies of postsecular rapprochement in the city', *Progress in Human Geography*, 37(1): 27–51.
Cloke, P., Baker, C., Sutherland, C. and A. Williams (2019) *Geographies of Postsecularity: re-envisioning politics, subjectivity and ethics*, London/New York: Routledge.
Corns, T. N., Hughes, A. and D. Loewenstein (2009) *The Complete Works of Gerrard Winstanley, Vol. 2*, Oxford: Oxford University Press, (references to this CHL followed by volume and page number, e.g. CHL i.485).
Erdman, D. and W. Blake (1988) *The Complete Poetry and Prose of William Blake*, newly revised edition, Berkeley: University of California Press, (references to this E followed by page number, e.g. E256).
Graham, E. (1996) *Transforming practice: pastoral theology in an age of uncertainty*, London: Mowbray.
Gurney, J. (2013) *Gerrard Winstanley: the digger's life and legacy*, London: Pluto.Habermas, J. (2010) 'An awareness of what is missing'. In Habermas, J., Brieskorn, N., Reder, M., Ricken, F., and J. Schmidt (eds.) *An Awareness of What Is Missing: faith and reason in a postsecular age*, Cambridge, MA: Polity Press, pp. 15–23.
Hill, C. (1983) *Winstanley 'The Law of Freedom' and other writings*, Cambridge: Cambridge University Press.
——— (1993) *The English Bible and the seventeenth-century revolution*, London: Allen Lane; Penguin Press.
Merton, T. (1981) 'Blake and the new theology'. In Hart, P. (ed.) *The Literary Essays of Thomas Merton*, New York, NY: New Directions, pp. 3–11.
Moltmann, J. (1977) *The Church in the Power of the Spirit: a contribution to messianic ecclesiology*, London: SCM Press.
Rowland, C. (1988) *Radical Christianity: a reading of recovery*, Cambridge, MA: Polity Press.
——— (2010) *Blake and the Bible,* London: Yale University Press.
——— (2014) 'Gerrard Winstanley: man for all seasons', *Prose Studies,* 36: 77–89.
——— (2017a) *Radical Prophet: the mystics, subversives and visionaries who strove for heaven on earth*, London: I.B. Tauris.
——— (2017b) 'Notes from an archivist of radicalism', *The Bible and Critical Theory,* 13(2): 1–12.

17

Pope Francis and the theology of the people

Rafael Luciani

Introduction

To understand Pope Francis' theological and pastoral options, and the turn of the Catholic Church during his Papacy, it is necessary to delve into the *Theology of the People*, born in the Latin American Church, and known by assuming the centrality of the option for the poor that comes from the Document of Medellín (the Second General Bishop's Conference of Latin America in 1968) though Aparecida (the Seventh General Bishop's Conference of Latin America in 2007). This option is practised by Francis through his view of a 'Poor Church and for the Poor', inspiring not only his process of ecclesial reforms but also his geopolitics (cf. Luciani 2017: Chap. 3). Elements of this theological and ecclesial frame challenge assumptions about postsecular society, while integrating social and civil movements with the Church's evangelizing mission in the world. This novelty in Francis' papacy leads towards a major dialogue and collaboration between the church and all those people and movements who struggle for human rights, democracy, and social justice.

The theology of the people

The theology of the people, also called theology of culture, is a branch of Latin American liberation theology developed in Argentina by theologians Lucio Gera and Rafael Tello. After the conference of *Medellín* in 1968, the Argentine bishops adopted this way of doing theology in 1969; however, its origins date back to 1966, with the creation of COEPAL (Comisión Episcopal de Pastoral), which coined the term 'people' as the existence of a common culture, rooted in a common history, and one committed to the common good. COEPAL proposed internalizing the spirit of Vatican II and assumed the task of centralizing a common form of being in the church by promoting *collegial structures* (cf. Conferencia Episcopal Argentina 1966) that favoured the defence of human dignity and the promotion of a liberated religion. Following the spirit of the Council's *aggiornamento* ('bringing up-to-date'), the bishops committed themselves to creating a reform of thought and the rules governing the structures of the church. In the end, they hoped for a 'more lively self-knowledge, reform, dialogue with our Christian brothers and sisters, and openness to today's world: the four aims of the Council' (cf. Conferencia Episcopal Argentina 1966).

Lucio Gera (1924–2012), author of *Sobre el misterio del pobre* (see Azcuy et al. 2007), among many other writings, was the one who shed light on this branch of Latin American theology. For him, the theology of the people did not seek to change social and political structures themselves, but as a consequence of the discernment of the mission and identity of the ecclesiastical institution from an *option for the poor*, expressed in a strong religious discourse that would drive the socio-political dialogue and promote a pastoral praxis. It is a theology informed by the praxis of social justice as the value of those faithful to Jesus (cf. Politi 1992). In that way, Gera understands culture as a place of meditation, for knowledge of reality, and specifically, popular or poor people's culture, an environment where you can meet the poor and share their lifeworlds.

Gera thinks of the poor as a people ('pueblo'), as a collective subject of history, with its own cultural *ethos*, whose soul, or religious heart, always focuses on hope, from the limited experiences and material deprivation in which they live (Gera 1998). This notion demands an insertion in the world of its own values of popular life, for its later theorization and evangelization. The latter, evangelization, will not be reduced to its social promotion, and not be understood as merely doctrinal formation, but implies, above all, those actions of *inculturation in the world as it is, social recognition*, and *promotion of the cultural richness of each people-culture*. This commitment will result in the integral promotion of the human subject, the promotion of socio-political dialogue, and the practice of social justice within the framework of a religion that frees people to show them the blessed face of history.

Since the 1970s, the future Pope Francis has had a very clear image of this vision, combined with an understanding of unity between the Christian political tradition, the secular dimension of the world, and the pastoral action of the church. His understanding was made clear in his opening speech of the Jesuits' provincial congregation in 1974 (Bergoglio 1982), where he explained how the Christian praxis—as much religious as socio-political—must focus on fraternal solidarity, social justice, and the common good, rather than notions of homeland, revolution, conservative or liberal that are exclusive to any dissent or alternative. Thus, his theological-pastoral option could not be reduced to an analysis of the economic and socio-political in light of a Marxist method, as occurred in other branches of liberation theology (cf. Scannone 2014). He understood that the point of departure had to be the real connection with the people and the study of their culture or common *ethos*. This identity or *ethos* is found both in what appears to be *de facto*, an obstacle for the integral development of the people (socio-economic, political, and religious), and must be safeguarded in the face of all external influences, such as new forms of economic and cultural colonialisms.

One opts for the poor from the poor themselves, respecting their own culture and way of being, to recognize it effectively and affectively as a true subject of a historical process of development and liberation. It is precisely their way of being, their cultural autonomy and newness that can evangelize us. It is as the theologian Víctor Fernández states:

> We can see in poor some deeply Christian values: a spontaneous attention to the other, an ability to devote time to others and come to the aid of another without calculation of time and sacrifices, as illustrated, with a more organized life, difficult to give in to other time, attention and resignations spontaneously, willingly, and selflessly.
>
> *(Fernández 1998: 139)*

To know this reality, theology of the people assumes four instances to take into consideration in order to discover the very *ethos* of the popular reality: 'The revaluation of popular Catholicism, the contributions of the social sciences and humanities, the experiences of popular-pastoral, and academic theological reflection that accompanies and guides the pastoral praxis' (Galli 1991). Many misunderstandings and critics from enlightened socio-cultural contexts or from high-income Western industrial countries tend not to understand nor accept that the meaning of the notion of *culture* in Latin America is different; it cannot be separated from a spirit of transcendence and religious meaning that is related to daily life. Here, again, Víctor Fernández explains:

> It is customary to say that a theology of the people opts for the ignorant masses, lacking of culture and critical thought. That which the theology of the people defends is very different. It means to consider the poor not as the mere object of a liberation or an education, but as individuals capable of thinking with their own categories, capable of legitimately living faith in his/ her own way, capable of creating paths for popular culture. Additionally, that they express or view life in a different way, does not mean that they don't think or have a culture; it's simply a different culture.
>
> *(Rodari 2014: 65)*

By assuming a shared life with the people, the church in Latin America gains its credibility in the midst of the present world, and is able to contribute to the processes of global change, because :

> together with the various sectors of society, she supports those programmes which best respond to the dignity of each person and the common good... convictions which can then find expression in political activity.
>
> *(Francis 2013: 241)*

People and culture: a popular and relational option, but not populist

This experience and bond with the daily world of popular life has no connection to populist, nationalist, or localist visions. Following Latin American theology, 'poor is a theological category' (Francis 2013: 198). And specifically, it is Christological, because in the poor, 'we see the face and the flesh of Christ, who became poor to enrich us with his poverty (2 Co 8, 9)'. The poor are the flesh of Christ (Francis 2015a). This option is thus a condition, *sine qua non,* for the Christian life because the proposal of Jesus, which is the Kingdom of God, is not that of a private relationship and intimacy with God (Francis 2013: 183) but that of a relationship with the others and in the world of the others as it is, because its society is the setting to build fraternity, peace, justice, and dignity for all (Francis 2013: 180).

The option for the poor is not a choice only for each poor individual whom we meet on the journey; it is also a *structural option*, which brings with it a socio-economic transformation and a change of mentality of the whole of Christian life. It is a position which undertakes the defence of the planet, starting with the most vulnerable people and those suffering; a knowing of the people-as-poor as a whole, because 'people in every nation enhance the dimension of life by acting as committed and responsible citizens, not as a mob swayed by the powers that be' (Francis 2013: 220). As Pope Francis reminds us, isolated individuals are not saved. but by '*the social relations existing between men*', that each one is *in relations with* the people where they live (Francis 2013: 178).

When Francis uses the notion of the people, he does so in three ways: people-as-poor, people-as-nation, and people-as-faithful. The people-as-poor are the marginalized and excluded, in the gutters of socio-political and economic participation. In the 1970s in Argentina, the 'shirtless and workers' would fall into this category. Such a notion belongs to those who do not have real possibilities for development, and whose life conditions are marked by misery and exclusion. These conditions constitute the majority of humanity and the most affected by the present system of global development.

The people in this understanding are situated in the *place for the interpretation of its culture,* not the reason of an illustration studied from other forms of thought or external ideologies to its own lifeworlds and beliefs. This analytical approach from socio-cultural hermeneutics understands popular religion as a suitable means for the understanding of its struggles, hopes, and desires, like the strength that accompanies and gives meaning to its day to day. The resource of agency of the historical-cultural analysis, and not just the social-analytical, benefits this aspect of Latin American theology. Hence, for Latin Americans, popular culture obtains a theological status, because the 'way we listen to God the Father is how we should listen to his faithful people. If we do not listen in the same way, with the same heart, then something has gone wrong' (Francis 2015d).

This understanding is not to idealize the life of the poor, but to recognize their extreme suffering, the fruit of injustice caused by socio-economic structures that are undesired by God. The aim is to commiserate, affectively and effectively, and to assume the struggle for the cause of a better, more humane, and inclusive world. The poor live a daily, mental sorrow, as a consequence of exclusion and the powerlessness that they are subject to, which are not even the results of frustration or violence but from an experience of religion. These people, mostly poor, want to become *people-as-nation* but they live and understand, above all, as *people-as-faithful* because religion gives meaning and hope, while encouraging a continual fight against the dominant culture. Assuming the paradigm of popular culture, where even human interdependence is preserved, is to capture the lifeworld that flows forth. It is a socio-cultural approach that can be described as:

> rootedness in the barrio, the land, the office, the labor union, this ability to see yourselves in the faces of others, this daily proximity to their share of troubles – because they exist and we all have them – and their little acts of heroism: this is what enables you to practice the commandment of love, not on the basis of ideas or concepts, but rather on the basis of genuine interpersonal encounter. We need to build up this culture of encounter. We do not love concepts or ideas; no one loves a concept or an idea. We love people.
>
> *(Francis 2015a)*

Popular culture is the fruit of an ongoing process that has been built by a massive exodus from the countryside to the cities. From this process of rural to urban migration, in this process itself, people have formed their own distinctive cultural traits. However:

> This has not been a planned exodus. Each one left on his/ her own account. Each had to raid a piece of land to build their house, to find a job, to arrive at a place of relative stability and specialization, fought for basic needs, sought to bring their family forward, even putting some of their children through university, found loyal friends, and little by little built an understanding of what had happened of what was happening, and through general concepts of the world and life. Does this way, of tangible, human production of life, not constitute a culture?
>
> *(Trigo 1985: 90–1)*

This option for the people-as-poor is that which permits the construction of the true common good and achieves the necessary momentum to reach a superior unity at the national level. It is at this national level where the influence of extreme ideologies, including Marxist and liberal, socialist and capitalist, can be surpassed. These extreme ideologies only seek to destroy memory and identity and homogenize societies, failing to take into consideration the cultural differences that exist as true values of humanization, of development (Francis 2013: 220).

Francis is not posing populism. We are not facing a populist theologian, since, 'the term people is distinguished from the word mass because it presupposes a collective subject capable of generating its own historical processes' (Rodari 2014: 65). As described by O'Farrell (1976: 17), to speak of people 'represents a concrete entity or, better put, a historical subject and collective or political, able to take the good of all as a common value and enduring'. And as subject, 'each people is the creator of their culture and the protagonist of their history' (Francis 2013: 122). It is not, therefore, about taking a populist attitude of distributing handouts and converting the people-as-poor into an object; not political nor pastoral. As the Venezuelan Jesuit Pedro Trigo SJ sustains, 'the theology of liberation proposes, as a historical novelty based on the Gospel, the constitution of the people as a *historical subject* both in society and in the Church' (Trigo 1985: 89), because people is that which shares a way of life and a political project that longs for the common good (Scannone 2015).

It is understood, then, for 'people', a group of people, a collective subject that shares a common culture, that is united by the same history, and that has a project of shared life (Díaz 2015); this is their way of life despite the negative socio-economic conditions they suffer and the historical forces that are still causing poverty and misery inspired by the fetishism of the markets (Francis 2013: 204). The option for the poor involves building the common good and fighting against poverty, in all its causes. It is a choice that springs from the Golden Rule: 'Do unto others as you would like others to do to you' (Mt 7:12). Or, as Francis has explained:

> Let us treat others with the same passion and compassion with which we want to be treated. Let us seek for others the same possibilities, which we seek for ourselves. Let us help others to grow, as we would like to be helped ourselves. In a word, if we want security, let us give security; if we want life, let us give life; if we want opportunities, let us provide opportunities. The measure we use for others will be the measure which time will use for us.
>
> *(Francis 2015c)*

Called to be citizens in the womb of the people

In the 1970s, Argentina lived between socio-political conflict and severe divisions on the inside of the Catholic Church due to Peronism. Bergoglio proposed the promotion of a *greater unity* at this juncture, understanding that the common good is more important than each position and each individual option, which he refers to as the 'parts'. Absolutizing the individual vision destroys dialogue and all possibility of reaching the common good, leaving out the possibility of a real option for the poor and for their development. In this context, he is building the theme of greater unity of the *people-as-nation*, starting with the common good. This appears to be central in the theology that inspires him and in his pastoral practice. But this implies that the church should incarnate herself in the reality of the people-as-poor, to know their culture and understand and respect their religiosity, to let herself be evangelized and changed by them, and to bet on their cultural recognition and socio-economic development. Inspired by Mons. Pironio, Bergoglio also understood that 'evangelization is in direct relationship with the human promotion and the full liberation of all peoples' (Pironio 1975: 169). But the first step was to capture the essence of popular culture, to understand their lifeworld. That weight of the culture is its *soul*:

> Our people have soul, and because we can speak of the soul of a people, we can speak of a hermeneutic, of a way of seeing the reality, of an awareness. Today, in the midst of conflict, this people teaches us not to pay attention to those that seek to distil reality in ideas, not to serve intellectuals without talent, nor ethicists without kindness, but we must appeal to the depths of our dignity as a people, appeal to our wisdom, appeal to our cultural reserves.
>
> *(Bergoglio, 2005: 6)*

To be able to walk in this direction, Bergoglio knew that he needed a change of certain 'mentalities'. He proposes the following. First, and with all realism, he understands that we will not arrive at unity while there still exists the temptation to forget disputes and not accept them. This type of attitude he calls 'abstract spiritualism'. Second, unity will not be achieved if economic and public policies are applied to Christian ends, as an example of the ideological visions that want to be imposed on the most poor and vulnerable, for those groups that are in power, whether political, economic, or religious in nature. This mentality is given the name of the temptation of 'functionalist methodologism' and 'abstract ideologies'. Third, the position of 'ethicists' or moralizing should be avoided, or to say that 'they isolate the conscience of processes and make formal rather than actual projects'. This type of mentality he called 'the moralizing of the priests'. This is how it was explained during his exposition in the *VIII Jornada de Pastoral Social* in Buenos Aires in 2005.

Bergoglio adds to these criteria that which would speak to the processes of participation in the public life and appear in the mid-1970s. At that time, Bergoglio claimed that 'unity prevails over conflict, the whole is greater than the part, and time is greater than space'. Almost 40 years later, in 2010, as the Cardinal of Buenos Aires, he returned to these criteria at the conference he gave on the bicentennial of Argentinian Independence and there added a fourth criterion for discernment: 'reality over idea'. At that conference, he maintained the treatment of these ideas because 'they help resolve the challenge of being citizens and belonging to a society' (Bergoglio 2010). Now, as the Pope he returns to this vision as much in the encyclical *Lumen Fidei* (nn. 55.57) as in the apostolic exhortation, *Evangelii Gaudium* (Francis 2013: 217–37).

His invitation is as such to re-establish social bonds and generate a culture of encounter capable of resisting the increasingly fragmented culture, promoted by globalization. But the 're-establishment of social bonds' becomes a choice for the majority of humanity, even those living in poor conditions, because in order to solve the problems of the world there must be a solution for those in poverty. The option is thus to build citizenship and democracy. The conversion of the church springs from its commitment to promote and build citizenship and democracy in the world, making it so that each individual and institution commits itself to the development of the poor (people-as-poor) so that all may be subjects (people-as-nation) and not objects or recipients. The role of the church in this process answers that she is the 'People of God incarnate in the peoples of the earth' (Francis 2013: 115).

A peripheral hermeneutic of social processes and movements

The option for the poor is a *structural* element or ongoing condition in the life and mission of the entire church, and hence capable of bringing about 'real changes' (cf. Francis 2014) in society. Such a stance necessarily affects relations between the centre and the peripheries, between Rome and local churches, between the church and society at large. Under this scheme, changes of whatever nature cannot be driven from the centre but rather must come *from the peripheries*, whether existential and social or political and religious. Francis explains it in the following terms:

> the great changes in history were realized when reality was seen not from the center but from the periphery. It is a hermeneutical question: reality is understood only if it is looked at from the periphery, and not when our viewpoint is equidistant from everything.
>
> *(Spadaro 2014: 3)*

This means a shift from the security of our comfort zones towards the place where the excluded, the discarded of society, are living. It is from there and with people that the truth of reality can be understood. The displacement required implies a conversion, going to and being with the poor, where they live, knowing first-hand what people are experiencing, and not letting ourselves give in to the temptation presented by the prevailing system which permits us to think about the reality of others but without knowing them or experiencing their lifeworlds, their conditions. Nominalism that pervades academia brings such a temptation to bear, producing specialists who have never had a real connection with the persons or the situations about which they write. Such was the explanation given by Francis in a 2014 interview with the Jesuit Antonio Spadaro:

> It is not a good strategy to be at the center of a sphere. To understand we ought to move around, to see reality from various viewpoints. We ought to get used to thinking. I often refer to a letter of Father Pedro Arrupe, who was General of the Society of Jesus. It was a letter directed to the *Centros de Investigación y Acción Social* (CIAS). In this letter, Father Arrupe spoke of poverty and said that some time of real contact with the poor is necessary. This is really very important to me: the need to become acquainted with reality by experience, to spend time walking on the periphery in order really to become acquainted with the reality and life-experiences of people. If this does not happen we then run the risk of being abstract ideologists or fundamentalists, which is not healthy.
>
> *(Spadaro 2014: 4)*

Without this view from the periphery, the proclamation of the gospel will be unsubstantial, unattractive, unable to get to the root of the true problems in order to remedy them (Francis 2013: 199). It is a matter of changing our location, of opting to live on the peripheries, and assuming the world as it is. This option does not consist of providing instruction and religious formation or setting up charitable works of social assistance. It requires going beyond mere assistance to establishing a *loving attention* that regards poor people as subjects in a horizontal relationship, treating them equally. The loving attention that Francis talks about entails a change in life-direction, of the way everything is done, so that others—the poor—may have a decent life. We see this kind of attention being given by those:

> priests and pastoral workers [who] carry out an enormous work of accompanying and promoting the excluded throughout the world, alongside cooperatives, favoring businesses, providing housing, working generously in the fields of health, sports and education.
>
> *(Francis 2015a)*

Their incentive is not from an ideology or social programme, but from the heart of the gospel, from the right that we all have to live well.

This theologico-pastoral path calls to live the Church's mission from within the *new historical processes of social change that serve the struggles of the poor majorities* for a better world, one that is more brotherly/ sisterly and liveable here and now. We are not talking about processes offered by the ecclesiastical institution through its pastoral action. What is being proposed is that the ecclesial institution recognize, take on, and promote those processes—including those of non-Christian persons and groups—that have a humanizing aim in society as a whole, and that are moving against the tide of the contemporary process of financial and cultural globalization. One of the most novel and specific modes for embodying this vision is put into practice when Francis invites, accompanies, and promotes social movements that are leading processes in history. He said as much at the *First World Meeting of Popular Movements*:

> Such proactive participation transcends the processes of formal democracy. Moving toward a world of lasting peace and justice calls us to go beyond paternalistic forms of assistance; it calls us to create new forms of participation that include popular movements...They are a real sign of the inclusion of the excluded in the building of a common destiny.
>
> *(Francis 2014)*

These movements, many of them secular, reveal that 'the poor not only suffer injustice, they also struggle against it' (Francis 2014: 1), and the duty of the church, in fidelity to the Reign of God, lies in 'accompanying them suitably on their path of liberation' (Francis 2013: 199). Indeed, the Church must do so *from within the particular ways in which these movements understand and carry out their struggles.* This is where *the primacy of the periphery over the centre* becomes real: in the transition from one model of church that has been accustomed to determining the directions and modes of change to another model of church—*people of God*—that recognizes what already exists in secular history in order to take it on and strengthen it. In other words, Francis recalls that 'the Church cannot and must not remain aloof from this process in its proclamation of the Gospel' (Francis 2015a), because 'respectful cooperation with popular movements can revitalize these efforts and strengthen processes of change'.

Popular or social movements show us that the church ought to be at the service of everyone, not only of its members, for if it wishes to really foster a culture of encounter it cannot limit itself to gathering those who are already part of its structure, nor believe that anyone who is not in it must eventually become part of it. The periphery has its own life, its own particularity and dynamic, which must be respected as something that contains the presence of the Word. At the *First World Meeting of Popular Movements*, Francis expressed this insight in the following terms:

> I know that you are persons of different religions, trades, ideas, cultures, countries, continents. Here and now you are practicing the culture of encounter, so different from the xenophobia, discrimination and intolerance that we witness so often. Among the excluded, one finds an encounter of cultures where the aggregate does not wipe out the particulars.
>
> *(Francis 2014)*

The fundamental soteriological principle of a church going out of itself lies in understanding that 'those who enter into communion with God are those who are in communion with human beings' (Gutiérrez 1988: 65). It does not go out to occupy new spaces; we are not in a situation of initial evangelization. The reason for going out is to bring about new alternative and horizontal processes in history, and to do that by first going where people are and becoming familiar with their lifeworlds. This thinking lies at the heart of a proper Latin American theology of processes in history (Gera 2005) which seeks to identify the major events of an era, those that mark its historical direction, and to discern in them certain signs of the times that reveal the meaning of salvation (*ibid.*: 270), guiding people toward universal brotherhood and sisterliness. These new processes are being led authentically by the social and popular movements that struggle for the 'active participation of great majorities' (Francis 2014)— something very actual and needed today.

Just as for John XXIII, in *Pacem in Terris,* these processes included those promoting workers and women, decolonization and peace between peoples, for Francis these are all *movements* that, in the face of the current prevailing direction of globalization, denounce its consequences—that is, exclusion and inequality—while at the same time propose a more humane world, an alternative to the current system. This mediating, socio-political role of popular movements, is what the church had lost sight of during the so-called ecclesial winter before the arrival of Francis.

Conclusion

To conclude we must say that the Latin American theology of the people or theology of culture accepts that

> the world has a specific secular dimension, which has its own consistency and relative autonomy. The secular dimension finds expression within the common notions of civilization or culture. It is in this secular realm that the Spirit of God acts and is present, not only in people of good will be taken individually, but also in society and in history, peoples, cultures, religions.
>
> *(Gera 2005: 265–6)*

To take on secular history with all its depth and truth means believing that the Spirit is present in it, in all these movements, forces, and processes in history, beyond the institutional church, and that it bears in itself a saving orientation greater than that which is immediate and palpable.

Such a vision helps convert the church, helps it to go out of itself and accompany these processes of humanization or *growth in humaneness*. In other words, the vision is to work together towards achieving concrete expression of the universal brother/sisterhood of all peoples and their cultures, able to look beyond beliefs and creeds. This effort implies joining forces to demand better democracies, social inclusion, and respect of human rights. Francis assumes these efforts by supporting social and civil movements, because these movements collaborate to build a more inclusive and just society. His Papacy, inspired by a Latin American ecclesial and theological identity, is a testimony of how secular society and religious institutions can join forces for the common good of all people, based on an ethics of democratic citizenship, ecclesial sinodality and social participation.

Further reading

Berryman, P. E. (1973) 'Latin American liberation theology, *Theological Studies*, 34: 357–95.

The author presents the social, ecclesial, and theological debates that gave birth to liberation theology, developing its key notions and initial contributions of the involvement of the church in the changes of society.

Boff, L., Elizondo, V. and M. Lefébure (1984) *The People of God Amidst the Poor*, Edinburgh: T. & T. Clark.

This volume of the Journal *Concilium* presents a series of articles, written by some of the founders of liberation theology, on the notion of 'People of God' as understood by Catholic theology after the Vatican II Council.

Comblin, J. (2004) *People of God*, New York, NY: Orbis Books.

This Latin American theologian explains how the church as 'People of God' is the most striking image to emerge from Vatican II, opening the church to the role of the laity, the dignity of the poor, and the political changes in our history.

Cox, H. (2016) *The Market as God*, Cambridge, MA: Harvard University Press.

Inspired by Pope Francis' *Evangelii Gaudium*, Harvey criticizes the business theology of supply and demand in our globalized world. In continuity with Latin American liberation, he explains how the market is omniscient and omnipotent, and in which sense it proposes a secular theology.

Metz, J. B. (2007) *Faith in History and Society*, New York, NY: Crossroads.

Metz articulates the emerging social and political issues that have arisen from current technological and economic processes. He deals with the question of whose progress and at what cost. He proposes a political theology for our times, engaging religious beliefs with the struggle for a just world.

Pasquale, F. (2015) 'The concept of periphery in Pope Francis' discourse: a religious alternative to globalization?' *Religions*, 6: 42–57.

Pasquale analyzes the role of religions as agencies that defend the common good, questioning political nationalisms and universalisms. The author uses the concept of periphery, as understood by Pope Francis, to look into the world from the edge, from the poor, rather than from the centre.

Scannone, J. C. (2016) 'Pope Francis and the theology of the people', *Theological Studies*, 77: 118–35.

This important Latin American theologian presents the influence of the Argentine theology of the people in Pope Francis' Apostolic Exhortation *Evangelii Gaudium*, and how this helps to understand Francis' view on people, church, the poor, and the common good.

Xavier, J. (2010) 'Theological anthropology of *gaudium et spes* and fundamental theology', *Gregorianum*, 91: 124–36.

Xavier shows how the Vatican II Council relates theology to anthropology and history, where anthropology plays a mediatory role in a dialogue between faith and reason in our modern world.

References

Azcuy, V. R., Galli, C. and M. González (eds.) (2006) *Escritos Teológico-Pastorales de Lucio Gera II: de la Conferencia de Puebla a nuestros días*, Buenos Aires: Ágape.

——— (eds.) (2007) *Escritos Teológico-Pastorales de Lucio Gera I: del preconcilio a la Conferencia de Puebla (1956–1981),* Buenos Aires: Ágape.

Azcuy, V. R. and P. Scervino (eds.) (1998) 'Ministerio peregrino y mendicante: Lucio Gera 50 años de sacerdocio', *Nuevo Mundo*, 55: 37–63.

Bergoglio, J. M. (1982) *Discurso de Apertura de la Congregación Provincial XIV: 18 de febrero de 1974. Meditaciones para religiosos*, Buenos Aires: Ediciones Diego de Torres.
——— (2005) *Ponerse la Patria al Hombro*, Buenos Aires: Claretiana.
——— (2010) 'Hacia un bicentenario en justicia y solidaridad 2010–2016: nosotros como ciudadanos, nosotros como pueblo' ('Towards a bicentennial in justice and solidarity 2010–2016: we as citizens, we as a people'), speech, Buenos Aires, 16 October.
Conferencia Episcopal Argentina (1966) *Declaración* "la Iglesia en el período postconciliar". See www.episcopado.org/documentos.php?area=1&tit_gral=Documentos%20históricos, accessed online 29-05-2018.
Díaz, B. (2015) 'Alberto Methol Ferré: una influencia fundamental en el pensamiento del papa Francisco', *Cuadernos del Claeh*, 101: 63–85.
Fernández, V. M. (1998) 'El *sensus populi*: la legitimidad de una teología desde el pueblo', *Teología*, 72: 133–64.
Francis (2013) *Apostolic Exhortation "Evangelii Gaudium" of the Holy Father Francis to the Bishops, Clergy, Consecrated Persons and the Lay Faithful on the Proclamation of the Gospel in Today's World*, Vatican City: Vatican Press.
——— (2014) *Address to the Participants in the First World Meeting of Popular Movements*, Rome, 28 October.
——— (2015a) *Address to the Participants in the Second World Meeting of Popular Movements*, Bolivia, July 9.
——— (2015b) *Apostolic Visit to Paraguay*, July 11.
——— (2015c) *Apostolic Visit to the U.S.*, Washington, September 25.
——— (2015d) *Meeting with Clergy, Religious, and Seminarians*, Bolivia, July 9.
Galli, C. (1991) 'Evangelización, cultura y teología: el aporte de J. C. Scannone a una teología inculturada', *Teología*, 58: 189–202.
Gera, L. (1998) 'Entrevista al cumplir sus 50 años de sacerdocio'. In Azcuy V. R. and Scervino P., (ed.), *Ministerio peregrino y mendicante. Lucio Gera 50 años de sacerdocio, Nuevo Mundo* 55: 37–63.
——— (2005) 'La teología de los procesos históricos', *Revista de Teología*, 87: 259–79.
Gutierrez, G. (1988) *Líneas Pastorales de la Iglesia en América Latina*, Lima: CEP.
Luciani, R. (2016) 'Los signos de los tiempos como criterio hermenéutico fundamental del quehacer teológico', *Atualidade Teologica*, 52: 37–57.
——— (2017) *Pope Francis and the Theology of the People*, New York, NY: Orbis Books.
O'Farrell, J. (1976) *América Latina: ¿cuáles son tus problemas?*, Buenos Aires: Ediciones Patria Grande.
Pironio, E. F. (1975) 'La evangelización del mundo de hoy en América Latina: exposición presentada en el Sínodo Episcopal de 1974, *Teología*, 25–26: 155–65.
Politi, S. (1992) *Teología del Pueblo: una propuesta argentina a la teología latinoamericana 1967–1975*, Buenos Aires: Ediciones San Antonio de Padua.
Rodari, P. (2014) 'Conversaciones con Víctor Manuel Fernández', *Iglesia Viva*, 259: 55–68.
Scannone, J. C. (2014) 'Papa Francesco e la teologia del popolo', *La Civiltà Cattolica*, 3930: 571–90.
——— (2015) 'La teología argentina del pueblo', *Gregorianum*, 96: 9–24.
Spadaro, A. (2014) 'Wake up the world! Conversation with Pope Francis about the religious life', *La Civiltà Cattolica*, 2014-I: 3–17.
Tello, R. (2008) *La Nueva Evangelización: escritos teológico-pastorales I*, Buenos Aires: Ágape.
——— (2011) *Pueblo y Cultura Popular*, Buenos Aires: Ágape.
Trigo, P. (1985) 'Teología de la liberación y cultura', *Revista Latinoamericana de Teología*, 4: 83–93.

18

Interrogating the postsecular

Elaine Graham

Introduction

Are we justified in arguing, with Gregor McLennan, that the postsecular represents 'a dramatic sea change in our metaphysical proclivities' (McLennan 2011: 15)? If so, what does the postsecular actually signify? What is achieved, if anything, by the category?

For some scholars, such as James Beckford (2012), the terminology of 'postsecular' is incoherent and adds nothing to our understanding of the current state of religion in the public square. Even though it may possess a degree of rhetorical power, it achieves little beyond telling us what we already knew (which no one beyond Europe and North America needed reminding): that religion has not disappeared from the world.

For others, however, the postsecular represents a refreshing and welcome paradigm, 'a renovation or renegotiation of the secular, a corrective designed to open it up to the impulses of religious inspiration and insight' (Barbieri 2014: 19). As an explanatory or diagnostic concept, it goes beyond the descriptive to highlight the strengths and weaknesses of previous conceptual frameworks, such as secularity and secularization, and offers useful signposts towards new directions in the study of religion and society. But in order to justify its use, if only as a heuristic category, we may need to clarify and differentiate its meanings. Postsecularity calls into question and renders complex the precepts and features of secularity; but, like its successor term, it too is multidimensional. It operates at levels of polity and governance (not least in terms of who are regarded as critical actors in civil society); epistemologically; in terms of the *habitus* and mind-sets of personal beliefs and values; as well as at the level of social expressions of faith and associated practices (a shift from organized, creedal religion to more diffuse and ambient forms of spirituality).

For all their differences, both Beckford and Barbieri honour this multidimensionality by constructing working typologies of the postsecular—sixfold, in both cases (Beckford 2012; Barbieri 2014). In a similar spirit, I will also attempt some kind of classification of different meanings of the postsecular, charting (broadly) its *empirical* deployment to describe and circumscribe the 'new visibility' of religion in Western society (after); its use as a tool of *genealogical* or ideological critique (against); and as a *phenomenological* consideration of new social forms of religious belief and practice, often entailing a rejection of reductionist and essentialist accounts of religion itself (beyond).

This analysis leads me to emphasize above all what I will term the 'reflexive' nature of the postsecular. Above all I see the postsecular as a kind of 'third space' (Baker 2009) in which we are inescapably reminded both of the fragility of religion in Western societies that persist in being functionally secular and theologically sceptical, as well as of the enduring capability of religion to motivate and curate powerful repositories of social capital. There may be good reasons, empirically and theoretically, to contest the supremacy and inevitability of secular*ism*, but the enduring consequences of seculariz*ation* still condition the terms by which religious bodies negotiate the public sphere. The postsecular is an attempt to navigate that novel, if contradictory and agonistic, territory.

Scoping the postsecular

Conventionally, secularization theory argued that due to modernization, religion would decline and disappear, especially from public life. While scholars may disagree on matters of emphasis, the consensus is that this process has been differentiated and multidimensional, and embraces the following elements:

- Evacuation of religion from the public realm, such as the constitutional separation of church and state.
- Different dimensions of dynamics of belief and unbelief: political (separation of state and church) (Casanova 1994); social, cultural (use of religious symbols and reasoning in public) (Brown 2009); transcendental (horizons of meaning, ultimate sites of reference).
- Decline in religious affiliation and observance (Berger 1999; Bruce 2002).
- A shift in what Charles Taylor terms the 'conditions of belief', such that faith is relativized and construed as optional (Taylor 2010).
- An epistemological and existential orientation towards temporal causation and evolution rather than a narrative of salvation history. Barbieri contrasts this as the difference between living in a universe rather than a cosmos (Barbieri 2014: 131).

Since the last quarter of the twentieth century, however, events have conspired to challenge the narrative of secularization as global, inevitable and unilinear. Even in Europe, which remains statistically speaking the least religious in the world (Pew Forum 2012), migration and globalization have enriched religious diversity and required Europeans to reconsider the basis of their own political culture in the founding values of Christianity and Christendom.

The term postsecular first achieved prominence in response to work by Jürgen Habermas in the early years of the twenty-first century (Habermas, 2008b; Butler et al. 2011). Noting the re-emergence of religion as a force in global politics and civil society, Habermas reflected that it may be time to reassess the precepts of post-Enlightenment liberal democracy, whereby the truth-claims of religious worldviews are bracketed out of public discourse in order to maintain, pragmatically speaking, a broad-based consensus towards the nature of political debate. Like most late twentieth-century political philosophers, Habermas assumed that religious discourse could never co-exist with, or impinge upon, public reason (Rawls 1971; Dillon 2012).

Habermas began to question this consensus in response to a number of events: the emergence of radical Islam as a global political force following the attack on the World Trade Center in September 2001; his reflections on the impact of advanced biotechnologies, in which he began to ask on what basis an appeal to the integrity and inviolability of 'human nature' could be founded (Habermas 2008a), as well as his growing concerns about the moral underpinning of the global financial crisis of 2007–2008. The latter exposed the attenuation of public communicative reasoning around the proper regulation of the global economy and suggested that, for the sake of their own survival, democratic institutions everywhere needed a wholescale rejuvenation of public values. Habermas thus began to call for a rethinking of the segregation of religious sources of reasoning from the secular nature of public debate, to the extent of suggesting that the former ought to be incorporated into a renewed vocabulary of civic virtue. He concedes to the possibility that those 'who are neither willing nor able to divide their moral convictions and their vocabulary into profane and religious strands must be permitted to take part in political will formation even if they use religious language' (Habermas 2008b: 28–9). Can the state incorporate religious reasoning in the name of healthy democratic debate; can it claim legitimacy if it does not?

This turn on the part of a highly influential social theorist represented a way of coming to terms with the unexpected resurgence religion in the public realm, especially in matters legal, educational, constitutional, as well as its continuing—indeed increasing—vitality as a stakeholder in local civil society (Dinham 2012; Hjelm 2015). Yet as critics have noted, much remains unresolved in Habermas' manifesto: most notably, how far religious groups themselves might be prepared to moderate their convictions in the midst of public debate; to what extent a culturally pluralist majority would grant legitimacy to faith-based interventions; and, indeed, whether this is simply a solution which sees 'religion' solely in functional terms as providing legitimation for economic, political, and cultural institutions that are beyond repair (Welker 2010; Dillon 2012). If religious voices and values are permitted, on what terms are they accommodated? How can transparency and participation be guaranteed? How far can a pluralist public understand, let alone tolerate, the particularities of religious teaching?

> If [religions] have a public role, it is time ... to challenge the "public reason" arguments of Rawls and Habermas which render faith in liberal non-religious language ... It is time to advance faith-based social action for faith-based reasons in faith-based terms.
>
> *(Dinham 2012: 69)*

Illustration

The agonies of postsecular society can be encapsulated in an incident which occurred on a French beach in August 2016. In the French tradition of laïcité, a number of local courts in Mediterranean seaside towns had forbidden Muslim women from wearing the burkini, a swimsuit designed to conform with Muslim guidelines on modest dress in public (Amrani 2016). Speaking to the newspaper *La Provence*, Prime Minister Manuel Valls called some women's preference for the burkini an 'archaic vision ... an expression of a political project, a counter-society, based ... on the enslavement of women' (Wright 2016). The central Council of State over-ruled the prohibition, but many politicians and commentators continued to defend the ban. But of course, this was always going to be about more than beachwear.

The matter of the visibility of Muslim women bathers at the seaside occurred against the background of cultural tension after a series of attacks by self-identified Muslims on a number of public buildings and events in France during 2015–2016, beginning with a gun attack on the offices of *Charlie Hebdo* in January 2015, another armed assault on the Bataclan theatre in Paris in November 2015, a lorry attack on a crowd celebrating Bastille Day in Nice in July 2016, killing 86 people, and the murder in Normandy later the same month of Fr. Jacques Hamel, an 85-year-old Catholic priest as he said Mass. Since then, of course, other European cities—Brussels, Manchester, London, Barcelona, Ansbach—have experienced terrorist attacks by Muslim-identified perpetrators. The 'new visibility' of religion is, thus, for many, far from innocuous or benevolent. The fear of violent attacks and furious debates about immigration from the Islamic world currently raging throughout Europe and the USA are the other side of the coin of stories of the value of religious capital for social cohesion.

This incident, then, goes right to the heart of the unease we might associate with the postsecular. The values of secularism, enshrined in France's prohibition on any demonstration of religious allegiance or identity, conflict directly with the cultural traditions of many French Muslims for whom the outward and visible marks of religious affiliation (far from being incidental to their sense of identity as individuals and as citizens) are absolutely fundamental. It epitomizes the agonistic clash between the irresistible force of secular 'enlightenment' and the immovable object of religious observance. As a force behind mass mobilization for change, faith-based activism is returning to the public stage and demonstrating its potential for imagining, realizing, and sustaining 'new creative alternatives and sustainable futures' (Braidotti 2008: 19). Yet many secular liberals experience a certain anxiety over the re-emergence of religion in the public domain, fearing a 'clash of civilizations' in the face of resurgent Islam but also an exposure of the limits of an agreed moral vocabulary of human rights and freedoms.

Manifestations of the postsecular

> The idea of the post-anything signals uncertainty in the social sciences.
>
> *(Lyon 2010: 648)*

We may now be gaining an impression of the postsecular as a space in which 'the presuppositions of the secularisation thesis no longer apply' (Boeve 2012: 145). Modernization clearly has not led to the effacement of religion; rather, currents of modernization and secularism can and do co-exist with continuing attachment to traditional, religiously derived, forms of belief, and identity. Rather than heading towards an inexorable *telos* of secularity, secularization might rather be viewed as a more diffuse and disaggregated processes of 'detraditionalisation, individualisation and pluralisation' (Boeve 2008).

As for the *post*secular, it inevitably evokes associations with other intellectual and political movements of the past 40 or 50 years: postmodernism, poststructuralism, posthumanism, postcolonialism. (And since 2016, perhaps, albeit less elegantly, we are having to come to terms with post-truth politics and media representations.) Yet there is still a need to clarify its meanings and utility. 'Post' may perform a number of functions. Most plainly, it may denote a successor phase or paradigm: that which comes after something or other. But here, of course, it still begs the question: after what—and what *exactly*?

More broadly, it may flag up a critical corollary to the very coherence or taken-for-grantedness of a particular movement or term: what do we mean by 'modernity'; were we *ever* modern? Or possibly an invitation to break apart a dominant 'master narrative' in favour of new perspectives and paradigms: such as asking what new 'subaltern' political and cultural opportunities and voices are liberated by the end of formal colonial rule in the Global South. Alternatively, 'post' may implicitly be asking whether the end of a particular movement and the transition into a new era occur because the aims of the original have been fulfilled; or because they have been found wanting? It may also mean a reversal altogether.

As one way of drawing out some of the diverse meanings of the term, I have devised a threefold typology (see Table 18.1) in which ideas of the postsecular: (1) open up new 'readings' of religious agency in contemporary global civil society ('after'); (2) challenge

Table 18.1 Threefold typology of meanings of the postsecular

After	Policy and practice New visibility Resurgence if not revival	Whose project was the secular, and who is now allowed a voice? Religious literacy Pluralism Habermasian polity
Against	Were we ever secular? Historical genealogy Paradigms and categories	Whose project was 'the secular'? At what cost does it come? Deconstruction of categories Hermeneutics of suspicion towards grand narratives
Beyond	Religion ... but not as we know it ... New manifestations Postsecular habitus	What new configurations are emerging, and what difference will they make? Reflexivity towards the possibility of faith and of none Emphasis on phenomenological dimensions of faith

Source: author

the very coherence and hegemony of secularism and secularization and invite us to take a 'genealogical' reading ('against'); (3) invite new theorizing about religion in relation to lived experience and as a part of wider cultural imaginaries and social practices ('beyond').

1 *'After': the return of religion to the public square*

This corresponds with Habermas' use of the term as signalling the potential impact of re-emergent religious voices and activists into public life. It admits there may be post-metaphysical grounds for democratic values such as justice, human rights, and common good which could then alter the rules of engagement for what might be considered an 'overlapping consensus' (Rawls 1987).

This dimension of the postsecular reflects the realities of empirical social change. In late modernity, the fragmentation of the nation-state (as the epitome of a technical-rational, bureaucratic form of power) has opened up new public spaces for religion (Dinham 2012). The retreat of state welfare reveals a plethora of voluntarist forms of civil society, including faith-based organizations. Once relegated to private morality, now the political potency of religion as source of social capital is once more apparent (Cloke and Beaumont 2013; Bretherton 2014; Tracy 2014; Spencer 2017).

2 *'Against': postsecular as genealogy*

This second 'genealogical' approach (Asad 1993) regards the secular as a historically and geographically contingent phenomenon that arose in the West in response to particular political problems to do with freedom of religion. The secular is 'neither singular in origin nor stable in its historical identity' (Asad 2003: 24). The key question here is, were we ever secular? And a secondary question, whose project was it?

A related critique of the secular is that of an explicitly theological critique of modernity and the precepts of Enlightenment, and a rejection of the hegemony of secular reason. It represents a kind of 'theological post-modernism, in which the tools of continental philosophers are placed in the service of a confessional agenda'. (Barbieri 2014: 143). The 'post' involves the restoration of true religion—Christendom, theocracy, orthodoxy—as return to a prelapsarian unity of spiritual and temporal.

These postsecular theologians set out to expose the hubristic nature of any claims to the sovereignty of reason and inevitability of progress, pointing out the 'shadow side' of Enlightenment and humanism (Smith 2004).

For its sociological, postcolonial, and theological critics, then, secularism is understood to have displaced Christianity in its claim to be normative, transcendent, and universal, as in non-contingent, and representing a point of reference which stands apart from, or independent of, any particular context. The very category of the secular is a construct, dependent on the imposition of the reification of religion and the religious as oppositional to a state of secularity. It exposes the ideologies at work in the assumption that the condition of secularity, indeed the very concept, may be seen as 'the ground from which all understanding of non-Enlightenment traditions must begin' rather than as a phenomenon with particular historical and ideological 'genealogies' 'albeit [ones] that have largely shaped our modern world' (Asad 1993: 200).

> We need to see secularism as a presence, as something, and therefore in need of elaboration and understanding. Whether we see it as an ideology, a worldview, a stance toward religion, a constitutional approach, or simply an aspect of some other project ... secularism is something we need to think through, rather than merely the absence of religion.'
>
> *(Calhoun 2010: 34)*

Historically speaking, the secular represents a moment in political history whereby some differentiation of jurisdiction occurred between the religious realm of ecclesiastical rule and that of civil statecraft. Even conceptually, Asad's analysis ought to alert us to the ways in which religion and the secular are not 'categories that are embedded in the nature of things' (Barbieri 2014: 121), and thus also to the ways in which, methodologically as well, such terms are put to use—including and especially in ideologies of political domination and violence. All the more so, then, in relation to social scientific method: not to assume that these terms denote preexistent, reified essences, but to ask how they function as heuristic devices, as things to think with. The questions then become, 'What are we studying when we study "religion"?' 'From what conceptual vantage-points is this taking place?' What does 'imagining the post-secular' enable us to do that other frameworks or methodologies cannot?

3. *'Beyond': Epistemological and phenomenological re-enchantment*

This dimension of postsecularity traces the turn to religion within contemporary continental philosophy, and ways in which the sacred, the transcendent, discipleship, communion, redemption have become 'things to think with' constructively and critically, especially when it comes to formulating alternative economies of grace to consumer capitalism, militarism, and nihilism (Crockett 2011). It also pays attention to implicit political theologies at work in global society—how religion provides the rationale for radical movements of resistance and revolution. Or revealing the hidden divinities and objects of worship—the 'postmodern sacred' (Mcavan 2012) at work within allegedly secular culture, within ideologies of the market, austerity, celebrity, nationalism, and so on (Lynch 2007).

On example of this can be seen in the 'Belief and Beyond Belief' festival in London running throughout 2017, which aims to address 'the seemingly innate need for humans to find meaning for their lives and a sense of where they fit into the universe, with all its mystery and majesty'. It has included debates on virtue ethics, mindfulness and well-being, the morality of abortion, interfaith dialogue, concepts of good and evil, religion and violence, as well as the influence of religion on the arts.

> The seemingly innate need for humans to find meaning for their lives and a sense of where they fit into the universe, with all its mystery and majesty, is a constant in all periods of human history... Even in an era when religion is said to be on the decline, our headlines are dominated by it, often in extreme forms.
>
> *(Southbank Centre 2017)*

Such an event reflects the drift away from organized, creedal religion but—to echo one of the staple arguments of secularization revisionists—this does not necessarily mean that interest in spirituality, religious practices, and teachings necessarily diminishes in equal proportion. However, is it the case that without the organizational and historical force of traditional religious institutions behind them, such new manifestations are necessarily less sustainable and visible in the long-term, and hence less capable of mobilizing the density of faith-based social capital that distinguishes the new visibility of religion in civil society?

This emphasis on religion not as cognitive belief, or religious magisterium, but as an indication of the ways in which everyday life is always already suffused with questions of meaning and the sacred, prompts Gregor McLennan (2011) to call for a new, postsecular methodology within the social sciences in which 'religious motivations, practices and modalities of belief [have] to be reconsidered, to say the least, phenomenologically'. The emphasis turns to the lived experience of religious actors in order to understand, non-reductively, how faith builds a world (or a 'social imaginary') of meaning and value (Lyon 2010: 654). These social imaginaries are not simply intellectual or cognitive, but actually forged by the *habitus* of faith: embodied religious practices, relations with material artefacts, traditions of reading sacred texts, and so on (Bell 2000; McCutcheon 2000).

'Post' in this respect takes us *beyond* the master narrative of secularization, away from a bifurcation of belief and non-belief, religion and atheism, faith and reason, towards a paradigm in which there is a plurality of ways in which the relationships and boundaries between transcendence and immanence, spiritual and material, are articulated. This constitutes a change of conceptual framework—and possibly, too, the emphases within our historical narratives of causation and consequence. The question here, then, is what conceptual frameworks and trajectories best help us locate ourselves and our futures?

Reflexivity and dialogue

This renewed attention to the very meanings and usages of the term 'religion' begins to connect with the voices who call for greater understanding and literacy towards religion in all aspects of public life, as a corollary not only of its greater currency and influence but also precisely because, given the declining affiliation to organized forms of religion, the lived experience of religious faith is increasingly unfamiliar to large sections of the population. Thus, arguments are emerging which call for programmes of religious literacy as an educational essential, even a measure of responsible citizenship (Carr 2012; Dinham and Francis 2015).

> The need to speak of religion in front of a diverse audience, the inability to appeal to the connivance of co-religionists, the necessity to objectify and explain the worlds of representations and attitudes proper to a given religion, alone constitute a position that marries religious belief to citizenship in a pluralist democracy. It enforces

> the recognition from the start that the religious worldview under discussion is not an all-encompassing symbolic structure for all society—even if it is the majority religion—but one orientation among many. Such an approach must inevitably clash with all religious self-descriptions that refuse a historical perspective ... and especially with all understandings of religion that insist on forcibly applying their own norms to the whole of society.
>
> *(Jean-Paul Willaime, quoted in de Vries 2013: 220)*

This is in part a question of how we construct and inhabit our own phenomenological world, and whether or not that is determined by some kind of *habitus* of faith. In *A Secular Age*, Charles Taylor (2007) aims to chart the movement from a culture in which belief in God was axiomatic to one in which it was an exception. The gradual 'disenchantment' of the world associated with the 'disembedding' of individuals from an imaginary populated by sacred beings and practices, transforming into an 'immanent frame' dominated by scientific materialism, human self-sufficiency, and instrumental rationality (*ibid.*: 4). Even those claiming a religious faith are shaped by this world-view.

As we have seen, however, the contemporary stand-off between belief and unbelief has a history, evolving from earlier paradigms. For Taylor, the secular reflects the transition 'from a society in which it was virtually impossible not to believe in God, to one in which faith, even for the staunchest believer, is one human possibility among others' (Taylor 2007: 3). He is speculating as much about the quotidian reality of people's taken-for-granted existential milieu as he is about the relations between church and state, or formal statistics of religious affiliation. It is about the shift in consciousness whereby human fulfilment and meaning are identified as coming through sources and life-chances that have nothing to do with transcendence and everything to do with human autonomy: 'a secular age is one in which the eclipse of all goals beyond human flourishing becomes conceivable' (*ibid.*: 19). As 'buffered selves' we have a clear understanding of what lies beyond us and an informed sense of how our emotional, moral, and cultural subjectivities are socially constituted, not divinely sanctioned. The secular age is thus characterized by an inability to revert to a naivety regarding the possibilities of belief: everyone, even the most devoutly religious, recognizes that to believe or not to believe are not givens but choices.

> We all learn to navigate between two standpoints: an "engaged" one in which we live as best we can the reality our standpoints opens us to; and a "disengaged" one in which we are able to see ourselves as occupying one standpoint among a range of possible ones, with which we have possible ways to coexist.
>
> *(*ibid.: *12)*

The state of secularity has separated people from a sense of enchantment. From the process of disenchantment emerged the imperative of having to find ways of ordering the everyday world in other ways, without reference to a horizon of transcendence or divine agency. Yet the postsecular might, arguably, denote a further turn towards reflexivity, in that we have to acknowledge the persistence of faith in one and the same breath as realizing that culturally and existentially, the world operates unproblematically within an immanent frame. We cannot unmake the prospect of non-belief; for those of faith, it must either be accommodated within a pluralist world-view or opposed in the name of a one, true faith. Yet even that is merely a denial of the presence of the other, however heretical. It is out of this *habitus* of pluridoxy and co-existence that identities and beliefs must be shaped.

> To be reflexively religious is to live with an intense awareness of one's own religious way as one among numerous increasingly proximate, and, to some extent, increasingly plausible, other ways.
>
> *(Hogue 2010: 360)*

Conclusion

What does the idea of the postsecular achieve? Arguably, it serves as a corrective to the liberal 'firewall' that segregates religion from the public square, serving as an enduring reminder of the inescapable 'publicness' (Tracy 2014) of religion. It enables us to recognize the constructedness of the category of the secular and the historical particularity of Western modernity, and challenges linear or monolithic trajectories of secularization. On balance, therefore, I would choose to retain the terminology of postsecular to indicate continuity *and* change: the sense that while the hegemony and taken-for-grantedness of the secular can be deconstructed, we live powerfully in its shadow and remain heavily conditioned by its assumptions. It represents, essentially, a deep *reflexivity* towards the unique and paradoxical nature of the ebbs and flows of faith and scepticism in which we live.

Overall, postsecular theorizing has exposed the extent to which modern Western thought often took secular humanism as a simplistic panacea for progress, universal reason, and peace. Yet this resulted in the effacement of many other cultural identities and the undermining of other systems of governance and political tradition—something that has come home to roost in the rise of radical Islam (Roy 2007). What would it mean to rethink the secular imperative within studies of religion and society as a less imperious and reified, more provisional and heuristic category? That may entail distinguishing between 'imperial secularity' and 'the genuine potential of the secular as a category to do justice to all religious particularities' (Smith and Whistler 2010: 16). Is it possible to envisage the emergence of some kind of paradigm that honours the diversity, vitality, and complexity of global religious life, and its stubborn endurance as a source of social order and social change beyond privatized piety?

Western modernity's conflation of radical, progressive free-thinking and religious scepticism in opposition to the reactionary forces of religious orthodoxy may now have to be dissolved, enabling us to pursue, without contradiction, examples of public life in which appeals to transcendent values and sacred texts serve as animating well-springs for innovative and inclusive social movements. Yet at the same time, since we cannot disinvent the traditions of humanism and religious doubt, there must be space for confronting those occasions in which religion is used to sanction abuses of power and impose its own forms of totalitarianism and absolutism. This heterogeneity of approaches—of religion as independent variable, as *habitus* and lived experience, and religious affiliation and practice mutating, waxing and waning in different directions—may be the most significant new window of opportunity opened up by theories of the postsecular.

Further reading

Gorski, P. S., Kim, D. K., Torpey, J. and J. VanAntwerpen (eds.) (2012) *The Post-Secular in Question: religion in contemporary society*, New York, NY: New York University Press.

An interdisciplinary collection considering theoretical and methodological dimensions within the scientific study of religion of the debate over the postsecular.

Graham, E. (2013) *Between a Rock and a Hard Place: public theology in a post-secular age*, London: SCM Press.

Offers an anatomy of the phenomenon of the postsecular and its ramifications for the ways in which religious values are mediated into the public domain.

Habermas, J. (2008a) *Between Naturalism and Religion: philosophical essays*, Oxford: Oxford University Press.

In a series of essays, Habermas considers the implications of the new visibility of religion for theories of communicative action, liberal democracy, and the nature of civil society.

Hjelm, T. (ed.) (2015) *Is God Back? considering the new visibility of religion*, London: Bloomsbury.

A collection of case studies from Europe and the Middle East examining how the re-emergence of public religion is manifested and contested across a range of contexts in politics, culture, and social policy.

References

Amrani, I. (2016) 'France's burkini ban exposes the hypocrisy of its secularist state', *The Guardian*, 24-08-2016, accessed online 14-06-2018.

Asad, T. (1993) *Genealogies of Religion: discipline and reasons of power in Christianity and Islam*, Baltimore, MD: Johns Hopkins University Press.

——— (2003) *Formations of the Secular: Christianity, Islam, modernity*. Stanford, CA: Stanford University Press.

Baker, C. R. (2009) *The Hybrid Church in the City: third space thinking*, London: SCM Press.

Barbieri, W. A. (2014) 'Introduction'. In Barbieri (ed.) *At the Limits of the Secular: reflections on faith and public life*, Grand Rapids, MI: Wm B Eerdmans, pp. 1–25.

Beckford, J. A. (2012) 'Public religions and the postsecular: critical reflections', *Journal for the Scientific Study of Religion*, 51(1): 1–19.

Bell, C. (2000) ''Pragmatic theory'. In Rothstein, M. and T. J. Jensen (eds.) *Secular Theories on Religion: current perspectives*, Copenhagen: Museum Tusculanum Press, pp. 9–20.

Berger, P. (1999) 'The desecularization of the world: a global overview'. In Berger, P. (ed.) *The Desecularization of the World: resurgent religion and world politics*, Grand Rapids, MI: Wm B Eerdmans, pp. 1–18.

Boeve, L. (2008) 'Religion after detraditionalization: Christian faith in a postsecular Europe'. In Ward, G. and M. Hoelzl (eds.) *The New Visibility of Religion*, London: Continuum, pp. 187–209.

——— (2012) 'Religious education in a post-secular and post-Christian context', *Journal of Beliefs & Values*, 33(2): 143–56.

Braidotti, R. (2008) 'In spite of the times: the postsecular turn in feminism', *Theory, Cuture and Society*, 25(6): 1–24.

Bretherton, L. (2014) *Resurrecting Democracy: faith, citizenship, and the politics of a common life*, Cambridge: Cambridge University Press.

Brown, C. (2009) *The Death of Christian Britain*, 2nd edition, London: Routledge.

Bruce, S. (2002) *God is Dead: secularization in the west*, Oxford: Blackwell.

Butler, J., Habermas, J., Taylor, C. and C. West (2011) *The Power of Religion in the Public Sphere*, edited by Eduardo Mendieta and Jonathan VanAntwerpen, afterword by Craig Calhoun, New York, NY: Columbia University Press.

Calhoun, C. (2010) 'Rethinking secularism', *Hedgehog Review*, Issue Fall: 34–48.

Carr, D. (2012) 'Post-secularism, religious knowledge and religious education', *Journal of Beliefs and Values: studies in religion and education*, 33(2): 157–68.

Casanova, J. (1994) *Public Religions in the Modern World*, Chicago, IL: University of Chicago Press.

Cloke, P. and J. Beaumont (2013) 'Geographies of postsecular rapprochement in the city', *Progress in Human Geography*, 37(1): 27–51.

Crockett, C. (2011) *Radical Political Theology: religion and politics after liberalism*, New York, NY: Columbia University Press.

Dillon, M. (2012) 'Jürgen Habermas and the post-secular appropriation of religion: a sociological critique'. In Gorski, P. S., Kim, D. K., Torpey, J. and J. VanAntwerpen (eds.) (2012) *The Post-Secular in Question: religion in contemporarysSociety*, New York, NY: New York University Press, pp. 249–78.

Dinham, A. (2012) *Faith and Social Capital After the Debt Crisis*, London: Palgrave Macmillan.

Dinham, A. and M. Francis (2015) *Religious Literacy in Policy and Practice*, Bristol: Policy Press.

Habermas, J. (2008b) 'Notes on post-secular society', *New Perspectives Quarterly*, 25(4): 17–29.

Hogue, M. (2010) 'After the secular: toward a pragmatic public theology', *Journal of the American Academy of Religion*, 78(2): 346–74.
Lynch, G. (2007) *Between Sacred and Profane: researching religion and popular culture,* London: I.B. Tauris.
Lyon, D. (2010) 'Being post-secular in the social sciences: charles taylor's social imaginaries', *New Blackfriars,* 91(1036): 648–62.
McAvan, E. (2012) *The Postmodern Sacred: popular culture spirituality in the science fiction, fantasy and urban fantasy genres,* Jefferson, NC: McFarland.
McCutcheon, R. (2000) 'Critics not caretakers: the scholar of religion as public intellectual'. In Rothstein, M. and T. Jensen (eds.) *Secular Theories on Religion: current perspectives,* Copenhagen: Museum Tusculanum Press, pp. 167–82.
McLennan, G. (2011) 'Postsecular cities and radical change: a philosophical sea change?'. In Beaumont, J. and C. Baker (eds.) *Postsecular Cities: space, theory and practice*, London: Continuum, pp. 15–30.
Pew Forum on Religion and Public Life (2012) *The Global Religious Landscape: a report on the size and distribution of the world's magoe religious groups as of 2010,* Washington, DC: Pew Research Center.
Rawls, J. (1971) *A Theory of Justice*, Cambridge, MA: Harvard University Press.
——— (1987) 'The idea of an overlapping consensus', *Oxford Journal of Legal Studies*, 7(1): 1–25.
Roy, O. (2007) *Secularism Confronts Islam,* New York, NY: Columbia University Press.
Smith, A. P. and D. Whistler (2010) 'What is continental philosophy of religion now?'. In Smith, A. P. and D. Whistler (eds.) *After the Post-secular and the Postmodern: new essays in continental philosophy of religion*, with foreword by Pamela Sue Anderson, Newcastle upon Tyne: Cambridge Scholars Publishing, pp. 1–24.
Smith, J. K. (2004) *Introducing Radical Orthodoxy: mapping a post-secular theology,* Grand Rapids, MI: Baker Academic.
Southbank Centre (2017) *Belief and Beyond Belief,* www.southbankcentre.co.uk/whats-on/festivals-series/belief-and-beyond-belief, accessed online 03-08-2017.
Spencer, N. (2017) *Doing Good: a future for Christianity in the 21st Century*, London: Theos.
Taylor, C. (2007) *A Secular Age*, Cambridge, MA: Belknapp Press of Harvard University Press.
——— (2010) 'The Meaning of secularism', *Hedgehog Review,* Issue Fall: 23–34.
Tracy, D. (2014) 'Religion in the public realm: three forms of publicness'. In Barbieri, W. A. Jr. (ed.) *At the Limits of the Secular: reflections on faith and public life*, foreword by Charles Taylor, Grand Rapids, MI: Eerdmans, pp. 29–50.
Vries, H. de (2013) 'A religious canon for Europe? policy, education and the postsecular challenge', *Social Research*, 80(1): 203–32.
Welker, M. (2010) 'Habermas and Ratzinger on the future of religion', *Scottish Journal of Theology,* 63(4): 456–73.
Wright, R. (2016) 'A court overturns a Burkini ban, but not its mindset, *The New Yorker*, 26-08-2016, accessed online 14-06-2018.

19

Postsecularity and urban theology

Chris Shannahan

Introduction

The onward march of rationalism has not led to the death of religion. While it may have changed and, in some parts of the Global 'North', the number of people self-defining as non-religious continues to increase, faith remains a central element in many local communities and a key player in civil society politics.[1] The old secularist orthodoxy is increasingly under fire as a new postsecular social contract is hammered out. In this essay I consider the impact that postsecular discourse has had on theology. Shaped by a liberative approach to theological reflection, I will suggest that academic debates about postsecularity largely fail to reflect the dynamic diversity of the contemporary city and do not engage with the multiple struggles for social justice that characterize an urban landscape in the grip of an ongoing 'age of austerity' and the resurgent xenophobia that has accompanied the election of Donald Trump as US President and the Brexit phenomenon in the UK. I will argue that debates about postsecularity need to be people-centred rather than process-driven if they are to foster a social ethic of mutuality capable of resourcing a culturally credible, urban theology of liberation.

The rise and fall of secularism

The work of Max Weber defined social scientific debate about the public role of religion for most of the twentieth century. Weber (1930) argued that as societies modernize, religion would become less publicly important as people increasingly turned to science to answer life's big questions. Following its high watermark in the work of Wilson (1966) and Berger (1967), support for Weber's secularization thesis began to wane. Sociologists like Bruce (2002, 2011) still affirm Weber's hypothesis that the public decline of religion is the inevitable consequence of the differentiation of the world into public and private spheres, which emerged from the religious rationalism of the Protestant Reformation (Walsham, 2008: 499), but his is now a minority view.

Berger (1967) argued that although religion would continue to comfort some people in an unfriendly world, institutional secularization would ultimately be accompanied by a similar shift in consciousness. Just before the turn of the century, however, Berger (1997: 994) suggested that he, like many other sociologists of religion, had been mistaken because, 'Most of the world today is certainly not secular. It's very religious'. Writing a few years earlier, Davie (1990, 1994) contended that, while church membership had declined significantly in the postwar era, this did not necessarily mean that people had swapped faith for secularism. Her subsequent work (Davie 2016), like that of Lynch (2002, 2005, 2007) and Partridge (2002, 2004), invites us to be wary of such a simplistic assumption.

Early in the twenty-first century, Heelas and Woodhead (2005) explored the changing face of faith in the town of Kendal in the North of England, suggesting we are at a watershed moment in the history of religion in Europe. Heelas and Woodhead (2005: 3–4) spoke of a subjective turn towards individualized forms of spirituality: 'The goal is...to have the courage to become one's own authority...to forge one's own inner-directed...life'. Partridge (2002: 240) argues that the disenchantment to which Weber points paved the way for a resurgent interest in 'alternative spiritualities' in Europe and North America. For Partridge (*ibid.*: 241), however, 'Secularisation theorists are too wedded to their grand metanarrative of a linear, irreversible... decline [in] organised religion...to recognise the importance of new forms of spirituality'.

Lynch (2002) discusses the experience of 'Generation X'—a loose term used to designate a rejection of institutional religion and consequent search for individualized existential meaning.[7] For Lynch (*ibid.*: 27), Generation X was born as, 'religious and social certainties began to dissolve and become more fluid'. In this context, argued Lynch (*ibid.*: 31), 'meaning is not something that can be found in pre-packaged forms...but...has to be sought in a personal way'. Lynch (2007), like Partridge, examines the recontextualization of ancient earth-centred spiritualities and aspects of Buddhist philosophy and Jewish and Sufi mysticism by largely left-of-centre members of Generation X who are equally unpersuaded by organized religion and secularism.

Building on her earlier work, Davie (2016: 162) argues that while Weber forwarded his analysis of religion in the midst of a rapidly changing Europe, these contextual concerns 'turned into theoretical assumptions, with the strong, but unsubstantiated, implication that secularisation would necessarily accompany modernization whenever and wherever it occurred'. For Davie (*ibid.*: 163), the growing critique of the secularization thesis 'amounts to nothing less than a paradigm shift' within sociology. This new academic climate emerged as theorists sought to keep pace with a changed social context within which faith communities have assumed an increased 'significance as markers of identity' (Davie 2016: 172) as religious faith has become deprivatized (Casanova 1994: 211ff).

Two further factors have challenged secularist assumptions that faith should be excluded from the public sphere. First, the growth of the British-Muslim community has impacted on debates and policy formation in relation to social cohesion and civic action (O'Toole et al. 2015). Second, migration from West Africa has led to the rapid growth of new, often Pentecostal, Christian denominations (Burgess 2009). Davie (2016: 175) summarizes, '[R]eligion continues to influence almost every aspect human society—economic, political, social and cultural'. The secular orthodoxy, which excluded religion from the public sphere, has lost its credibility as faith groups have continued to be important players in civil society politics, locally, nationally, and internationally (Dinham et al. 2009; Dinham and Shaw 2012).

Looking for a new consensus

The work of Jürgen Habermas has had a major influence on the emergence of postsecular discourse. Habermas (1991: 130) argues that we need to move beyond a secular-religious binary if we are to understand how religion can 'help us express our best moral intuitions'. He suggests that, 'Religious traditions have a special power to articulate [*these*] moral intuitions, especially with regard to vulnerable forms of communal life' (Habermas 2006: 10) and points out that, 'religious organizations are increasingly assuming the role of "communities of interpretation" in the public arena of secular societies' (Habermas 2008: 20). Ironically, however, Habermas (2006: 4) echoes the influential but widely contested (Shannahan 2017) 'clash of civilisations' thesis popularized by Huntington (1993), implying that religious diversity threatens social cohesion. Furthermore, like Castells (2010), Habermas (2006, 2008) seemingly equates the influence of religion almost exclusively with conservative social movements, rather than with progressive networks like Drop the Debt, Stop the War, or Sojourners.

Sketching out the postsecular landscape, Habermas (2006: 5ff) suggests that, 'For all their ongoing dissent on...world views and religious doctrines, citizens are meant to respect one another as free and equal members of their political community'. However, he still appears to suggest that the needs and perceptions of a secular state and citizens who are people of faith are in conflict. Habermas (2006) acknowledges that a new social contract should not expect faith communities to translate every religious statement into 'a universally accessible language' (2006: 8), while also asserting that, '[G]iven that in the liberal state only secular reasons count, citizens who adhere to a faith are obliged to establish a kind of "balance" between their religious and their secular convictions'. Habermas (2006: 9) is concerned that the separation of religion and politics in the public sphere does not cause 'an undue mental and psychological burden for those citizens who follow a faith' but insists that people of faith must acknowledge the 'priority that secular reasons enjoy in the political arena'. For Habermas, it seems, the public sphere should be shaped around secularist principles, which cast doubt on whether postsecularity reflects an existential shift or a pragmatic example of realpolitik.

Postsecularity and theology

The postsecular turn in theology coincided with a resurgent interest in what Hoelzl and Ward (2008) called the 'new visibility of religion' in civil society politics (Scott et al. 2004; Baker 2007; Bretherton 2011, 2014; Baker and Skinner 2014; Shannahan 2014). What impact has postsecular discourse had on theological discussions about the political role of faith groups on an increasingly fluid cultural landscape? A brief look at the 'radical orthodoxy' (RO) movement and recent developments in public theology can help us to respond to this question.

The memorable, 'Once there was no secular' with which John Milbank (1990: 9) began *Theology and Social Theory: beyond secular reason*, served as a pithy preface to the emergence of the RO movement almost a decade later. The publication of *Radical Orthodoxy: a new theology* (Milbank et al. 1999) was the academic equivalent of hurling a rock into a pool. Almost 20 years later, the ripples are still disturbing theological waters. Milbank et al. (1999: 1) nailed their colours to the mast in the first paragraph, 'For several centuries now, secularism has been...constructing...a world in which the theological is either discredited or turned into a harmless leisure-time activity...And yet...today the logic of secularism is imploding'. RO is a postmodern bricolage, mingling, 'exegesis, cultural reflection and philosophy in a complex...collage' (Milbank et al. 1999: 2) but is deeply critical of modernity and the Enlightenment secularism that followed in its wake. This postsecular theological movement seeks 'a return to patristic and medieval roots' because, for Milbank et al., the 'fragments' (1999: 23–4) of presecular Augustinian orthodoxy can challenge rationalist secularism.

RO captured the imagination but struggles to speak in a dialogical manner to a postsecular age. As Doak (2007: 372) notes, 'Milbank challenges Christian theology to recover its proper position as the master discourse that positions all other discourses'. Shakespeare (2000: 167) argues that RO, 'seems incapable of admitting any good whatever in secular or non-Christian discourses', thereby (*ibid.*: 168) fostering an 'endless conflict between mutually exclusive discourses all seeking to occupy the hermeneutical high ground'. For many, it is Milbank's name that is most publicly connected with RO, but it is the less acerbic work of his compatriot Ward that brings the movement's broad themes into a sharper dialogue with the particularities of the modern city. Contemporary urban theologians would benefit from reflection on Ward's (2000: 36ff) contention that in the face of the 'rise of market-driven consumerism, cities become increasingly secular places…' and that, as a result, 'faith becomes increasingly privatised'. Ward's conflation of consumerism and secularisation is overly generalized, and his insistence that faith is inevitably confined to the private sphere fails to reflect the dynamic diversity of the twenty-first century city. However, his wrestling with the specificity of urban life presents RO in a more provisional light than the vision articulated by Milbank. Ward's (2000: 55ff) reflections on deindustrialization and his apparent movement beyond Christian exclusivism could stimulate a creative dialogue between RO advocates and urban theologians who are critical of the movement. However, it is unclear how the concerns of RO engage with life in an 'age of austerity' or grassroots interfaith dialogue in superdiverse but splintered cities where the movement's grand themes fail to resonate.

A second theological response to postsecular discourse is found in public theology, the contemporary cousin of the older discipline of political theology, and in debates about the nature of theology and its place in the public sphere. Although many commentators dismiss him as a Nazi apologist, the first use of the term 'political theology' is found in the work of Carl Schmitt (1922, 1985). However, political theology finds its roots not in Nazi Germany but in the writings of the Hebrew Prophets and the description of Jesus' ministry in the Gospels. In an academic context we can date the birth of the political theology to the work of three German theologians: Johann Baptist Metz (1969), Dorothy Sölle (1974), and Jürgen Moltmann (1967, 1974); to the first generation of Latin American liberation theologians: Gustavo Gutiérrez (1974), Leonardo Boff (1978), and Juan Luis Segundo (1975); and to the pioneer of US black theology, James Cone (1969). While liberation theology, black theology, and urban theology are contextual theologies that engage in depth with lived experience, much political theology has confined itself to debates about macro issues, a critical dialogue with political philosophy and a theological critique of institutionalized political and economic power (Scott and Cavanaugh 2004; Stackhouse 2004).

A fixed distinction between political theology and public theology is misleading because both disciplines explore the theological implications of faith-based engagement in the public sphere. However, the rise of public theology has signalled a theological shift from institutions and structures to networks and neighbourhoods, state to civil society, and a secular landscape to an uncertain postsecular settlement. Martin Marty (1974) was probably the first to use the term 'public theology', which Stackhouse (2004: 284ff) suggests was quickly adopted for three reasons. First, it symbolized a shift to a broader understanding of the public sphere and faith-based civil society politics. Second, the term resonated with attempts to find a credible way of talking about faith in an increasingly secularized and diverse public sphere. Third, it echoed an older social gospel tradition, which emphasized, 'the responsibility of the ecclesial community to engage in [*the*]…constant reformation of the social order' (*ibid.*: 284).

Public theology seeks to articulate a theology for a postsecular age.[3] The focus it has on individuals and local communities puts it in a better position than political theology or RO to provide a theological voice capable of engaging dialogically with the central players in an increasingly fluid public sphere—government, economic structures, and civil society. However, while the Global Network for Public Theology now includes researchers in Nigeria, Hong Kong, Indonesia, South Korea, Brazil, and Tonga, Cady (2014) suggests that public theology has, to date, been largely confined to theologians in the USA, Europe, Australia, and South Africa.[4] In an interwoven world, does the implicit dominance of the Global 'North' and an enduring link to the church limit the capacity of public theology to engage with the existential concerns of communities in the Global 'South' or hinder its development in inherently multifaith communities?

The work of Elaine Graham exemplifies public theology's attempt to respond to the postsecular shift in academic discourse. Graham has sought to articulate a culturally credible postsecular Christian apologetics. She suggests that 'the world appears to be…entering an unprecedented political and cultural era' (Graham 2013: xiii) in which the separation of religion and politics that was a central plank of modernity is becoming increasingly anachronistic. However, while Graham (*ibid.*: xvi) notes that secularization is not 'uniform, inevitable or irreversible', she suggests that we are not witnessing a religious revival or a 'return to Western Christendom'. For Graham (*ibid.*: xvi), public theology represents 'a quest for a new voice in…a public debate that is more fragmented, more global and more disparate' than anything we have known before.

Graham (2013: xx) argues that public theology revolves around two fundamental concerns. First, 'it privileges…the societal meanings of faith in contrast to forms of religious belief and practice that confine faith to private and pietist intentions'. Second, she contends that public theology is 'less concerned with defending the interests of specific faith communities than generating informed understandings of the theological and religious dimensions of public issues'. This may be true, but public theology continues to frame its discourse within a Christian worldview. It is therefore reasonable to ask whether the discipline can credibly engage with the search for meaning in the multifaith cities of the twenty-first century within which an increasing majority of people have no connection whatsoever with the institutional church.

The emergence of public theology exemplifies the need for Christian theologians in a postsecular age to rethink old debates about the relationship between Christianity and culture, which find their roots in Richard Niebuhr's (1951) *Christ and Culture*. Niebuhr's ideal types continue to raise questions about the role of faith in civil society politics and the relationship between theology and culture. Graham (2016: 25), however, argues that the church needs to recover an 'apologetics of presence', not 'as a weapon of conversion, but a gesture of solidarity and reconciliation'.[5]

Influenced by its Christian Realist roots, public theology is better placed to resource the development of an inclusive postsecular theological narrative than RO. However, its broad-brush approach, implicit linkage to the institutional church, and insufficiently developed dialogue with the social sciences cast doubt on its ability to articulate the everyday realities and existential questions of the millions for whom the 'age of austerity' has become a seemingly unending fact of life.

Bursting the postsecular bubble

Debates about postsecularism arguably reflect the existential questioning of a small metropolitan intellectual élite, rather than the concerns of most city dwellers. For many people, religion has never played any discernible role in their lives. Stark (1999) and Swatos and Christiano (1999) suggest that debates about secularization and postsecularity are premised on a misreading of history, which imagines the past as a religious 'golden age'. For some city dwellers, faith is a submerged part of their cultural hinterland that surfaces during times of communal or personal crisis (Reddie 2001; Wilkinson, 1993). For others, faith continues to act as an affirming identity marker in a society that demonizes them (Modood and Ahmad 2007). And for other people, talk of the 'new visibility of religion' is mistaken. Faith never went away but has undergirded progressive social movements for more than a century and remains the basis of their political engagement (Shannahan 2014).

I have summarized the growing critique of the secularization thesis and the emergence of theological responses to postsecularism and suggested that they fail to meet the needs of socially excluded communities. I want now to ask some awkward questions of postsecularism's cheer-leaders by drawing on two case studies that burst the postsecular bubble.

Social exclusion and urban youth spiritualities

From 2009–2012, I spent most of my time working alongside socially excluded urban youth on a large housing estate in the city of Birmingham in the UK. This ethnographic project explored how the experience of social exclusion impacted the ways in which unemployed young men thought and spoke about identity, community, and spirituality. At the time, the estate was among the 5% most multiply-deprived communities in England and Wales.[6] Five years later, it remains the most multiply-deprived parliamentary constituency in Birmingham and among the 10% most deprived in England and Wales. Three things became clear during fieldwork. First, the dominant public discourse surrounding unemployed young men stigmatized them and robbed them of their agency. Second, none of the young men had any connection with organized faith groups, which were viewed as hypocrites who talked about social justice but did little about it. Third, academic analyses of 'belief' did not capture the contextualized nature of their existential questioning (Shannahan 2012).

At first glance it might appear as if these young men were signed-up secularists, but first impressions can be misleading. Their lives disrupt neat arguments about the secular-postsecular spectrum, raising questions about its usefulness in socially excluded urban communities. Debates about the decline of religion, its new visibility in civil society politics, or the emergence of a postsecular co-existence between secularists and people of faith did not resonate with young men in Bromford. Their rejection of religion was of peripheral importance because church had never been a part of their lives anyway. However, their critique of the church should not be equated with a rejection of the possibility of God as the words of one 19-year-old make clear, 'I believe in God, but He doesn't live round here!' (Shannahan 2012: 322).

A generation ago, Tricia Rose (1994: 21) argued that 'Hip-Hop emerges from the de-industrialisation meltdown where social alienation, prophetic imagination and yearning intersect'. Hip-Hop provides young men from Bromford with the vocabulary to articulate their own hopes, dreams, and fears. It seemed natural, therefore, to work with the Birmingham graffiti artist, Mohammed Ali, to find a way in which the young men could give voice to their hopes and dreams. Together, we created a huge graffiti cube entitled 'Bromford Dreams'.

As the young men created the cube they made it clear that they wanted to include two figures in prayer—one Muslim and one Christian—for two reasons. First, because, as they said, 'Everyone prays sometimes', and second, as a slap in the face of the far-right English

Defence League that had begun organizing on the estate. Rap music enabled the young men to articulate their own individualized spirituality which revolved around a contextualized Holy Trinity: Jesus, rap musician Tupac Shakur, and Malcolm X. These iconic figures were not seen in dogmatic terms but as heroes standing in solidarity with the marginalized. Truth was not validated by outsiders but in relation to its capacity to embody an ethic of hope and agency in the face of social exclusion. Academic debates about secularization, postsecularity, and 'belief' fail to capture the raw spirituality of the young men I got to know. They do not fit into the neat sociological or theological categories that have been created in other places and by other people. Theirs is a non-dogmatic spirituality of immanence and solidarity. It provides us with a window onto the complexity of meaning-making for young men for whom the debates of academics have no significance whatsoever.

Community organizing and faith-based politics

Community organizing was born in South Chicago just before the outbreak of World War II, through the work of Saul Alinsky. Influenced by his Chicago Area Project which supported young people caught up in gang violence, Alinsky spearheaded the development of the Back of the Yards Neighborhood Council, which first met in July 1939. The success of its first assembly enabled Alinsky to establish the Industrial Areas Foundation (IAF) the following year. Other, explicitly faith-based community organizations like the Gamaliel Foundation and the PICO National Network followed in the late 1960s and early 1970s. Such organizations premised their campaigning on a Biblical commitment to social justice rather than Alinsky's more utilitarian approach. The Gamaliel Foundation makes this distinction explicit in its mission statement:

> Gamaliel's organizing...draws on struggles for justice by people of faith stretching back thousands of years...Our work draws on...Christ's life and teaching, the Torah, the Qur'an, Catholic social teaching, the founding principles of American democracy and the US civil rights movement.[7]

However, it is the model of community organizing created by Alinsky and developed by the IAF that took root, in an adapted form, in the UK in the early 1990s, with the creation of the *Citizens Organising Foundation* (renamed *Citizens UK* in 2009). Bretherton (2014: 34) suggests that Alinsky sought to develop a 'common life politics', rather than a model of activism, revolving around the specifics of class or ethnicity. Bretherton (*ibid.*) notes that Alinsky's model of community organizing was shaped by a commitment to networked, bottom-up work, the emphasis within the US Labour Movement on mutuality and cooperation, and common good thinking within Catholic Social Teaching. The work of the IAF during Alinsky's life time was characterized by a pragmatic rather than a principled engagement with faith groups (Shannahan 2014: 25ff). However, since Alinsky's death in 1972, the IAF has placed a stronger emphasis on the animating force of Jewish and Christian spirituality (Chambers and Cowan 2003; Bretherton, 2014: 41ff).[8]

The community organizing model that took root in the USA and was to catapult former organizer Barack Obama into the White House in 2009 crossed the Atlantic at the end of the Thatcher decade (Bretherton 2014: 76ff; Shannahan 2014: 33ff). Influenced by his engagement with US organizing, Neil Jameson *established The East London Citizens Organisation* and the *Citizens Organisation Foundation/ Citizens UK*, which, by 2017, involved more than 24,000 people from over 300 civil society organizations. Although not explicitly faith-based,

Citizens UK relies heavily on the support of churches, mosques, gurdwaras, and synagogues, bridging the gap between the model adopted by the IAF and that exemplified by the Gamaliel Foundation in the USA.[9] Bretherton (2014: 76) describes community organizing in the UK as a practice that 'mediates the relationship between "faith" and democratic citizenship', and both Shannahan (2014) and Bretherton (2014) argue that it draws on a centuries-long heritage of socially progressive faith-based activism.

A generation ago, the urban theologian Kenneth Leech (1997: 140) anticipated the ecclesiological implications of postsecular theological discourse, suggesting that the calling of the church was to be 'a creative minority within society', rather than the dominant voice seemingly imagined by advocates of RO. A framing of the 'new visibility of religion' in civil society politics within Leech's agitating minority ethic could animate the postsecular witness that public theologians are searching for.

Postsecular discourse has been articulated within an academic sub-culture. However, its carefully honed analysis is disrupted by the particularities and gritty political activism of community organizing. Fusing *Qur'anic* and Biblical commitments to equality and human worth with the comparable values of trades' unionists, community groups and students' unions, community organizing side-steps disengaged debates about secularism and postsecularity. Its reweaving of the fabric of society bursts the secular/ postsecular bubble, focusing instead on building broad-based campaigns that put food on the table and a roof over someone's head.

Conclusion

As the world becomes an increasingly urban planet, it is becoming clear that the future of faith will probably be decided in large cities. In this essay I have suggested that debates about secularization, re-enchantment, and postsecularity fail to engage with the complex realities of urban life. The secular/postsecular narrative does not reflect the more nuanced discourse in diverse urban communities where faith continues to be an important, if sometimes submerged, feature of life or where it never made much of an impact in the first place. Life in the twenty-first-century city is not reducible to a neat secular/ postsecular either-or but is characterized by fluidity and change. It is in this liminal 'third space' (Bhabha 1994) that new meanings and identities are forged, as Beaumont and Baker (2011: 3) recognize in their exploration of the 'flows of interaction between diverse groups' in the contemporary city. Disengagement from organized religion in socially excluded urban communities does not neatly equate with secularism, on the one hand, or re-enchantment on the other, but can give rise to new hip-hop-infused spiritualities of solidarity and resistance. The resurgent visibility of religion within civil society politics need not reduce faith groups to welfare delivery agencies in a postsecular 'age of austerity', but can stimulate bottom-up models of prophetic faith-based activism that speak truth to power. It is to this agenda that urban theology must turn as the twenty-first century draws to the end of its second decade.

Notes

1 In 2014, in all, 22% of people in the USA self-defined as non-religious, see http://www.pewforum.org/religious-landscape-study/. In 2017 in the UK, the British Social Attitudes survey indicated that 53% of people described themselves as non-religious, see https://www.theguardian.com/world/2017/sep/04/half-uk-population-has-no-religion-british-social-attitudes-survey

2 The popularizing of the term 'Generation X' is often attributed to the novelist Douglas Coupland and his 1992 novel *Generation X: tales for an accelerated culture.*

3 See http://www.theosthinktank.co.uk/ accessed 17 September 2017. The work of the Theos think-tank exemplifies the broad-based approach of public theology in the twenty-first century.
4 See note #3.
5 See https://chestercathedral.com/wp-content/uploads/2015/10/A-New-Apologetics-Cathedral.pdf accessed 17 September 2017.
6 See file:///C:/Users/ac0971/Downloads/Index_of_Deprivation_2015.pdf accessed 25 September 2017.
7 See http://www.gamaliel.org/AboutUs/History.aspx accessed 29 September 2017.
8 See http://www.industrialareasfoundation.org/content/mission accessed 29 September 2017.
9 See http://www.citizensuk.org/ accessed 29 September 2017.

Further reading

Shannahan, C. (2010) *Voices from the Borderland: cross-cultural urban liberation theology in the twenty-first century*, London: Equinox.

A wide-ranging critique of differing models of British urban theology since its inception in the 1970s, which Shannahan uses to develop a new, cross-cultural, and postsecular approach to urban theology.

Woodhead, L. and R. Catto (eds.) (2012) *Religion and Change in Modern Britain*, London: Routledge.

This multidisciplinary edited volume charts the changing and increasingly fluid nature of religion in the UK and enables the reader to wrestle with a turn away from secularism towards more provisional postsecularist analyses of the search for meaning in the twenty-first century.

References

Baker, C. (2007) *The Hybrid Church in the City: third space thinking*, Aldershot: Ashgate.
Baker, C. and H. Skinner (2014) *Faith in Action: the dynamic connection between spiritual and religious capital*, Chester: William Temple Foundation.
Beaumont, J. and C. Baker (eds.) (2011) *Postsecular Cities: space, theory and practice*, London: Continuum.
Berger, P. (1967) *The Sacred Canopy: elements of a sociological theory of religion*, Garden City, NY: Doubleday.
——— (1997) 'Epistemological modesty: an interview with Peter Berger', *The Christian Century*, October 29: 972–78.
Bhabha, H. (1994) *The Location of Culture*, New York, NY: Routledge.
Boff, L. (1978) *Jesus Christ Liberator: a critical Christology for our time*, Maryknoll, NY: Orbis Books.
Bretherton, L. (2011) 'A postsecular politics? Inter-faith relations as a civic practice', *Journal of the American Academy of Religion*, 79(2): 346–77.
——— (2014) *Resurrecting Democracy: faith, citizenship and the politics of a common life*, New York, NY: Cambridge University Press.
Bruce, S. (2002) *God is Dead: secularization in the west*, Oxford: Wiley-Blackwell.
——— (2011) *Secularization: in defence of an unfashionable theory*, Oxford: Oxford University Press.
Burgess, R. (2009) 'African Pentecostal spirituality and civic engagement: the case of the Redeemed Christian Church of God in Britain', *Journal of Beliefs and Values*, 30(3): 255–73.
Cady, L. E. (2014) 'Public theology and the postsecular turn', *International Journal of Public Theology*, 8: 292–313.
Casanova, J. (1994) *Public Religions in the Modern World*, Chicago, IL: University of Chicago Press.
Castells, M. (2010) *The Power of Identity*, Oxford: Blackwell Publishing.
Chambers, E. T. and M. A. Cowan (2003) *Roots for Radicals: organizing for power, action and justice*, London: Continuum.
Cone, J. H. (1969) *Black Theology and Black Power*, San Francisco, CA: Harper Collins.
Davie, G. (1990) 'Believing without belonging: is this the future of religion in Britain?', *Social Compass*, 37(4): 455–69.
——— (1994) *Religion in Britain since 1945: believing without belonging*, Oxford: Blackwell.
——— (2016) 'Resacralization'. In Turner, B. S. (ed.) *The New Blackwell Companion to the Sociology of Religion*, Oxford: Blackwell, pp. 16–179.

Dinham, A., Furbey, R. and V. Lowndes (eds.) (2009) *Faith in the Public Realm: controversies, policies and practices*, Bristol: The Policy Press.

Dinham, A. and M. Shaw (2012) 'Measurement as reflection in faith-based social action', *Community Development Journal*, 47(1): 126–41.

Doak, M. (2007) 'The politics of radical orthodoxy: a Catholic critique', *Theological Studies*, 68(2): 368–93.

Graham, E. (2013) *Between a Rock and a Hard Place: public theology in a post-secular age*, London: SCM Press.

——— (2016) 'A new apologetics: speaking of God in a world troubled by religion', Chester Theological Society, Chester Cathedral Lecture, 23 January.

Gutiérrez, G. (1974) *A Theology of Liberation*, London: SCM Press.

Habermas, J. (1991) *The Structural Transformation of the Public Sphere: an inquiry into a category of bourgeois society*, Cambridge, MA: MIT Press.

——— (2006) 'Religion in the public sphere', *European Journal of Philosophy*, 14(1): 1–25.

——— (2008) 'Secularism's crisis of faith: notes on post-secular society', *New Perspectives Quarterly*, 25: 17–29.

Heelas, P. and L. Woodhead (2005) *The Spiritual Revolution: why religion is giving way to spirituality*, Oxford: Wiley-Blackwell.

Hoelzl, M. and G. Ward (eds.) (2008) *The New Visibility of Religion: studies in religion and cultural hermeneutics*, London: Continuum.

Huntington, S. (1993) 'The clash of Civilisations', *Foreign Affairs*, 72(3): 22–49.

Leech, K. (1997) *The Sky is Red: discerning the signs of the times*, London: Darton Longman & Todd.

Lynch, G. (2002) *After Religion: Generation X and the search for meaning*, London: Darton, Longman & Todd.

——— (2005) *Understanding Theology and Popular Culture*, Oxford: Blackwell.

——— (2007) *The New Spirituality: an introduction to progressive belief in the twenty first century*, London: I.B. Tauris.

Marty, M (1974) 'Reinhold Niebuhr: public theology and the American experience', *The Journal of Religion*, 54(4): 332–59.

Metz, J. B. (1969) *Theology of the World*, New York, NY: Seabury Press.

Milbank, J. (1990) *Theology and Social Theory: beyond secular reason*, Oxford: Blackwell.

Milbank, J., Pickstock, C. and G. Ward (eds.) (1999) *Radical Orthodoxy: a new theology*, London: Routledge.

Modood, T. and F. Ahmad (2007) 'British Muslim perspectives on multiculturalism', *Theory, Culture & Society*, 24(2): 187–213.

Moltmann, J. (1967) *Theology of Hope: on the ground and implications of a Christian eschatology*, London: SCM Press.

——— (1974) *The Crucified God: the cross of Christ as the foundation and criticism of Christian theology*, London: SCM Press.

Niebuhr, H. R. (1951) *Christ and Culture*, New York, NY: Harper & Row.

O'Toole, T., Meet, N., DeHannas, D. N., Jones, S. H. and T. Modood (2015). 'Governing through prevent? Regulation and contested practice in state–muslim engagement', *Sociology*, 50(1): 160–77.

Partridge, C. (2002) 'The disenchantment and re-enchantment of the west: the religio-cultural context of contemporary western Christianity', *Evangelical Quarterly*, 74(3): 235–56.

——— (2004) *The Re-enchantment of the West: alternative spiritualities, sacralization, popular culture and occulture*, vol 1, London: T&T Clark.

Reddie, A. G. (2001) *Faith Stories and the Experience of Black Elders*, London: Jessica Kingsley.

Rose, T. (1994) *Black Noise: rap music and black culture in contemporary America*, Middletown, CT: Wesleyan University Press.

Schmitt, C. (1985) *Political Theology: four chapters on the concept of sovereignty*, trans. G. D. Schwab, Chicago, IL: University of Chicago Press.

Scott, P. and W. Cavanaugh (eds.) (2004) *The Blackwell Companion to Political Theology*, Oxford: Blackwell Publishing.

Segundo, J. L. (1975) *The Liberation of Theology*, Maryknoll, NY: Orbis Books.

Shannahan, C. (2012) '"NEET" believers? An analysis of "belief" on an urban housing estate', *Culture and Religion*, 3(3): 315–35.

——— (2014) *A Theology of Community Organizing: power to the people*, London: Routledge.

——— (2017) 'Zombie multiculturalism meets liberative difference: searching for a new discourse of diversity', *Culture and Religion*, 17(4): 409–30.
Shakespeare, S. (2000) 'The new romantics: a critique of radical orthodoxy', *Theology*, 103(813): 63–177.
Sölle, D. (1974) *Political Theology*, Philadelphia, PA: Fortress Press.
Stackhouse, M. (2004) 'Civil religion, political theology and public theology: what's the difference?', *Political Theology*, 5(3): 275–93.
Stark, R. (1999) 'Secularization R.I.P', *Sociology of Religion,* 60(3): 249–73.
Swatos, W.H and Christiano, K.J. (1999) 'Seculraization Theory: The Course of a Concept', *Sociology of Religion*, 60.3: 209-228
Walsham, A. (2008) 'The Reformation and the "disenchantment of the world" reassessed', *The Historical Journal*, 51(2): 497–528.
Ward, G. (2000) *Cities of God*, London: Routledge.
Weber (1930) *The Protestant Ethic and the Spirit of Capitalism*, London: Allen & Unwin.
Wilkinson, J. (1993) *The Church in Black and White*, Edinburgh: St Andrews Press.
Wilson, B. (1966) *Religion in Secular Society: a sociological comment*, London: Watts & Co.

Part III

Theory, space, social relations

20

Postsecular plasticity

Expansive secularism

Gregor McLennan

Introduction: frame, spin, temperament

Critics of the concepts of postsecularity and postsecularism typically object to the terminology's analytical vagueness, its distance from concrete empirical and political problems of the day, and its embrace of theoretical and normative positions that are substantively different from one other. These charges are weighty. Yet, such critiques tend to assume that, by comparison, the older-style notions of secularity and secularism, religion and theology, remain stable. This is no longer the case, if it ever was. And it should be acknowledged that postsecular thought—across its many variants—has had considerable impact in disrupting conceptual rigidity and normative polarization across this spectrum (McLennan 2010a). Accordingly, new articulations of the philosophy-politics-religion relationship often resist easy classification as secular(ist), or postsecular(ist), or religious—seeing as opposition to strict secularism is frequently formulated as a defence of *religiosity*, even when evidence of devout belief is slight.

In this contribution, I pull together diverse postsecular articulations which fit that description, construing them for discussion purposes as constituting 'expansive secularism'. One reason for formulating this category is my sense that the portrayals of 'strict' secularism that kick-start many manifestos along the spectrum not only amount to rather crass caricatures, but can mislead us into thinking that postsecularism equates to *anti*-secularism. Not infrequently, however, the headline denunciations give way to revisionist rather than abolitionist thinking, and only seldom—in the social science literature at least—are they theological in any serious sense (McLennan 2010b).

A second rationale is to explore the distinction made by two leading philosophers between two types of secularist outlook on the world. Famously, in *A Secular Age*, Charles Taylor defines the 'immanent frame' as a matter of 'exclusive humanism', and generally pits immanence, both conceptually and normatively, against the alternative frame of 'transcendence'. But Taylor does not go all the way with this founding contrast, because he allows that the immanent frame itself can be subdivided into two 'spins', one closed to transcendence, the other not (Taylor 2007: 544–9). Taylor's reasoning in making this move is slightly curious and not entirely convincing given the drift of his argument. Accordingly, the second immanent variant is barely mentioned through all the supportive commentary on Taylor's *magnum opus* (e.g. Warner et al 2010). This neglect is in part due to the Taylorians' rejection of the very terms postsecular and postsecularism, a refusal which arguably runs against the underlying logic of the 'secularism re-framed' project, and which seems to be motivated chiefly by a determination not to be seen in the same camp as Habermas (Warner et al. 2010: 22–3). Be that as it may, we can push further than Taylor and his followers into the space of the second immanent spin.

Coming from the tradition of mainstream analytical philosophy, Thomas Nagel notes the way in which, even in that generally secular discourse, the 'religious temperament' is getting a more sympathetic airing than before. Nagel defines the religious temperament broadly, as trying to 'bring into one's life a recognition of one's relation to the universe as a whole', thus living 'in harmony with the universe and not just in it' (Nagel 2010: 5–6). Like Taylor, who acknowledges that his whole project 'tilts towards the believer' (Taylor 2007: 7), Nagel thinks that what he calls 'inside-out' secularists will be inclined to dismiss this rather fuzzy, questing mind-set, as explicitly religious intellectuals cannot. But Nagel accepts that 'outside-in' secularists also possess the appropriate attitude. So, again, it makes sense to scan the field of postsecularism to see what outside-in perspectives can be identified, and then critically evaluated.

Expansive, exhilarating

For convenience, what I am calling expansive secularism is drawn inductively from a small number of diverse authors and publications. Taken together, they share four important features: a critique of narrow secularism, an appreciation of religion, an affirmation of naturalism, and a more qualified attachment to both materialism and humanism. Bruce Ledewitz's *Hallowed Secularism* (2008) embodies something of the old democratic Left tradition. Spurred by the revival of religion to improve upon secular-humanist norms, Ledewitz finds that a lot more can be taken from the structure and practice of religious faith than previously might have been allowed (in this vein, see also Saxton 2006). Behind Rosi Braidotti's otherwise self-explanatory title 'The Residual Spirituality in Critical Theory' (2014) lies a hinterland of feminist, postcolonial understandings of cultural and political solidarity, underwritten, for Braidotti, by a neo-vitalist philosophical perspective (also present in William Connolly's postsecularism and Jane Bennett's writings on the enchanted modern). Simon Critchley's *The Faith of the Faithless* (2012) fronts up as a series of 'experiments in political theology', in effect being a series of lengthy and not especially experimental engagements with texts by, among others, St. Paul, Rousseau, Heidegger, Schmitt, Levinas, Benjamin, and Badiou. In mining the political theology and critical theory borderlands (for more of which see de Vries 2006), Critchley searches for arguments and talismans with which to ward off the corrosive thesis—John Gray (2007) is the featured proponent—that overtly religious and Left-liberal secular people alike are entirely deluded in their anticipations of a progressive, ultimately wholesome world. Roberto Mangabeira Unger's *The Religion of the Future* (2014) does not belong to any more general tradition or circuit; he refers to few authors and almost never engages with texts. Reviewers of Unger have spotted something of the Renaissance humanists in him, something of Sartre, something of Freire. To me, both the content of Unger's proposals and his fiery, stentorian tones echo those of Feuerbach. The result, depending on one's mood, is either exhilarating and distinctive or else over-the-top and idiosyncratic. Unger's is a deliberately, incantatorially repetitive mode, constantly rehearsing a handful of core ideas and ringing phraseologies that resound throughout his body of work on law, social theory, experimental democracy, and the Left alternative, all geared to preparing us for our prospective 'ascent' to a higher, better life. What, then, are the overlaps; what constitutes expansive secularism?

Secularism critiqued

For Ledewitz, conventional secularism yields insufficient grounds either for subjective meaning or social peace. Like Charles Taylor in *A Secular Age*, Ledowitz regards mundane secularism as a matter of moral and experiential *flatness*, when what we need is personal and collective *fullness*. Likewise, Critchley also thinks radical politics requires a religious sense of fullness, and fully buys into the desecularization of the world thesis, whereby the secular age has effectively passed, and we are faced by 'a new situation' of metaphysical conflict, political violence, and religious war. All the more important, Critchley thinks, not to accept a stand-off between secularism and faith. Braidotti and colleagues identify the 'myth of secularism' as dangerous and mistaken, especially 'the axiom that equates secularism with emancipation'. Specifically, feminism today 'cannot avoid a head-on collision with the very secularism that has historically been its point of reference'. Unger for his part accepts that a vital distinction is to be made between the sacred and profane, but this is very different from that between the religious and the secular, which in privatizing belief negates the necessary promise of religion 'from the perspectives of its enemies' (Unger 2014: 52–3). The expansive thus shares with the 'secularism re-framed' discourse a militant sense that secularism must be dramatically redirected, if not quite abandoned.[1]

Religion embraced

It is not just that secular hostility must cease; any acceptable secularism must honour the fact that religion serves functions, generates motivations, and cherishes values beyond the reach of secularism to date. In that way, Braidotti (following Saba Mahmood) praises the piety of Muslim women as a progressive, feminist modality. She holds that 'transformative becoming' in a political, materialist sense requires a strong spiritual drive and horizon. And she wishes to counter some of the pessimistic, Lacanian moves within what we might call post-secular Leftism (Slavoj Zizek, for example, though he scorns the postsecular label), by forwarding an affirmative view of human desire as plenitude. Critchley incorporates Christian *love* (agape) conceived as acts of 'absolute spiritual daring' that bring into being new forms of subjectivity and solidarity. The indispensable aspect of *faith* (even for the faithless) is that it binds us to a committed enactment of self in relation to the kind of 'supreme fictions'—not least supreme *political* fictions—that make 'infinite demands' on us, demands we try to meet while knowing that we never fully can. Mystical anarchism is Critchley's term for this syndrome whereby we can be born again, while remaining in full grasp of our general limitations and even our situated wretchedness. For Ledewitz, religion's progressive values are objectively necessary for 'living life abundantly', and this should be secularism's goal too. To that end, religion gives us a heightened sense of personal encounter, encourages social responsibility, nurtures hope, and orientates us to the unconditional, gesturing to a grander universalism beyond the confines of blinkered relativism. The dialectic of immanence and transcendence encapsulates the intense panorama of Unger's 'religion of the future'; this necessarily braced and demanding relationship is what, in his view, has been diminished not only in standard secularism but in the three main religious pathways over many centuries. To underplay the irreparable flaws and existential terrors in the human situation either by offering abstract spiritual detachment (Buddhism for example) or by glorifying conventions and roles in order that we cleave to them in comfort and security (Confucianism perhaps) is grievously to belittle our potential greatness, in fact our *god-likeness*. The Christian form of Unger's third religious category—what he calls 'struggling with the world'—comes closest to grasping the nature and scale of the task, though it has too often backed off. What we must do, unblinkingly, is realize transcendence not by escaping the world, but by pouring our inexhaustible energies and talents into our immanent, improvised, earthly projects and into our relationships with others. That way, through personal and political transformation, we prophetically bring the future into the being of the present. Philosophy, politics, and art are also indispensable, but it is religion that uniquely infuses them, such that 'the whole of individual life and social experience be penetrated by the vision' (this being Unger's main definition of religion).[2]

Naturalism, historicism

If religion, in a generic sense, is undoubtedly essential within the expansive frame, its effects do not take place, and relate to no realms, divinities, or existences, outside the natural and human universe. As in Nagel's conception of outside-in-ness, religion is a more a matter of *temperament* than personal devotion, ideational system, or sacred ritual. Thus, his notion of living 'in harmony with the universe and not just in it' corresponds to what Ledewitz means by religion imparting a sense of holiness, and a 'sense of oneness with all reality'; and to Braidotti's notions of 'intimacy with the world' and 'connections with human and non-human others'.

What we have here is neither conventional religion nor straight secularism, but religion through 'a secular lens' (Ledewitz). That lens comes fitted with an unbelieving filter. Thus, for Braidotti, spirituality is residual, but decidedly non-theistic. Critchley is also 'atheistic', even if not 'triumphally so', because the faith of the faithless is about the immanentizing of the transcendental, religion without God. Unger too is 'without faith', accepting that his is a 'strange theology', being devoted to a Godless religion 'that does not yet exist'. The new worldview also needs to be constrained by modern analytical reasoning, even if the expansives are hardly rationalists *a la* Bertrand Russell. Thus, Unger holds that customary religion and theology simply cannot answer the 'three scandals of reason'. One is the suspension of belief in natural and historical causality in deference to supernatural interventions, however supportively intended. Nor is it feasible to promulgate an eternalist universalism in the face of today's thoroughgoing explanatory contextualism, according to which the rise and fall of all 'sacred plots' quite evidently reflect the historical particularities of the social groups, structured societies, and cultural presuppositions that produced them. Finally, Unger reports that the very idea of God as a being and as a presence, whether conceived as personal or impersonal, singular or multiple, beneficent being or entrancing *non*-being, is strictly incoherent, and it is futile to construct an elaborate apologetics around this idea.[3]

Materialism, humanism

In line with that naturalistic horizon, and in supercharged style, Unger talks of the utter historicity, the death-boundness of ourselves, our social contexts, the natural world, and even nature's laws. *Everything*, without exception, is subject to constant emergence, change, and decay. Material, temporal, and singular causality comprise the *only* Absolute there is. For Critchley too, everything and everyone is radically contingent, with all the agonism that such a condition unleashes. For Braidotti (and all the others), metaphysical *monism* therefore becomes the only viable ontological stance. However, monism here is not equivalent to reductive or eliminative materialism, because it is closely attuned to historicity, possibility, affect, and complexity. Nevertheless, this metaphysical stance contrasts with the sort of *cognitive dualism* that remains definitive of religious philosophy, despite many concessions to 'secular' reasoning. Dualism divides reality into the higher plane of the divine/ immaterial and the physical, mundane sphere of human endeavour and scientific materiality, both of which are held to exist necessarily, to be knowable, and to interact with one another, even if the causality behind this interaction is ultimately mysterious (Scruton 2014).

Humanism is also significantly qualified. Certainly, much attention is devoted to the primacy of the human situation, the facts of human experience, the vastness of human potential, and the necessity of human commonality. Such humanism is important for Braidotti's feminist-vitalist sense of affirmation and solidarity, and for Unger's conception of our raw powers of context-making and context-breaking: we work within constraints, yes, but we readily transcend and transform them. Humanity still can 'rise', show itself to be infinitely aspiring and experimental, ever responsive to the claims of the (human) other. We are not socially determined, nor are we merely natural creatures. At the same time, this is not sentimental or completely 'exclusive' humanism, in Taylor's term. Braidotti follows Actor Network Theory and the 'new materialism', making assemblages of human, non-human, and posthuman agencies and affectivities the relevant focus. Critchley is impressed by Gray's puncturing of progressivist illusions, such that only hope can prevail over pervasive cynicism. Unger (2014: 221) warns against facile Prometheanism. Ledewitz thinks it is folly to be 'happy clappy' about humans, given their record (2008: 162).

Nomenclature

Is 'expansive secularism' the appropriate term for this stream of thinking? Although it aligns with Taylor's reframing current in castigating sectarian secularism, its atheistic and naturalistic lean is so pronounced, even as it claims to be taking religion on board, that the parallel with the reframers cannot go all the way. The expansives are not *political* secularists, but philosophically and sociologically the suggestion is that pluralization and globalization cannot but steadily undercut the force of scriptural, monotheistic, priestly, and doctrinal religiosity, and the nature and remit of 'spirit' outreaches a purely religious interpretation. These inclinations braid with some other updated statements of postsecular and secular commitments. For example, one manifesto for 'continental philosophy of religion' which seeks to be 'faithful to the postsecular event' warns sternly against the 'new imperialism' of the 'theological postsecular' (Smith and Whistler 2010: 11, 14). To similar effect, the introduction to a special issue of *The European Legacy* immediately sets itself against 'post-rationalist', 'inflationary' readings of the postsecular according to which we are not witnessing the death of God but rather the death of secular anticipations of the death of God (Ungureanu and Thomassen 2015: 103–4). There are postsecularists, in other words, whose receptivity to the functional and affective attractions of religion does not extend to theology or to any sense that religions are 'propositionally' valid.

Meanwhile, the editors' introduction to the recent *Oxford Handbook of Secularism* conveys an image of secularism that does not match the identikit picture reproduced in many postsecular and social-theological polemics. Although some basic tenets can still be identified, 'it seems harder to find a clear definition of the term' across the growing number of books with secularism in their titles. Indeed, secularism is 'multipronged and multifaceted'; *not* reducible to assertions that non-belief is growing (even though it is); *not* entailing that secularization *necessarily* follows from modernization and/or science (even though those things strongly contribute); and *not* arguing that religion's disappearance is inevitable (even if it might decline or modulate in various ways). So what we are now dealing with is 'polysecularity', in ways that can readily include religious people and the concerns expressed by 'religion's admirers' (Zuckerman and Shook 2017: 6, 11).

Only one of Zuckerman's and Shook's six foundational secularist tenets is normatively laced—the other five simply advise taking great care in studying trends towards or away from religion. According to the main point (2017: 10), secularism is about 'explanatory justifications for secularization in society and personal secularity for individuals, and the promotion and improvement of partially or fully secular living'. But the value of secular living in this formula remains 'undefined', these authors insist, being subject to the assessment of specific, concrete proposals—which, as per the aforementioned quote, need only be '*partially*' secular proposals. As for 'advocacy', this neither requires nor prescribes a 'formulaic method'. Within those permissive guidelines, the expansives we are considering are secularists rather than—or perhaps as well as—postsecularists. Certainly, many of the perceived deficits of strict secularism are ameliorated.

As well as rendering the whole idea of secularism more provisional and complex than often allowed, expansive secularism helps counter a familiar anti-secularist tactic of portraying modern science, or mundane relationships, or democratic life, or technological being, as utterly spiritless and unimaginative, at least when cut adrift from divine anchorage. Thus, Ledewitz lists the glories of scientific endeavour as key to secular spiritual renewal, Braidotti's neo-vitalism has a certain proximity to the freethinking side of the life sciences, and Unger's views about natural causality and total historicity are learned enough to enlist an established theoretical physicist as co-author (Unger and Smolin 2014). Of course, there are perfectly good scientists who are religious, but the scientific joys and pains of opening into new entities and forces do not, as is often alleged, exemplify the religious quest as such, nor do they justify an *idealist* style of ontological monism. This is indicated in its way by the avalanche of popular science works sweeping through the bookstores, written by non-religious specialists who enthuse about and illuminate all manner of questions around creativity, infinite macro- and micro-worlds, uncanny ideas and events, irreducible beauties, and philosophical dilemmas that perhaps can never be resolved. Indeed, if science and philosophy are undergoing a new lease of metaphysical life, then that only tells us that religion and theology have lost their long-term monopoly on wonder and argument about what it might be that transcends us (see for, example, the 'speculative realism' of Brassier 2007; Meillassoux 2008; De Landa and Harman 2017).

On the downside, in terms of both appropriate classification and intellectual rigour, the expansives—with the paradoxical exception of Unger—portray religion, or at least the religious temperament, in almost uniformly positive light. This is understandable in a way, seeing as they are aiming for an inclusive solidarity, and a new kind of affirmation, contrasting with the New Atheist campaign to present religion in general as nothing but harmful. All the same, and twisting Taylor's trademark vocabulary directed against secularism, it is negligent to overlook the numerous 'cramped horizons' and repressive 'closures' that are effected in the name of religions.

A second area of questioning is ongoing across all the secularism and postsecularism debates, namely secularism's Euro-Christian-centredness. The critical observations here—made by variously deconstructionist, postsecular, and 'Southern' authors (and perhaps most thoroughly by Talal Asad [2003])—are twofold. One is that it is above all in the Western Christian tradition that such a sharp and double-sided intertwining of 'religion and secularity' occurs. Another is that only this tradition, and only when it goes secular, erects a distorted *singular* concept of 'religion', highlighting the primacy of propositional doctrine, such that it can then be demarcated from, and judged in relation to, 'non-religious' cultures and disinterested theoretical systems. Among the expansives listed, Braidotti is the obvious exception to that charge, seeing as she makes it herself. On the other hand, even self-consciously 'global' feminist discourse cannot totally jettison the modernist cast with which it is partially breaking. And vitalist philosophy itself, to which Braidotti appeals, has a distinctly modernist-Northern heritage, even if it is designed to go beyond that. (But then again, all philosophical insights and modes can be adapted beyond their original formative context.) It should be noted at this point that one of the most substantial postsecular thinkers along these lines, Judith Butler, would probably distance herself from both straight-laced *and* expansive variants of secularism as schematized here. Embedding her reflections in a diasporic, unsettled, Jewish-exemplified set of concerns, Butler is not after any convenient patching up of issues and perspectives. Arguably, this is because, more than any other postsecular writer, Butler identifies the violence-conducive *statism* of secularity and secularism as their principal feature, and their disgrace (for a resume of Butler's evolving positions, Dickenson and Morgan 2015). This is not something that stands out in the same way for the expansive secularists we are looking at here.

Viability

In terms of theoretical consistency and coherence, the expansive secular imaginary faces strenuous challenges. I will home in on Simon Critchley and Roberto Unger in that regard. The former insists that transformative political theology requires quasi-religious faith in supreme fictions of the (earthly) sacred. Except that we secular-minded mystical anarchists also realize that our supreme causes, hopes, and dreams are, precisely, *fictions*. In other words, we understand the fictional status of supreme entities and principles, 'but we believe in it nonetheless'. *Prima facie*, this assertion carries little psychological or socio-political plausibility. Perhaps what is meant is that all political notions of progress and hope summon up and operate within myth-like structures—a common refrain within postsecular thought. But how fully conscious we are in treating the myths that guide us *precisely as myths* is another matter. Myths tend to inhabit political ideas in a background, figurational way; they don't provide the objects of attention and command as such. As for faithfully adhering to explicit political doctrines or traditions, such as democracy, or feminism, these are better regarded as ideals than as myth or fiction in any tight sense. As such, we can readily accept that they may not be practically achievable or guaranteed in full, or any time soon, but work on the basis that philosophical and political commitment, together with activism, can take us a considerable way towards making them real. It is not a matter of elaborate make-believe, which is what Critchley's 'fiction' implies, even if derived from Rousseau.

In that respect and in general, the self-positioning of Critchley's reflections as exercises in political *theology* is essentially gestural, which is to say, *analogical*—obliquely rather than literally intended. The same might be said of much of Hent de Vries's important (2006) collection under that same heading, and a brief digression can indicate why. Appearing around the time of *A Secular Age*, de Vries out-guns even Taylor and the reframers with the strength of his blast against received secularism. The 'modern critique of religious conviction', de Vries asserts, is itself a religion, a chant, a counter-sacredness, a desperate appeal against forces and movements it cannot contain, a vain hope that the spectre of religion will become zombie-like. The battle against 'theological truth', it follows, is 'utterly misplaced', because religious sources of progressive politics today are 'benign', and it is folly to try to bleach out their motivational drive, embedded as they are in 'transcendent-transcendental motifs' (2006: ix–x). But the force of the extraordinary paragraph or two in which those claims are rolled out—almost without any argument as such—considerably weakens by the time we get to the end of de Vries's mammoth prefatory statement for the book. Of the authors extracted as political theologists, almost none fit that label without stretching it beyond precision. Thus, Laclau (also Lefort in his own way) is summarized as theorizing the endless series of equivalences that can be produced in the logical space where 'the social' comes unstuck and the absolute and eternal meet the finite and particular; Benjamin and Schmitt are presented not as political theologians in any doctrinal sense, but in terms of the lure and continual resurfacing of theological figures of thought, which is a significantly different matter; an extract from Butler then makes the point that Benjamin is principally concerned with the commandment of ethical address (something that secularist fans of Benjamin would support); Nancy shows how politics retreats, *as politics*, when religion and theology continue to be a real presence in political-institutional forms; and Detienne and several other luminaries do no more—and no less—than exemplify the postsecular inclination to look comparatively and closely at the myriad things that pass for religion in relation to political and social formations. Now very little of this in any direct sense theological, even politically theological, though neither is it normatively secularist as such. It therefore comes as no surprise when de Vries sums up (2006: 88) his nomenclature in an inordinately lateral way: political theology is thinking that seeks to grasp and quiz the elusive absolute that governs or drives or aspires or destabilizes or terrorizes the public domain. He also—rightly in my view—deduces from this that it may be better to move on from political theology either to a new metaphysics of the political, or to the politics of the everyday, problematizing religion every bit as much as secularism. Tomoko Masuzawa's jacket blurb underlines this not-religious, postsecular mind-set: political theology is 'rough terrain', featuring numerous efforts to dissociate, calibrate, and realign the domains of the political and the religious.

The brunt of that digression applies fully to Critchley's adoption of the terminology of political theology, little of which justifies the casual conflation of faith *qua* hopeful belief in human betterment with faith in divine or personified or absolute entities and forces. A final difficulty with Critchley's faithless religion then directly follows. It concerns his positive response to the idea of being 'born again' and wishing to recruit this syndrome to radical democratic transformation. Now arguably, in full-value religious transfiguration, we are not especially perturbed by the prospect of totally upgrading our unworthy, sinful worldly selves for a deeper, new, transcendentally infused identity. In fact, that is precisely what we seek. But does that craving, that degree of conversion experience, equally absorb critically minded, religiously unbelieving anarchistic activists? Do we—does Critchley personally?—really want to leave our or his current situated subjectivity completely behind, entrusting it to higher moral personalities that we must accept are beyond our ken? Understandably, Critchley quickly backs off from this almost embarrassing incongruity, cautioning us against 'purist', indeed 'puerile' notions of a 'blessed alien land', a wholly new 'absolute beginning', and 'life without law' (2012: 91–3, 202–3).

As for Unger's commitment to a thoroughly naturalistic, single-order, time-bound universe, this sits more comfortably with a secular orientation towards science than it does with religious expressions of science's ultimate purpose. He also goes to great lengths to lay bare abiding contradictions in the human condition, in the conviction that no religion of the future can take off without these being fearlessly acknowledged. On the one hand, we are nothing less than embodied spirit, indomitable context-breakers, enemies of compromise, creatures of infinite capacity and plasticity. Thus, in pursuit of a new divinity—our own—we throw ourselves into acquiring the highest 'virtues of connection, purification, and divinization' (2014: Ch. 7). It is these virtues that Unger says will steer the radical shift to a new level of civilization and (presumably) sustain it forever. It might seem that such demands, and their transitional role, could only be met by an elite of saints, but Unger says we all have this saintliness in us.

The trouble is that Unger also presents us (currently) as somnambulant, over-conventionalized, and even 'mummified' social beings, so the road to reinvention is sure to be extremely rocky. Accordingly, the virtue of universal solidaristic connection is constantly undercut by the urge he claims we have for 'self-standing personality', the assertion of which involves rather *anti-social* traits and even a degree of personal cruelty (for example, towards our loved ones when, perforce, we must move on). Moreover, the compendious virtues of purification (comprising simplicity, enthusiasm, and attentiveness) are sharply offset by a core part of human nature, namely our *insatiability*: the unquenchable thirst we have for experience, meaning, and fulfilment, which Unger thinks—like the Lacanian Leftists—always overshoots its partial realizations. As for the virtues of divinization, 'analogous to the theological virtues of Christian doctrine', that is, faith, hope, and charity (2014: 283), even these are constantly challenged by the definitive *groundlessness* that is our unavoidable plight as limited beings in a finally unknowable, ever-changing life and universe.

These contrasts and combinations make for a dense congeries of intrinsic tensions, the 'prophetic' resolution of which looks too demanding to be credible, in the normal scheme of things. Of course, this is exactly Unger's point: the normal scheme of things is too slow, too embedded, ever to come around to the necessary degree of radicality. What we need, and at high speed, is a jolting transformation into the altogether quicker, 'awakened' selves that are required for the new religion and new society. And it is *prophecy* that stands as the proper medium for that; prophecy, the 'self-fulfilling' nature of which is so often taken by liberals to be fatal, yet which, as Unger sees it, is precisely prophecy's unique strength. We just need to commit to it, trust it, practise it. Well, this *does* begin to look like a political theology. But it is extremely hard to see how sheer exhortation alone is going to suffice to fill out the political dimension of the concept. If this really is a political manifesto for a new type of social personality, people as they are right now somehow have to be brought on board. In other books, Unger's theses on socio-economic transformation and political reform manage a more programmatic blend of radical edge and social realism. For example, he insists that the apparently tame proposal that everyone should be able to live a *petit bourgeois* life is, in effect, revolutionary. This seems to me spot on, and just the way to recruit a wide social constituency for more fundamental transformation. Yet, on religion, there is no strategic, which is to say no political, aspect: the pitch of the appeal is searing, and the demands on the individual personality are total. This raises serious doubts about how the routinized masses come into the transitional scenario, and, relatedly, who exactly will be the saints and prophets who can be entrusted to lead the way. Although Unger's focus on social personality is original and invigorating, his paean to human divinity harbours the considerable dangers of elitism, populism, and irrationalism.

Braidotti and Ledewitz face less fateful conundrums, but these exist nonetheless. The universalism intrinsic to vitalism notably weakens when full affirmation of 'life itself' is withdrawn from people and syndromes that contemporary vitalists do not much like (violence, disease, racism, misogyny, and much besides). Hallowed secularism is more easily rendered into a non-contradictory amalgam of secular values and religious virtues, but whether a deontological, functional appreciation of faith amounts to significantly more than good-hearted humanism is, for all Ledewitz's protests, still open to question.

Conclusion: what next?

I have been selectively weaving a path through core issues around secularism, postsecularism, and religion. These issues are inescapable in social theory at present, actively modifying what Taylor has usefully called 'the modern social imaginary'. One upshot is that even if—as I have suggested—it has been unfairly 'framed' somewhat in the 'post' literature, secularism *does* need rethinking and representation, and the authors that we have been examining help considerably in that regard. But I have also sought to convey that there are tangible theoretical and normative problems within this current, whether individually expressed or taken as a whole. These difficulties are not only conceptual; they are to do with political and rhetorical interpellation as well, because at a certain threshold of inflationary overkill, latent tensions become all but intractable contradictions. Both the advantages and conundrums of expansive secularism are worth pinpointing additionally because we can anticipate that more variants of that ilk will be coming on stream as the fulcrum of the postsecular plank slides towards the secularist end, given the striking emergence of 'no religion' as a (cross-) cultural identity-marker at the present time. As we have also noted, a more deflationary, modest form of secularism is under way too, and perhaps what is most needed now is a conceptually elaborate, globally attuned, and, yes, appropriately 'expansive' statement of that.

Notes

1 For the quotes and emphases in this paragraph: Ledewitz (2008: 1–2, 42); Critchley (2012: 8, 24); Braidotti et al. (2014: 1, 5); Braidotti (2014: 252); Unger (2014: 52–3).
2 Braidotti (2014: 251, 255); Critchley (2012: 10, 13, 15, 18, 20); Ledewitz (2008: 1, 56–9); Unger (2014: 53, 121, 55).
3 Ledewitz (2008: 5, 196); Braidotti (2014: 251, 256); Critchley (2012: 18–20); Unger (2014: 202–3, 225–8).

Further reading

Molendijk, A. L., Beaumont, J. and C. Jedan (eds.) (2010) *Exploring the Postsecular: the religious, the political, the urban*, Leiden: Brill.

A wide-ranging collection, covering the principal meanings and debates around postsecularism, theoretically, politically, and empirically, especially in geography and sociology.

Calhoun, C., Juergensmeyer, M. and J. VanAntwerpren (eds.) (2011) *Rethinking Secularism*, Oxford: Oxford University Press.

High-quality set of discussions of secularism following in Charles Taylor's footsteps, though somewhat uncritical in its editorial orientation and selection.

The European Legacy, 20 (2) (2015).

Special issue of this journal on 'Post-secularism: Between Public Reason and Transcendence', with a good range of perspectives on the nature and future of postsecular ideas and thinkers.

Scott, D. and C. Hirschkind (eds.) (2006) *Powers of the Secular Modern: Talal Asad and his interlocuters*, Stanford, CA: Stanford University Press.

Important collection on the impressive and forceful, but also frustratingly elusive, contribution of Foucauldian anthropologist Asad.

References

Asad, T. (2003) *Formations of the Secular: Christianity, Islam, modernity*. Stanford, CA: Stanford University Press.

Braidotti, R. (2014) 'The residual spirituality in critical theory: a case for affirmative postsecular politics'. In Braidotti, R., Blaagaard, B., de Graauw, T. and E. Midden (eds.) *Transformations of Religion and the Public Sphere: postsecular publics*, Houndmills: Palgrave: 249–72.

Braidotti, R., Blaagaard, B., de Graauw, T. and E. Midden (2014) 'Introductory notes'. In Braidotti, R., Blaagaard, B., de Graauw, T. and E. Midden (eds.) *Op cit*: 1–13.
Brassier, R. (2007) *Nihil Unbound: enlightenment and extinction*, Houndmills: Palgrave.
Critchley, S. (2012) *The Faith of the Faithless*, London: Verso.
De Landa, M. and G. Harman (2017) *The Rise of Realism*, Cambridge, MA: Polity Press.
Dickenson, C. and S. Morgan (2015) 'Dwelling in diaspora: Judith Butler's post-secular paradigm', *The European Legacy*, 20(2): 136–50.
Gray, J. (2007) *Black Mass: apocalyptic religion and the death of utopia*, New York, NY: Farrar Straus and Giroux.
Ledewitz, B. (2008) *Hallowed Secularism: theory, belief, practice*, New York, NY: Palgrave.
Meillassoux, Q. (2008) *After Finitude: an essay on the necessity of contingency*, London: Continuum.
McLennan, G. (2010a) 'Spaces of postsecularism'. In Molendijk, A. L., Beaumont, J. and C. Jedan (eds.) *Exploring the Postsecular: the religious, the political, the urban*, Leiden: Brill, pp. 41–62.
——— (2010b) 'The postsecular turn', *Theory, Culture & Society*, 27(4): 3–20.
Nagel, T. (2010) *Secular Philosophy and the Religious Temperament*, Oxford: Oxford University Press.
Saxton, A. (2006) *Religion and the Human Prospect*, New York, NY: Monthly Review Press.
Scruton, R. (2014) *The Soul of the World*, Princeton, NJ: Princeton University Press.
Smith, A. P. and D. Whistler (2010) 'Editors' introduction: what is continental philosophy of religion now?'. In Smith, A. P. and D. Whistler (eds.) *After the Postsecular and the Postmodern*, Newcastle: Cambridge Scholars Publishing, 1–24.
Taylor, C. (2007) *A Secular Age*, Cambridge, MA: Harvard University/Belknap Press.
Ungureanu, C. and L. Thomassen (2015) 'The post-secular debate: introductory remarks', *The European Legacy*, 20(2): 103–8.
Unger, R. M. (2014) *The Religion of the Future*, London: Verso.
Unger, R. M. and L. Smolin (2014) *The Singular Universe and the Reality of Time: a proposal in natural philosophy*, Cambridge: Cambridge University Press.
Vries, H. de (2006) 'Introduction: before, around, and beyond the theologico-political'. In Vries, H. de and L. E. Sullivan (eds.) *Political Theologies: public religions in a post-secular world,* New York, NY: Fordham University Press, pp. 1–88.
Warner, D., VanAntwerpen J. and C. Calhoun (eds.) (2010) *Varieties of Secularism in a Secular Age*, Cambridge, MA: Harvard University Press.
Zuckerman, P. and Shook, J. R. (2017) 'The Study of Secularism', In Zuckerman, P. and J. R. Shook (eds.) *The Oxford Handbook of Secularism*, Oxford: Oxford University Press, 1–17.

21

Dialogue with religious life in Asia

Lily Kong and Junxi Qian

Introduction

Over the past decade or so, debates on the advent of a postsecular present have moved center-stage in social and cultural theory. These exchanges have made commendable achievements in analyzing the increasing relevance of religion to public politics and policies. Theorization of the postsecular makes for a hefty intellectual agenda, and a comprehensive review of this agenda is clearly beyond the scope of a single contribution, addressed, more appropriately, in a collective way by this Handbook. Nevertheless, we would like to contribute a few ideas to help delineate the broader permutations of the postsecular turn in social sciences, which serve as the intellectual context for our arguments in this piece.

The gist of the postsecular treatise criticizes the assumption of the privatization of religion in modernity, one central component of the secularization thesis, and argues for the persistence or resurgence of religions in public cultures and the public sphere. In Europe, encounters between different religious traditions due to massive human migrations in an era of global mobilities, and the consolidated positions of various Rightist and Fundamentalist religious movements, have meant that the relations between the state, the public, and the religions need to be carefully managed and balanced, rather than defined *deus ex machina* as those of privatization and differentiation. In Habermas' (2006, 2008) formulations, liberal democracy once held that religious languages must be translated into secular languages for faith actors to converse in the public sphere; this is, however, no longer so. In a postsecular context, there is a relationship of mutual learning between the secular and the religious; secular citizens are expected to be more self-reflexive, both cognizant of and responsive to religious utterances, while religious actors become more sensitized to secular worldviews and concerns. In this sense, the cognitive dissonances and burdens of understanding between the religious and the secular need to be dealt with by secular and religious actors in equal terms. Concurrently, Habermas has attempted to chart a middle path between secular philosophies and religions as potential resources for the normative principles of rights, justice, freedom, emancipation, and so forth (Habermas et al. 2010; also Köhrsen 2012; Ascione 2017). Eventually, scholars' growing interest in the study of religions in public life has brought together various terms such as public religion, public theology, and political theology, encompassing a rich variety of research foci, from political integration and pluralism, to politics of identity and difference, and to conflict and violence in the contemporary world (de Vries and Sullivan 2006; Graham 2013).

In our discipline of geography, the notion of the postsecular has nurtured a small but vibrant area of inquiry. On the one hand, the postsecular has been understood in terms of people becoming more conscious about, even proud of, the religious and spiritual dimensions of everyday life. Increasingly, people turn to theological discourses and interpretations as well as spiritual experiences to constitute the everyday, lived, and embodied subjectivities. Religions supply ordinary people with a system of vocabularies to make sense of, and negotiate, secular processes, such as changing political economic conditions, gender, national identity, mobility and migration, and multicultural encounters (Olson et al. 2013; Oosterbaan 2014a, 2014b; Saint-Blancat and Cancellieri 2014; Gökarıksel and Secor 2015, 2017).

On the other hand, in a postsecular context, the focus of religious actors and organizations is undergoing a transition from transcendence to immanence, from faith-by-dogma to faith-by-praxis, reflected in faith actors' engagements in civil services, social welfare, and activism (Cloke and Beaumont 2013). In this process, faith-based organizations bring theo-ethics to bear on professional and voluntary participation in civic affairs, which addresses the lethargy of the neoliberal state in provision of public services, and in so doing fills the gulfs in social justice (Cloke et al. 2005; Beaumont 2008; Beaumont and Baker 2011; Lancione 2014; May and Cloke 2014; Bolton 2015; Williams 2015; Cloke et al. 2016). Religious people evidence an assemblage of motivations, acts, and discourses that bring together different religious communities to fight for a common ethical cause. In sum, this growing body of geographical scholarship has supplied a vivid portrayal of a new ecology of social life, which is increasingly enchanted by religious impulses, now diffuse in the textures of everyday life.

In sum, the postsecular agenda involves concerted and sophisticated efforts to reconstruct the historical narratives of modernity; it represents a sort of 'strong theory' reversing secularist epistemologies and yearning for anti-secular impulses (McLennan 2010). However, as scholars located outside Europe, observing the rapid societal changes in numerous parts of Asia, we suspect that the strength of the postsecular thesis paradoxically seems to be its simultaneous weakness. First, notwithstanding its strong and assertive language, the discourses of the postsecular are largely preoccupied with historical and contemporary specificities of Europe, and incompatible with the experiences of societies where secularity took root in different ways from modern Europe (Camilleri 2012). In many parts of Asia, for example, secularization was less a 'naturally' occurring process than a top-down civilizing and modernizing mission—postcolonial political elites 'learned' the doctrines of secularity from the West, and orchestrated campaigns of secularization at the local scale. In other words, secularity was very much a legacy of Western imperialism and the global triumph of Anglo-European modernity. In this sense, religions in these societies have always been 'public' since the incipient stage of modernization, always within the rubric of postcolonial state politics, discourses, and policies. Postsecular discourses, as currently formulated in the mainstream Anglophone academia, tend to lose sight of state construction of secularity in non-European contexts, even though the historical emergence of secularity and laicity in Europe was largely a political project overseen by the state, as well.

Second, postsecular discourses underscore a historicity that reifies the 'newness' of the rapprochement between the secular and the religious; they imply a recent past in which the privatization of religion was dominant and largely unchallenged. This framing of historicity, however, veils the fact that in many parts of the world, the privatization of religion has never been such a desired quest or an empirical reality as in modern Europe (Kong 2010; Köhrsen 2012). Rather, secularity and religiosity have been produced and managed in local and specific ways, traversing the boundaries between the public and the private.

Third, postsecular discourses have been deficient in considering the ontological difference between non-Western religions and Christianity, obfuscating the fact that the religious-secular divide was not a given, but derived largely from Judeo-Christian theologies (Asad 2003; Taylor 2007). In contrast, the boundaries between transcendence and immanence, other-worldly pursuits and this-worldly engagements, have been historically vague, porous, and fluid in many belief systems of Asia. Buddhism, for example, has long emphasized the accumulation of good karma through this-worldly service and care, and involvement of Buddhist actors in suffering alleviation, care, and social welfare has been common in Chinese communities and societies for centuries.

Bearing in mind these observations, this contribution ventures to suggest an alternative framing of the postsecular. Above all, we suggest that the postsecular remains a useful concept to theorize the abiding and vibrant presence of religions in the public sphere. But for this thesis to make more sense to contexts beyond Europe, two assumptions need to be rejected: (1) the epistemology of secularity and religiosity as oppositional to each other; (2) the historicity of a linear transition from a secular past to a postsecular present. We propose that the secular and the religious are not antithetical to, but mutually implicated in, each other—secularity restricts the expression of religiosity, but also creates new meanings and conditions for the existence of religion; vice versa, the persistence or resurgence of religion in the public sphere compels people to rethink, negotiate, and maintain the reach and limit of secularity. Secularization is not simply a *deconstructive* force, but also a *reconstructive* one.

It is in the spirit of this hermeneutic that this contribution brings religion back onto the canvas of modernity, while resisting the temptation of announcing the end of the 'secular age' (Taylor 2007). We argue that a fuller theorization of the crossing-over between the secular and the religious, the public and the private, cannot be realized unless we attend more closely to the regional and historical contingencies of religiosity in modernity—the construction of the religious-secular divide is intrinsically negotiable and fluid. On the one hand, the postsecular does not necessarily imply the decline of secularity; rather, it captures the ongoing dynamics and competitions between the religious and the secular. Secularity and secularism are still highly relevant to state governance and governmentality across the globe. As Wilford (2010) suggests, religiosity and secularity have distinctive spatialities and scales, which may or may not overlap. Hence, the effects of religious revival do not necessarily supersede those of secularization. On the other hand, the postsecular does not imply, in a linear-temporal sense, the transition from one configuration of modernity to another; instead, it is better suited as an epistemological manifesto for the incomplete nature of any modernizing project. In this vein, scholars need to question and de-essentialize the concepts of secularization, secularity, and secularism, rather than dismantle or devalue them (McLennan 2010).

To support our arguments, we present some empirical discussions to delineate two parallel trajectories of the entanglement of religion, secularity, and modernity, both of which deviate from conventional wisdom on the postsecular in notable ways. First, we discuss the state-imposed secularization in China since the demise of the imperial monarchy in 1911, and the recent 'religious revival' in the era of Reform and Opening (1979-). We illustrate how the state construction of secularity creates undulating, not ossified, conditions for the exclusion but also inclusion of religion in the public sphere. Second, we consider the persistent visibility of religion in public politics and social life in the city-states of Singapore and Hong Kong, where religious vibrancy and state secular interventions co-constitute landscapes of modernity, and the public-private boundaries are constantly being redrawn. In this process, the state and the public advance their respective agendas and purposes; obviously, the general secularization of society and the flourishing of public religions are very much co-existent. We are mindful, however, of the fact that there is no such thing as 'Asian religious studies' (Kong 2015), and the cultural and historical idiosyncrasies of religions as practiced in Asian societies defy any generalization. To include 'Asia' in the title simply indicates our hope that those from Asian societies other than (Mainland) China, Singapore, and Hong Kong may be sympathetic to our thesis, given comparable trajectories of subjection to colonial influences and the postcolonial construction of 'our modernity' (Chatterjee 1997).

China: the making of the secular and postsecular as state projects

The secular-religious divide was historically unknown to traditional Chinese belief systems. The renowned Chinese-American sociologist C.K. Yang (1961), for example, characterized Chinese religiosities as 'diffused religions'—cosmological-spiritual views and discourses were so inextricably interwoven with the fine-grained textures of social, economic, and cultural life that this relationship of interpenetration undermined any dichotomous understanding of the secular and the religious in China. The differentiation of the religious from the secular was in fact thanks to the Chinese state's and political elites' absorption and dissemination of Western discourses and theories. The concerted effort to suppress religious expressions and practices in the public sphere started in the Republican era (1911–1949), only to reach its apogee in the Cultural Revolution under the rule of Mao (1966–1976). China's humiliation by Western colonial powers in the late nineteenth and early twentieth century prompted a culture of *self-critique* among Chinese intellectuals and elite. They identified traditional Chinese cultures, in which religions were syncretically embroiled, as the primary reasons for China's 'backwardness' and blindness to progress. As Mayfair Yang (2011: 10) poignantly criticizes, Chinese elite's embrace of key elements of European Enlightenment resulted in 'repeated waves of cultural self-laceration, religious destruction and state campaigns of secularization'.

Accepting European modernity as the epitome of progress and civilization, indigenous elite in China initiated an ambitious project of social engineering by means of enforced secularization. Intellectuals and elite borrowed from Meiji Japan the word *shukyo* (the Japanese translation of the Western term 'religion', later translated into Chinese as '*zongjiao*' 宗教). This notion was then conceptualized scientifically as a coherent system of theosophy, scriptures, rituals, clergy, and religious sites (M. Yang 2008; Ashiwa and Wank 2009). Despite the fact that the boundaries between Buddhism, Taoism, Confucianism, and folk religions were highly obscure in Chinese cultures, the concept of religion had given ontological existence to institutional religions (Buddhism, Taoism, Islam, Catholicism, and Protestantism, excluding Confucianism and folk religions). Traditional folk religions, in particular, were devalued as irrational superstitions to be denounced and eradicated. Indeed, superstitions became the foil against which rational and scientific knowledge of religion could take root (Duara 1991a, 1991b). Even institutional religions were increasingly relegated to the sphere of private faith, and their significance in the day-to-day social and cultural life had to be surrendered to secular reason, science, and industrial modernity (M. Yang 2008).

On the top of this top-down secularization campaign, the post-1949 communist regime added an extra layer of Marxist evolutionism and atheism. Under the radicalism of Mao, the state consistently criticized religions as 'opiate of the masses' and accomplices of internal feudalism and external imperialism. The state also disgraced clergies as an 'exploitative class' parasitic on the produce of workers and peasants. Eventually, the communist state launched perhaps the ironhanded iconoclastic campaign that human history had ever seen (M. Yang 2008; F. G. Yang 2012). The cumulative effect of the anti-superstition movement in the Republican era and atheist purges under Maoism was 'one of the most dramatic secularization processes in the modern world' (M. Yang 2011: 7).

The successive episodes of state-enforced secularization represented an extreme version of the secularization thesis. In this sense, it is in principle justifiable to speak of a postsecular present in China, given the extraordinary revival of religious beliefs and practices in China after the Communist Party initiated the Reform and Opening in 1979. While atheism remains an official ideology in the public sphere, many private citizens are suspicious of atheist values, or at least in search of alternative political, cosmological, and cultural imaginaries. The Chinese state also recognizes that even in a socialist country where religions cannot be used as an instrument of class exploitation, people still need a spiritual domain to negotiate disasters and misfortunes. This revised rationale is part and parcel of the state's own self-reflexivity of modernity. Notably, the relaxation of religious restriction and limited practice of religious freedom are reflective of the broader redirection of the state's cultural policies, from the total negation of traditional cultures to flexible accommodation and selective use of traditional elements as resources of cultural governance (Qian and Kong 2018).

Religions, once stigmatized and targeted for eradication, are well positioned in this new agenda, manifested in at least three ways. First, religions have been incorporated into the state's mission of constructing 'spiritual civilization' and a 'harmonious society'. The state emphasizes religions as sources of moral values and ethics that are conducive to the maintenance of social stability. Both the spiritual guidance and philanthropic work conducted by faith actors may be co-opted by the state to enhance the governance of a rapidly diversifying and polarizing society. Religion can also provide anchors of spiritualities and collective identities for people negotiating the fluidity and uncertainty of the market economy. For example, in 2016, President Xi Jinping remarked in an address to UNESCO that Buddhism and other traditional religions were important components of the Chinese civilization, and of high 'spiritual value' to the Chinese people. Second, religions give the state a new edge of *soft* political power in both domestic and international affairs. For example, the Chinese state's promotion of Mazu, a maritime guardian goddess, constitutes a symbolic force that asserts the cultural affinity between Taiwan and China as the 'Motherland', and facilitates the rapprochement cross the Taiwan Strait (Chau 2011; Zhang 2017). In the international arena, Confucianism has been an ideological cornerstone of the alleged 'peaceful rise' of China and a 'harmonious' world order envisaged by the Chinese state, epitomized by the controversial project of establishing Confucius Institutes worldwide. Third, religions have been exploited as resources and assets for regional development. The selling of spiritualities and religious cultures is indeed indispensable for countless initiatives of tourism development. Not only have historical religious sites and shrines been opened to tourism for generating local revenues, but religions increasingly figure visibly in flagship projects of regional development and regeneration, exemplified by the widely criticized commodification of Shaolin Monastery (少林寺) and the expansion project of Famen Temple (法門寺) in Shaanxi Province.

To conclude this section, we suggest that, if there is a Chinese postsecularity at all, it can challenge and enrich the theorization of the postsecular in at least two ways. First, postsecular discourses in Europe tend to imply the empowerment of religions, and religious claims as quasi-independent utterances in the public sphere. In contrast, in China, both secularity and postsecularity have in fact been carefully orchestrated by the state, and the increased visibility of religions in postreform public cultures does not entail the straightforward empowerment of religions vis-à-vis secular forces. Precisely because of the centrality of the state in defining the relative importance of secularity and religiosity, the secular and the postsecular are not mutually oppositional, but largely two sides of one coin—the cultural and ideological administration of the society by the state.

Second, the state construction of secularity and religiosity is not simply geared towards the privatization of religions. Rather, it is a highly tortuous, uncertain agenda, moving back and forth between multiple priorities of governance, and creating ebbing and flowing spaces of visibility and invisibility for religions. Postsecularity may be as much contingent on the power of secular state as secularity. In his religious economic interpretation of religious decline and revival in China, F. G. Yang (2012) contends that religious regulation restricts the 'religious market' and suppresses market mechanisms of supply of and demand for religious goods. We take a different view. Instead of equating state regulation with the *decrease* of religiosity, we introduce sensitivity to the *nature* and *quality* of religious regulation. Just as the Chinese state intervenes actively to *create* (rather than reduce) market mechanisms for China's economic development (Wu 2008), the state never simply restricts the religious market, but oftentimes creates new elements and dynamics to shape, define, and direct the religious market (just think of the grandiose Buddhism- and Taoism-themed mega-projects proliferating in China!). Above all, in the theorization of the postsecular, we need to give nuanced considerations to the variegated and ambivalent positions of the modern state as a secularizing force and agent.

Singapore and Hong Kong: coexistence of secularity and religiosity in the public sphere

The case of modern China represents a transition from a highly secularized past to the postsecular present. Such a linear temporality may not be uncommon worldwide, but it is not necessarily so. As Kong (2010: 764) points out, in numerous contexts across the world, 'the engagement of sacred and secular was not "re-emerging" but rather continuing'. The postsecular discourse would be problematic, if sensibilities of 'resurgence' and 'revival' are universally applied. The postsecular does not have to imply religions as the backlash against the disenchantment and pathologies of modernity; rather, religions may be a spiritual force that has persisted in the public sphere over the course of modern transition, and indeed been constitutive of secular modernity. Thus, our proposition in this section is that the relationships between secularity and religiosity are multiple and context-specific, much richer than being unitarily about privatization and differentiation. The state, the people, and faith actors can harness public religions to their own benefits and interests, be they state-building, governance of diversity, participation in social welfare, promotion of specific faiths and theological ideas, or reinforcement of civic agency. We would like to illustrate this point of view through the lens of the postcolonial societies of Singapore and Hong Kong.

A quick stroll through the streets of Singapore and Hong Kong will easily impress one with the sheer diversity and quantity of religious spaces, sites, and shrines (Christianity, Buddhism, Taoism, folk religions, and more in Singapore than Hong Kong, Islam, and Hinduism). In both cities, myriad scenes of vibrant religious life enrich the landscapes of cosmopolitan urban modernity. Both Singapore and Hong Kong are global hubs of commerce and finance which, thanks to tectonic economic changes since the 1960s, are now among the major world cities. Religions have traditionally figured prominently in the cultural landscapes of both cities, closely intertwined with the social and economic conditions of urban existence—state-led urbanization, capitalist development, individualism, migration, social polarization, and so on. Concurrently, however, due to British colonial rule and the advent of European modernity, religions are certainly no longer diffused systems encompassing everyday life, but have differentiated from the state, the market, economy, and science (Goh and van der Veer 2016). In this specific configuration of modernity, it is particularly interesting to consider relationships between the state, the public sphere, and religions.

For both cities during British rule, the secularity of the state did not rule out partnerships with religions for the effective and cost-minimizing management of the society. For example, Christianity enjoyed patronage of the state, and was used as an instrument for culturally assimilating and co-opting indigenous elites (Kwong 2002; Goh 2016). Religions also played an indispensable role in the provision of social services and welfare (much earlier than the recent proliferation of faith-based care and services in the public sphere of Europe). In Hong Kong, before compulsory state education was implemented in 1971, the major providers of public education were actually religious groups. Even in 2004, the three mainline Christian churches—Catholic, Anglican, and Methodist—operated 40% of all primary and secondary schools in Hong Kong, in addition to the various schools operated by Buddhist and Taoist organizations (J. Tan 1997). In this sense, faith-based services helped the state reduce the cost of secular government and constituted the laissez-faire economy in Hong Kong. In the context of Singapore, many of the earliest schools were established by religious groups, and they remain active in the provision of pre-tertiary education, though overtaken in proportion by state schools after decolonization.

In the postcolonial contexts, religions have remained visible and vibrant in the public sphere of both Singapore and Hong Kong. But the very nature of publicness has charted very different courses. In Singapore, public discourses of religions are essentially constructed and chanted by the state. Neo (2016) suggests that the constitutional order of Singapore is only 'quasi-secular' because the state is openly entangled with religious affairs and is explicit about its desire to regulate religions by means of public policies.

Alongside Neo, we summarize the Singapore state's approach to religions with reference to two doctrines. The first is the principle of equal treatment and harmony. The state holds in high value the four Ms of multiracialism, multilingualism, multiculturalism, and multireligiosity, in conjunction with a fifth M, meritocracy. In this ideology, political discourses emphasize state neutrality and equal treatment of all faiths—no privilege shall be conferred on the basis of race, language, culture, or religion (Kong 2015; note exceptions elaborated in Kong 1993). The mantra of equality is reinforced by prioritizing racial and religious harmony. Religious groups are exhorted to identify foremost with a unified nation of Singapore, while intolerance towards other faiths is not approved by the state (C. Tan 2008).

The second doctrine to which the state adheres is hierarchy; that is, secular reason is prioritized over religions (Neo 2016). Insofar as the state welcomes the presence of diverse religions in public cultures, religious concerns cannot override secular purposes. Even when religious teachings are valorized by the state, it is to prioritize secular goals, such as when religious adherents are exhorted to strive for economic progress and development by drawing on religious teachings (Kong 2012). The prioritization of secular reason and objectives has been elucidated by scholars with reference to, for example, the strong emphasis on secular multiculturalism and national identity in faith-based schools (Kong 2005). Insights have also been drawn from the availability and locations of religious spaces and sites. In a global city where land is of extreme scarcity, religious groups face perennial challenges for the supply of spaces and buildings. State-led urbanization and zoning, however, fall short of the expectations of religious people, for the state prefers a functionalist approach that prioritizes centralized decision-making, pragmatism, efficiency, and order (Kong 1993, 2002). As a result, we occasionally witness the impromptu socialization of urban sites, which defies the binaries of private/ public, legal/ illegal, sacred/ profane (Kong 2002).

In contrast, religious organizations and faith actors in post-1997 Hong Kong are not content with being the objects of state discourses and directions, but much more active in the shaping of secular politics. On the one hand, Buddhist, Taoist, and Confucian groups, which received little state patronage during colonial rule, naturally expect that their cultural affinities with China will help them gain the favour of Beijing and restore privilege in postcolonial Hong Kong. Unsurprisingly, these groups are generally in rapport with the post-handover state in Hong Kong, and actively promote the Chinese national identity and the revival of Chinese cultures (Kwong 2002). On the other hand, Catholic and Protestant groups have developed a strong public identity as well, but focus on the democratization of Hong Kong and the thorny issue of religious freedom in the face of the Chinese state's anti-Christian mindset. Divergent from the thriving of Pentecostal Christianity in Singapore, which focuses on the cultivation of individual self and spiritual communities, Christianity in Hong Kong features much more strongly liberal theologies (Goh 2016). Both Catholic and Protestant churches are vocal in the advocacy of human rights, democracy, rule of law, freedom, and a robust Hong Kong local identity (to fend off the political assimilation of Beijing) (Li et al. 1998; Nedilsky 2014). Christian concepts and theories have been reimagined to inspire political mobilization, giving rise to a vibrant political theology (Leung 2009; Kung 2010; Jung 2016). The heavy involvement of Christian groups in the 2014 Umbrella Movement is but the latest culmination of this spirit of public engagement (Bosco 2016; Chow and Lee 2016).

Conclusion

We begin this concluding section by summarizing our contribution to the postsecular debate. We acknowledge that they have not been developed to any degree of sophistication, given space constraints; yet, we hope that they cast some light on a global comparative perspective for the study of religion and the postsecular.

To theorize postsecularity, a comprehensive understanding of the ways in which secularity took roots in different contexts is needed. In Asia, secularity was historically complicit with the power-knowledge complex of Western colonialism and European modernity. The fin-de-siècle encounters of non-Western societies with secularity involved complex operations of power—unchallenged authority of Western discourses and epistemologies, and subsequently the power of indigenous elites to institute a secular age locally. In China, the taking-root of secularism cannot be understood without critically reflecting on the symbolic and material violence exercised by the state in successive campaigns of exorcism and iconoclasm, in the name of progress and modernity. In Singapore and Hong Kong, secularity took root as a result of colonial rule, and became one of the founding political principles of the postcolonial Singapore state, albeit less so in Hong Kong. In all these contexts, however, secularizing missions are never capable of absorbing religious impulses completely and cannot march forward without being challenged or contested. The dramatic renaissance of religions in post-1979 China is a vivid testimony to this, while in Singapore and Hong Kong, the state has adopted more pragmatic approaches that regulate, rather than eradicate, public religious cultures. The interweaving of the secular state and public religion ensures that the particular formation of the postsecular in these contexts defies easy comparison with the European experience.

The manifestations and consequences of public religions in the postsecular condition are contingent and indefinite. In both China and Singapore, the visibilities of religions in public cultures are overseen, managed, and approved by the state, but postsecularity means very different things in the two societies—China is postsecular in the sense that the state has orchestrated and closely controlled the transition from overwhelming secularism to the revival of religions in public cultures, with the purpose of harnessing the benefits brought by faiths and religious cultures. In Singapore, postsecularity does not imply a comparable temporal transition, but the persistent entanglement of religious affairs and political cultures of the state. In Hong Kong, akin to Singapore, the term postsecularity makes sense in so far as it contains no connotation of having evolved from a monolithically secular age as such. But this does not mean that postsecularity is a given status quo, a *fait accompli*, rather than an ever-changing process. Indeed, the public theologies of Buddhist, Taoist, and Christian groups in Hong Kong have all changed due to the Handover—for Buddhists and Taoists, their primary public roles have transitioned from social welfare to alliance with Beijing and the postcolonial local state, while for Christians, philosophies of public engagements have changed from those aligned with colonial state interests to more liberal, anti-hegemonic, and resistant ones.

Above all, we do not propose to dismiss the conceptual scaffold of the postsecular even though the thesis is currently largely preoccupied with European experiences, which represent a somewhat exceptional case of secularization rather than a global norm (Davie 2007). Our proposition is that the prefix 'post' does not imply a linear temporality, but paradigmatic changes to social scientific theories of modernity (McLennan 2010)—secular modernity is not a totalizing project but always punctuated by religiosities, and it is imperative that we do not treat the religious dimension of modernity in reductive ways. The postsecular may manifest itself as an enforced, fully fledged secularity mutating into a postsecular present (as in the case of China), or the abiding and relative peaceful coexistence of secularity and religiosity in the public sphere (as in Singapore and Hong Kong). A key caveat, though, is that the nature of public religion is not singular but needs to be construed and expounded against specific contexts. After all, we need to develop theorizations of *multiple postsecularities*, rather than a single postsecularity.

Further reading

Palmer, D. (2017) *Dream Trippers: global daoism and the predicament of modern spirituality*, Chicago, IL: University of Chicago Press.

Palmer analyzes spiritual trips centered on Chinese Daoism in a context of religious revival in China, and situates religiosity at the interface of spiritual pursuit, market economy, and the state.

Qian, J. and L. Kong (2018) 'Buddhism Co. Ltd? Epistemology of religiosity, and the reinvention of a Buddhist monastery in Hong Kong', *Environment and Planning D: Society and Space*, 36(1): 159–77.

The work examines the contestation between Buddhism and urban development processes, seeing religion as a powerful public actor that shapes both the market culture and state governance in Hong Kong.

Sinha, V. (2016) 'Marking spaces as 'sacred': infusing Singapore's urban landscape with sacrality', *International Sociology*, 31(4): 467–88.

Sinha's work illustrates vividly how the distinctions between the secular and the religious, the private and the public are constantly transgressed and relativized in the context of urban Singapore.

References

Asad, T. (2003) *Formations of the secular: Christianity, Islam, modernity.* Stanford, CA: Stanford University Press.

Ascione, G. (2017) 'Dissonant notes on the post-secular: unthinking secularization in global historical sociology', *Journal of Historical Sociology*, 30(2): 403–34.

Ashiwa, Y. and D. L. Wank (eds.) (2009) *Making Religion, Making the State*, Stanford, CA: Stanford University Press.
Beaumont, J. (2008) 'Faith action on urban social issues', *Urban Studies*, 45(10): 2019–34.
Beaumont, J. and C. Baker (eds.) (2011) *Postsecular Cities: space, theory and practice*, London: Continuum.
Bolton, C. (2015) 'Tracing faith-based service landscapes: the contours of messiness at the open door community in Atlanta', *Urban Geography*, 36(2): 221–35.
Bosco, J. (2016) 'The sacred in urban political protests in Hong Kong', *International Sociology*, 31(4): 375–95.
Camilleri, J. A. (2012) 'Postsecularist discourse in an "age of transition"', *Review of International Studies*, 38(5): 1019–39.
Chatterjee, P. (1997) *Our Modernity*, Rotterdam and Dakar: Sephis & Codesria.
Chau, A. Y. (ed.) (2011) *Religion in Contemporary China: revitalization and innovation*, London: Routledge.
Chow, C. and J. Lee (2016) 'Almost democratic: Christian activism and the Umbrella Movement in Hong Kong', *Exchange*, 45(3): 252–68.
Cloke, P. and J. Beaumont (2013) 'Geographies of postsecular rapprochement in the city', *Progress in Human Geography*, 37(1): 27–51.
Cloke, P., Johnsen, S. and J. May (2005) 'Exploring ethos? Discourses of "charity" in the provision of emergency services for homeless people', *Environment and Planning A*, 37(3): 385–402.
Cloke, P., Sutherland, C. and A. Williams (2016) 'Postsecularity, political resistance, and protest in the Occupy Movement, *Antipode*, 48(3): 497–523.
Davie, G. (2007) *The Sociology of Religion*, London: Sage.
Duara, P. (1991a) *Culture, Power, and the State: rural north China, 1900–1942*, Stanford, CA: Stanford University Press.
——— (1991b) 'Knowledge and power in the discourse of modernity: the campaigns against popular religion in early twentieth-century China', *The Journal of Asian Studies*, 50(1): 67–83.
Goh, D. (2016) 'Secular space, spiritual community and the hybrid urbanisms of Christianity in Hong Kong and Singapore', *International Sociology*, 31(4): 432–49.
Goh, D. and P. van der Veer (2016) 'Introduction: the sacred and the urban in Asia', *International Sociology*, 31(4): 367–74.
Gökarıksel, B. and A. J. Secor (2015) 'Postsecular geographies and the problem of pluralism: religion and everyday life in Istanbul, Turkey', *Political Geography*, 46: 21–30.
——— (2017) 'The post-Islamist problematic: questions of religion and difference in everyday life', *Social & Cultural Geography*, 18(5): 645–64.
Graham, E. (2013) *Between a Rock and Hard Place: public theology in a post-secular age*, London: SCM Press.
Habermas, J. (2006) 'Religion in the public sphere', *European Journal of Philosophy*, 14(1): 1–25.
——— (2008) 'Notes on post-secular society', *New Perspectives Quarterly*, 25(4): 17–29.
Habermas J., Brieskorn, N., Reder, M., Ricken, F. and J. Schmidt (eds.) (2010) *An Awareness of What Is Missing: faith and reason in a postsecular age*, Cambridge, MA: Polity Press.
Jung, P. Y. (2016) 'The Eucharist as transformable spirituality: reflections on protestant spirituality and a recent civil disobedience movement in Hong Kong', *Spiritus: A Journal of Christian Spirituality*, 16(2A): 14–38.
Köhrsen, J. (2012) 'How religious is the public sphere? A critical stance on the debate about public religion and post-secularity', *Acta Sociologica*, 55(3): 273–88.
Kong, L. (1993) 'Negotiating conceptions of "sacred space": a case study of religious buildings in Singapore', *Transactions of the Institute of British Geographers*, 18(3): 342–58.
——— (2002) 'In search of permanent homes: Singapore's house churches and the politics of space', *Urban Studies*, 39(9): 1573–86.
——— (2005) 'Religious schools: for spirit, (f)or nation', *Environment and Planning D: Society and Space*, 23: 615–31.
——— (2010) 'Global shifts, theoretical shifts: changing geographies of religion', *Progress in Human Geography*, 34(6): 755–76.
——— (2012) '"Managerialist" constructions of religion: the geographical and historical contingencies of religion in Singapore', *International Symposium on Concepts of Religion between Asia and Europe*, Museum Rietberg (Park-Villa Rieter), University of Zurich, Zurich, 1–3 November.
——— (2015) 'Disrupting "Asian Religious Studies": knowledge (re)production and the co-construction of religion in Singapore', *Numen: International Review for the History of Religion*, 62(1): 100–18.

Kung, L. Y. (2010) 'The publics, church and Hong Kong: the narrative nature of public theology', *Logos & Pneuma: Chinese Journal of Theology*, 32: 85–115.
Kwong, C. (2002) *The Public Role of Religion in Post-Colonial Hong Kong*, New York, NY: Peter Lang.
Lancione, M. (2014) 'Entanglements of faith: discourses, practices of care and homeless people in an Italian City of Saints', *Urban Studies*, 51(14): 3062–78.
Leung, B. (2009) 'The Hong Kong Catholic Church: a framing role in social movement'. In Kuah-Pearce, K. E. and G. Guiheux (eds.) *Social Movements in China and Hong Kong: the expansion of protest space*, Amsterdam: Amsterdam University Press, pp. 245–58.
Li, K. M., Cheung, K. H. and K. S. Chan (1998) 'The social role of Catholics in Hong Kong society', *Social Compass*, 45(4): 513–31.
May, J. and P. Cloke (2014) 'Modes of attentiveness: reading for difference in geographies of homelessness', *Antipode*, 46(4): 894–920.
McLennan, G. (2010) 'The postsecular turn', *Theory, Culture and Society*, 27(4): 3–20.
Nedilsky, L. V. (2014) *Converts to Civil Society: Christianity and political culture in contemporary Hong Kong*, Waco, TX: Baylor University Press.
Neo, J. L. (2016) 'Secular constitutionalism in Singapore: between equality and hierarchy', *Oxford Journal of Law and Religion*, 5(3): 431–56.
Olson, E., Hopkins, P., Pain, R. and G. Vincett (2013) 'Retheorizing the postsecular present: embodiment, spatial transcendence, and challenges to authenticity among young Christians in Glasgow, Scotland', *Annals of the Association of American Geographers*, 103(6): 1421–36.
Oosterbaan, M. (2014a) 'Public religion and urban space in Europe', *Social & Cultural Geography*, 15(6): 591–602.
——— (2014b) 'Religious experiences of stasis and mobility in contemporary Europe: undocumented Pentecostal Brazilians in Amsterdam and Barcelona', *Social & Cultural Geography*, 15(6): 664–82.
Saint-Blancat, C. and A. Cancellieri (2014) From invisibility to visibility? The appropriation of public space through a religious ritual: the Filipino procession of Santacruzan in Padua, Italy', *Social & Cultural Geography*, 15(6): 645–63.
Tan, C. (2008) 'Creating "good citizens" and maintaining religious harmony in Singapore', *British Journal of Religious Education*, 30(2): 133–42.
Tan, J. K. (1997) 'Church, state and education: Catholic education in Hong Kong during the political transition', *Comparative Education*, 33(2): 211–32.
Taylor, C. (2007) *A Secula Age*, Cambridge, MA: The Belknap Press of Harvard University Press.
Vries, H. de and L. E. Sullivan (eds.) (2006) *Political Theologies: public religions in a post-secular world*, New York, NY: Fordham University Press.
Wilford, J. (2010) 'Sacred archipelagos: geographies of secularization', *Progress in Human Geography*, 34(3): 328–48.
Williams, A. (2015) 'Postsecular geographies: theo-ethics, rapprochement and neoliberal governance in a faith-based drug programme', *Transactions of the Institute of British Geographers*, 40(2): 192–208.
Wu, F. L. (2008) 'China's great transformation: neoliberalization as establishing a market society', *Geoforum*, 39(3): 1093–96.
Yang, C. K. (1961) *Religion in Chinese Society*, Berkeley: University of California Press.
Yang, F. G. (2012) *Religion in China: survival and revival under Communist rule*, Oxford: Oxford University Press.
Yang, M. (ed.) (2008) *Chinese Religiosities: afflictions of modernity and state formation*, Berkeley: University of California Press.
Yang, M. (2011) 'Postcoloniality and religiosity in modern China: the disenchantments of sovereignty', *Theory, Culture and Society*, 28(2): 3–45.
Zhang, J. J. (2017) 'Paying homage to the "Heavenly Mother": cultural-geopolitics of the Mazu pilgrimage and its implications on rapprochement between China and Taiwan', *Geoforum*, 84, 32–41.

22

Four genealogies of postsecularity

Kristina Stoeckl and Dmitry Uzlaner

Introduction

For the Russian formalist literary critic Victor Shklovsky (1893–1984), the purpose of art was to turn the familiar object into something unfamiliar and strange, thus enabling the spectator to behold the object anew. A piece of art, a theatre performance, or a cinematographic scene has the power 'to make the stone *stony*', he wrote, meaning that they break up habitual perceptions and open the gaze to the essence of an object (Shklovsky 1991 [1925]). The liberating effect of distancing and estrangement is valid not only in the sphere of art, but also applies to social theories. Social theories develop in specific contexts, frequently with the purpose to give sense and order to a confusing reality in a particular time and space. It is once we take them out of their original context that theories and concepts reveal their potential to flourish or, on the contrary, their limitations and contradictions.

Postsecularity is one such concept that is put to the test when taken outside of the Western context where it originally emerged. In our contribution to the *Handbook*, we look at postsecularity through the lens of its perception in the Russian scholarly context. This exercise in distancing oneself from the original—Western academic—context of the theory has the purpose to free the imagination for a critical analysis. The Russian lens, we will argue, allows us to discern four different genealogies that inform postsecularity: the sociological, the normative, the postmodern, and the theological. All of these genealogies have to a certain extent produced their own postsecularity, but due to the high specialization of Western academia, they don't usually intersect and therefore rarely clash. In the Russian context, on the contrary, all four genealogies are received and interpreted simultaneously. The Russian lens therefore reveals the differences and commonalities between existing approaches to postsecularity.[1]

Postsecularity: a variety of approaches to a complex phenomenon

Since the late 1990s, debates on postsecularity have multiplied, developing the concept into different disciplinary directions: sociology, normative political theory, philosophy, and theology. In front of such proliferation, the sociologist James Beckford came to the conclusion that the notion of postsecularity was problematic in itself and that it offered no help in explaining important recent trends connected to religion and secularity (Beckford 2012: 16–7). Beckford claimed, in particular, that 'it is not easy to reconcile the idea that the secular has somehow come to an end with the idea that postsecularity represents a refinement—or a more productive phase—of secularity' (2012: 12). Beckford is right in his observation that the concept of postsecularity has proliferated into different directions that are not easily reconcilable with each other, but he underestimates the underlying unity of the existing approaches.

Several authors have offered their own meta-analyses of postsecularity, pointing to multidimensional interpretations while attesting to the underlying unity of existing approaches (McLennan 2007, 2010a, 2010b; Uzlaner 2011, 2013; Cistelecan 2014; Molendijk 2015; Parmaksız 2016; Fordahl 2017; Bradotti 2008). Arie Molendijk, for example, comes to the conclusion that 'the emergence of the "postsecular" refers to very real phenomena, the most important being the "intertwinement" of the secular and the religious in sometimes new forms' (2016: 110). George McLennan stresses the importance of understanding postsecularity as 'intra-secular rather than anti-secular' (2010a: 19), as critical exploration of faults and limits of secularity rather than straight rejection of it. Clayton Fordahl emphasizes hyper-reflexivity and anxiety over taken-for-granted conceptions regarding religion and secularity (2017: 564–5), while Bradotti analyses the impact of postsecular theory on feminism (Bradotti 2008). Umut Parmaksız, finally, praises postsecularity for challenging 'the natural status ascribed to the secular, interrogating seculanormativity in the social, political and cultural realms' (2016: 111). This attitude is echoed by Morteza Hashemi, who also sees in different manifestations of postsecularity 'a form of challenging the conception of the secular as the neutral' (2016: 474), that is, as the neutral foundation which is beyond doubts and which is the shared background for all further processes and transformations. Our contribution adds to this line of argumentation. We take the existence of different interpretations of postsecularity not as a weakness in the theory, but rather as an indicator that the complex underlying experience of revision of the secularization thesis requires an equally complex response.

In the Western context, different academic disciplines—sociology, normative political theory, philosophy, and theology—cultivate different understandings of postsecularity. In the well-ordered world of Western academia, these understandings do not usually intersect. In the Russian context, the situation is different. Genealogies of theoretical argument as distinct as the political liberalism of John Rawls and Jürgen Habermas are brought into interplay with the postmodernism of Slavoj Žižek or the radical orthodoxy of John Milbank. In the Russian perception, postsecularity from the beginning referred to a general paradigm shift—affecting different disciplines and working in different social, political, legal, and cultural contexts. What mattered was not concrete positions and nuances but the general understanding that 'religion is back' and that it is starting to refashion the theoretical and empirical landscape formed by previous decades and even centuries of secularization. It is this holistic Russian approach to the concept of postsecularity that provokes the distancing or estrangement effect that was hailed by Shklovsky as the essence of art, and which allows also the social theorist to cast a fresh look on the theoretical debates that define postsecularity.

The sociological genealogy

The first scholarly article to systematically introduce the term postsecular into Russian was 'A Postsecular Age: Religion and Culture Today' by Alexander Kyrlezhev (2004, published in English in 2008). It took another couple of years before the term gained wider prominence and a more systematic reception set in, mostly with the works of Dmitry Uzlaner

(2008, 2011, 2013), Kyrlezhev (2011, published in English in 2012, 2013, 2014), and few others (cf. Morozov 2008; Shishkov 2010a; Horujy 2012). The authors who actively introduced the term into the Russian academic debate were sociologists of religion, philosophers, and theologians. The concept of postsecularity had an immediate sociological appeal in the Russian situation, characterized for over a decade already by religious revival and a renewed public role of the Russian Orthodox Church. Kyrlezhev defined secularization as 'the age of war against religion and of the attempts to find substitutes for it' (Kyrlezhev 2008: 25). He identified the post-Soviet religious revival with postsecularity.

In the West, also, the sociological genealogy of postsecularity was connected with empirical claims regarding the demise of the secularization thesis and the widely shared agreement that modern societies are experiencing a 'return of religion'.[2] Postsecularity, in this context, had a great rival, the concept of desecularization introduced into the debate by Peter Berger (1999). However, postsecularity and desecularization tried to catch different dynamics in the situation of religious resurgence, with postsecularity stressing the transformative nature of the interaction of religion and secular modernity; and desecularization focusing more conventionally on religious restoration and religious-secular conflicts. In the Russian context, the difference between the two theoretical approaches was evident, with postsecularity-studies trying to detect changing modes of interaction between religion and Russian society on the level of practices and discourses (Uzlaner 2014), and desecularization studies focusing on the renewed role of Orthodoxy as a public religion and on church-state relations (Karpov 2010; Shishkov 2010b).

The aim of sociological studies of postsecularity has, on the whole, been the desire to show the transformative nature of religion in secular societies and to highlight the novelty of the postsecular religious constellation, rather than simply to argue a 'return of religion' (Rosati and Stoeckl 2012). Rosati, in his study on religion in post-Kemalist Turkey, highlighted novel aspects of religious practices in contemporary Turkey, for example the creation of non-traditional sanctuaries of prayer and commemoration (Rosati 2015). Geographers, theologians, and urban sociologists, such as Molendijk et al. (2010), Beaumont and Baker (2011), Herman et al. (2012), and Cloke et al. (2019), have, likewise, stressed the transformative role of religion in, what has been called, 'postsecular spaces of engagement', which differ from the spaces traditionally occupied by religious organizations.

All of these studies share the intuition that postsecular society is a place where the negotiation of the relation between different religions and between religions and secular worldviews is not a confrontation between self-contained ideological universes, but an encounter that unsettles each of the actors involved through a process of self-reflexivity. Against the grain of the concept of desecularization, these studies argue that what is happening in a situation of religious resurgence is *not* the return to a presecular, premodern religious status quo ante, but a novel constellation of pluralism. Such pluralism applies also to religions themselves, which are no longer studied as monoliths, but as multivocal bodies, as a part of highly pluralized societies, with which they interact in multiple ways.

The normative genealogy

The normative genealogy of postsecularity is connected with the academic discipline of political theory and with political liberalism. The postsecular political liberal agenda has been shaped in particular by Rawls (1993) and by Habermas (2006) responding to Rawls. It has a clear normative dimension and has inspired a debate in political philosophy about 'reflexive' forms of secularism (cf. Ferrara et al. 2010; Calhoun et al.

2011; Gorski et al. 2012). Postsecular political liberalism holds that an ideology of secularism is not an integral part of liberalism and that secularism as a political ideology discriminates against religious citizens. All citizens must in principle be free to enter into public debates from within the framework of their 'comprehensive doctrines', provided that they are ready to deliberate over political norms in a reasonable fashion and in the view of a consensus that can become valid for all (the 'overlapping consensus').

Habermas' contribution to this debate is informed by his previous work on communicative action and deliberative democracy. The norms that underlie our modes of political coexistence, and this is Habermas' most basic position, do not lie out there in 'principles from nowhere' nor do we need to abandon the idea of general guiding norms in the light of a multiplicity of moralities and beliefs; agreement on 'principles valid for all' can, instead, emerge in the process of communication and deliberation, which can be the fruit of a mutual learning process and general consent. Habermas himself describes this kind of reasoning as 'post-metaphysical', because it affirms the validity of moral and political principles not by indication of some transcendental point of reference, but through an immanent deliberation process. The equality of public deliberation is threatened, however, when the secular public discourse renders it difficult for religious citizens to voice their arguments. Habermas responds to this particular problem with the assertion that not only should religious citizens be asked to translate their claims into the language of secular public discourse, but also the non-religious citizens are asked to play their part, namely, to scale down their secularist aspirations. Such a reciprocal work of translation should give rise to what he calls 'the complementary learning process' (2006). Habermas' concepts of translation and of the complementary learning process are premised on the idea that religions undergo a process of modernization in response to the challenges of religious pluralism, modern science, positive law, and profane morality. This notion of 'modernization of religious consciousness' has been accused of a secularist and ethnocentric bias by some commentators (Leezenberg 2010; Habermas 2011; Maclure and Taylor 2011).

In the Russian reception of postsecularity, the Habermasian normative understanding of postsecularity is strikingly absent. It is in particular the idea that religions in postsecular societies undergo a modernization process that is rejected. Kyrlezhev argued instead: 'Religion in the postsecular age does not need to adjust; there is no need for it to make itself "modern," because the Western world has already moved on to the "postmodern" phase' (2008: 31). Even though Habermas was duly introduced to the Russian debate later on (Uzlaner 2011), the Habermasian version of postsecularity has remained of secondary importance. One reason for this neglect could be that the Habermasian genealogy implicitly presupposes a set of conditions, namely, the institutional context of liberal democracy and the existence of an academic theology or academic philosophy of religion (Stoeckl 2014b), that are largely absent in the Russian context. Through the Russian prism, the limitations of the Habermasian normative genealogy of postsecularity stand out clearly. The persuasiveness of this political liberal genealogy, which has arguably been the motor of the debate on postsecularity in the West, has no parallel in Russia. The Russian readers were not really interested in the normative genealogy of postsecularity from the start and identified their authoritative source for postsecularity elsewhere: in the postmodern genealogy.

The postmodern genealogy

The Russian reception linked postsecularity to postmodernity. 'The start of the postsecular age coincides with the start of the postmodern age', Kyrlezhev wrote, because 'postmodernism gives freedom to religion as religiosity'. 'Postmodern indifference' was the ground on which 'postsecular attitudes [could] grow' (2008: 25). The Western key authors in the Russian reception of postsecular social theory were theologians like John Milbank (his 'Beyond secular reason: theology and social theory', 1993) and—for some of them—John Caputo (his chapter 'How the secular world became post-secular' in 'On religion', 2001: 37–66), or philosophers and critical theorists such as Charles Taylor, Slavoj Žižek, and Talal Asad (see Uzlaner 2011 for the first Russian collected volume on postsecularity).

The postmodern genealogy of postsecularity—though some of the mentioned authors might object to the label 'postmodern'—is connected with the linguistic turn in philosophy, the constructivist turn in the social sciences, and with critical theory. Based on Lyotard's definition of the postmodern condition (Lyotard 1984), the key component of postmodern secularity is reflexivity. Heeding the lesson of the linguistic turn, the modern secular self-understanding of the social and political sciences appears as one 'grand narrative' and secularism as a form of authority (Connolly 1999). Postmodern philosophers discover religion as a reservoir of meaning, on which they draw independently of specific religious traditions of interpretation (Derrida and Vattimo 1998; Žižek 2000; Nancy 2008); critical theorists in all disciplines scrutinize the inbuilt biases in the theories and concepts they are working with. The postmodern perspective perceives postsecularity as a constitutive element of a postmodern turn which marks a rethinking of key tenets of modernity and rejection of religion as the most important of these tenets (Caputo 2001: 37). This leads to 'contamination of philosophy with theological thinking' or to the 'theologisation of philosophy' (Smith and Whistler 2010: 2).

This postmodern dimension of postsecularity is closely connected to genealogical studies about the modern construction of the religion-secular binary, which shows how both modern religion and modern secularity as two incompatible dimensions are not some eternal entities, but artificial constructs resulting from theoretical and political efforts dating back at least to the fifteenth century (Despland and Vallee 1992; Asad 1993, 2003; Molnar 2002; Dubuisson 2003; Masuzawa 2005; Fitzgerald 2007; Hurd 2008; Cavanaugh 2009; Nongbri 2013; Sullivan et al. 2015).

Postsecularity becomes, in this genealogy of argumentation, the study of religion and society *after* the rejection of strong assumptions of the secularization thesis and *after* the deconstruction of conceptualizations of religion in the Western social sciences. In this sense, the postmodern genealogy of postsecularity is also a genealogy of postreligion, if by religion we understand the modern concept of 'religion' which is being deconstructed.

It is important to note, however, that not all theoretical literature and empirical studies associated with the postmodern genealogy of postsecularity use the term 'postsecular'. Some even reject the label. For example, Slavoj Žižek criticizes the term as closely connected to 'a new reenchantment of the world' (Žižek and Milbank 2009: 255–6). It is in the Russian reception that the postsecular nature of Žižek's thinking is highlighted. Žižek, Giorgio Agamben, and Alain Badiou with their turn to St Paul and political theology are read as founders of an alternative postsecular philosophy, alternative both to the political liberal postsecularity of Habermas and to 'the postsecular version of deconstructionism' (Žižek and Milbank 2009: 256) as represented by Jacques Derrida or John Caputo (Uzlaner 2011: 10).

McLennan has described the range of postsecular approaches that we are presenting as four separate genealogies as a continuous spectrum, ranging from 'religious' approaches such as Radical Orthodoxy (RO) (Milbank 1993; see Chapters 11 and 12) to secular-materialist positions such as held by Žižek. In McLennan's scheme, Habermas and Derrida sit in a 'middle range' (McLennan 2010a: 4). McLennan's arrangement comprises, in our systematic approach, the normative, philosophical, and theological genealogy. The three meta-theoretical outlooks that he identifies—genealogical, neo-vitalist, and postcolonial anti-historicist—do not overlap with our threefold disciplinary distinction but add an additional layer that identifies strategies vis-à-vis secularism across the board of different approaches.

The theological genealogy

The Russian reception of postsecular social theory highlights a fourth genealogy of postsecularity, which exists somewhat in isolation from the sociological study of the phenomenon: theology. This fourth genealogy is logically connected to the reflexive turn in philosophy, for which both the normative and the postmodern genealogy are representative. It constitutes its reverse side: secular philosophy reflects upon its secularist bias and opens itself towards religion as a result (what has been called 'theological turn', see Janicaud 1991); theology also reacts to this postmodern turn perceiving it as a chance to re-enter into postsecular philosophy and social theory on its own grounds (Blond 1997). As James K. Smith in his reflections on postsecular theology makes clear:

> the theoretical foundations for the secular have been systematically dismantled. So if we are witnessing the advent of the postmodern ..., then we should also be seeing the advent of the *post*-secular. And insofar as twentieth-century Christian theology ... allied itself with the Enlightenment project, resigning itself to an 'apologetic' project of correlation with secular thought, the demise of modernity must also spell the demise of such theology.
>
> *(Smith 2004: 33)*

The postmodern postsecular turn is perceived as a kind of emancipation for Christian theology, which is now free to develop without constant looking back at secular philosophy and ontology (see Smith 2006). Postmodern philosophers, in turn, look to theology as 'a key "site of resistance" against the alienations of what is perceived as a singularly Western modernity' (Žižek and Milbank 2009: 255–6). This leads to the blurring of the boundary between theology and philosophy and to the expansion of theological-philosophical discussion, where secular philosophers engage in theological reflections and theologians reinterpret key philosophical concepts along theological lines.

It is important to stress that there is a bifurcation in the theological treatment of postsecularity depending on whether the authors stand in the postmodern or the normative lineage of postsecularity. We can, in fact, speak of two theological genealogies of postsecularity: one theological postmodern, the other theological normative. For both, modern philosophy has become self-reflexive to the point that it overcomes its secular bias and opens itself towards religion and theology.

The theological postmodern genealogy moves in the direction of a rather radical rethinking of the distinctions between religion and secularity, faith and reason, philosophy and theology, and so on; this is a search 'for another modernity' or for alternative versions of modernity beyond the secular and liberal model. Theological postmodern postsecularity is far from unanimous—this is a pluralistic, burgeoning conflict field ranging from Milbank's 'Radical Orthodoxy' (Milbank 1993), to Caputo's spiritual deconstruction (Caputo 2001; Caputo and Vattimo 2009), to Michel Henry and Jean-Luc Marion's theological turn in phenomenology (Staudigl and Alvis 2016), to Žižek's 'theology of the death of God' (Žižek and Milbank 2009: 110–233).

The theological normative genealogy, on the other hand, remains in the mainstream of Western philosophy: it limits itself to accurate dialogue and complementary learning without attempts to renew foundational distinctions of modernity. This theological normative genealogy has also produced interdisciplinary debates between philosophers and theologians, whose aim, however, has not been to go beyond foundational aspects of modernity, but rather to reconnect secular modernity with its religious component. Habermas' famous dialogue with Cardinal Ratzinger (Pope Benedict) (Habermas and Ratzinger 2006), and publications of Christian theologians discussing Habermas' ideas (cf. Habermas et al. 2010; Mrówczynski-Van Allen et al. 2016), put theology into the centre of postsecular processes of 'translation' between secular and religious arguments. To this field we may add sociologists of religion who study theological discourses from the angle of postsecular theory and examine how church doctrine and secular realities intersect (Stoeckl 2014a; Dillon 2018).

Conclusion

At the time when a handful of Western scholars were arguing over the question whether a postsecular social theory was meaningful at all, or tried to operationalize the theory for concrete empirical studies, or discussed the normative implications of it, their Russian colleagues took the term at face value. Applying postsecular social theory to the Russian situation promised new avenues for thinking about religion, state, and society beyond the dichotomy of Soviet atheism and Orthodox religious restoration. By working out this research agenda, the Russian reception of postsecular social theory made visible and tangible what Western debates on postsecularity tend to conceal, namely, the existence of four separate genealogies. However, through the Russian lens we also glance that we are dealing not with unconnected debates, as Beckford (2012) remarked critically, but with a more general paradigm shift that manifests itself differently across different contexts. This shift is essentially about the questioning of the modern secular paradigm, according to which secular reason and a secular public sphere were the stable, neutral, taken-for-granted ground on which all other processes evolve. From the postsecular perspective, this ground becomes now the site of intense debate and even conflict, a space of 'uncertainty concerning the religious-secular configuration' (Kyrlezhev 2004: 101). Our identification of four genealogies of postsecularity tries to advance an orderly exploration of this new space of uncertainty.

Every research on postsecularity connects to one of the aforementioned four genealogies; actually, most studies connect to more than one. The four genealogies that we have identified in this essay function as epistemological backdrop for the selective gaze of scholars; they determine the methodology, the processes, and the material that come under scrutiny. The Russian lens brings these four genealogies as well as their unity into focus. The Russian readers are less professionally biased in their reception of Western debates than Western academics, who tend to stay within the bounds of their disciplines. After the closed world of Soviet censorship, since the 1990s, the Russian humanities and social sciences have caught up with Western scholarship all at once. For that reason, they tried to grasp the picture in general while worrying much less about subtleties in debates, chronological phases, and incompatibilities of schools of thought. The Russian reception of postsecular social theory moved from the very beginning along the lines of a broad 'postsecular turn', embracing all four genealogies of postsecularity and interpreting them as a rethinking of secularism as a kind of a new openness to religion and theology.

This reception has produced a highly original debate on postsecularity, but one which has been difficult to 'translate back' into the Western context. This essay is actually an attempt to do precisely that. In the mindboggling situation of post-Soviet religious life in Russia, where theological reform exists side by side with religious fundamentalism, where the boundaries between the secular and the religious are in constant flux, and where fierce battles of ideas (and not only ideas) are being waged over Russian identity, culture, and politics, postsecular social theory has a considerable emancipatory and critical potential, because it starts from the assumption that pluralism, and not the dichotomous secular-religious divide, is the basic feature of society. But the term has also not been immune to abuse, and for several years already one can witness representatives of the Russian Orthodox Church make use of 'postsecular theory' in order to bestow scientific credentials to their claims for privilege in the secular Russian state. In this context, it is even more important that postsecularity as a research programme is upheld as a critical endeavour.

Notes

1 The purpose of this essay is theoretical. The overview on the Russian reception of the postsecular, which is mostly of interest to an expert readership, is kept purposefully short. For an extensive overview and analysis of the Russian reception of Western theories of secularization, desecularization, and the postsecular, see Stoeckl and Uzlaner (2019). The authors acknowledge support by the grant ERC-STG-2015–676804 for the writing of this essay.

2 Many sociologists of religion have rejected the idea that the situation of increased public visibility of religion, the heightened focus on religion in politics and society, and rising levels of religious pluralism warrants the necessity to add the postsecular qualifier (cf. Turner 2010; Gorski et al. 2012). Critics of the postsecularity concept argue that the (renewed) vitality of religion under modern conditions is a novelty only in a restricted European context and common reality in most other parts of the world (cf. Eder 2002; Burchardt, Wohlrab-Sahr, and Middell 2015).

Further reading

Calhoun, C., Mendieta, E. and J. VanAntwerpen (eds.) (2013) *Habermas and Religion*, Cambridge, MA: Polity Press.

This edited volume contains one of the most up-to-date and comprehensive discussion of Habermas' engagement with religion and with his ideas about postmetaphysical thought and postsecular consciousness. It is therefore the ideal starting point for the study of the normative genealogy of postsecularity.

Connolly, W. E. (1999) *Why I am not a Secularist*, Minneapolis: University of Minnesota Press.

Connolly's book is a classic for postsecular social theory and a foundational text for the postmodern genealogy of postsecularity.

Rosati, M. and K. Stoeckl (eds.) (2012) *Multiple Modernities and Postsecular Societies*, Farnham: Ashgate.

This edited volume offers an exemplary series of case-studies in the sociological genealogy of postsecularity, comprising cases from Turkey, Iran, Nepal, Russia, and Nigeria. The introduction and the first chapter on multiple democracies make clear how the normative perspective can intersect with sociological studies on postsecularity.

Smith, J. K. (2004) *Introducing Radical Orthodoxy: mapping a post-secular theology*, Grand Rapids, MI: Baker Academic.

This theological work, directed to a Christian readership, explains the foundations and trends of the theological genealogy of postsecularity in its postmodern key. The author is a representative of the RO first propagated by John Milbank.

References

Asad, T. (1993) *Genealogies of Religion*, Baltimore, MD: The Johns Hopkins University Press.

——— (2003) *Formations of the Secular: Christianity, Islam, modernity*, Stanford, CA: Stanford University Press.

Beaumont, J. and C. Baker (eds.) (2011) *Postsecular Cities: space, theory and practice*, London: Continuum.
Beckford, J. (2012) 'SSSR Presidential Address: public religions and the postsecular: critical reflections', *Journal for the Scientific Study of Religion*, 51(1): 1–19.
Berger, P. L. (1999) 'The desecularization of the world: a global overview'. In Berger, P. L. (ed.) *The Desecularization of the World: resurgent religion and world politics*, Washington, DC: Ethics and Public Policy Center, pp. 1–18.
Blond, P. (ed.) (1997) *Post-Secular Philosophy: between philosophy and theology*, London/New York, NY: Routledge.
Bradotti, R. (2008) 'In spite of the times: the postsecular turn in feminism', *Theory, Culture & Society*, 25(1): 1–24.
Burchardt, M., Wohlrab-Sahr, M. and M. Middell (eds.) (2015) *Multiple Secularities Beyond the West: religion and modernity in the global age*, Boston, MA: de Gruyter.
Calhoun, C., Juergensmeyer, M. and J. VanAntwerpen (eds.) (2011) *Rethinking Secularism*, Oxford: Oxford University Press.
Caputo, J. D. (2001) *On Religion*, London/New York, NY: Routledge.
Caputo, J. D. and G. Vattimo (2009) *After the Death of God*, edited by J. W. Robbins, New York, NY: Columbia University Press.
Cavanaugh, W. T. (2009) *The Myth of Religious Violence: secular ideology and the roots of modern conflict*, Oxford: Oxford University Press.
Cistelecan, A. (2014) 'The theological turn of contemporary critical theory', *Telos*, 167 (Summer): 8–26.
Cloke, P., Baker, C., Sutherland, C. and A. Williams (2019) *Geographies of Postsecularity: re-envisioning politics, subjectivity and ethics*, London/New York: Routledge.
Connolly, W. E. (1999) *Why I am not a Secularist*, Minneapolis, MN: University of Minnesota Press.
Derrida, J. and G. Vattimo (1998) *Religion*, Stanford, CA: Stanford University Press.
Despland, M. and G. Vallee (eds.) (1992) *Religion in History: the word, the idea, the reality* ("La Religions dans L'histoire: le mot, l'idee, la realite"), Waterloo: Wilfred Laurier University Press.
Dillon, M. (2018) *Postsecular Catholicism: relevance and renewal*, Oxford: Oxford University Press.
Dubuisson, D. (2003) *The Western Construction of religion: myths, knowledge and ideology*, Baltimore, MD: The Johns Hopkins University Press.
Eder, K. (2002) 'Europäische säkularisierung: ein sonderweg in die postsäkulare sesellschaft?' *Berliner Journal für Soziologie*, 12: 331–44.
Ferrara, A., Kaul, V. and D. Rasmussen (2010) 'Special issue: postsecularism and multicultural jurisdictions', *Philosophy and Social Criticism*, 36(3–4): 265–80.
Fitzgerald, T. (2007) *Discourse on Civility and Barbarity: a critical history of religion and related categories*, Oxford: Oxford University Press.
Fordahl, C. (2017) 'The post-secular: paradigm shift or provocation? *European Journal of Social Theory*, 20(4): 550–68.
Gorski, P., Kyuman Kim, D., Torpey, J. and J. VanAntwerpen (eds.) (2012) *The Post-Secular in Question: religion in contemporary societies*, New York: New York University Press.
Habermas, J. (2006) 'Religion in the public sphere', *European Journal of Philosophy*, 14(1): 1–25.
——— (2011) 'Dialogue. Jürgen Habermas and Charles Taylor', In Butler J. and J. VanAntwerpen (eds.) *The Power of Religion in the Public Sphere*, Columbia: Columbia University Press, pp. 60–9.
Habermas, J. and J. Ratzinger (2006) *The Dialectics of Secularization: on reason and religion*, San Francisco, CA: Ignatius Press.
Habermas, J., Reder, M. and J. Schmidt (2010) *An Awareness of What is Missing: faith and reason in a post-secular age*, Hoboken, NJ: Wiley.
Hashemi, M. (2016) 'A post-secular reading of public sociology', *Social Compass*, 63(4): 461–77.
Herman, A., Beaumont, J., Cloke, P. and A. Walliser (2012) 'Spaces of postsecular engagement in cities', In Beaumont, J. and P. Cloke (eds.) *Faith-Based Organizatinons and Exclusion in European Cities*, Bristol: Policy Press, pp. 59–80.
Horujy, S. S. (2012) 'Postsekulyarizm i antropologiya ("Postsecularism and anthropology")', Chelovek. *ru*, 8: 15–34.
Hurd, E. S. (2008) *The Politics of Secularism in International Relations*, Princeton, NJ: Princeton University Press.
Janicaud, D. (1991) *Le Tournant Théologique de la Phénoménologie Française*, Combas: Éd. de l'Éclat.
Karpov, V. (2010) 'De-secularization: a conceptual framework', *Journal of Church and State*, 52(2): 232–70.

Kyrlezhev, A. (2004) 'Postsekulyarnaya epokha ("The postsecular age")', *Kontinent*, 120: 252–64.
——— (2008) 'The postsecular age: religion and culture today', *Religion, State, and Society*, 36(1): 21–31.
——— (2011) 'Postsekulyarnoe: kratkaya interpretatsiya ("The postsecular: a brief interpretation")' *Logos*, 3(28): 100–6.
——— (2012) 'Postsekulyarnaya kontseptualizatsiya religii: k postanovke problemy ("Postsecular conceptualization of religion: towards a formulation of the problem")', *Gosudarstvo, Religiya, Tserkov' v Rossii i za Rubezhom*, 2(30): 53–68.
——— (2013) 'Sekulyarizm i postsekulyarizm v Rossii i v mire ("Secularism and post-secularism in Russia and in the world")', *Otechestvennye Zapiski*, 52(1): 175–92.
——— (2014) 'A post-secular conceptualization of Rreligion: defining the question', *State, Religion and Church*, 1(1): 7–22.
Leezenberg, M. (2010) 'How ethnocentric is the concept of the postsecular?' In Molendijk, A. L., Beaumont, J. and C. Jedan (eds.) *Op cit.*, pp. 91–112.
Lyotard, J.-F. (1984) *The Postmodern Condition*, Manchester: Manchester University Press.
Maclure, J. and C. Taylor (2011) *Secularism and Freedom of Conscience*, Cambridge, MA: Harvard University Press.
Masuzawa T. (2005) *The Invention of World Religions, or, How European Universalism Was Preserved in the Language of Pluralism*, Chicago, IL: University of Chicago Press.
McLennan, G. (2007) 'Towards postsecular sociology?' *Sociology*, 41(5): 857–70.
——— (2010a) 'The postsecular turn', *Theory, Culture & Society*, 27(4): 3–20.
——— (2010b) 'Spaces of postsecularism'. In Molendijk, A. L., Beaumont, J. and C. Jedan (eds.) *Op cit.*, pp. 41–62.
Milbank, J. (1993) *Theology and Social Theory: beyond secular reason*, Oxford: Blackwell.
Molendijk, A. L. (2015) 'In pursuit of the postsecular', *International Journal of Philosophy and Theology*, 76(2): 100–15.
Molendijk, A. L., Beaumont, J. and C. Jedan (eds.) (2010) *Exploring the Postsecular: the religious, the political and the urban*, Leiden/Boston, MA: Brill.
Molnar A. (2002) 'The construction of the notion religion in early modern Europe', *Method & Theory in the Study of Religion*, 14(1): 47–60.
Morozov, A. (2008) 'Has the postsecular age begun? *Religion, State & Society*, 36(1): 39–44.
Mrówczynski-Van Allen, A., Obolevitch, T. and P. Rojek (eds.) (2016) *Beyond Modernity: Russian religious philosophy and post-secularism*, Eugene, OR: Pickwick.
Nancy, J.-L. (2008) *Dis-enclosure: the deconstruction of Christianity*, New York, NY: Fordham University Press.
Nongbri, B. (2013) *Before Religion: a history of a modern concept*, New Haven, CT/London: Yale University Press.
Parmaksız, U. (2016) 'Making sense of the postsecular', *European Journal of Social Theory*, 21(1): 98–116.
Rawls, J. (1993) *Political Liberalism*, New York, NY: Columbia University Press.
Rosati, M. (2012) *The Making of a Postsecular Society: a Durkheimian approach to memory, pluralism and religion in Turkey*, edited with a foreword by A. Ferrara. Farnham: Ashgate.
Rosati, M. and K. Stoeckl (eds.) (2012) *Multiple Modernities and Postsecular Societies*, Farnham: Ashgate.
Shishkov, A. (2010a) 'Osmyslenie poniatiya "postsekulyarnoe" v russkoyazychnoi periodike za poslednee desyatiletiye' ("Making sense of the "postsecular" in Russian language publications for the last ten years"), *Bogoslov.ru*, (28 April 2010), see www.bogoslov.ru/text/733826.html, accessed online 20–03–2018).
——— (2010b) 'Nekotorie aspekty desekulyarizatsii v postsovektskoj Rossii' ("Some aspects of desecularization in the post-Soviet Russia"), *Gosudarstvo, Teligiya, Tserkov' v Rossii i za Rubezhom*, 2(30): 165–77.
Shklovsky, V. (1991 [1925]) *The Theory of Prose*, Bloomington, IN: Dalkey Archive Press.
Smith, A. P. and D. Whistler (2010) *After the Postsecular and the Postmodern: new essays in continental philosophy of religion*, Cambridge, MA: Cambridge Scholars Publishing.
Smith, J. K. (2004) *Introducing Radical Orthodoxy: mapping a post-secular theology*, with a foreword by John Milbank, Grand Rapids, MI: Baker Academic.
——— (2006) *Who's Afraid of Postmodernism? Taking Derrida, Lyotard, and Foucault to church*, Grand Rapids, MI: Baker Academic.

Staudigl, M. and J. W. Alvis (2016) 'Phenomenology and the post-secular turn: reconsidering the "return of the religious"', *International Journal of Philosophical Studies*, 24(5): 589–99.
Stoeckl, K. (2014a) *The Russian Orthodox Church and Human Rights*, London/New York, NY: Routledge.
——— (2014b) 'The theology blind spot', *Immanent Frame Blog of the Social Science Research Council* (13 February 2014): http://blogs.ssrc.org/tif/2014/02/13/the-theology-blind-spot/, accessed online 20–03–2018.
Stoeckl, K. and D. Uzlaner (2019) 'The Russian postsecular". In Schneider, C. (ed.) *Theology and Philosophy in Eastern Orthodoxy*, Eugene, OR: Wipf and Stock.
Sullivan, W. F., Hurd, E. S., Mahmood, S. and P. G. Danchin (eds.) (2015) *Politics of Religious Freedom*, Princeton, NJ: Princeton University Press.
Susen, S. (2015) *The 'Postmodern Turn' in the Social Sciences*, London: Palgrave Macmillan.
Turner, B. S. (2010) 'Religion in a post-secular society'. In Turner, B. S. (ed.) *The New Blackwell Companion to the Sociology of Religion*, Oxford: Wiley-Blackwell.
Uzlaner, D. A. (2008) 'V kakom smysle sovremennyi mir mozhet byt' nazvan postsekulyarnym' ("In which sense can the contemporary world be called postsecular"), *Kontinent*, 136: 16.
——— (2011) 'Vvedenie v postsekulyarnuyu filosofiyu' ("Introduction to postsecular philosophy"), *Logos*, 83(3): 2–32.
——— (2013) 'Kartografiia postsekulyarnogo' ("Cartography of the postsecular"), *Otechestvennye Zapiski*, 1: 17.
——— (2014) 'The Pussy Riot case and the peculiarities of Russian post secularism', *State, Religion and Church*, 1(1): 23–58.
Žižek, S. (2000) *The Fragile Absolute—or, why is the Christian legacy worth fighting for?* London/New York: Verso.
Žižek, S. and J. Milbank (2009) *The Monstrosity of Christ*, edited by Creston Davis, Cambridge, MA: MIT Press.

23

Beyond salvaging solidarity

Umut Parmaksız

Introduction

Over the last two decades, the 'postsecular' has become an increasingly popular concept in various disciplines. For some (Milbank, 1992a, 1992b), the concept represents the emergence of a new age, an era that requires the rethinking of most fundamental social and political categories that we use to make sense of social and cultural life. For others (McLennan 2011; Bader 2012), there is not much clarity as to what, if anything, the concept offers to social and normative political theory. In this contribution, I intend to show that despite its eclectic nature and various shortcomings, the concept has much to contribute to normative thinking and theory. But to reveal this potential, I argue that we need to go beyond the instrumental, solidarity-oriented direction that most scholars of the postsecular have taken the concept and redirect it towards a fuller, more substantive affirmation of religion which challenges what I call the modern 'seculanormative truth-regime'.

In what follows, in the first section I will demonstrate that the postsecular represents a loose but still recognizable normative position concerned with salvaging modern solidarity through a recourse to religion. Within this literature, I will pay particular attention to Habermas, not only because his thought exemplifies to a large extent the postsecular moment but more importantly for my purposes, he recognizes that religion can be the bearer of truth-content, but fails, in my view, to appreciate this insight fully. I will then point out that the postsecular literature problematizes secularity for the wrong reasons, that is its incapacity to stimulate solidarity and/or provide humanity with a world of meaning that can rival the religious, and hence fails to provide a novel and substantial normative critique of secularity. In the second section, using Heidegger's insight into truth, and Taylor and Kompridis' insights into world disclosure, I draw the outlines of a hermeneutical approach that can lead to a more substantive postsecular affirmation of religion's truth-content, without abandoning rational public discourse and opening a space for religious mysticism and revivalism.

Salvaging solidarity with religion?

Ever since the emergence of the 'postsecular' as a concept in social theory in mid-1960s, the focus within the literature has centred on problematizing secular modernity for its incapacity to stimulate solidarity, humane action and responsibility towards others, and many scholars have presented aspects of religion as the remedy to this problem (Parmaksız 2018). This fundamental, deep anxiety about secularity's capacity to foster solidarity, which I will refer to as 'solidarity anxiety', is present in the thinking of many scholars. For example, Greeley (1966), who first uses the concept in a social theory context, argues for a postsecular society that overcomes the pathologies of secular modern societies through the warmth of religion. In his analysis, Greeley (1966: 125) identifies the crisis of the age as the modern man's desire to have 'best of both possible worlds'; that is, 'the intimate support of *gemeinschaft*' while enjoying 'the freedom, the rationality, and the technological flexibility of a *gesellschaft* world'. In this respect, the loss of community is a theme that occupies a fundamental place in his post-Christian age construal. Borowitz (1970), writing from a narrower Jewish perspective, holds secularity responsible for the Holocaust and calls on Jewry to return to the Jewish man of the old tradition.

In the 1970s and the 1980s, the term pops up in various instances without much explanation as to what it means. Morris (1970: 18) employs the phrase 'postsecular world' with a Christian motivation to denote a 'rediscovery of the supernatural', and Graham (1973: 12) seeks the influence of 'Oriental religions' to support Christianity in appealing to the 'postsecular world', where a new religious sensibility is pushing traditional religion and materialism to the margins. For Peursen (1989: 39–40), the postsecular era is defined by 'an intensified urge for meaning', and Martin's (1992: 5–6) 'apocalyptic' declaration that humanity is about to forever lose its sense of community is followed by a call for a 'postsecular community' that takes its inspiration for solidarity from 'pre-secular communities' but is nonetheless regulated by the 'real achievements of modernity'.

In the 1990s, the concept comes to be used increasingly in relation to the marginalization of religion in the public sphere. Gedicks (1991), for instance, focuses on the effects of secular power and advocates the formation of a public sphere where both the religious and the secular can coexist without either serving as a locus of power. In his significant contribution to debates on secularism, Connolly (1999: 3) criticises 'modern secularism' for forfeiting 'some of the very resources needed to foster a generous pluralism' for the sake of a 'public sphere in which reason, morality and tolerance flourish'. Taking a poststructuralist route, Connolly (1996: 259) calls for a postsecular ethics that is 'alert to' its own 'fragility' 'and open to the play of difference in cultural life'. In that, the core of Connolly's criticism is essentially concerned with restoring vitality to the politics of becoming that Connolly believes is necessary to 'sustain the creative tension' that in turn sustains democratic politics. His engagement with the secular problematic is predetermined with a goal of decentralizing the secular, which he thinks stifles this creative dimension.

Undoubtedly, it is with Habermas that the concept is propelled from the fringes of social theory into the mainstream of social and political theory (Habermas 2008a). In the early 2000s, Habermas appropriates the term and applies it to his evolving views on religion's role in public life. To make sense of Habermas' adoption of the postsecular, there is a need to make sense of Habermas' approach to religion and what has exactly changed in his thinking.

Habermas' earlier texts on religion can be understood as an effort to modify Durkheim's sociology of religion in light of communicative action theory (Habermas 1984, 1987). For the early Habermas, as with Durkheim, religion plays a crucial role in providing society with its initial collective identity. Habermas combines this Durkhemian insight with his communicative action theory and argues that there occurs a secularization through the 'linguistification of the sacred', which refers to the process through which the collective identity, constructed by/on religion, is secularized through attempts to legitimize its authority through language. It is through language that religion is transformed into worldviews that acquire a legitimating role, and by way of communicative action, they establish their authority—an insight which, according to Habermas (1987: 57), Durkheim did not sufficiently appreciate. In this earlier model, secularization is conceptualized as the transformation of the consensus imposed by religious authority into the secular through which 'the aura of rapture and terror that emanates from the sacred, the spellbinding power of the holy, is sublimated into the binding/bonding force of criticisable validity claims' (Habermas 1987: 77). In spirit of this insight, Habermas (2002: 79) argues that 'after the collapse of the religious worldviews' what is to be 'salvaged' is 'nothing more and nothing other than the secular principles of a universalist ethics of responsibility'. In this functionalist view, religion plays an important but transitional role in the ladder of social evolution. After this initial constructive role, it disappears into the subjective and the private realm of conscience.

Habermas (2008b: 131) has in the last two decades amended this position by adopting the view that religious traditions, having a 'special power', are of value for public life through their capacity to penetrate a realm that can reveal moral intuitions that other forms of inquiry cannot. What is perhaps most striking about this change is that Habermas (2008b: 132) grants that religious language can be the carrier of *truth-content* and obliges secular citizens to 'open their minds' to learn from these religious discourses. In a telling paragraph, Habermas (1992: 15) lays bare the fears that have led to this rethinking and the core of his new-found interest in religion when he argues that 'without the transmission through socialization and the transformation through philosophy of any one of the great *world religions*', the 'semantic potential' that religion inhabits 'could one day become inaccessible'. He goes on to argue that 'if the remnant of the intersubjectively shared self-understanding that makes human(e) intercourse with one another possible is not to disintegrate, this potential must be mastered anew by every generation. Each must be able to recognize him-or herself in all that wears a human face' (Habermas 1992: 15). For Habermas, it seems the solidarity that is necessary to engage in communicative action is dependent on our willingness to incorporate this semantic potential of the world religions into public discourse. Hence, instead of separating the religious and the secular, as envisaged by his earlier conception of secularization, Habermas now argues that in the postsecular society, secularization 'functions less as a filter separating out the contents of traditions than as a transformer which redirects the flow of tradition' (Habermas 2010: 18).

While there is much to be said about the shift in Habermas' thinking, for my purposes what ought to be emphasized is that even though Habermas recognizes and wants to appropriate the potential truth-content of religious discourse and moves away from a Weberian privatization of religion perspective, he is still mainly operating in the imaginary of classical social theory and more specifically Durkheimian paradigm. The shift in Habermas' stance, therefore, despite at times resembling a radical re-evaluation, is still within his earlier functionalist framework, which affirms religion through its solidarity creating function. The main difference between the early and late Habermas is that instead of attributing religion a role *only* in the constitution of solidarity, Habermas later attributes a new postconstitutive role to religion in the sustainment of solidarity.

Despite its eclectic use, the postsecular represents a thin but recognizable normative stance that predominantly involves an affirmation of religion for the sake of fostering solidarity and peaceful coexistence in modern societies, while almost exclusively ignoring (Connolly) or leaving undeveloped (Habermas) the truth-content dimension. The notion, when construed as such, is misdirected for at least three reasons.

Firstly, the general anxiety over solidarity and the fears regarding us losing our sense of humanity are exaggerated. This is because modern societies are more resilient and resourceful in their capacity to sustain solidarity through interdependence and identification than the proponents of the postsecular acknowledge. These two forms of solidarity, interdependence and identification, can work to check and balance the extreme corrosive effects of each other and they provide a relatively stable ground to reflect on the shortcomings and limitations of the other. As interdependence can foment resentment and create a sense of domination, the sense of belonging that emanates from a common identity can remind us about our obligations and responsibility to our peers that go beyond a future expectation of reciprocity. When identity becomes too entrenched, flirting with the metaphysical, interdependence can work to grind away its thick borders and open it up to a critical examination and reformulation. The complex interdependence that modern societies exhibit has the capacity to erode and reveal the artificiality of race, religion, ethnicity, nation, family, and other categories of association or identity, and thereby open up possibilities of establishing new instances of solidarity.

Secondly, even if we were to concede that modernity suffers from a deficit of solidarity, it is not clear as to what extent religion can really be a remedy for this problem. Much of the conviction that religion can be fundamental to sustaining solidarity is based on the Weberian view that secularization and disenchantment of the world have stripped humanity from the web of meaning it was comfortably residing in, creating 'specialists without spirit, sensualists without heart' (Weber and Parsons 2010: 124). While for Marx this was a welcome development as 'man' was finally face to face with the reality of its existence, for others this implied the restructuring of the social world on the Hobbesian principle of *homo hominu lupus*, which essentially dismantled links of community and solidarity, reducing human existence to a solitary atomism. While this is a widely held conviction in social and political theory, it is not so easily defensible in empirical terms. In fact, when we look at empirical data, the evidence suggests that countries with higher levels of non-religious population, such as Norway, Sweden, Finland, and the Netherlands, exhibit the highest levels of social trust, while some strongly religious societies are very untrustful of each other (Breggren and Bjornskov 2011). Research also shows that members of different religious traditions tend to exhibit different levels of trust towards strangers; while societies with individualized religions like Protestantism exhibit high levels of trust, societies with hierarchical religions such as Catholicism and Islam exhibit lower levels (Bjornskov 2006; Welch et al. 2007; Breggren and Bjornskov 2011). Hence, the nature of how one is religious also affects trust. On the other hand, research also shows that rather than religiosity, income equality is a better determinant of reciprocal trust in societies, and as equality increases, trust in each other tends to increase as well (Bjornskov 2006). This is not to say that religion cannot be a contributor to the sustenance of solidarity, however, whether it is the solution to a solidarity crisis is questionable. Moreover, the evidence demonstrates that the assumption that non-religious societies are incapable or less capable of generating meaning and good will towards others is devoid of convincing empirical proof.

Lastly, to the extent that the postsecular is located within a solidarity and peaceful coexistence-oriented framework of rethinking the public sphere, it does not represent a substantially new or unique normative perspective. There is already a plethora of works that problematize secularism with the aim of reforming it to make it respond better to the diversity of complex societies. For instance, Maclure and Taylor's (2011:15) 'open-secularism' calls for a 'radical reformulation of secularism' intended to reshape our understanding of secularism on the basis of an overlapping consensus that will guarantee 'peaceful coexistence'. Other political theorists, such as Modood (2010), have defended the normative supremacy of 'moderate' Western secular traditions as a starting point to reconsider the relationship of state and religion on a more accommodative basis. Bhargava (2011) similarly asserts that secularism, while problematic in its insistence that state and religion ought to be separated, can still be salvaged by rehabilitating it to work with what he calls a 'principled distance' towards religions. Bader (2012), on the other hand, radically questions the normative utility of the concept and argues that liberal democratic politics inhabits within itself principles which make a distinct emphasis on secularism unnecessary. Hence, he argues that the concept of postsecular is redundant and cannot do any substantial theoretical work.

These critiques of secularism demonstrate that it is possible to envisage a pragmatic critique of secularism without engaging with the wider and more grandiose questions that the postsecular literature aspires to address. In that, one is justified to ask what the postsecular brings new to the way we ought to think about religion and politics in modern societies.

Despite its various shortcomings, I hold that the concept has the potential to mount a substantial critique of the existing state of affairs regarding religion and politics. But if the postsecular is to do any substantive theoretical work in how we think about religion in the modern public sphere, its normative focus needs to shift from solidarity to truth and it needs to represent a thicker outlook, which takes into account the possibility of religion bearing truth-content that goes beyond solidarity sustainment.

I argue for this thicker, truth-content of religion-oriented understanding not only because the concept is theoretically redundant when conceptualized as contributing to solidarity sustainment. More importantly, we must not forget that for many who wish to use religious discourse in public debate, the issue is not about the sustainment of solidarity but more about the expression of a truth that they deem to be manifested in the religious. Unlike the scholars of the postsecular for whom the value of religion resides in its capacity to mend modernity and take it to a new level by enhancing its solidarity fostering capabilities, for many religiously motivated citizens the value of their convictions is intrinsic to them being the expression of a truth.

The postsecular beyond solidarity

One of the most significant consequences of secularization as a social and cultural transformation has been, to use Foucault's terminology, the emergence of a new 'regime of truth' (Foucault 2002: 131), which has had effects in the way we think about religion and politics. Like other social and cultural regimes, this truth-regime is an evolving phenomenon, and it has the capacity to manifest itself in different ways under different circumstances. But in the realm of politics, I hold that there have been two major developments that have defined the present truth-regime.

Firstly, this truth-regime has at its most stringent disqualified religious traditions from access to truth, or greatly restricted their capacity to generate truth-content. With Kant's intervention in epistemology, metaphysics, and along with it much of religion, has been cast into the territory of the unknowable, and the authority on deciding what constitutes legitimate knowledge, which had largely been enjoyed by religious institutions, has been transferred to institutions of science. This new normative order, underpinned by Enlightenment thought, has solidified the conviction that in politics secular language has primacy over religious language as it has the capacity to exhaust the meaning generated by the religious. In other words, it is assumed that whatever meaning that the religious can manifest, the secular can also convey, perhaps even better. This truth regime has been canonized in what Eberle and Cuneo (2008) refer to as the 'doctrine of religious restraint', that is, the principle that religious citizens ought to restrain themselves from using religious language and justifications in public debates.

Secondly, truth in general has been cast as antithetical to politics. In the context of Europe, this change took place following the trauma of fascism and disenchantment from socialism with emergence of a more pragmatic vision of politics. Thinkers, such as Arendt, Berlin, Voegelin and Popper, coming from different philosophical backgrounds, have developed a family of critiques that has emphasized the open-endedness of political life against what they see as the stopping/stifling power of truth. Arendt (2005: 303) succinctly puts the 'trouble' that these thinkers see with truth, 'truth peremptorily claims to be acknowledged and precludes debate, and debate constitutes the very essence of political life'. As such, there has been a strong drive to expunge truth from the public sphere, and construe politics as an activity not occupied with truth.[1]

This truth-regime constitutes an important part of the *seculanormative* thinking (Parmaksız 2018) that dominates much of contemporary political theory. Much like social and political theory has construed secularity as the absence of the religious, the natural substratum that we are left with when religion disappears, secular language has been construed as the stable substratum that all other forms of justifications or reasons can essentially be grounded on, having the capacity to exhaust the realm of meaning that religious discourse can bring into public debate. In rethinking secularism, I hold that the postsecular makes normative sense when it problematizes this truth-regime and questions to what extent the seculanormative exclusion of religion's truth-content is necessary.

While there is the need for an approach that affirms the possibility of truth-content in religious discourse, postsecular thinking ought to also stay clear of essentialist conceptualizations of religion that attribute religion a *sui generis* status. We must avoid the inclination to attribute a 'special power' to religion (Habermas) or suggest to re-establish the 'metaphysical' or 'religious' as an essential, 'instinctual' aspect of being human which relates to an existential 'constitutive choice' (Meyer 1995). Conceding religion such an all-encompassing role goes beyond the warrant of postsecular thinking and dangerously flirts with religious revitalization at the expense of the non-religious. We cannot simply roll back the immanentist achievements of modernity, which have established a sense of publicness, equality, and freedom, and needless to say, any affirmation of religious traditions cannot mean *carte blanche* approval. In that, what is to be affirmed is the potential and not necessarily the actuality of religion's truth-content, the occurrence and value of which would be subject to open and unrestricted debate in the public sphere. In that what is necessary is a postsecular thinking that operates within a politically immanentist framework.

Such a rethinking of religion's potential for truth-content, which avoids the pitfalls of a revivalist perspective, can be realized through a hermeneutical approach that makes use of Heidegger's insights into truth and world-disclosure. To start with, what makes world-disclosure helpful in challenging the predominant truth regime is that it complicates our understanding of truth-content, by enabling us to think of truth in an ontological framework that connects the self, intersubjectivity, and knowledge. Heidegger's insight about language and its relation to the ontological preunderstanding, which makes things intelligible in the first place, not only challenges notions of unsituated knowledge and subjectivity, but it also complicates our understanding of universally criticizable validity and truth claims. Heidegger in his search for the essence of truth has grasped that in order for the question of truth to arise, we need to have been disclosed a preunderstanding of being through language in the first place. Hence, there can be no truth without disclosure of a world. This insight has two consequences: (1) it enables us to conceive as to how religious traditions can be the carriers of truth-content; and (2) it casts a shadow on any tradition that claims to have an all-encompassing capacity to be the carrier of truth-content.

Before I explain how both of these aspects relate to religious traditions, there is a need to clarify Heidegger's conceptualization of truth because it is the basis of much controversy. While some (Wrathall 1999) have taken Heidegger's insight as essential to making sense of our predicament, Habermas (1990) has argued that Heidegger's linking of disclosure and truth does not leave any space for the use of critical reason. For Habermas (1990: 154), disclosure has been elevated to an almost mystical level and construed as a 'truth-event' to the extent that it 'uproots propositional truth' and devalues 'discursive thought', leaving no critical space for context transcendence.

Yet the extent to which Habermas accurately represents Heidegger's views on truth is doubtful. Heidegger famously acknowledged that it was a mistake to equate disclosure, which is the condition of intelligibility, with truth, which is made possible through disclosure, as they are on different ontological levels. While this is a late but welcome acknowledgement of a major mistake, it is a terminological clarification rather than a fundamental shift in his understanding of truth. Wrathall (1999) convincingly argues that Heidegger always intended propositional truth and disclosure, that is truth as unconcealment, to mean different notions, and that his approach does not deny the possibility of construing truth in a propositional sense. Quite to the contrary, Wrathall (1999: 70) argues that Heidegger is concerned with 'clarifying' the 'foundations', that is the conditions which make truth as a correspondence relation meaningful to us, 'rather than abandon it altogether'. In that, the Heideggerian approach does not deny criticizable validity-claims or elevate truth to an uncriticizable level, nor does it assert that we can only talk about truth in a fuzzy and intangible sense. Rather it provokes us to be reflexive about our truth-claims and recognize their indebtedness to an underlying understanding of the world.

As much as disclosure and truth ought to be deconflated, Heidegger's examination still proclaims that disclosure inhabits truth-content albeit on a different ontological level. Heidegger (1969: 27) holds that disclosure manifests through language a 'primordial truth', which arises through the unconcealment of the world to us. In this sense, we can conceptualize religious traditions as carrying truth-content as they can unconceal a primordial truth about the ontological reality of being human.

In making sense of this primordial truth as unconcealment, the concept 'human condition' can provide us with a broad reference point. The notion of the human condition inhabits both a claim to universality and context transcendence, on the one hand, through its reference to the human and not just a particular community of humans, while embracing contextuality, as it is concerned with conditionality, on the other hand. From this broad perspective, interpretations and how we make sense of ourselves can be construed as disclosing the contours of the human condition, thereby helping construct the intersubjective world—what Taylor (1995: 112) calls 'the space we share between us'. This view challenges the idea, which Habermas also adheres to, that world-disclosure makes sense when it is related to the aesthetic realm and when taken to be expressive of a truth that relates to the self.[2] On the contrary, Taylor (1995) establishes that Heidegger's position stands firmly against subjectivist 'intrapsychic' accounts that construe expression as a revelation of an inner being, or Platonist accounts that understand expression as being representative of a higher order. Rather he asserts 'disclosure is not intrapsychic but occurs in the space between humans' and perhaps more importantly, 'it helps to define the space that humans share' (Taylor 1995: 112).

Religious commitments can be more than just a frame for subjectivity or residues of tradition, but by forming an indirect relationship with backgrounds, they have a potential to give body and structure to an otherwise fuzzy and indiscrete milieu. In that, they have not only a capacity, in a positive sense, to disclose a particular aspect of what being human can potentially mean, but also in the negative sense, they disclose the possibilities and limits of the human condition. Hence, they do not just have the capacity to tell a truth about a particular self and the subjective web of meaning it is located within, but they can also indirectly provide access to and 'unconceal' a primordial truth about the fabric of the human condition.[3] Every possibility is also a recognition of a capacity that being human inhabits.

An important consequence of this affirmative view is that, contra Habermas, what makes religious language valuable might not be what is translatable in it, but more so its poetic aspect, that is, precisely that which is lost in translation. Such a perspective can go beyond the limited, solidaristic, 'moral insight'-oriented affirmation of postsecularism, and open up a space for the affirmation of religious discourse for its value in itself. While Habermas seems to recognize this possibility, his adherence to the solidarity-centred reading of religion and deep scepticism towards world-disclosure as a notion keeps him from doing justice to his own acknowledgement.

Heidegger's account of truth does not only allow us to conceptualize how religious traditions can be the carriers of truth-content, albeit in a primordial human-related sense, but it also provokes us to question to what extent any tradition can be the carrier of unbounded, deworlded meaning and truth-content. The dependence of truth-claims to a predisclosed world understanding challenges the infinite capaciousness of secular language and its primacy over the religious—an assumption taken for granted by Habermas, Rawls, and others. Such an unsettling of the supremacy of secular language rightly provokes the question as to how we can make judgements to determine the value of various different disclosures within a framework of reason. Given various historical examples, Habermas is right in being cautious that the infatuation with the poetic and world-disclosure can lead to an uncritical affirmation of the unconcealed 'world' with all its problems and taken-for-granted inequalities.

The answer to this problem lies within world-disclosure itself, more specifically, in understanding how world-disclosure is related to rationality. In this regard, Taylor (1995) has been influential in instigating the view that world-disclosure does not deny reason but it ought to be understood as 'a new compartment of reason', which complements its critical capacity. Kompridis (2006) follows and develops this line of thought, laying out how world-disclosure can be understood as 'another voice of reason' to advance critical thinking. Drawing again on Heidegger, Kompridis distinguishes between two kinds of world-disclosure; 'pre-reflective disclosure', which corresponds to unconcealment or clearing of being in Heidegger's terminology, and 'reflective disclosure' which involves re-evaluation, questioning of the background that we are thrown into, opening up the possibility of critical engagement. In this sense, disclosure in its reflective active form, Kompridis (2006: 35) explains:

> can critically introduce meanings, perspectives, interpretive and evaluative vocabularies, modes of perception, and action possibilities that stand in strikingly dissonant relations to already available meanings and familiar possibilities, to already existing ways of speaking, hearing, seeing, interpreting, and acting.

World-disclosure can also have an active world-making capacity, and it is not merely manifestive. It is from the start postured in relation to an 'other', and as to the extent that we recognize disclosure as possibility, we do so in relation to an actuality or other possibilities.

Such expressions of alternative critical possibilities provoke us to evaluate the value of the commitments and meanings that we take for granted and in doing so they entail a normative stance.

In many ways, in a liberal political context, these alternative disclosures provide the most effective ammunition against other world-disclosures, and they can be employed argumentatively to unground ontologies that enable problematic truth-claims. In this respect, Kompridis (2006: 118–20) demonstrates how the neglect and aestheticization of world-disclosure have robbed the critics of world-disclosure from benefitting from its argumentative utility and power. This rejection has led to the exclusion of various legitimate modes of thinking and argumentation from rational argumentation, such as transcendental arguments, dialectical arguments, and hermeneutical arguments, all of which have been crucial to Western philosophical tradition. For political purposes, hermeneutical arguments are particularly relevant as they involve:

> creative redescriptions of prior practice and cultural paradigms, [that] reconstruct how we came to have the practices that we have, how we came to be the kind of beings that we are, having the pre-reflective commitments that we do, preoccupied with our particular concerns and anxieties'—all of which would be beneficial to the emergence of a critical communicative reason.
>
> *(Kompridis 2006: 118)*

As such, hermeneutical arguments can be understood as indirect arguments against truth-claims enabled by prereflective disclosures. Instead of a head on engagement, world-disclosing arguments enable us to make a case against such disclosures by way of unsettling their foundation.

The acknowledgement that disclosure can take different forms allows us to devise how we can distinguish between 'good' and 'bad' world-disclosure—a problem that has troubled the critics of world-disclosure. Kompridis (2006: 35) argues that a differentiation between 'good and bad disclosure' can be achieved:

> by distinguishing between disclosures of the world that more fully and generously create the conditions for reflective disclosure from those that create conditions that obscure their own status as disclosures.

This difference affirming future-oriented approach allows us to appreciate as valuable those unconcealments that are conducive to the unconcealment of other primordial truths. Not all truth-content is equally valuable or politically relevant. To the extent that we affirm world-disclosure as an unconcealment of truth, we must also disaffirm disclosures that fundamentally harm the manifestation of disclosures. This means that because disclosure takes place through language and reflective disclosure is realized through critical thinking, those practices and norms that harm and inhibit the development of these capacities would need to be debilitated. Political norms must protect the conditions that enable reflective world-disclosure, and at times this may mean restricting and regulating certain other world disclosures. While such restrictions may be presumed to be harmful to diversity, on the contrary it protects diversity as any commitment to difference necessarily implies a commitment to the conditions that make difference possible. These restrictions provide a level of protection against the slide to social, cultural, and political relativism that an affirmation of world-disclosure can be used to justify.

Conclusion

In conclusion, I should clarify that my intention is not to argue that religion inhabits truth-content without which we would be deprived of something essential to human existence. This is what some of the postsecular scholars (Greeley 1966; Borowitz 1970; Blond 1998; Dooley 1999) seem to be convinced of and hence assert as the solution to the pathologies of modern, secular societies. What I instead suggest is that we need to fully affirm, not just in a limited sense, the potential of religion bearing some truth-content.

Such an affirmation of religion's capacity will not only require a rethinking of religious truth and its place in public debate, but it will undoubtedly be accompanied by a general rethinking of the exclusion or marginalization of truth from politics in general. I take the postsecular problematization of the modern truth-regime, I advocate here, to be more than just a contributor to this larger project of bringing truth back into the political, because historically the modern scepticism to truth has its roots in the scepticism to religious truth. The challenge is finding ways of incorporating truth, religious or not, into the political without fundamentally undermining open debate, which forms the bedrock of modern public discourse.

This is a challenge that will require devising new ways of governing non-seculanormative public spheres, and in this endeavour a question of central importance will be determining what kinds of language can be admitted to debate with potentially coercive consequences. 'Religious' and 'secular' can no longer be adequate or exhaustive yardsticks for determining legitimate forms of argumentative language use. What needs to be done is to rethink, from a postsecular perspective, what these categories stood in place for and squeeze out what principles can be salvaged. Only then can we set up new standards that transcend these two categories while preserving both the inherent capacity of language to manifest truth and the argumentative character of modern public spheres.

Notes

1 It is now common to speak of a 'post-truth' era. I find that a more appropriate terminology to describe the current move away from factual debate would be a change in the 'truth-regime'.
2 See Kompridis (2006) for a critique of Habermas' aesthetization of world-disclosure in general, and Rothberg (1986) for a similar critique which takes issue with Habermas' confinement of the religious to the aesthetic and the subjective realms.
3 The 1981 World Value Surveys (WVS) provide a glimpse into our conviction about the truth-content of religion. According to this survey, in the 16 Western countries that the survey was conducted, including USA and Canada, 90.6% said one or more religions had some truth and meaning (World Values Survey Association 2009). Unfortunately, this question has since been removed from the survey, and we do not know if this affirmative outlook has changed in the last three decades.

Further reading

Cooke, M. (2006) 'Salvaging and secularizing the semantic contents of religion: the limitations of Habermas's postmetaphysical proposal', *International Journal for Philosophy of Religion*, 60(1–3): 187–207.

Cooke criticizes Habermas' postmetaphysical mode of reasoning, which she defends is merely another form of secular reasoning. She argues instead that public debate ought to be based on non-authoritarian mode of reasoning, whether it be secular or religious.

Elkins, J. and A. Norris (2012) *Truth and Democracy*, Philadelphia: University of Pennsylvania Press.

This book is a collection of essays, written by prominent authors, which explore important questions about truth and politics. It provides a comprehensive examination of the relation of truth to opinion, agreement, authority, justification, deliberation, and public reason.

Estlund, D. (1993) 'Making truth safe for democracy'. In Copp, D., Hampton, J. and J. Roemer (eds.) *The Idea of Democracy*, Cambridge: Cambridge University Press.

Estlund deals with epistemic questions regarding truth, knowledge, access to truth, and how truth can be incorporated into democratic normative thinking.

Ferrara, A. (2009) 'The separation of religion and politics in a post-secular society', *Philosophy & Social Criticism*, 35(1–2): 77–91.

Ferrara defends the view that the postsecular means a rethinking of the relationship between politics and religion, which would have repercussions for how democratic equality in the public sphere is understood.

References

Arendt, H. (2005) 'Truth and politics'. In Medina, J. and D. Wood (eds.) *Truth: engagements across philosophical traditions*, Oxford: Wiley-Blackwell, pp. 295–314.

Bader, V. (2012) 'Post-secularism or liberal-democratic constitutionalism', *Erasmus Law Review*, 5(1): 5–26.

Berggren, N. and C. Bjørnskov (2011) 'Is the importance of religion in daily life related to social trust? Cross-country and cross-state comparisons', *Journal of Economic Behavior & Organization*, 80(3): 459–80.

Bhargava, R. (2011) 'Rehabilitating secularism'. In Calhoun, C., Juergensmeyer, M. and J. VanAntwerpen (eds.) *Rethinking Secularism*, Oxford: Oxford University Press, pp. 92–113.

Bjørnskov, C. (2006) 'Determinants of generalized trust: a cross-country comparison', *Public Choice*, 130(1–2): 1–21.

Blond, P. (1998) *Post-Secular Philosophy*, London: Routledge.

Borowitz, E. B. (1970) 'The postsecular situation of Jewish theology', *Theological Studies*, 31(3): 460–75.

Connolly, W. E. (1996) 'Suffering, justice, and the politics of becoming', *Culture, Medicine and Psychiatry*, 20(3): 251–77.

——— (1999) *Why I Am Not a Secularist*, Minneapolis: University of Minnesota Press.

Dooley, M. (1999) 'The politics of exodus: Derrida, Kierkegaard and Levinas on "hospitality"'. In Perkins, R. (ed.) *International Kierkegaard Commentary: works of love*, Macon, GA: Mercer University Press.

Eberle, C. and T. Cuneo (2008) 'Religion and political theory', *Stanford Encyclopedia of Philosophy*, acccesed online 15-06-2018.

Foucault, M. and J. Faubion (2002) *Power: the essential works of Michel Foucault, 1954–1984*, London: Penguin.

Gedicks, F. M. (1991) 'The religious, the secular, and the antithetical', *Capital University Law Review*, 20(1): 113–45.

Graham, A. (1973) 'Contemplative Christianity'. In Needleman, J., Bierman, A. K. and J. A. Gould (eds.) *Religion for a New Generation*, New York, NY: Macmillan, 9–14.

Greeley, A. M. (1966) 'After secularity: the neo-gemeinschaft society: a post-Christian postscript', *Sociology of Religion*, 27(3): 119–27.

Habermas, J. (1984) *Theory of Communicative Action, Vol. I*, Boston, MA: Beacon Press.

——— (1987) *The Theory of Communicative Action, Vol. II*, Boston, MA: Beacon Press.

——— (1990) *The Philosophical Discourse of Modernity: twelve lectures*, translated by Frederick G. Lawrence, reprint edition, Cambridge, MA: The MIT Press.

——— (1992) *Postmetaphysical Thinking*, Cambridge, MA: The MIT Press.

——— (2002) *Religion and Rationality*, Cambridge, MA: The MIT Press.

——— (2008a) 'Notes on post-secular society', *New Perspectives Quarterly*, 25(4): 17–29.

——— (2008b) *Between Naturalism and Religion*, Cambridge, MA: Polity Press.

Habermas, J. (2010) *An Awareness of What Is Missing: faith and reason in a post-secular age*, with Brieskorn, N., Reder, M., Ricken, F. and J. Schmidt, English edition, translated by Ciaran Cronin, Cambridge, MA: Polity Press.

Heidegger, M. (1969) *The Essence of Reasons*, Evanston, IL: Northwestern University Press.

Kompridis, N. (2006) *Critique and Disclosure*, Cambridge, MA: The MIT Press.

Maclure, J. and C. Taylor (2011) *Secularism and Freedom of Conscience*, Cambridge, MA: Harvard University Press.

Martin, B. (1992) *Matrix and Line*, Albany, NY: SUNY Press.

McLennan, G. (2011) 'Spaces of postsecularism'. In Beaumont, J. and C. Baker (eds.) *Postsecular Cities: space, theory and practice*, London: Continuum, pp. 41–61.

Meyer, W. J. (1995) 'Private faith or public religion? An assessment of Habermas' changing view of religion', *The Journal of Religion*, 71(3): 371–91.
Milbank, J. (1992a) 'Problematizing the secular: the post-postmodern agenda'. In Berry, P. and A. Wernick (eds.) *The Shadow of Spirit: postmodernism and religion*, London: Routledge, pp. 30–44.
——— (1992b) 'The end of Enlightenment: postmodern or postsecular'. In Geffré, C. and J.-P. Jossua (eds.) *The Debate on Modernity*, London: SCM Press, pp. 39–48.
Modood, T. (2010) 'Moderate secularism, religion as identity and respect for religion', *The Political Quarterly*, 81(1): 4–14.
Morris, P. D. (1970) *Metropolis: Christian presence and responsibility*, Notre Dame, IN: Fides Publishers.
Parmaksız, U. (2018) 'Making sense of the postsecular', *European Journal of Social Theory*, 21(1): 98–116.
Peursen, C. (1989) 'Towards a post-secular era', *The Ecumenical Review*, 41(1): 36–40.
Rothberg, D. J. (1986) 'Rationality and religion in Habermas' recent work: some remarks on the relation between critical theory and the phenomenology of religion', *Philosophy & Social Criticism*, 11(3): 221–43.
Taylor, C. (1995) *Philosophical Arguments*, Cambridge, MA: Harvard University Press.
Weber, M. and T. Parsons (2010) *The Protestant Ethic and the Spirit of Capitalism*, London: Routledge.
Welch, M. R., Sikkink, D. and M. T. Loveland (2007) 'The radius of trust: religion, social embeddedness and trust in strangers', *Social Forces*, 86(1): 23–46.
World Values Survey Association (2009) *World Values Survey 1981–2008*, official aggregate v. 20090901, Madrid: ASEP/JDS.
Wrathall, M. (1999) 'Heidegger and truth as correspondence', *International Journal of Philosophical Studies*, 7(1): 69–88.

24

Christianity and the Indian diaspora

Robbie B. H. Goh

Introduction: Christianity as 'abjected' religion in India

Christianity forms a rather small part of the religious composition of India, with just over 2% of the population being Christian, whereas more than 80% is Hindu, and just over 13% is Muslim; other minority religions include Sikhism (just under 2%), and Buddhism and Jainism each with less than 1% share of the population (Census of India 'Religious Compositions'). Religious politics in India largely centre on the relationship between the two main religions of Hinduism and Islam. The 1947 Partition was a political settlement intended to create separate homelands for these two religious communities, although resettlement was only partially successful, and intercommunities tension continues in India to the present day (Alam 1993; Krishna 2005; Fernandes 2007).

Among the small minority religions, Christianity stands out for a number of reasons. Firstly, it is a religion which is associated with socio-economic marginals in India, including the Dalits and Scheduled Castes, and the tribal peoples outside of the Indo-Aryan majority. It is also the only religion in India in which women outnumber men (albeit by a small margin); men form close to 52% of the population of India overall, and likewise around 52% of the Hindu and Muslim populations (and 53% of the Sikh and 51% of the Buddhist and Jain populations), but men form just under 50% of the Indian Christian population (Census of India 'Religious Compositions'). Despite its long history in India, particularly among the 'Thomas' Christians in Kerala state, Christianity is often viewed with suspicion as a 'foreign' religion associated with evangelicals outside of India who are intent on proselytizing Indians to Christianity (Goh 2018: 38–43). While Indian Christians for the large part do not experience the same kind of consistent and violent persecution faced by their counterparts in (for example) Pakistan and Bangladesh, they do experience periodic episodes of communal violence, particularly the isolated Christian communities in rural parts of North India. They also experience what might be called (following Julia Kristeva) a socio-psychological 'abjection', an endemic discrimination which perpetually associates them with socio-economic abjects, or cast-offs, and limits some of their more overt religious expression, such as evangelism (for those who identify as evangelical Christians) and large-scale rallies especially featuring foreign evangelical speakers (Kristeva 1982; Goh 2018).[1]

A significant factor in the abjection of Christianity must be the role of 'Hindutva' politics in India. In the face of the increasing tensions between Hindus and Muslims in pre-Partition India from the 1920s onwards, Hindutva ('Hindu-ness') became mobilized by Hindu propagandists to 'appear to include—yet subordinate—[religious] minorities' within an India defined as effectively a Hindu nation (Pandey 1993: 245; Copland 2007: 263–5). This has meant, among other things, the teaching in schools run by the Rashtriya Swayamsevak Sangh (RSS) and Bharatiya Janata Party (BJP) of an 'extremely *virulent communal* view of Indian history', and the promotion of Hinduism at the expense of the other religions in India (Hasan 2007: 237–8; Mukherjee et al. 2008: 21). For Christians in India, the rise of Hindutva politics has meant a greater wariness with regard to evangelical activities, or even the appearance of evangelism; Christians have been arrested for allegedly converting Hindus to Christianity, and there have also been reports of Hindu hardliners forcibly 'reconverting' Indian Christians back to Hinduism (Burke 2015; Ghatwai 2017).

Diaspora, secularizing forces, and religious-cultural maintenance

In general (and as might be expected), moving out of India leads to a weakening of ties and cultural markers with the Indian homeland, with the weakening intensifying under certain conditions. Such conditions could be the greater degree of assimilation and settlement in the host country, with the greater socio-economic success that fosters attachment to the host country, and with successive generations living in the host country. The cultural loss can take the form of weakening of competence in and use of vernacular languages, negative attitudes to India and a lack of desire to visit or return, the weakening of racial/ religious/ caste endogamy, or the loss of adherence to religious practices and beliefs (Vertovec 2000; Bhatia 2001; Siddique 2004; Hiralal 2015). The weakening of cultural-religious ties with India—manifested, among other things, in a loss of competence in Hindi, and an increasing lack of interest in and knowledge of Hinduism—is acknowledged among its diaspora and is a cause for concern for both the Indian government and the Hindu religious organizations. The aggressive work of Hindu, Sikh, and Muslim religious organizations among the Indian diasporic communities indicates their awareness of the cultural-religious loss inherent in migration out of India and long-term settlement in host countries, particularly those (for example, the USA, UK, Canada, Trinidad) where secularizing tendencies and liberal individualistic values dominate society (Vertovec 2000; Mehta 2004; Wilke 2006; Madsen and Nielsen 2014).

Indian Christianity offers a particularly interesting case here. Unlike Indian-origin religions like Hinduism and Sikhism, Christianity has a stronger influence outside of India than inside it. While Hinduism and Sikhism are by no means strongly cephalic religions, nevertheless their main spiritual and cultural influences are based in India, so that migration out of India is usually also a movement (to varying extent and dependent on other factors as well) away from the main spiritual influences of their respective religions. The situation is somewhat different for Indian Muslims who migrate: there are a number of host countries where religious influences (albeit with some differences compared to Islam in India) are as strong, if not stronger than, in their home country. This is true not only of the Gulf Cooperation Council (GCC) nations to which many Indians move for employment and higher salaries, and where obviously Islam is the official religion and dominant social influence, but also other predominantly Muslim countries like Indonesia and Malaysia, and even other countries (for example, in the UK and Western Europe) where by many accounts localized support systems for Muslim migrants thrive (Jenkins 2007). One of the respondents in my study was 'Josiah' (a pseudonym), a teenage boy from a Telugu-speaking background, whose family was from Andhra Pradesh but had moved to Muscat, Oman, when 'Josiah's' father found work there. 'Josiah' (who was Christian, like his family) attended the largest Indian school in Muscat, where there were about equal numbers of Indian Christian and Indian Hindu students, with a minority of Indian Muslims. 'Josiah' reported to me that, in the Muslim-dominant culture of Oman, the minority Indian Muslim students in his school were emboldened to evangelize to the non-Muslim students. 'Josiah' himself had been the recipient of some of this aggressive Muslim evangelical attention, and he reported to me that a number of his fellow students from Christian, Hindu, and even Jain backgrounds had converted to Islam as a result. These examples can also be considered as instances of cultural-religious loss, not a succumbing to secularism, but to the influences of an alternative religion that very likely would not have had the same conversion influence had the subject not left his or her religious community in India.

Indian Christians face contradictory influences in the diaspora. Many of the popular migration destinations for Indians—including the USA, UK, Canada, and Australia, and, to a lesser extent, Asian destinations like Hong Kong and Singapore—are also places which might, to varying degrees, be called Christian 'hubs' (Goh 2016). These are either (in varying combinations of these factors) Christian-legacy countries, where Christianity maintains a general social influence although not the dominance it historically enjoyed; where Christianity flourishes under a relatively tolerant political regime and enjoys some measure of aspirational status because of associated educational, cultural, or economic capital; or where a powerful Christian cultural industry in the form of large and popular megachurches, Christian artistes, Christian publishing, or other media can be found. There are, of course, Christian hubs (for example, South Korea) which are not highly popular destinations for Indian migrants.

As with Islam in the Muslim-dominated societies, so likewise Christianity in many of these host societies exerts a degree of cultural-religious influence. Sinha (2006: 99) points out that 'Indian Christians are a small minority in Hindu-dominated India, but form a part of the majority population in the West, a situation quite unlike that of other Indian religious groups overseas'. My qualitative research indicates that this 'majority' Christian influence in countries like the USA, UK, Canada, and Australia does exert a form of reinforcement of religious identities among Indian Christian residents there, although this influence is manifested in rather complex ways. Rather than a simple augmentation of a singular Christian identity—a direct contrast to the abjected position many Christians experience in India—Christian majority culture exerts contrary push-pull pressures on Indian diasporic Christians, depending on cultural parameters such as mainline denominational networks, megachurches, Christian schools, Christian social circles, as they intersect with socio-economic factors like age, extent of settlement in host country, gender, and education. At the same time, such Christian-legacy countries also exert a secularizing influence in a number of ways: younger, especially third-generation onwards, Indian Christians face pressure from non-Indian and non- (or nominal-) Christian peer groups towards behaviour such as excessive drinking, drugs, and sexual promiscuity—behaviour which directly contravenes the conservative Christian values of their families. There is also peer pressure towards eclectic social relations, including close friendships with non-Indians and non-(or nominal) Christians. These close social relations also form a basis for behaviour such as decreasing church attendance, delinking from church social circles which are usually exclusively Indian, and, in some cases, dating and even marrying outside of the Indian Christian community. The social influences of Christian-legacy countries can be more effectively secularizing because of the pervasiveness of a nominal Christianity whose apparent religious affiliation may conceal liberal values, and secular (non-evangelical, relaxed in terms of observances) tendencies are quite opposed to the conservative Christianity of many Indian Christian migrant families. Several of my respondents, especially those in their late teens and early 20s, and living in the USA, Canada, the UK, and Australia, reported such instances of peer pressure, and even older respondents expressed anxieties at cultural influences that were perceived as inducing their children to depart from the conservative Christian values of their families and communities.

Communal Indian Christianity: insulating factors

Secularizing and liberalizing cultural influences on Indian Christian settlers, even in Christian-legacy host countries are qualified primarily due to two reasons: the strong communal and vernacular nature of Christianity in India, and the relatively limited role of Pentecostalism in India.

Indian Christianity is generally characterized by a strongly communal organization, whose main parameter is a regional identity defined by language or dialect spoken, and whose secondary parameters include caste and socio-economic markers. This is especially true of the small rural Christian communities largely composed of agrarian working-class families. As is generally the case with Christianity in India, in these rural Christian communities it is the Dalits, backward castes, and tribal people, who dominate, since Forward Caste Hindus have historically been less willing to convert to Christianity and thus abandon their caste privileges. Usually lacking the education and other capital that would confer any real social and geographical mobility, such small Indian Christian communities perpetuate an identity which is very much bound by language, class, and location. Migration, when it does occur, is usually the short-term labour migration practised by the young men of this community, who go to work typically in GCC countries (but also to places like Singapore, Hong Kong, and South Korea where there are shortages of unskilled labourers); there they live generally isolated lives among other young men from the same state and speaking the same language (e.g., Telugu among those from Andhra Pradesh). Without any possibility of permanent settlement in the host country, such migrants generally plan on returning to their village and Christian community in India. If they maintain their religious practices in the host country, it will be in their communal group, largely homogeneous in terms of language, class, and caste.

Even among more highly skilled migrants who move with their families with the option and intention of staying in the host country, the communal clustering is strong. In large cities with a relatively large Indian community overall (such as Los Angeles and Chicago in the USA, or Vancouver and Toronto in Canada) and a corresponding Indian Christian community, it is still the norm for Indian Christians to form small churches based on language and regional origin, and to measure their congregations in terms of the number of 'families' it consists of, rather than the number of individuals, which is the more common practice among churches in the USA and elsewhere (Goh 2018: 64–66). Liturgies (prayer, worship, and preaching or teaching) are reliant on the vernacular language, which makes entry into the communal church difficult, if not impossible, by someone who does not speak that language; evangelism and growth of the church are usually targeted specifically at Indian migrants in the area who belong to that linguistic-regional community. Caste and socio-economic distinctions also apply in Indian churches in the diaspora: one of my respondents, David Dorapalli, was born in Andhra Pradesh but spent a significant part of his life in

Chennai, before coming to study at the Chicago Theological Seminary where I interviewed him. Fluent not just in the Telugu of his birth but also in Tamil and English, Dorapalli expressed a liberal and progressive mind-set that was opposed to Indian Christianity's caste organization. Yet he admitted to me that the church he chose to attend in Chicago—the CSI (Church of South India) Wesley Church Chicago (Telugu congregation)—was a church that was quite conscious of caste distinctions. He explains this seeming contradiction between his avowed principles and the church of his choosing, by his desire to reach out (including with religious influence) to non-Christian Indians while in Chicago, which for him means other Telugu-speakers. The evangelical imperative, for Dorapalli, necessitated a home church that was Telugu-speaking; evangelism in this model accords with the communal and caste boundaries familiar from church organization back in India.

The limited role of Pentecostalism in India must be understood, firstly, in contradistinction to its expansive and even explosive growth in countries like Brazil, Chile, Nigeria, Kenya, South Korea, the Philippines, and elsewhere. While evangelical Pentecostal churches and movements can certainly be found in parts of India, Pentecostalism's much more muted position there can be attributed to a number of factors: firstly, there is the sheer small size of Christian believers overall in India, being less than 3% of the total population, largely dispersed—with the exception of a few states like Kerala, Tamil Nadu, and the small states of Northeast India—into small communities in remote areas, and generally constrained by Hindu and to a lesser extent Muslim religious activism. Secondly, even among Indian Christians in Orthodox and mainline denominational churches, there is a marked condescension and outright hostility towards Pentecostals, who are characterized as 'ostentatious', 'boorish', 'uncouth', 'uneducated', and other disparaging terms (Bauman 2015: 92). Combined with the vernacular organization of many of the Indian diasporic churches, this suspicion and disparagement of Pentecostalism means a generally conservative mind-set that was not receptive to the popular and sweeping Pentecostal movements familiar in (for example) many of the Latin and African Christian diasporic communities.

These two main factors combine with others—the social, cultural, and linguistic issues that generally inhibit most Indian migrants from fully integrating into their host society—to create barriers between the majority of Indian communal churches, and their surrounding environments including nominal Christian values, the expansive megachurches, and the extensive mainline denominations in Christian legacy countries. This is particularly true of first-generation Indian Christian migrants, older individuals for whom the communal church has become the primary or almost exclusive social network, and individuals whose educational and/ or linguistic background prevent deeper interactions with their host society. It is less true of highly educated Indians (particularly from the larger urban centres in India) who have the socio-economic and linguistic capital to fit easily into their host societies, and of younger Indian diasporic Christians particularly those in their late teens or early adulthood. For these latter groups, the structure of the typical Indian communal church is largely seen as restrictive, standing in the way of the more interracial social interactions and larger Christian religious interactions in which they are interested. The urban, cosmopolitan, and often fluently multilingual professionals probably represent the most inclusive end of the Indian diasporic Christian spectrum: several of my respondents fitted this profile—a banker working in Hong Kong who wished to be referred to as 'VS', and a professor living in Trinidad named Roydon Salick—and were married to non-Christians (Hindu women, in both cases) while negotiating religious aspects of the household. Both 'VS' and Roydon assiduously maintained their own Christian identity and regular practices; 'VS' and his wife agreed that their son would be exposed to both Hinduism and Christianity and make his own choice, while Roydon and his wife agreed to have their two children brought up as Christians.

Younger diasporic Indian Christians had their own difficulties with the Indian communal church model, although among my respondents this did not manifest itself in the kind of religious openness that tended to occur in the cosmopolitan professional class. The young respondents I spoke to—those in their late teens and early 20s—typically did not share their parents' and grandparents' attachment to their communal church. Most found it too small and confining, and struggled both with the vernacular language in which the liturgy was conducted, as well as with the conservatism of liturgy, theology, and values. Some were concerned with their parents' intentions to matchmake them with a potential partner from within the small church circle. Having been born, or spent most of their formative years, in the liberal society of these Christian-legacy countries, my young respondents did not share their elders' alignment of race, vernacular identity, and religion; more than one respondent reported that they 'do not see Christianity in racial terms', much less vernacular or caste ones.

These factors were in some cases sufficient to push young Indian Christians out of their parents' communal churches. However, the enduring influence of those communal churches, even in young Indian Christians who expressed their discontentment, can be seen in the fact that many remained in their parents' churches for the sake of unity with their families and so as not to upset their parents. Almost all the young respondents who expressed dissatisfaction with their communal churches insisted on anonymity to avoid any possibility of appearing disrespectful to their parents and their church community. Only a few respondents had actually left their parents' communal churches: of these, none had lapsed from their Christian faith, although a number had struggled with periods of spiritual 'dryness' and crises of faith, and had instead moved to other churches. The overwhelming majority moved to what they called 'international churches', large churches with evangelical-Pentecostal leanings, contemporary worship services featuring bands with electronic instruments, and with multiracial (although usually white-dominated) congregations. In other words, young respondents who left their parents' communal churches, chose churches almost diametrically opposed to that of their parents' small, conservative, vernacular, and racially homogeneous churches. While not many of my respondents could make the break with their parents' churches, it is, of course, significant that those who did, did so not by abandoning Christianity but by making a move that was on the one hand a reinforcement of their Christian faith, and on the other hand a marked break from their specific identity as Indian Christians.

Religious-cultural reinforcement and/as postsecularity in the Indian Christian diaspora

It can thus be seen that, with few exceptions, the insulating and self-reinforcing nature of Indian Christianity in the diaspora—the close association of a conservative Christianity and a vernacular Indian identity—serves quite effectively to rebuff the forces of secular and liberalizing influences, and indeed even the modernization of religious rituals and observances to keep up with the times. Instead of a rapid dilution of Indian Christianity in the diaspora, the conservative and communal nature of this Christianity has been manifested and preserved through a number of strategies:

1 Religious endogamy
2 Adaptive Pentecostalism
3 Transnational communalism

Religious endogamy: One of the most striking findings of this project was the extent to which religious endogamy was preserved amongst Indian Christians in the diaspora. Apart from a small number of well-educated cosmopolitan individuals, all other respondents had married Christian spouses. Even those who had married outside their race (as in the case of a Gujarati woman living in England who had married a white British man) or outside their communal group (as with a Punjabi woman and a Tamil man married in Malaysia), both spouses were Christian at the point of marriage. This was true across the different communal groups, and across the different host countries (not merely the Christian-legacy Western ones) in which my respondents had settled. Religious endogamy was thus an even stronger operative force than vernacular, caste, or even ethnic endogamy.

Women respondents tended to be more insistent on religious endogamy than men, perhaps in recognition of the fact that the husband would wield authority in the typical Indian household, and a non-Christian husband would make the wife's religious practice difficult if not impossible. Evangelical-Pentecostal individuals were, not unexpectedly, the most insistent on religious endogamy, not just to another Christian, but in particular to another evangelical Pentecostal—what a number of respondents called 'knowing Jesus in the heart', that is, a committed and confessional Christian faith, not merely a nominal one. One female respondent—a Malayalee professional who had been brought up in the Kerala Orthodox church, but who had personally become converted to Pentecostalism while studying outside of Kerala—had struggled against her family's insistence on matchmaking her to another Orthodox Malayalee, and insisted on a spouse who was not just a nominal or legacy Christian, but one with a personal and confessional form of Christianity. The respondent finally got her way, marrying a fellow professional from the Church of South India (CSI) who satisfied her religious criterion, even though her church community maintained a general prejudice against CSI Christians who were believed to be backward-caste converts to Christianity. Another female respondent, originally from Andhra Pradesh but who was living in Oman when she reached marriageable age, held out against efforts to matchmake her to a fellow Christian from Andhra Pradesh for fear that she would end up married to a nominal Christian. She conducted her own search for a marriage partner, including long-distance via email, and finally found a partner (from the Christian community in Andhra Pradesh) who fulfilled her criterion of 'knowing Jesus in the heart'.

Religious endogamy is in part ensured by the cohesiveness of the Indian communal church in the diaspora. Since for the majority of Indian diasporic Christians the communal church—homogeneous in religion, regional origin, language, and often caste as well—is the main social-religious community, this ensures that marriage is usually not just to another Christian, but one from the same vernacular group. In the case of transnational marriages, this too is often fostered by communal Christian links, thus again ensuring both religious and vernacular endogamy.

However, even among those who had their dissatisfactions with the Indian communal church and its social ties, religious endogamy was a strong imperative. Thus, the younger Indian Christians that I spoke to, although rejecting the racial and vernacular homogeneity and the theological-liturgical conservatism of their parents' communal churches, uniformly expressed the conviction that their marriage partners had to be Christian. While most of these respondents were 'race-blind' in their attitude to Christianity and the kind of church they would like to be in, most seemed to recognize that marrying a non-Indian would be more of a stumbling block for their parents and community. However, they were certainly open to relationships with Indian Christians outside of their vernacular group and communal church, and several of my respondents had actually married Christians from other vernacular groups. The one non-negotiable criterion, for practically all of these respondents, was that their prospective spouses had to be Christians with a personal faith conviction rather than a nominal or legacy form of Christianity.

Adaptive Pentecostalism: A number of Indian communal churches in Christian-legacy countries, while in basic structure the antithesis of large Pentecostal churches, have nevertheless been able to adapt some features of Pentecostalism. These include contemporary worship bands featuring electronic instruments (and singing both vernacular songs and the popular English songs of the contemporary Christian media industry), and the work of supernatural healing and signs (such as revealing knowledge of the supplicant's inner emotional state or personal history, and offering redress or healing through prayer). These adapted forms of Pentecostalism can thus sit (with varying degrees of comfort) within the more common Indian communal church model: the use of the vernacular is preserved, although supplemented by the use of English or Hindi in some of the more popular worship songs. The use of Pentecostalism's supernatural ministry can be tailored to the acceptance level of particular communal churches, and can even be tailored to evangelical outreach to the surrounding (non-Christian) Indian community, relying on the generally greater receptivity to the offer of prayer for healing and personal problems as compared to outright evangelical overtures.

Other types of adaptive Pentecostalism include a 'private' or 'sub-group' Pentecostalism, in which an individual continues within a communal church structure while maintaining evangelical-Pentecostal beliefs and practices, often in a small coterie of other similar individuals in the church. This strategy is employed partly because Pentecostals may not have the option of joining an overtly Pentecostal church (either because it is not convenient or possible to do so in the areas where these individuals are, or because of family and community reasons), and partly because such individuals hold fast to the possibility of influencing the more conservative Christians in such churches to Pentecostal beliefs and practices.

Related to point 2, one form of adaptive Pentecostalism that has had transnational implications was the use of communal Christian networking to pursue evangelical work (with underlying Pentecostal flavours) back in India. In other words, while the Indian communal church structure continued to limit evangelical work and growth in numbers in the host country, some groups had adapted strategies which mobilized transnational groups of communal Christians towards evangelism back in India. The most extensive example of this would probably be the Kerala Pentecostals in the USA and Canada, who proved adept at mobilizing financial and other resources towards evangelical and social relief missions in parts of India. Thus, while the (Keralite) communal identity was still adhered to, this was complemented by the Pentecostal affiliation, and mobilized not to reinforce small and isolated Keralite communal churches, but rather towards work in India. A similar transnational-communal network operated amongst Punjabi evangelicals, whose work focused on Punjab state in India. Other communal Christian groups affirmed transnational ties among those groups, without necessarily committing to mission and social work in India.

These practices of religious endogamy, adaptive Pentecostalism, and transnational communalism collectively serve to counteract and even reverse the secularizing and fissiparous forces faced by the Indian Christian diaspora in many host countries.

Conclusion

The interesting example of the Indian Christian diaspora cautions against a blanket notion of a global Christianity strongly and uniformly driven by Pentecostalism. At the same time, it also cautions against the intuitive notion of a weakening of cultural-religious identity in the diasporic subject, a weakening that is generally thought to intensify over generations in the host country.

In host countries where religions other than Christianity dominate the social landscape (such as Islam in the GCC countries), the evangelical influences of those religions do exert a conversionary force on diasporic Indian Christians. While this is not a secularizing force, from the point of view of the Christian diasporic subject, it works in the same manner to weaken originary religious identity and observances. In Christian legacy countries, the social influence is more complex. On the one hand there is a wider Christian network than in India, together with more freedom of religious practice than is available for many subjects in India. On the other hand, Christian legacy societies are also the ones where secularizing influences are the strongest. Even the Christian networks pose a challenge to the majority of diasporic Indian Christians, whose religious practice is closely tied to conservative practices in small vernacular churches. To embrace and integrate into non-Indian Christian networks in such countries is to be potentially open to influences of liberal and nominal Christianity, or Pentecostal forms of worship and practices, which would mean the loss of the characteristic religious identity of the Indian Christians.

In responding to these forces, Indian Christianity in the diaspora offers an interesting example of postsecularity, in which the characteristic features of tight communal and familial ties, religious endogamy, and intracommunal transnational links are mobilized to reinforce religious identity. Unable (for cultural, linguistic, and other reasons) to plug directly into the wider Christian environment of Christian-legacy host countries, such Indian Christian diasporic communities are nevertheless able to adapt selective Pentecostal elements to vitalize communal Christian links and capitalize on communal links to strengthen Pentecostal methods and actions, including transnationally. In this sense, while in many ways different from other Asian Christian diasporas such as the Chinese or Korean ones, the Indian Christian diaspora does point to a model of what might broadly be called ethnic Christian postsecularity in the diaspora, where small vernacular Christian groups are generally isolated from the surrounding host society, and thus preserve their cultural and religious forms against the forces of change posed by other religious influences, secularizing modernity, or even other forms of Christianity more liberal or markedly Pentecostal than the common Indian model. This ethnic Christianity, while by no means monolithic or unchanging, is a form of religious preservation against the dilution or modification of religious identity and practice, even as it has to come to terms with the different conditions of being Christian in the Indian diaspora as opposed to being Christian in India.

Note

1 From 2008 to 2010, I conducted fieldwork in parts of India (primarily Uttar Pradesh, Bihar, and around Nagaland), also in 10 other countries and regions with significant Indian diasporic populations (namely Singapore, Hong Kong, Australia, Qatar, Oman, Switzerland, the UK, Trinidad and Tobago, Canada, and the USA). That fieldwork has gone into the book *Protestant Christianity in the Indian Diaspora* (Goh 2018). The present paper utilizes some of the fieldwork but necessarily summarizes and focuses on its implications for thinking about postsecularity. A more nuanced and comprehensive account of Christianity in the Indian Diaspora can be found in my book.

Further reading

Bergunder, M. (2008) *The South Indian Pentecostal Movement in the Twentieth Century*, Grand Rapids, MI: Wm. B. Eerdmans Publishing.

There has not been a great deal of work on Pentecostalism in India, and Bergunder's book is a noteworthy exception. His account is particularly useful in understanding the role of Pentecostalism amongst the Kerala diaspora, and to a certain extent the Telugu-speaking and Tamil-speaking ones as well.

Jacobsen, K. A. and S. J. Raj (ed.) (2008) *South Asian Christian Diaspora: invisible diaspora in Europe and North America*, Farnham: Ashgate.

A collection of essays on different Christian communities (of different denominations) in various parts of Europe and North America. There is some good fieldwork and qualitative interview data in the essays.

Levitt, P. (2007) *God Needs No Passport: immigrants and the changing American religious landscape*, New York, NY: The New Press.

Good account of the role of religion in the immigrant experience, including the significance of religious social networks.

References

Alam, J. (1993) 'The Majlis-E-Ittehad-Ul-Muslimeen and the Muslims of Hyderabad'. In Pandey, G. (ed.) *Hindus and Others: the question of identity in India today*, New Delhi: Viking, pp. 146–76.

Bauman, C. M. (2015) *Pentecostals, Proselytization, and Anti-Christian Violence in Contemporary India*, Oxford: Oxford University Press.

Bhatia, T. K. (2001) 'Media, identity, and diaspora: Indians abroad', *Diaspora, Identity, and Language Communities* (*Studies in the Linguistic Science*), 31(1/Spring): 269–87.

Burke, J. (2015) 'India investigates reports of mass "reconversion" of Christians', *The Guardian*, 29 January, accessed online 11-08-2017.

Census of India "Religious Compositions", see www.censusindia.gov.in, accessed 19-01-2010.

Copland, I. (2007) 'Crucibles of Hindutva? V. D. Svarkar, the Hindu Mahasabha, and the Indian princely states'. In McGuire, J. and I. Copeland (eds.) *Hindu Nationalism and Governance*, New Delhi: Oxford University Press, pp. 253–82.

Ghatwai, M. (2017) 'Madhya Pradesh: two held for taking 11 tribal kids for "religious conversion"', *Indian Express*, 23 May, accessed online 11-08-2017.

Goh, R. B. H. (2016) 'Christian capital: Singapore, evangelical flows and religious Hubs', *Asian Studies Review*, 40(2): 250–67.

——— (2018) *Protestant Christianity in the Indian Diaspora: abjected identities, evangelical relations, and Pentecostal visions*, Albany, NY: State University of New York Press.

Hasan, M. (2007) 'The BJP's intellectual agenda: textbooks and imagined history'. In McGuire, J. and I. Copeland (eds.) *Hindu Nationalism and Governance*, New Delhi: Oxford University Press, pp. 226–52.

Hiralal, K. (2015) 'Changing caste identities in the Indian diaspora: a South African perspective'. In Kumar, P. P. (ed.) *Indian Diaspora: socio-cultural and religious worlds*, Leiden: Brill, pp. 158–76.

Jenkins, P. (2007) *God's Continent: Christianity, Islam, and Europe's religious crisis*, New York, NY: Oxford University Press.

Kristeva, J. (1982) *Powers of Horror: an essay on abjection*, New York, NY: Columbia University Press.

Madsen, S. T. and K. B. Nielson (2014) 'Hindutva and its discontents in Denmark'. In Gallo, E. (ed.) *Migration and Religion in Europe: comparative perspectives on South Asian experiences*, London/New York, NY: Routledge, pp. 155–70.

Mehta, S. R. (2004) 'The uneven "inclusion" of Indian immigrants in Mauritius'. In Jayaram, N. (ed.) *The Indian Diaspora: dynamics of migration*, New Delhi: Sage Publications, pp. 186–99.

Mukherjee, A., Mukherjee, M. and S. Mahajan (2008) *RSS, School Texts and the Murder of Mahatma Gandhi: the Hindu communal project*, New Delhi: Sage Publications.

Pandey, G. (1993) 'Which of us are Hindus?' In Pandey, G. (ed.) *Hindus and Others: the question of identity in India today*, New Delhi: Viking, pp. 238–72.

Siddique, C. M. (2004) 'On migrating to Canada: the first generation Indian and Pakistani families in the process of change'. In Jayaram, N. (ed.) *The Indian Diaspora: dynamics of migration*, New Delhi: Sage Publications, pp. 78–102.

Sinha, V. (2006) 'Religious traditions in the diaspora'. In Lal, B. V. (ed.) *Encyclopedia of the Indian Diaspora,* Singapore: Editions Didier Millet/National University of Singapore, pp. 94–101.
Vertovec, S. (2000) *The Hindu Diaspora: comparative patterns*, London: Routledge.
Wilke, A. (2006) 'Tamil Hindu temple life in Germany: competing and complementary modes in reproducing cultural identity, globalized ethnicity and expansion of religious markets'. In Kumar, P. P. (ed.) *Religious Pluralism in the Diaspora*, Leiden: Brill, pp. 235–68.

25

Resisting the transcendent?

Christopher Baker

Introduction

The idea of the postsecular, and subsequent developments of the concept of postsecularity, have struggled to gain critical purchase and acceptance in critical urban and human geography. This places it at odds with other disciplines where there is a wider acceptance of the term—for example, religious studies, political philosophy, anthropology, cultural studies, and international studies (Baker and Dinham 2017). However, critical urban and human geography, or critical urban theory for short (henceforth CUT),[1] has historically shared a greater scepticism of the term along with sociology and sociology of religion, with which it seems to share a number of misgivings.

First, there appears to be an ongoing suspicion, derived from Marxist and sociological materialist analysis, that religion and belief are largely irrelevant social and political actors in the public sphere, and that even if they are not, there is little particularly good that can come from the impacts that they create (Lancione 2014; Hjelm 2015). Second, even those who recognize the increased salience of religion and belief in the current modernity, appear suspicious that the 'postsecular project' is an attempt to generate a narrative of the twenty first century whereby religion somehow replaces the secular in a linear or teleological fashion (Beckford 2012; Martin 2016). Third, there is a line of thought that interest in religion and belief is driven by policy panics over cohesion and integration on the one hand, and radical and violent extremism on the other. The degree of interest in the supposed power of religion and belief is therefore not empirically matched by the policy hype surrounding it (Bruce 2011; Dinham and Francis 2015). Notwithstanding these and other critiques, part of the potential challenge of the postsecular concept is that it unambiguously challenges many disciplines to unlearn their presuppositions and working assumptions. There is natural resistance to this challenge on the grounds of both time and intellectual credibility. I will nuance some of these stances by means of my thesis that proposes that undergirding, or perhaps running parallel to this *normative* ontology of scientific and empiricist certitude within CUT, there is an *operant* ontology of the sacred, the mystical, and the enchanted buried deep within the traditions of this discipline.

In this scenario, I am indebted to the work of public theologian Helen Cameron who, in her work with several others (Cameron et al. 2010), comes up with four categories of interrelated knowledge: the operant, the formal, the normative, and the espoused. This framework traces the complex and nuanced ways in which practical knowledge and applied wisdom are held in a creative tension with the authoritative and normative prescriptions of a particular discipline, ideology, or theoretical construct. Within Cameron's exemplar, theology, the boundaries of acceptable or recognized knowledge and behaviour are laid out in the formal and normative theology categories. Formal theology refers to 'the theology of the academic theologians and the dialogue with other disciplines', while normative theology describes 'the theology the group names as authoritative and will allow to challenge its operant and espoused theologies' (*ibid.*: 13). The operant and espoused categories refer to the practical and nuanced ways in which these formal and normative sets of propositional beliefs get 'cashed out' in the day-to-day public sphere. Operant refers to 'the theology embedded in the actual practices of a group (what we do)', whilst espoused is 'the theology embedded in a group's articulation of its beliefs (what we say we do)'. Espoused theology, based on reflection of actual praxis, can be fairly congruent with normative and academic assumptions of the discipline, but it can, of course, be deeply subversive and/ or incongruent with them.

I now delineate this relationship between formal/normative and operant/espoused forms of knowledge with a fourfold, largely chronological, schemata outlining the engagement by CUT with ideas and practices of religion and belief (as proxy terms for deep ontologies of transcendence) as indicative of what we might broadly call a 'turn' towards postsecularity.

In the beginning was absolute space—Lefebvre's ontological category of space production

The Production of Space (Lefebvre 1991) was a landmark volume that came towards the end of Lefebvre's long career as a Marxist philosopher and sociologist. Among other things, he had been a taxi driver and fought for the French resistance in World War II. According to David Harvey (2000: 100), at the heart of his intellectual project was an attempt to infuse material Marxist philosophy with art, literature, and poetry, as well as a deep appreciation for the sociological encounters of everyday life encountered in both rural and urban situations. Harvey defines this plea for recognizing the emancipatory value of the body and human creativity a subversion of 'the mechanistic view... the Cartesian/ Newtonian conception of [how] spacetime [is] produced'. This lived and vibrant quality of space, was for Lefebvre, the key to developing a new critical theory (updating Engels' Cottonopolis theory of the industrial city, for example) of how space was produced, and was an important endeavour in showing how hegemonic, bourgeois ideologies associated with capitalism constrain the political and economic agency of the proletariat.

Lefebvre's new critical urban theory was constructed around the fluid interplay of three different ways of looking at how urban space is produced: perceived space (*l'espace percu*), conceived space (*l'espace concu*), and lived space (*l'espace vecu*). Broadly conceived, lived space reflects the tactics for everyday living deployed by the working class whereby social and working relations are reproduced. Conceived space refers to prescriptive vision of the city laid out by planners and capitalists that is designed to disrupt and thwart the ability of the working class to own and reproduce the city. Perceived space is by all accounts a tricky category, but one which according to Merrifield (2006: 111), 'mediates' between lived and conceived space because it represents the values and the meaning that are attached to patterns of movement around the city, monuments, landmarks, and so on that interconnect people with places and create new patterns of interaction. Despite, or perhaps because of, its creative imprecision, this typology of space production has had a huge impact on several eminent

Marxist and post-Marxist geographers and cultural commentators, including David Harvey, Soja, Jamesan, Lebas, Shields, Swyngedouw, and Brenner.

An *a priori* category of space that Lefebvre (1991) uses to critically inform his deployment of these other three categories, however, is his notion of *absolute space* (or *l'espace absolu).* These are spaces of ontological and sacred depth that derive their power from natural energy, but then get overlaid by architectural, cultural, and political readings. For a time, their preeminence as foundational spaces of meaning and power lies precisely in the fact that they combine all these elements. Thus, they are spaces of 'symbolic mediation' between human and divine power. Lefebvre's exposition is worth quoting is full because of its centrality to not only his argument, but the argument I am outlining in this chapter.

> Absolute space, religious and political in character, was the product of the binds of consanguinity, soil and language, but out of it evolved a space which was relativized and historical. Not that absolute space disappeared in the process; rather it survived as the bedrock of historical space and the basis of representational spaces (religious, magical and political symbolisms) ...[and] embodied an antagonism between full and empty.
>
> *(48)*

Within Lefebvre's (1991) typology, the power of absolute space becomes irrevocably corrupted and subsumed by material capitalism, or what he calls 'abstract' space. Abstract space emerges fully in the medieval period when towns and cities became more systematically constructed according to the logic of accumulation which he defines as 'the accumulation of ... knowledge, technology, money, precious object, works of art, symbols' (*ibid.*: 49). This accumulation, Lefebvre suggests, means that space loses its ability to reproduce itself socially. It becomes disconnected from social relationships and thus turns into an 'abstracted' commodity—as indeed does time, nature, human sexuality and sensuality, as well as meaningful public spaces. Capitalism with its capacity to distort labour and social relations assumes a subjectivity that 'dissolves and incorporates' other subjects. Yet this subjectivity is occluded; 'it appears as an impersonal pseudo-subject, the abstract 'one' of modern social space, and – hidden within it, concealed by its illusory transparency—the real subject, namely state (political) power' (*ibid.*: 51).

Lefebvre (1991) highlights how each century, from Roman and Greek Empires to the twentieth century, moves human society inexorably from 'absolute' space to 'abstract' space. Under this narrative, religion or at least Christianity, is characterized as a religion obsessed with death and darkness, with pilgrimages to crypts of dead saints a key characteristic. This 'cryptic' faith is 'decrypted' by the emergence of modernity in the form of the medieval town with its bustling energy conducted in the light and transparency of secular activity. It is in the twelfth century, in the epic intellectual battle of two theologians Bernard de Clairvaux and Peter Abelard, that the decisive *intellectual* battle for modernity is won by the latter. Yet Lefebvre is clear in his high regard for both thinkers, and sometimes appears to argue against his own historical and teleological logic with regard to abstract space, by asserting that preexisting or absolute space in fact 'underpins' 'representational' spaces. Modern knowledge forgets this arrangement at its peril as it 'falls into a trap when it makes representation of space (i.e. the conceptualised space of scientists, planners and urbanists) the basis for the study of life for in doing so it reduces lived experience' (*ibid.*: 90).

Elaborating Lefebvre's thinking on 'absolute' space reminds us that he took the historical interplay between the religious and the secular, the theological and the philosophical, with a concerted intellectual approach, and on occasions seems to suggest that that tension is still ongoing and properly unresolved. The historically contingent (yet still ongoing) relationship between 'absolute' space and 'abstract space' is a central oppositional pull that drives the argument of the book.

Yet, it is hard, if not impossible, to find reference to, or critical development of, Lefebvre's use of *absolute* space in any of the followers of the Lefebvrian tradition. Their analysis remains by and large fixated with the *abstract* and accumulative production of space. This is why, within the globalized flux of the twenty-first century, which has brought the practices of religion and belief back again into sharp relief, their analysis struggles to find the purchase it once did in explaining the modern urban city in which we now live. Gregor McLennan (2011: 29) puts his finger on this issue in an outstanding chapter for the *Postsecular Cities* collection. He reflects on David Harvey's 2008 book *The Right to the City,* a substantive reworking of an earlier volume entitled *Social Justice in the City,* written in 1973. Despite its interest in emerging new civic solidarities as expressions of new forms of social justice, Harvey makes no reference at all to the role of religion and religious social movements in either volume. As a critical Marxist, McLennan suggests that his natural response should be one of relief that Harvey's 'daring and unflinching' case for a traditional Marxist analysis of global inequality represents a moment when 'the tables have been decisively turned once again, with normal (critical-materialist) service resumed'. But McLennan is too honest to slip uncritically back into a comfortably reassuring mind-set. For him, Harvey's inability or unwillingness to include religious activism and phenomena in his analysis into the 'multiple signs of rebellion... against the burgeoning processes of creative destruction that have dispossessed the masses to any right of the city whatsoever' (*ibid.*: 29) is 'puzzling' and inaccurate. For McLennan, it's hard to see how any serious debate about what 'kind of social ties, relationship to nature, lifestyles, technologies and aesthetic values we desire', can avoid 'explicit political and philosophical discussion concerning the place of religion in public life and identity' (*ibid.*: 29).

'The echoes return slow'[2]—the emergence of geographies of re-enchantment

A counternarrative to this normative materialist Marxist and post-Marxist critique on religion and belief begins to emerge in the late 1990s. It is now gathering pace and ontological intensity, and appears willing to expand the terms under which religion and belief can be said to 'co-produce' the urban with the secular. This turn to what I am calling 're-enchanted geographies', emerges under the impetus of three strands within CUT which include the psychoanalytic revolt and the liberation of political desire and jouissance, the feminist turn towards religion and belief, and the emergence of a fully fledged ontology of secular enchantment based on the idea of 'cosmopolitics', inspired by the work of Bruno Latour and Isabelle Stengers.

The post-oedipal turn in CUT

There are a series of Marxist and post-Marxist theorists who are seeking to replenish the wellsprings of political socialism, particularly after the failure of the Paris uprising of 1968, with an 'immanentist' appeal to what in effect is a sublimated language of quasi-spiritual ecstasy. In this camp, I am thinking particularly of the work of Andy Merrifield and Mark Purcell. They do not go down the path of some of their fellow post-Marxist theorists, for example Žižek (2003), Badiou (2003), and Hardt and Negri (2000), for whom the political replenishment of the leftist cause as global movement lies in a creative engagement with religious and theological ideas, and the ideational role of the early Christian church as the bringer down of 'empire' and the establishment of a new universal 'brotherhood'.

Merrifield and Purcell's 'immanentist/psycho-analytical' rather than the 'theological' genealogy of post-Marxist theory derives inspiration from the radical reinterpretation of the Freudian psycho-analytical tradition that was so clearly linked to Marxist materialist critiques of religion and bourgeoisie morality. This re-interpretation starts with Lacan, and then is driven forward in spectacular fashion by Deleuze and Guattari in their book *Anti-Oedipus* (1980/1988). Postanalytical critical geographers are drawn to Lacan's new emphasis on the importance of desire (or *jouissance*) and a union with an external Real or Other (or at least external to the unconscious of the individual). Lacan's formulation was that 'desire is the desire of the Other', which he came to symbolize, in his complex analytical algebra, as the '*petit object a*'. The 'a' in this formulation is the first letter of *autre,* or 'other', and it represents the *cause* of the desire, rather than the end point of desire, which is the Autre, or Other, with a capital A. Referring to Lacan's tripartite system of psycho-analytical enlightenment including the Imaginary, the Symbolic, and the Real, Caldwell suggests that 'as a force beyond both the Symbolic and the Imaginary the "*objet petit a*" is the residual part of the Real that resists completion' (2009: 23).

Deleuze and Guattari are concerned to break through what they see as the hermetically sealed world of the Freudian framework, where desire is introvertedly and neurotically located within the universe of the individual analysand and often predicated on a manifestation of a 'lack' (for example, the lack of a penis). This repressed and internally driven energy instead needs releasing externally for the sake of political revolution. Deleuze and Guattari's tactic is to free the 'objet petit a' from its subordination to a lack (i.e., a lower case manifestation of an upper case Real or Other) into a primordial energy that 'transforms and is transformed through the ways in which it is organised' (Caldwell 2009: 23). In collapsing the ontology between the Imaginary, Symbolic, and Real, Lacanian-influenced geographers redirect the energy of the desire for a transcendent Real to within an immanent frame and horizon. This tactic, they believe, releases a flow of energy that creates machinic or 'real' expressions of that desire that is then enabled by other machines ad infinitum. In this way, the subject never becomes a unified object but is always in the process of fluid expression. Thus, the idea of desire and joy (or what Lacan defines as *jouissance*) as a form of transformational, externalized, and enchanted life-drive (rather than internalized death-drives) begins to populate post-Marxist critical geography.

One example of this genre would be Merrifield's (2011) book *Magical Marxism*. In an editorial based on its contents, he argues that Marxist studies move away from tired and dry categories of materialist and class-war analysis which are cognitive, self-referencing, and out of touch with the aspirations of ordinary people. As he pithily suggests, 'Over the years we have let the spirit of Descartes crush that of Rabelais' (Merrifield 2009: 385). Rather, he pleads, Marxism must move into a more affective register, where the essence of the search for social justice and economic empowerment engages with everyday life and new forms of utopian imagination, including those inspired by poetry. This passage is typical of what one might call a re-enchanted Marxist materialism.

> The emphasis on poetry is a crucial one, not least because Magical Marxism's best adherents are perhaps lyric poets, people who don't necessarily write poetry but who somehow lead poetic lives, who literally become-poets, as Deleuze might have said, who internalize powerful feelings and poetic values, spontaneous values with no holds barred. The key point here is that Marxists make life a poem, adopt a creative attitude towards

> living. Poetry, accordingly, becomes something ontological for Magical Marxists, a state of Being-in-the-world, the invention of life and the shrugging off of tyrannical forces that wield over that life.
>
> *(Merrifield 2009: 382)*

Other post-Marxist commentators like Castells and Purcell see the tumultuous events of the Arab Spring and the emergence of Occupy in 2011 as epoch-defining events that liberate the potential of this magical and enchanted materialism. In his book, *The Deep-down Delight of Democracy,* Purcell (2009), channelling the likes of Spinoza, Ranciere, Mouffe, as well as Deleuze and Guattari, talks about new forms of economic and spatial production, based on desiring. These forms of 'desiring-production' are likened to flows that escape the capitalist-state machine's attempts to control them; they are

> accelerated flows...that seek out other flows and enter into connections with them. In the best case... they will produce large aggregates of free connected desire, and those aggregates, as they grow, will begin to trace out something they (Deleuze and Guatarri) call "*a new land*".
>
> *(47)*

Purcell is clear, however, that this new land is an affective state of mind—a new political consciousness that has been liberated by the Occupy phenomenon—that he hopes will survive despite retrenchments by the state and the market to restore 'business as usual'. The new land is a metaphor, not a physical space (87). What will keep the flame alive is not dead ideology or musty tomes. Rather it is the effect of joy, of jouissance. As well as the power of ideas, Purcell says, 'we need to really *feel* what it is we desire' (*original emphasis*) (119).

The city of spirit—the feminist turn back to religion and the sacred

Much feminist critical geography has centred on ideas of cities of difference or ordinary cities (Fincher and Jacobs 1998; Robinson 2002), which normalize and diversify the ways in which urban spaces are planned, conceived, and experienced, not least by women, ethnic minorities, young people, LGBTQ, and disabled constituencies. Clara Greed (2011), however, outlines how the portrayal of religious women in urban contexts within both feminist and CUT studies has largely been ignored. There were clear overlaps between religious identities and first-wave feminism during the late nineteenth and early twentieth centuries in which women of faith were at the forefront of suffragette and social reform movements. Some of the early second-wave feminist theorists in the late 60s and early 70s were also theologians and philosophers or were religiously educated (*ibid.*: 107). But since that period, a secular, Marxist, and Enlightenment tradition within Western feminism has predominated which has foregrounded a critique of religion as patriarchal and anti-modern, and therefore deeply inimical to women's agency and empowerment. Over time, however, this model has also become perceived as too normative and privileged in favour of white, Western women, ignoring, for example, how 'third-world women' are multiply oppressed on the grounds of race, gender, and class. Crenshaw (1989) coined the concept of intersectionality to describe these multilayered experiences of oppression, domination, or discrimination, to which have now been added the categories of sexuality and disability. According to Sarah Salem (2013), religious identities have now also become part of the intersectionality debate as part of the postcolonial challenge to normative, white Western feminism.

There is now an emerging discourse, particularly from Muslim women scholars, challenging the traditional assumption that religious identities stifle both feminine agency and autonomy. They present the complex and often problematic ways Muslim women must negotiate contemporary urban pluralism within non-theocratic states. Banu Gökarıksel and Anna Secor (2014), for example, reflect on how devout Sunni Muslim women inhabit the postsecular geographies of Istanbul.

> Women navigate a gendered moral order made taut with the tension of political and cultural contestation, and in the intimate micro-geographies of offered prayer mats, rejected pastries, a visit to the house of an Alevi aunt or a refusal to visit a friend's place of worship.
>
> *(28)*

Clara Greed (2011) is sceptical that the notion of the postsecular city eases the complexity of religious women's lives and the prejudice they encounter, even though the idea of the postsecular is, at one level, pointing towards a positive re-evaluation of religion and belief in the public sphere. Greed focuses on the growing cohort of black Pentecostal women ministers and church leaders who work in diverse, global cities like London.

> They are more likely to apply their belief in the miraculous to the political situation, and to intercede for the nations and the rulers, in the "public" realm as an extension of their beliefs in physical healing, casting out demons and the operation of the gifts of the Holy Spirit in the 'personal' realm...
>
> *(ibid.: 112)*

Their willingness and ability to span traditional public and private zonings of space means these women become 'political' and 'feminist' actors, not necessarily through conscious choice, but because they 'inevitably' get 'drawn into community politics and equality battles in dealing with local problems' (112). Greed suggests that despite this important political and community work on the ground, the contribution of these women remains 'politically invisible' to central government, despite an apparent 'concern for "faith communities" within the "diversity and equality agenda"' (112).

The new inclusion of religion and belief in the intersectionality debate highlights a further shift towards the religion and belief agenda within feminist critical urban theory. Hopkins (2009) suggests a new interface between religion, belief, and feminist geographies has been generated by the 'emotional turn' in human geography (see Bondi 2005). Such an alliance, suggests Hopkins (2009: 9), promotes shared understandings based on 'intimate, personal and embodied accounts of the salience of religion to people's everyday experiences, the emotions and feelings associated with particular religious places, events and times may also be better understood'.

Meanwhile, a less private and individualized feminist critical geography is drawing attention to the ways in which particular configurations of physical and human elements combine to create certain religious and spiritual spatial affects, and indeed, effects. Leonie Sandercock (2003) proposes a counter-hegemonic practice to the paternalistic and modernist norms to which she was exposed as a trainee planner. These norms were 'designed to triumph over both politics and nature with its rational decision making and problem-solving techniques, grounded in rigorous social analysis.... which would liberate societies form ideologies superstitions, prejudices' (*ibid.*: 31). As a reaction against the sterile and unsafe public spaces that this form of planning created (see, for example, the work of Jane Jacobs), Sandercock suggests that the hitherto marginalized voice of women, and particularly minority women, came to the fore in the 80s and 90s, creating a series of alternative imaginations of the city which Sandercock characterizes as the city of memory, the city of desire, and the city of spirit (*ibid.*: 33).

The 'city of spirit' category refers to the sense of individual and communal alienation that Sandercock (2003) perceives has been produced by the 'dead' landscapes of urban overcrowding, pervasive consumerism, and mass industrialisation. As an Australian citizen, living in British Columbia, she observes that several of her compatriots 'now go in search of comfort to aboriginal songlines or Native American sacred places' (*ibid.*: 225). She is painfully aware of the Orientalist paternalism inherent in such practices (*ibid.*: 226) but seeks to expand her understanding of 'spirit' in a way that fits a renewed sense of 'Western' and immanentist re-enchantment. This is her vision of a re-enchanted urban space.

> The nourishing of the spirit or soul needs daily space and has everyday expressions: a group of students in a coffee shop discussing plans for a protest; and elderly Chinese man practising his tai-chi on the beach or in a park; amateur musicians performing in front of cafes and museums; an old woman tending her flowers in a community garden; kids skateboarding among the asphalt landscaping of sterile bank plazas; lantern parades through city streets on the winter solstice.
>
> (ibid.: *227)*

This expression of a materially inclusive if somewhat ontologically nebulous urban enchantment contrasts with the more edgy and contested response to overtly religious symbols of dress worn by women of faith (e.g., niqabs, hijabs, crucifixes). But both responses are clear examples of postsecular engagement with notions of religion, belief, and spirit from within the tradition of critical feminist thinking.

Cosmopolitics—object-centred politics and the shift towards full spectrum secular enchantment

Sandercock's 'spiritual' vision of urban space becomes more fully blown in the work of what we might call 'second generation' assemblage theorists. In this line of enquiry initiated by Isabelle Stengers and Bruno Latour, every human and non-human element that constitutes the fabric of urban life is a fully blown 'actant' that possesses an ontologically real energy and personality that profoundly shapes the material structures around us. In an article for *City* journal on the theme of 'Assemblage and Critical Urbanism', Colin McFarlane (2011) develops Deleuzian assemblage theory as a critical tool for thinking about 'a more just and ecologically sound' theory of urban transformation. The Marxist materialist analysis that simply sees cities as outcomes of excess and unjust capitalist accumulation has occluded the role of many other actors and processes that also shape the urban. Instead, McFarlane sees the city as a decentred object which is 'relentlessly being assembled at concrete sites of urban practice ... as a multiplicity of processes of becoming, affixing societal networks, hybrid collectivities and alternative topologies' (*ibid.*: 205). He researches into the informal settlements in Mumbai as a case study of this set of intertwined practices. But he is also keen to attach a greater 'potentiality' to this assemblage theory. By potentiality, McFarlane means 'both the intensity and excessiveness of the moment—the capacity of events to disrupt patterns, generate new encounters with people and objects and create new connections in ways of inhabiting everyday urban life' (*ibid.*: 209).

Blok and Farias (2016: 5) take McFarlane's assemblage theory further up the ontological dial with reference to what they refer to as an 'object-centred politics' (OCP) which becomes the operating framework for their enhanced model. This OCP, also designated as 'ontological politics', has, I detect, four interrelated dimensions associated with it: the *notion of radical co-presence* that highlights a ontologically significant relationship between all the actants that combine to create new common urban assemblages; the new knowledge and experience of this radical co-presence often generates an *excess or surplus of meaning*; this surplus of meaning in turn generates *new political-ethical subjectivities* based on a common understanding of the singularity of moral intent and ontological depth that lies within the urban; this understanding in turn leads to the creation *of new political imaginations and practices of urban and civic engagement.*

This ontological and radical co-presence of multiple objects highlights the innate (or virtual) set of possibilities that are generated by these new and ever-fluxing assemblages. A surplus of knowledge and affect is generated by these ontological co-presences that impel us to make these potential realities more visible (Blok and Farias 2016: 5). These commentators are clear that this is not only a critical research task (*ibid.*: 5) but also a politico-ethical one. The search for an underlying unity (or commons) to the impact created by both human and non-human actants requires a visualization of hitherto new possibilities for creating the co-existence of these human and non-human actants.

This is the implicit moral task lying at the heart of *Cosmopolitics*. Following Latour's extrapolation of a *Dingpolitics* (or politics of things) (Blok and Farias 2016: 7), Blok and Farias suggest that the objects that co-construct our urban assemblages are not 'objects' in the standard sense of the word; rather, they are relationally intended. In shaping our 'shared, common public matters' they, therefore, come 'loaded with moral and political capacities' (*ibid.*: 7).The key question that emerges from this ontological worldview is, 'not first and foremost for *whom* these [enactments] function, but rather *how* shared urban realities are made and remade' (*ibid.*: 7).

In other words, OCP involves a radical *decentring* of our assumptions about how material reality is produced. It requires the formation of an inductive, but politically astute, sensibility; one might even say, spirituality. An OCP radically decentres the human subject. 'It is important to stress', say Blok and Farias, that 'this ontological multiplicity does not just point to the different furniture of human worlds, but to different ways of "being human," of assembling and enacting humanity' (Blok and Farias 2016: 7). Blok and Farias never directly allude to the spiritual quality of their work, but their call for a new political-ethical subjectivity derived from an ontological politics could potentially 'cross-over' into religious, spiritual, and no-religious sites of engagement: 'a politics of exploring and provisionally settling what does and does not belong to our common (urbanised) worlds' (Blok and Farias 2016: 7).

Blok and Farias (2016: 33) commission multiple case studies to flesh out their idea of OCP, ranging from austerity-economy tactics for sharing train tickets between strangers which are derived from excesses of 'social energy'. They also disrupt traditional transactions of neo-liberal economics, through to the truly apocalyptic disruption of urban systems by actants derived from water and air in the form of hurricanes and tsunamis. Put succinctly, the common thread linking these examples of OCP lies in the way they all reflect the fundamental politico-ethical question 'of how we can live together in ways that remain sensitive to the active inclusion and the making visible of all the heterogenous constituents of common worlds' (*ibid.*: 11).

In conclusion, Cosmopolitics and OCP have their roots in Actor Network Theory which begins as a 'techno-scientific' enquiry into objects, before widening the debate to explore the spaces in which these objects circulate. These tactics lead to exploring how these objects exert influence on one another and how they become entangled with one another. Blok and Farias then use the language of 'entanglements' (Blok and Farias 2016: 12) to shift the analysis from one located within techno-scientific frameworks, to one with more obvious moral-ethical and political dimensions. The journey towards secular enchantment, energized by 'cosmopolitics' thus makes the hermeneutical shift from the scientific to the ethical to the political, including the emergence of new politico-ethical subjectivities of what it is to be human and a citizen (Baker 2017).

Conclusion

In this essay, I have sought to highlight how, as part of the postsecular turn, those disciplines that hitherto had not paid much attention to religion, belief, and ontologies of transcendence and the sacred, are now starting to address them. The purpose of constructing this typology has been threefold. First, I have set out to show that ontologically real readings of the transcendent become more pronounced as we move deeper into this postsecular century. OCP, for example, represents a full-spectrum secular ontology reflecting an externally transcendent reality that actively shapes the material structures that surround us.

Second, the shift within CUT towards a deeper ontological view of reality suggests an unconscious unravelling of Lefebvre's foundational thinking about the significance of 'absolute space'. Although largely ignored by CUT in the 1980s and 1990s, a more symbiotic, if still edgy, relationship with notions of the transcendent (and by implication, religion, and belief) seems to have developed. In other words, the work of Blok and Farias symbolizes a return to the lost or sublimated category of the transcendent established by Lefebvre.

Finally, this typology attempts to understand the emerging relationship between CUT and notions of the transcendent, as part of the new space for critical debate and research opened by the 'postsecular turn'. My title for this essay suggests that the relationship has historically been characterized by a resistance on the part of a normative intellectual secularism to acknowledging the full impact and complexity of religion and belief on the spatial, and vice versa. However, my use of a question mark in my title is deliberate. In deploying Cameron's theory of knowledge development, I have suggested that the category of the transcendent has consistently emerged as an operant category, alongside and within the normative sanctions self-imposed by CUT on referring to it.

But perhaps active resistance or avoidance within CUT is giving way to recognition of the validity of religion and belief as a shaper of modernity. This recognition is still grudging and sceptical in some quarters, but increasingly intrigued and enchanted in others. Either way, a new relationship between the religious and the secular is being forged at both an empirical and theoretical level that requires us to rise to the challenge of new thinking and praxis if the multiple challenges facing our modern, urban world are to be progressively overcome for the benefit and flourishing of all.

Notes

1 I use the term Critical Urban Theory (CUT) as a shorthand concept for both critical theory that looks at the urban, as well as the tradition of critical human geography. In this sense I am deploying the term as argued by Neil Brenner (2009: 198) in an influential *City* article where he defines CUT as 'involving the critique of ideology (including social–scientific ideologies) and the critique of power, inequality, injustice and exploitation, at once within and among cities'.

2 This phrase comes from the title of a volume of the poetry of the twentieth-century Welsh poet R. S. Thomas (1988) *The Echoes Return Slow*. MacMillan, London.

Further reading

Becker. J., Klingan, K., Lanz, S. and K. Wildner (eds.) (2013) *Global Prayers: contemporary manifestations of the religious in the city*, Zurich: Lars Muller Publishers.

A fantastically visceral and materialistic reading how embodied religion shapes the smallest to the largest of urban spaces, and across multiple cities not usually theorized upon. Interdisciplinary and full of visual resources.

Berking, H., Steets, S. and J. Schwenk (eds.) (2018) *Religious Pluralism and the City: inquiries into postsecular urbanism*, London: Bloomsbury Press.

The latest critical enquiry into the nature of the postsecular from a predominantly secular/urban sociology perspective. Whilst critically regarding postsecular labelling, this volume does equally challenge the discipline of urban sociology to engage far more openly than it has done to date with the lived and complex realities of urban religion and belief.

Cloke, P. and A. Williams (2018) 'Geographical landscapes of religion'. In Baker, C., Crisp, B. and A. Dinham (eds.) *Re-imagining Religion and Belief for 21st Century Policy and Practice*, Bristol: Policy Press.

A critical and comprehensive up-to-date literature review of the multiple and innovative ways religion and belief are being theorized and researched in human geography.

References

Badiou, A. (2003) *St Paul: the foundation of universalism*, Stanford, CA: Stanford University Press.
Baker, C. (2017) 'Aiming for re-connection: responsible citizenship'. In Cohen, S., Fuhr, C. and J. Bock J (eds.) *Austerity, Community Action and the Future of Citizenship in Europe*, Bristol: Policy Press, pp. 255–77.
Baker, C. and A. Dinham (2017) 'New interdisciplinary spaces of religions and beliefs in contemporary thought and practice: an analysis', *Religions*, 8(2): 16, accessed online 18-04-2018.
Beckford, J. (2012) 'SSSR Presidential Address: public religions and the postsecular: critical reflections', *Journal for the Scientific Study of Religion*, 51(1): 1–19.
Blok, A. and I. Farias (2016) *Urban Cosmopolitics*, London and New York, NY: Routledge.
Bondi, L. (2005) 'Making connections and thinking through emotions: between geography and psychotherapy', *Transactions of the Institute of British Geographers*, 30(4): 433–48.
Brenner, N. (2009) 'What is critical urban theory?', *City*, 13: 198–207.
Bruce, S. (2011) *Secularisation: in defence of an unfashionable theory*, Oxford: Oxford University Press.
Caldwell, L. (2009) 'Schizophrenizing Lacan: Deleuze, [Guattari], and anti-Oedipus', *Intersections*, 10: 19–27.
Cameron, H., Bhatti, D., Duce, C., Sweeney, J. and C. Watkins (2010) *Talking about God in Practice*, London: SCM Press.
Crenshaw, K. (1989) 'Demarginalizing the intersection of race and sex: a black feminist critique of antidiscrimination doctrine, feminist theory and antiracist politics', *University of Chicago Legal Forum*, 140(1, article 8): 139–67.
Deleuze, G. and F. Guattari (1988) *A Thousand Plateaus: capitalism and Schizophrenia,* London and New York, NY: Continuum.
Dinham, A. and M. Francis (eds.) (2015) *Religious Literacy in Policy and Practice*, Bristol: Policy Press.
Fincher, R. and J. Jacobs (eds.) (1998) *Cities of Difference*, London and New York, NY: Guilford Press.
Gökarıksel, B. and A. Secor (2014) 'The veil, desire, and the gaze: turning the inside out', *Journal of Women in Culture and Society*, 40: 177–200.
Greed, C. (2011) 'A feminist critique of the postsecular city: god and gender'. In Beaumont J. and C. Baker (eds.) *Postecular Cities: space, theory and practice*, London: Continuum.
Hardt, M. and A. Negri (2000) *Empire*, Cambridge, MA: Harvard University Press.
Harvey, D. (2000) *Spaces of Hope*, Edinburgh: Edinburgh University Press.
Hjelm, T. (ed.) (2015) *Is God Back? Reconsidering the new visibility of religion*, London: Bloomsbury
Hopkins, P. (2009) 'Women, men, positionalities and emotion: doing feminist geographies of religion', *ACME*, 8: 1–17.
Lancione, M. (2014) 'Entanglements of faith: discourses, practices of care and homeless people in an Italian City of saints', *Urban Studies*, 51(14): 3062–78.
Lefebvre, H. (1991) *The Production of Space*, trans. Donald Nicholson-Smith, Oxford: Wiley-Blackwell.

McFarlane, C. (2011) 'Assemblage and critical urbanism', *City*, 15: 204–24.
McLennan, G. (2011) 'Postsecular cities and radical critique: a philosophical sea-change?'. In Beaumont, J. and C. Baker (eds.) *Postecular Cities: space, theory and practice*, London: Continuum.
Martin, D. (2016) *The Future of Christianity*, London and New York, NY: Routledge.
Merrifield, M. (2006) *Henri Lefebvre: a critical introduction*, London and New York, NY: Routledge.
——— (2009) 'Magical Marxism', *Environment and Planning D: Society and Space*, 27: 381–86.
——— (2011) *Magical Marxism: subversive politics and the imagination,* London: Pluto Press.
Purcell, M. (2009) *The Deep-down Delight of Democracy*, Malden, MA: Wiley-Blackwell.
Robinson, J. (2002) *Ordinary Cities: between modernity and development*, London and New York, NY: Routledge.
Salem, S. (2013) 'Intersectionality and its discontents: intersectionality as traveling theory', *European Journal of Women's Studies*, doi:10.1177/1350506816643999, accessed online 18-04-2018.
Sandercock, L. (2003) *Cosmopolis II: mongrel cities of the 21st century*, London: Continuum.
Thomas, R. S. (1988) *The Echoes Return Slow,* London: Macmillan.
Žižek, S. (2003) *The Puppet and the Dwarf*, Cambridge, MA: The MIT Press.

26

Architecture of radicalized postsecularism

Krzysztof Nawratek

Introduction

Architecture as an academic discipline or a profession is not particularly interested in post-secular thought. There are obviously buildings that were designed for religious functions, but there are very few examples of an interest in postsecularism as a particular way of thinking among architects and architecture theorist (Starkey 2006; Martin 2010), and it should be stressed that in architectural discourse postsecularism relates mostly to spiritualism. One can, however, point out some fundamental questions present in the architectural discourse which overlaps with postsecular thinking. There are at least two issues where architectural and postsecular thinking could meet: one is a spatial context, defined as the unknown or 'out there'; the other is problem of hierarchies and dependencies. As I will show in this contribution, space is always hierarchical; therefore, one can always describe any spatial configuration as a chain of unequal relationships between dominating and submissive spaces; one can find similar discussions in contemporary postsecular thought. These issues are not, however, in a centre of a mainstream postsecular discourse inspired by Jurgen Habermas' interventions at the beginning of the twentieth century.

There are two main strands of research investigating an issue of religion in urban space. One is influenced by Jurgen Habermas' intervention in the beginning of the twenty-first century (Habermas 2006, 2008) and focuses on the communication between secular and religious logics and actors. The other is more related to activities conducted by diverse religious groups and focuses on the influence of religious or religious-inspired practices in contemporary cities (Beaumont and Cloke 2012; Cloke and Pears 2016, 2017). However, there is a third strand growing (present to some extent also in aforementioned books), which is both more conceptual (focused, for example, on merging queer studies and religious studies; Bauman 2018) and more policy-oriented (for example, investigating religious values in a context of sustainable practices, Narayanan 2015). This third strand could be described as 'narrative oriented', and by engaging with phenomenological methodologies (Ahmed 2006) looks at urban spaces beyond reductive linguistic or pragmatic perspectives (Falahat 2014, 2018). The notion of 'radicalized postsecularism' is not coined against current postsecular discourse, but rather aims to engage more directly with the issue of secular, disenchanted universe.

From this point of view, 'radicalized postsecularism' could be seen not really as 'post' but rather as 'anti' secularism, which is a fundamental position preventing final disenchantment of the world. It deals with an impact of religious practices on a social and political life (what postsecular thinking does), but its main interest lays in questioning the existential and ontological consequences of humanity facing 'the Absolute Other'. Following Otto's work, radicalized postsecularism perceives the acknowledgement of 'mysterium tremendum et fascinans' (mystery of terror and awe) as the foundations of any religion.

'Radicalized postsecularism' questions Habermas' position and declares we have never been secular. Radical postsecularism is rooted in Rudolf Otto's claim that is presented in his seminal work, *The Idea of the Holy: an inquiry into the non-rational factor in the idea of the divine and its relation to the rational* (1917, 1958), which defines an essence of religious experience as a relationship with the numinous—'the Absolute Other'. Rudolf J. Siebert (2005) defines, 'the longing for the entirely Other' as a foundation for critical theory, while Rudolf Otto defines the relationship with the numinous ('the Absolute Other') as a foundation of all religions. From this perspective, 'radicalized post-secularism' focuses on a void between stabilized elements of human existence and 'the Absolute Unknown'.

Steven Engler and Mark Gardiner (2017: 618) suggest four approaches to the sacred; one of these perspectives is related to Otto's position claiming:

> that the sacred is accessible to language, but that it requires a specific mode of communication. This builds on the assumption that religious language talks about an unusual kind of thing and involves special kinds of experiences.

In this contribution I analyze a built environment (at a meta-scale of a city and micro scale of an architectural intervention) from a 'radicalized post-secular' position as a (potential) set of gestures able in the same moment to stabilize human ontological security and to engage with the unknown. The (potential) uniqueness of this architectural approach lays in an ability to use an 'open gesture', which is visual, spatial, and temporal, and at the same time to fundamentally engage *dialogue* (architecture does not exist without users) and *experimentation* (it is never fully able to be replicated).

Space as a hierarchical entity

From a perspective of architecture and urbanism, space is fundamentally a hierarchical entity. Bill Hillier's (2007) configurational theory of architecture and space could be seen as a foundational theory of architecture as spatial practice. Relationships between spaces could seem equal when seen from a neutral (Cartesian) perspective, but when human bodies and perception are taken into consideration, equality is replaced by hierarchy. Equality and equivalence are replaced by a constant process of segregating and uniting, as Robin Evans (2011) beautifully puts it:

> If anything is described by an architectural plan, it is a nature of human relationships, since the elements whose trace it records—walls, doors, windows, and stairs—are employed first to divide and then to selectively reunite inhabited space.

Before I will discuss this issue further, I need to define two notions I am using in this chapter. When I talk about points, I mean one geometrical point or set of points (defined by coordinates) creating Cartesian space. When I talk about space, I mean a (socially, economically, cognitively constructed) relationship between points. Spatial hierarchy is directly related to the way the human body experiences space and how it moves through space. If the body moves from point A to point B (and A doesn't equal B), it must travel a certain distance. To do this, it needs energy (therefore space/ distance itself is an obstacle), but it must also overcome any potential material (and/ or legal) obstacles because urban or architectural space, in general, is not empty; there are always (built, imposed, or pre-existing) objects and regulations influencing any movement from A to B. Even if there is nothing but a flat empty plateau, there is still a space between A and B, and the space itself is an obstacle—space is a 'thing' (measured as a distance) and a relationship at the same time (Nieuwenhuis 2014). It means that we can describe the (body) movement from A to B as a sequence of transfers between a set of points.

```
A - - - - - B
```

If the movement is linear and if to get from A to B the body must go through point X (and there is always X between different A and B), it means that point X controls access to point B. This position of control builds spatial hierarchy:

```
A - - X
        - - B
```

This hierarchy transfers point (set of points) X into space X, established in and by relationships with A and B. Hillier's theory focuses on access and visual connections, but it could be easily extended to other senses (for instance, sound, smell, tactility). I believe there is nothing to prevent us using this theory as a foundation for universal thinking on connectivity, understood as a physical (material/ spatial) access and as an ability to create connections in a more metaphorical sense (part of the narrative, feeling of belonging, and so forth). Visibility is obviously material (related to light and lack of elements obscuring the view), but it is also related to an ability to make a cognitive connection between architectural and urban phenomenon (material and spatial).

The tradition of seeing (organizing/ understanding) urban space through individually created patterns could be defined as mainstream architectural and urban thinking; think of Kevin Lynch's (1960) theory on the image of the city, of or Christopher Alexander's (1977) theory presented in his seminal book *The Pattern Language*.

It is safe to say that essentially architectural thinking should be defined as an attempt to organize space and spatial phenomenon as narratives. Sometimes these narratives are linear—one can argue that every human body shapes a topological (continuous) space from the birth to its death; there is no break in this particular space our body is shaping through spatial and temporal continuum. Sometimes, these narratives look more like collage or hypertext, when we consider people as an interacting multitude of topological spaces; while not all interactions are spatial (especially when using digital media), they always attempt to organize space and matter and to tell the story, to 'make sense' of seemingly unrelated elements. Configurational theory focuses on a movement; it is a dynamic exploration of the world, always attempting to change a given location. Configurational spatial theory in a radicalized postsecular perspective, however, goes beyond just a physical movement—it allows us to oscillate between spatial locations and non-spatial logics and narratives, because there is always 'The Absolute Other' looming out of the void. In fact, Hillier's and Hanson's (1989) work recognizes the non-spatial context that defines spatial relationships by making a distinction between inhabitants of the building and between visitors.

Configurational theory could help to understand differences between Georges-Eugène Haussmann's project for Paris and Ildefons Certa's plan for Barcelona. At the first glance, these two projects look similar (for example, they both use perimetric urban blocks and long linear transport cuts—boulevards in Paris and Las Ramblas in Barcelona), but their ideological background and socio-political consequences are quite different. Haussmann's project is convincingly described by David Harvey (2006) as authoritarian. First of all, it destroyed existing 'organic' spatial structure supporting existing social interactions. Then, his plan introduced wide boulevards as elements of 'infrastructure of control'. Ildefons Certa's plan for Barcelona is much more egalitarian and democratic. Firstly, it was designed as a new part of the city. It did not, therefore, destroy existing social connections, but introduced new conditions for new inhabitants. According to the original plan, these inhabitants should be from different social classes, but at the end they were rather middle- and upper-middle class who settled in this new Barcelona. Secondly, his plan allowed for relatively equal access to urban infrastructure. The equality is provided by spatial means—'sameness' of urban grid, lack of distinctively monumental spaces, and diversity of potential connections between points A and B. Squares in general are just 'empty perimetric block'; they are not designed to be monumental. Barcelona might be seen as a direct ideological opposite to Paris. A similar approach, focused on spatial hierarchies and accessibility, could be used to analyze any urban environment.

Spatial hierarchies and (in)justice

To talk about emancipation, first we need to discuss justice and injustice. Spatial justice is a relatively new term, developed by Edward Soja (the term has been used before by O'Loughlin 1973; Reynaud 1981; Pirie 1983; Soja 2009) in his seminal work *Seeking Spatial Justice* (2010); from a similar perspective this issue has been discussed earlier by David Harvey (1973) who used the term 'territorial justice'. Spatial (or territorial) justice is related to an access to urban infrastructure and resources, and therefore one can say it is rooted in a classic discussion on redistribution. More recently, the debate between Nancy Fraser and Axel Honneth (2003) on the importance of recognition and redistribution (defined, however, in a broader, multilayered sense) allows us to develop the discussion on spatial justice at a deeper, ontological level.

Nancy Fraser's perspective translates relatively easily into spatial analyses. It could be illustrated by putting her discussion of the 'two axes' of injustice (one related to lack of recognition, the other an unequal redistribution) in a context of the aforementioned Haussmann's project for Paris. Boulevards introduced in this project play two roles. On the one hand, the boulevards establish clear distinction between higher and lower classes. These streets provides transport corridors and spaces of ostentation to 'show off' (boulevards could be seen as 'salons of the city'); they also thus marginalize the lower classes who are clearly not welcome (but not physically excluded) in that space. On the other hand, these boulevards (as 'void' but also as long, linear facades) block access to the urban and social structures of the 'old Paris' hidden behind the facades.

According to configurational theory, a spatial condition is always built around hierarchies, and therefore (ontological) spatial injustice is immanent in any urban and architectural situation. Any emancipatory mechanisms must be located beyond (outside) the particular spatial condition itself. On this basis, 'radicalized postsecularism' may become a useful theoretical framework that allows to reintroduce critically discussion on justice (and emancipation) into the very core of architectural and urban thought. 'Radicalized postsecularism' should be seen as a 'two-step thinking'. In the first step, it puts any discussion in a wider context. There is always something 'out there' that is impossible to fully understand and describe. This 'something' is not emptiness but rather an 'active void'; there is always a risk that some unexpected forces and factors will come from that void. The second step is closer to a large body of thinking on the postsecular that focuses on diverse religious techniques (theological speculation, rituals, and ethical recommendations) stabilizing the known against the Absolute Other (Otto 1917, 1958).

Following Nancy Fraser's discussion on justice further, I'll now use her description of affirmative and transformative strategies in a spatial context:

> affirmative remedies tend to reify collective identities. Valorizing group identity along with a single axis, they drastically simplify people's self-understandings—denying the complexity of their lives, the multiplicity of their identifications, and the cross-pulls of their various affiliations (...) meanwhile, transformative approaches are solidaristic. Focused on expanding the pie and restructuring the general conditions of labour, they tend to cast entitlements in universalist terms.

From the affirmative perspective, point B needs to be considered as 'unique/special' to guarantee free access from point A through X to B. It means that point/space X has been 'forced' to reduce its 'sovereignty' and ability to control (decide) who and what is going through it, because of the importance and uniqueness of point B. The spatial (geometrical) relation remains the same, but the 'nature' (or 'status' in a social, legal, or cultural sense) of spaces, and therefore the relationship between spaces, have been changed. An external force (socio-political, cultural, and/ or economical context) decides on conditions of the existences and functions of point/ space X and point B.

A - - X - - B

A transformative strategy would question a linearity of a movement A to X to B, defining (creating) a set of points/spaces X, X1, X2, and so on, in order to diminish the controlling position of point/space X in providing access to point B.

A - X - B
- X1- B
- X2- B

A transformative strategy in an urban context can be illustrated by an underground transport system—it does not disrupt urban structure on the ground (or does it in a limited way by above-ground buildings of metro stations), yet it changes the conditions how above-ground spaces are accessed. For that reason, the meaning of urban structure and the narrative of the city are changed. From an architectural perspective, a transformative strategy could be illustrated by any building extension, especially providing new access (doors, staircases) to existing spaces.

Transformative strategies are only possible in an 'open world', where new spaces and new connections could be created (out of 'thin air'). A transformative strategy changes the *space* by multiplying its fragmentation and *spatial relationships* (by creating new spaces it creates new hierarchies and new connections) imposing additional conditions or restrictions on existing spaces. On this basis, transformative strategy is spatially active; it 'colonises' space 'out there', whereas the affirmative strategy is spatially passive as it allows the forces from 'out there' to define the status of spaces.

It is important to realize that in both strategies, affirmative and transformative, the hierarchical control of one point/ space over another has been questioned and diminished. In both cases, the status of the weakest point (B in our example) has remained the same, but the status of a 'controlling space' (point/ space X) has changed. In both strategies, the intervention of the external force or expanse to the external space was needed. An affirmative strategy needs a non-spatial external force to intervene; a transformative strategy needs 'emptiness' (or we can even say 'nothingness') for new spaces to be created. I find it intriguing that the transformative strategy, clearly preferred by Fraser, when translated into space, appears as a kind of expansionist strategy. There is an assumption of emptiness waiting to be filled in, to be colonized. Taking this perspective seriously, spatialized transformative strategy may lead to justification of colonial practices, when to achieve a social and spatial justice in one place there is a need to treat some other places as emptiness, 'invade' them, and colonize by new logics, spaces, and functions.

In the discussed scenario, the weakest point (B) is paradoxically the most important one—it is a goal of the journey. This is a fundamental issue—only because point B is important as an aim of the journey, it 'puts itself' in the weakest position, as being controlled by other points/spaces located 'along the way'.

There is a clear dialectical relationship between point B and point/ space X (they are interdependent—only because of the importance of space B, the space X matters; the 'controlling power' of space X is coming from the importance of space B, increasing power of space X decreases access of space B, it makes space B weaker), but the spatial hierarchies as such are contextual, not dialectical. Points B and point/space X are related, but the importance of point B is coming from beyond points A, X, or B; it is defined by the meaning and importance of point B defined by 'another', non-spatial logic which exists beyond any spatial relationships. The importance of space X is purely spatial. It means we have here two logics operating at the same moment—one is *immanent* and spatial (space X controls access to space B, space B makes space X important), the other is *transcendent* ('something out there' defines space B as important). This transcendent logic is not defined by spatial location but is actualized spatially (defines space B as a goal of the journey and a spatial relationship between spaces A, X, and B). These two logics are separated, but each is actualized in a context of the others (so they connect and interact with each other). While put into context of architecture and urban planning, the mechanism connecting these two logics is based on a set of hybrid (being spatial and non-spatial in the same time) gestures such as planning documents, project's drawings, budgets, images, and stories, but also emotions and kinship relationships between all actors involved into the process of designing, building, and then using of the object/ space/ building. These gestures aim to bridge the gap between these logics. Ultimately, the gap is always present, and its existence allows actualization to occur.

Their interrelationship is interesting—the actualization happens in space, but in each case this actualization has a certain non-spatial and non-material, ontological 'depth'. How could this 'depth' be defined? In the more pragmatic terms of the discussed case—who/ what decided on the importance of point B? It can't be done by the point B itself, and even so, this self-created importance must be known (recognized) beyond point B. There is a void between these two logics—they are intertwined when they interact with each other, but to understand 'why?' they interact in this particular space and time, we would need to contextualize the situation further. From an architectural perspective, it is always the architect or urban planner who decides on the spatial sequences and further contextualization would involve motives of the client/ investor, users, legal framework, economic and cultural context, and so forth. The process of contextualization in terms of going further beyond this particular spatial and temporal situation continues *ad infinitum*.

If points A and B are to be designed, their locations and the distances between them are going to be decided. If points A and B are given (pre-existing architectural or urban situation), the distance and 'geometrical' relationship is unchangeable, but the space between them (therefore their mutual relationship) is under dispute. The process of architectural design always deals with spatial sequences and spatial hierarchies; it is therefore always related to spatial (in)justice.

Dependency and emancipation

Giorgio Agamben's (2013) *Highest Poverty: monastic rules and forms-of-life* investigates the Franciscan concept of *usus moderatus*, which was the centre of a dispute between Franciscans and the Papacy. *Usus moderatus* (*usus facti*), sanctioned by Pope Nicholas III in bull Exiit qui seminat (1279), made a distinction between a right to use and a right to possess. The right to use is rooted in the Franciscans' dependency on the Papacy. This dependency allows a 'smaller' mode of existence (hence the Franciscans are called the Friars Minor) to act beyond the 'normal' (designed for humans / citizens) legal norms. There is a higher authority allowing Franciscans to be like:

> And just as a horse has a use of fact of the oats which it eats, and not have any lordship, so a slave and an expropriated religious has a simple use of fact to the bread, wine, clothing, and other things that are consumed by use.
>
> *(Oliger 1929; Robinson 2010: 37)*

Agamben's focus is on the mechanism by which the rule shapes monks (making them religious men), freeing them at the same time from the rule of the 'external' (secular/ civil) world. Agamben's analysis shows how life/ use could defy law and codification, how action and agency in a context of a 'smaller/ depended mode of existence' and a 'higher order' liberate a human being from the rule of law.

It is crucial to analyze carefully the process of shifting the scales of a legal framework here—a monk acts here on a scale of his own life, defined by the rule under the supervision of the Pope. The monk is dependent on the papacy (the Pope recognizes and actualizes the existence of the monk), but the papacy is also dependent on the monk as his authority is recognized and therefore actualized by the dependent being. The papacy legitimises the monk's actions but is itself also legitimized by the monk who references to the power higher than the Pope—God. One could argue that the church in that time embodied the higher power and that God was absent in a power structure, but I would insist that without faith in the Absolute Other the church would lose its legitimacy. The monk, by his direct engagement with a smaller life, is making himself a mediator between the Absolute Other and regulated, that is, linguistically and institutionally stabilized religion. The assumed existence of the highest power (God/ the Absolute Other) 'opens' (pushes it beyond) the dialectical relationship between 'superior' and 'minor' beings. As Jacob Taubes (2009) argues, from this perspective, theocracy could be seen as potentially the most rebellious and anti-authoritarian of all political regimes.

Could this reintroduction of hierarchy anchored in transcendence be a way to reshape contemporary critical urban and architectural theory? The Absolute Other forces all actors (points A, X, and B) into 'cooperation for recognition' (going beyond a Hegelian 'struggle for recognition' between Master and Slave) by creating a threat to them both. In architectural and urban discourses, the Absolute Other is the limitless (spatial, social, political, and so forth) context; the unknown power hiding 'out there'. The Absolute Other in a contemporary urban context could be defined as an ecological catastrophe, a cosmic disaster, or, on a much smaller scale, as the change of legal regulations influencing the city by its national state or international organizations. In general, it could be just about anything that influences in a dangerous or unpredicted way the city from beyond.

In any architectural or urban situation, existence of context should be assumed as given. There is always something—spatial, material, and legal/ cultural or social—beyond this particular situation. 'Radicalized postsecularism' accepts our limited ability to define this context. 'Radical postsecularism' has a symbiotic relationship with any scientific mechanism of producing knowledge—the process of stabilizing the unknown into the known—but recognizes the existence of a constant void between the known and the unknown and, crucially I would argue, the unpredictable influence of the unknown on any particular spatial situation.

Taking a 'radicalized postsecular' perspective, architectural and urban approaches to injustice could be twofold. It could be reactive, attempting to use the void between spatial and 'transcendent' logics. Spatial interventions could influence the situation—providing wider doors allows people with certain disability to access; providing uneven surfaces can prevent certain groups of users to comfortably use the space. These interventions are coming out of the gap between standards, regulations, budgetary requirements, and certain design decisions. In the process of materialization and spatialization of non-spatial logics lies a chance for designers to 'hack' the narrative. This transcendent logic is a hybrid of the known and the unknown (but active) contextual forces. It is not necessary to understand the nature of the external forces. It is fundamental to understand the consequences of these forces in the actions of the here and now. One way of seeking justice and emancipatory forces in architecture is rooted in a process of translating (or 'materialising') non-spatial logic into space. The other way requires architects to design solutions 'out of thin air' that are unpredicted and unexpected.

Architecture and urbanism seen from the 'radicalized postsecular' perspective attempts to use the void as a reactive and partially empty entity. From this perspective, architectural activities are focused on experimentation, on making spatial gestures, and then on reacting to a never fully predicted outcome.

Conclusions

Postsecularism is predominantly understood in the linguistic and postmetaphysical framework proposed by Jürgen Habermas, or as a return of spirituality and symbolic dimensions into discussion of the built environment. I argue that there is a shared interest between postsecular thought and architecture that goes much deeper and engages with ontological issues. This 'radicalized postsecularism', as I propose to call it, would help to radically contextualize any architectural situation in an ultimate attempt to re-enchant the world. The process of radical contextualization means that there is always 'something out there' to influence spatial situations, and that this 'something' (because it is unknown) is potentially in a dominant position towards 'the known'. This condition allows continuous reconfiguration of any spatial hierarchical structure.

Spatial hierarchy, I argue, is an eminent feature of any architectural and urban situation. This spatial hierarchy creates conditions for ontological spatial injustice, dependency, and oppression. Emancipation, therefore, comes from beyond space, yet is actualized spatially. The dialectical relationship between space B (important but controlled by space X) and space

X (gaining its importance as a gateway to space B) could be overcome by contextualization of this relationship. Non-spatial logic, the force coming from beyond, changes and disrupts any spatial hierarchy.

Nancy Fraser's affirmative and transformative strategies seem particularly suited for spatial (architectural and urban) interpretation. However, I argue that any meaningful discussion about spatial justice is possible only in the context of an open world where any spatial situation is contextualized beyond spatial conditions. The void ('abyss of the unknown') beyond the represented world is always in a potentially dominant position. On the one hand, there is always the possibility of actions drastically influencing the situation coming 'out of the void'. On the other hand, the void could be (at least partially) 'empty', therefore allowing new spaces to be created 'out of the represented world'. The void is a threat and an opportunity. This is a tenet of an architecture and urbanism in a 'radicalized post-secular' framework.

My argument is that to ontologically stabilize human existence in the built environment, the transformative approach, leading to a multiplication of spatial connections that leads to an increase of spatial complexity, is needed. Paradoxically, the ontological stabilization in architecture and urbanism leads to fundamental engagement and support for complexity and density, and at the same moment it forces architects and urbanists to take into account that this approach could be expansive and could lead to spatial conflicts. 'Radically post-secular' architecture and urbanism advocate spatial hyper-pluralism built around careful consideration of minor spaces and minor lives under the threat of the Absolute Other.

Further reading

Sandercock, L. (2003) *Cosmopolis II: mongrel cities in the 21st century*, London: Continuum.

I would argue that nobody writes more passionately and convincingly about diversity and openness of contemporary city than Leonie Sandercock.

McCullough, L. (2012) *Conversations with Paolo Soleri* (Conversations with Students), New York, NY: Princeton Architecture Press.

Paolo Solieri's work, both in writing and design, has been strongly influenced by theology of Pierre Teilhard de Chardin. This chapter does not refer directly to Solieri nor to the work of de Chardin, but is deeply rooted in their thoughts.

Yoon, J. (2010) *Spirituality in Contemporary Art: the idea of the numinous*, London: Zidane Press.

Jungu Yoon's book is an important attempt to re-examine the relationships between contemporary art and spirituality in a context of Rudolf Otto's work on the numinous.

References

Agamben, G. (2013) *The Highest Poverty: monastic rules and form-of-life*, Stanford, CA: Stanford University Press.

Ahmed, S. (2006) *Queer Phenomenology: orientations, objects, others*, Durham, NC: Duke University Press.

Bauman, W. A. (ed.) (2018) *Meaningful Flesh: reflections on religion and nature for a queer planet*, Goleta, CA: Punctum Books.

Beaumont, J. and P. Cloke (2012) *Faith-Based Organisations and Exclusion in European Cities*, Bristol: Policy Press.

Cloke, P. and M. Pears (eds.) (2016) *Mission in Marginal Places: the praxis*, Milton Keynes: Paternoster.

Engler, S. and Q. M. Gardiner (2017) 'Semantics and the sacred', *Religion*, 47(4): 616–40.

Evans, R. (2011) *Translations from Drawing to Building and Other Essays*, Pamela Johnstone (ed.), London: Architectural Association Publications.

Falahat, S. (2014) *Re-imaging the City: a conceptualisation of urban logic of the 'Islamic city'*, Wiesbaden: Springer.

——— (2018) *Cities and Metaphors: beyond imaginaries of Islamic urban space*, London and New York, NY: Routledge.
Fraser, N. and A. Honneth (2003) *Redistribution or Recognition?: a political-philosophical exchange*, London: Verso.
Habermas, J. (2006) 'Religion in the public sphere', *European Journal of Philosophy*, 14(1): 1–25.
——— (2008) 'Notes on post-secular society', *New Perspectives Quarterly*, 24(5): 17–29.
Harvey, D. (1973) *Social Justice and the City*, Athens, GA and London: University of Georgia Press.
——— (2006) 'The political economy of public space'. In Low, S. and N. Smith (eds.) *The Politics of Public Space*, London: Routledge.
Hillier, B. (2007) 'Space is the machine: a configurational theory of architecture', *UCL Space Syntax*, http://spaceisthemachine.com/ accesed online 15-06-2018.
Hillier, B. and J. Hanson (1989) *The Social Logic of Space*, Cambridge: Cambridge University Press.
Lynch, K. (1960) *The Image of the City*, vol. 11, Cambridge, MA and London: The MIT Press.
Martin, D. (2010) 'Inscribing the general theory of secularization and its basic patterns in the architectural space/time of the city: from presecular to postsecular?'. In Molendijk, A. L., Beaumont, J. and C. Jedan (eds.) *Exploring the Postsecular*, Leiden and Boston, MA: Brill, pp. 183–206.
Narayanan, Y. (2015) *Religion, urbanism and sustainable cities in South Asia*, London and New York, NY: Routledge.
Nieuwenhuis, M. (2014) 'Taking up the challenge of space: new conceptualisations of space in the work of Peter Sloterdijk and Graham Harman', *Continent*, 4(1): 16–37.
O'Loughlin, J. V. (1973) *Spatial Justice for the Black American Voter: the territorial dimension in urban politics*, Philadelphia: Pennsylvania State University Press.
Otto, R. (1958) *The Idea of the Holy: an inquiry into the non-rational factor in the idea of the divine and its relation to the rational*, New York, NY: Oxford University Press.
Pirie, G. H. (1983) 'On spatial justice', *Environment and Planning A*, 15: 465–73.
Reynaud, A. (1981) *Société, Espace et Justice: inégalités régionales et justice sociospatiale* (Society, Space and Justice: regional inequality and social area justice), Paris: Presses Universitaires de France.
Robinson, J. (2010) *Bonagratia of Bergamo: a tract on the poverty of Christ and the Apostles*. individual.utoronto.ca/jwrobinson/translations/bonagratia_de-paupertate.pdf
Siebert, R. (2005) 'The critical theory of society: the longing for the totally other', *Critical Sociology*, 31(1): 57–113.
Soja, E. (2009) 'The city and spatial justice' (La ville et la justice spatiale), *justice spatiale*, 01 September.
——— (2010) *Seeking Spatial Justice*, Minneapolis: University of Minnesota Press.
Starkey, B. (2006) 'Models, architecture, levitation: design-based research into post-secular architecture', *The Journal of Architecture*, 11(3): 323–28.
Taubes, J. (2009) *Occidental Eschatology*, Stanford, CA: Stanford University Press.

27

Islamophobia, apophatic pluralism, and imagination

Giuseppe Carta

Introduction: Islamophobia and the conflict over mosques

In the autumn of 2007, Italian politician Roberto Calderoli announced he was calling the *maiale day* ('the day of the pig') in response to the Bologna City Council's approval of the building of a long-awaited purpose-built mosque. Calderoli, vice-president of the Senate, former minister for institutional reforms, and coordinator of the *Lega Nord* ('Northern League'), declared he would 'walk his pig' over the designated area in order to desacralize the land and sanction the dominance of Christianity over the territory. Although widely stigmatized, the initiative succeeded: attacked by various media and neighbourhood associations, Bologna city council decided to review and halt the planning application (Conti 2016). This was not the first occasion in which a demonstration of this kind was proposed by Lega Nord and, indeed, not the last. Seven years earlier, in Lodi, near Milan, pig urine was spilt over the land where a mosque was to be constructed; after that, a priest consecrated the land and said mass (Saint-Blancat and Schmidt di Friedberg 2005). A similar protest was performed in Padua, only two months after Calderoli's contention (Guolo 2011). In all these instances, the activists achieved their goal: by 2018, the demands of the Muslim communities of Bologna, Lodi, and Padua for the consecration of a permanent sacred space are still pending.

Despite a rapidly expanding population of more than two million Muslims, 43% of whom are Italian citizens (Ciocca 2018), in Italy the number of mosques characterized by specific architectural features has remained remarkably low—only six throughout the whole national territory. Alongside these are an ever-growing number of *musallayat*, temporary prayer halls, estimated at around 1,200. As illustrated by the photographer Niccolò Degiorgis (2015), most of these are hidden in anonymous buildings lacking in symbolic denotations, including a plethora of former shops and commercial facilities, apartments, garages and car parks, sport halls, and former warehouses. Interiors are usually furnished with few liturgical elements—a carpet, a wooden *mihrab*, a tap and a sink used for ablution; the orientation to Mecca, *qibla*, is commonly indicated by tape lines applied over the carpet, and often there is no ventilation and only artificial illumination. Since such informal places hardly suffice the demand, often Muslims gather around their *musalla* and pray upon the pavements. *Musallayat* are ordinarily depicted by media and institutions as *moschee abusive* ('unauthorized mosques') and are frequently closed by authorities for the infringement of local building regulations. Nonetheless, repressive measures are not supported either by a set of regulations or by a manifest political will to offer planning directions to Muslims communities (Chiodelli and Moroni 2017). According to activists, for enjoying the constitutional right of religious freedom, Muslims would be forced paradoxically to violate current planning regulations, since these regulations neglect their specific requirements (see Figures 27.1 and 27.2).

Figure 27.1 Rome's Muslim association Dhuumcatu and CAIL practice the Friday congregational prayer facing the Colosseum, in response to the closure of several prayer halls by Rome's authorities

Source: Photo: Graziano Panfili—Ulixes Pictures, 21 October 2016.

Figure 27.2 Rome, Masjid E Rome, via Serbelloni

Source: Photo: Raffaele Petralla—Ulixes Pictures.

Conflicts over mosques represent a common pathway for the opposition to the presence of Islam and Muslims in non-Islamic countries to be channelled (Allievi 2010; Morgan 2016). Such tensions target Muslim groups' acts of territorialization, and, more immediately, their impact on the shape of the city, on the sensible materiality of the built environment, and on its *image*. Furthermore, the building of mosques is often stigmatized as integral to a harmful multiculturalist strategy that would undermine social cohesion and enhance urban revanchism (see Modood 2013; Fincher et al. 2014). Not always coinciding with anti-religious positions, but often paralleled by the call for a revival of Christianity, Islamophobia became a crucial factor in political affairs during the same years in which the academic debate on postsecularism was flourishing. Alongside being assimilated to the philosophical ideal type of the postsecular city (Beaumont and Baker 2011), the contemporary city might appear as a 'fundamentalist city', a space in which rights of minorities are discarded by explicit or latent forms of marginalization (AlSayyad and Massoumi 2010).

How could clashes over the presence and building of Islamic sacred spaces in non-Islamic cities be addressed? How can a postsecular sensibility help to ameliorate urban pluralism? This chapter tackles these questions by advancing two intertwined claims. First, I uphold that the philosophical project of the postsecular can better achieve its aims if conceived as a form of apophatic pluralism (see Rose 2013; Franke 2014). The second related point is that its register, rather than that of Habermas' consensual reason, must be that of imagination. My argument is that postsecular urbanism should stimulate the emergence of a form of non-establishmentarian and non-foundational pluralism at the level of the built environment. I contend that, while the promotion of interfaith encounters and of more-than-confessional shared spaces can be beneficial to many (see Sandercock and Senbel 2011), the first unescapable requirement for postsecular principles to be initiated should be that of guaranteeing the possibility for a plurality of incommensurable specificities to autonomously explore their own sense of sacrality—in other words, to produce and dwell their own space. The task of a renewed urbanism inspired by postsecular concerns and apophatic pluralism, I argue, is then that of desecularizing, and pluralizing, imagination.

Religious diversity and the postsecular city

The contemporary city is characterized by a nova effect of diverse beliefs, from established organized religions to non-religion, from orthodoxies to atheism—between these poles, a wide plurality of new religions, syncretic ones, and various degrees of unstructured spirituality (Woodhead 2016). Rather than a mere backdrop for diversity to appear, the urban is increasingly thought as an active force in the production of religious forms. At the same time, it is the everyday domain where theological and philosophical predicates can profitably interact and dialogue or rather clash and collide. As the space in which a variety of unfolding of the real and the social find expression as spatial demands and affective practices, the city constitutes the locus wherein principles and pragmatics of pluralism are needed the most (Oosterbaan 2014; Berking et al. 2018).

Modern thought has approached religious diversity through the socio-political doctrine of secularism. Secularism is classically framed around the articulation of two principles: on the one hand, the principle of functional and spatial separation of politics and religion; on the other, that of state neutrality, sanctioned by the primacy of secular reason over religious truths. According to Saba Mahmood (2016), there are two main problems with secularism. The first is that, rather than abstaining from expressing its preferences on religious matters, it actively regulates doctrines and practices. The second is that, by naturalizing as *rational* the epistemic situatedness of its evaluation on the real, it is particularly hostile towards religious minorities. As argued by Talal Asad (2003: 61), secularism lies on 'contradictory foundations: on the one hand elite liberal clarity seeks to contain religious passion, on the other hand democratic numbers allow majorities to dominate minorities even if both are religiously formed'. Rather than being a solution to their strife, secularism ultimately exacerbates the tensions it is called to manage, while polarizing religious difference and intensifying inequalities.

Emerging in the last two decades, the notion of the postsecular has been coined with the precise goal of philosophically deferring secularism. In a similar vein to postmodern and postcolonial theses (see Abeysekara 2006; Lloyd and Viefhues-Bailey 2015), the postsecular can be framed as twofold. First, by rejecting the divide between private faith and public reason, it aims at triggering a counter-hegemonic challenge towards the totalizing narrative of Western modernity. Second, it advances new ways for granting the right to difference and for rethinking the public realm through the integration of an undefined multiplicity of knowledges. Exceeding the scopes of liberal toleration and multiculturalism, postsecular scholars call for a revived interchange between metaphysics and the political, while fostering an intellectual climate in which religious contributions are relevant to the structuring of epistemologies, ethics, and social order. The task of postsecular theories is, then, to accommodate the religious by opposing the epistemic asymmetries brought by secularism, while cultivating a cohesive space for the political in which difference can be respected and flourish.

In recent years, postsecular challenges have reached critical geographers, as new distinctive orientations have developed. By exploring how 'secularization and "the secular" are cogenerated with "the religious" through space', geographers are increasingly concerned with how 'the ways that we study religion in and through society, politics, and history require an engagement which potentially disturbs the categories of "religious" and "secular" and their presumed meanings and affiliations' (Olson et al. 2013: 8). For Veronica della Dora (2018: 48), the notion of the postsecular should be primarily conceived as a 'critical lens and methodology', through which accessing the 'coexistence of different forms of belief and non-belief'. Following John Milbank's critique of the secular (2006) and Mircea Eliade's *homo religiosus*, Justin Tse (2014) stresses that every performance of place-making is a theologically constituted articulation of transcendence and immanence. Through the concept of *grounded theology*, Tse's contribution disrupts the disciplinary boundaries of geography of religion: rather than conflicting categories folded across distinct spaces, religion, and the secular intertwine to the point that 'it is not necessary to define the religious in geography, as if there were anything that could be considered outside the bounds of religious inquiry' (202).

One of the most prominent indications for research is then the contestation of the discreteness of the religious, for interrogating instead how different religions and secularities cohabit within a multitude of poetics and politics of the sacred. Such a theoretical perspective expands the geographical agenda beyond 'official sacred spaces' and institutional politics, opening the field to the perpetual unfolding of subjectivities and embodied performances. Postsecular geographies are commonly related with phenomena of blurring of narratives of belonging, with the formation of crossover identities, of shared practices of the everyday, and of the discursive and affective negotiation of the commons (see Cloke et al. 2019). Found in the emerging of theo-ethic praxis of generosity in place of dogma and sectarian confessional theologies, being religious or non-religious, scholars such as Paul Cloke and Justin Beaumont (2013) envisage the postsecular city as the space in which *hope* and *love* erupt as means for resisting and subverting the dominant framework of capitalism and urban indifference. The postsecular city, according to this view, is transformative and emancipatory precisely because it encourages the translation of vernacular registers into something contingently hybrid.

The shift from dogma to charitable praxis 'points to the embodied performances of identity—religious or otherwise—in which local lived spaces can come to represent the potential for new formations of tolerance and agreement in place of previous sectarian tendencies' (Cloke and Beaumont 2013: 33). The postsecular city is demanded to constitute a laboratory for practices of social justice and for the development of transformative social identities, allowing to 'transcend the excesses of postmodern heterogeneity and nihilistic relativism with something fresh and new' (Beaumont and Baker 2011: 4).

Nevertheless, I would like here to raise two critical remarks. The first is that postsecular geographies, by emphasizing forms of rapprochement between different constituencies, are often dismissive about how power relations might impact the theological orientation of such encounters. In an argument extensively elaborated by Chantal Mouffe (2013), consensual approaches inspired by Habermas' deliberative model, such as postsecularism commonly is, fail to come to terms with the irreducible conflictual dimension of the social and with the pluriversality of the world. While 'public spaces are always striated and hegemonically structured' (91), consensus-seeking obliterates the counter-hegemonic struggles of minorities at the risk of reinforcing majority inclinations and epistemic discrimination. Conditions of religious diversity call for theoretical and pragmatic tools for positively balancing the respect of theological particularities with the excesses of identity politics that would Balkanize cohabitation. It is precisely for this reason that it is fundamental to avoid dropping the voices of minorities to irrelevance, by paying special attention to how majoritarian knowledge, more established grounded theologies, or better-represented groups may channel dynamics of rapprochement.

The second point is that postsecular geographies are often dismissive towards the agency of the city, the physicality of its built environment, and its capacity of embodying and regulating social relations and hierarchies. As posed by Michele Lancione, (2014: 3064), 'postsecular scholars mainly understand the city as a back-drop scenario … Where is the "city"—with its more-than-human actants, contextual dynamics, power, affective atmospheres—in the "postsecular city"?' As blatantly exemplified by Islamophobia and the widespread rise of establishmentarian populism, the postsecular city seems yet to be fully approached. Clara Greed (2011), for example, stresses that

> postsecularism may "allow" for a greater acknowledgement of the importance of religion in city life, but this is not yet manifested in the design of the built environment … the archaeology of the past weighs heavily on the present nature of urban form and structure, and upon the subcultural values and attitudes of urban policy makers.
>
> *(109)*

The postsecular struggle for pluralism is still to be achieved: ongoing phenomena of marginalization of religious groups and knowledge in the city dramatically emphasize how 'postsecularism embeds ideals that are better understood as ethical projects towards which a society may strive; one does not, perhaps, arrive at post-secularism so much as one struggles with its demands' (Gökarısksel and Secor 2015: 28).

Interreligious dialogue and apophatic pluralism

How can pluralism be conceived? How can different forms of religiosity coexist and dialogue? The canonical way of framing interreligious confrontation is the threefold categorization of exclusivism, inclusivism, and pluralism (Schmidt-Leukel 2005). *Exclusivism* is the position according to which only one religion contains the means of salvation; while this religion is final, universal, and utterly true, all the others are false beliefs incapable of responding to any ultimate question. According to *inclusivism*, instead, while all religions may contain some valid traits, only one religion is depositary of absolute truth. The critique that can be moved against inclusivism is that it is naively paternalistic in its benign forms, and colonialist in its muscular version: while the religious other is not necessarily an enemy but someone we can at least tolerate, we anyway cannot but try and convert him, for his own sake.

Differently from exclusivist and inclusivist confessionalism, *pluralism* rejects the intrinsic superiority of any religion towards the others. It assumes that different religions are united by a substantial agreement on the subject matter, an ineffable living reality they all speculate on and approach towards. The main advocate of the pluralist position is John Hick (1980). His proposition is that of a theo-centric pluralism, in which all religions orbit around a transcendental divine principle: all religions are unified in their attempts to come to terms with the transcendental divine, yet the transcendent portrayed by religions' teachings is not to any extent the noumenical itself but only its phenomenal representation, drawn according to the means humankind acquires through the worldly experience. Hick considers interreligious dialogue not only as a viable tool for mutual understanding, but as an inescapable stage for religions to renovate their theological repertoire. In his view, interreligious dialogue is an open field in which even non-religious people can be profitably engaged: 'the religious pluralist does not, like the traditional religious exclusivist, consign non-believers to perdition, but invites them to try to produce a better explanation of the data' (1997: 163).

Many contest Hick's pluralist proposition by objecting that it establishes nothing but a form of liberal confessionalism, whose pretention of neutrality would insinuate a quite specific form of syncretic religion, Hick's religion (Howard-Snyder 2017). This critique is advanced by Gavin D'Costa (1990: ix), according to whom the pluralist project 'operate[s] in a curiously absolutist fashion, proposing to incorporate religions on the system's own terms rather than on terms keeping with the self-understanding of the religions'. Another critical account is formulated by the postsecular theorist John Milbank (1990). According to Milbank, interreligious dialogue is an illusion shaped by the false sociological notion of religion as a genus, a notion in reason of which theological claims are dissociated from the situated voices through which these express 'our biographical or transbiographical processes of coming-to-know' (177). Milbank claims that the lived realities we call religions are integral, non-reflected, autonomous constructions that 'entail incommensurable social projects' (179). By promoting their conflation and hybridization, Hick's pluralism would constitute another form of the deterritorializing power of secularism, by which difference is forfeited in the name of modernity's collated myth of governability.

A different form of pluralism is that of the *particularists*, who, by considering religious systems as inherently irreducible one to the other, evade the search for common ground as reductionist, incapable of rendering the specificity of the religious other's lived claims. Still, differently from confessionalists, particularists theorize that, first, different religions might share something, even if not on a doctrinal level, and, second, the incompatibility of their exoteric expressions does not preclude the possibility for paralleling their inward truths. On the contrary, it is maintained, an open dialogue between different traditions can happen only on the preconditional recognition of incommensurable difference. A distinguished proponent of such an orientation is the Islamic philosopher Seyyed Hossein Nasr (1989). Differently from Hick, Nasr affirms that we cannot access different religions' messages other than from within the integral unity of their doctrinal *and* methodical apparatus. The underlying unity of religions cannot be found in the contingent similarities their doctrines might have, as Hick believes, but only on what can be experienced through the contemplative practice of a fully religious life.

A third model for pluralism, proposed in recent years by Kenneth Rose (2013), is that of *apophatic pluralism*. The rationale for apophatic pluralism lies on a critique of the limits of language. According to this perspective, religion is the articulation of the unsayable: every religious doctrine is a transitory approximation towards a reality that cannot ever be fully grasped. This does not imply that cataphatic formulations are implausible and need to be rejected as false, but only that they are articulated in impermanent forms, to the point that in the long term these *might* be substantially transformed or substituted by others. Rose maintains that we need to respect the sovereignty of each theology, like particularists recommend, and that we must strengthen our epistemic openness to diversity by objecting to reductionist approaches to the religious. To be aware of the epistemic limitations of discursive reason entails our conversations to be as open as possible, and for us to be self-critical about our own assumptions. Rather than dismissing the finality and universality of religious claims, apophatic pluralism postulates that none of them constitute the sole truth.

William Franke (2014) argues that apophasis is the most authentic register for the postsecular. His philosophy of the unsayable conceives pluralism as unescapable and inherently favourable for a theology of peace:

> when different philosophies are repositioned and redefined as attempts to say what cannot be said, they reflect upon each other as reflecting a common … something/ nothing that they cannot say—except each in its own inadequate way … Different, even apparently contradictory, philosophies are revealed thereby as necessary to each other rather than as excluding and as having to suppress each other.
>
> *(149)*

Apophatic pluralism does not correspond to methodological atheism, for it disposes the idea that every truth claim within the nova effect leads to their unsayable core; it is not even agnostic, for it is

> rather the moving away by negation from any position in which knowing can rest secure that is most characteristic of apophaticism. It does not remain within the limits of any knowledge or even of any definite and achieved state of unknowing.
>
> *(199)*

Apophatic pluralism provides a strong basis for postsecularism precisely because it encourages the search for ultimate truths while perpetually trying to be self-reflexive about the positionalities from which enunciating any propositions. Differently from a secularist doctrine of mere neutrality, it postulates a non-substantive openness towards a plurality of incommensurable worldings as the only suitable strategy for enacting interreligious dialogue and building a shared city in cooperation. Nonetheless, it 'requires an unceasing, never satisfied effort of the imagination. This makes poetic imaginings of the cosmos and its Creator crucial for theological vision' (229). The unsayable cannot adequately be said nor thought but can be *imagined*. Postsecularism can be successfully pluralist, respectful of the particular and still aiming to forge a cohesive public realm, only when it self-reflectively opens its discursive formulations, its lived doctrines and narratives, to the perpetual unfolding of the unsayable, which can be found nowhere but in imagination.

Postsecular urbanism: pluralizing imagination

Katie McClymont (2015) holds that postsecular theories can contribute to a radical redefinition of urban planning. First, if its aim is to accommodate religious diversity, postsecular planning must go beyond approaches based on difference. For religion is essentially a register that we *all* dispose of, the postsecular city must be:

> open to all regardless of designated identity; you do not have to be an active part of any established religious community for it to be meaningful to you. It offers a different understanding of human experience, an engagement with values such as 'beauty' and 'awe' cannot be expressed instrumentally.
>
> *(538)*

Second, since the cultivation of postsecular religiosity transcends official sacred places, the scope of planning must exceed the provision of places of worship. For McClymont, planning should instead be thought as a tool for recognizing, engaging and even promoting municipal spirituality, a form of place-making which 'gives access to the transcendent, a potentially counter-hegemonic way of being, an alternative set of values underpinned by shared humanity' (542–43). The places in which this can be employed are those whose value transcends instrumental rationality for rather comprising the common good. Such places are exemplified by cemeteries, community assets, and nature—places whose 'claims to value are spiritually inclusive rather than for a defined denomination or established faith' (549). Ultimately, postsecular planning would aim to protect and promote the more-than-confessional spiritual significance of places that are 'for all faiths or none' (546).

Even though these arguments sound appealing, I have some concerns in fully subscribing such an agenda. On one side, while I agree about the need to capture and facilitate the unfolding of the religious beyond officially sacred space, I still contend that to adequately respond to the demands of specific communities of faith constitutes the prime, unescapable task for postsecular urbanism. If not weighted by a critical reflection on how the common is hegemonically shaped, the search for inclusivity and more-than-confessional spatiality can result, instead, theologically prescriptive and discriminatory. First, because religious groups have the right to be closed groups with clear-cut boundaries and restricted access—just like any other group which is not conventionally marked as religious. And secondly, because place matters—for different places are differently seized by different grounded theologies. To epistemically and politically support the struggle of marginalized minorities is a crucial step for establishing the right to religious difference, a principle coextensive to the project of postsecularism. To accommodate the spatial demands of specific religious groups is to oppose their social *and* theological discrimination: to do so, rather than defending only that specific community, is to defend the right to the city and the right to difference for *all*—even though the space planners would give shape to is not open to all.

On the other side, the idea that 'it is with the "invisible" that planning needs to engage if it is to meet the challenge of postsecularism' (McClymont 2015: 538) needs to be complemented with the principle that the *visible* is also an essential component of religion. The visible is essential for religious individuals because it provides the means through which the invisible is self-disclosed; it is essential for every *homo religiosus*, for everyone's everyday life is sensuously experienced within the world of bodies. To argue that what really matters for the territorialization of religion is the invisible is to actively regulate their own doctrinal and experiential apparatus. Altogether, I contend that to dismiss the importance of the visible is hazardous precisely because it validates the claim according to which the aesthetic quality of the place provided to a specific community through planning processes is irrelevant, and that what matters is only that a space, whatever space, responds to the functional requirements set by authorities. That, in other words, a *musallayat* equals a *masjid*, a mosque, and that the form in which these are realized is not relevant for their inhabiting and sacrality.

Form matters, and so too does beauty. While it is true that beauty cannot be expressed instrumentally, it is also true that beauty is instrumental to the thriving of the religious, for it is inseparable from spirituality (see Nasr 1987). Alongside the right to difference, we must fully maintain the right to beauty. What is key, for planning and for the postsecular project, is not to dematerialize beauty, but to decolonize it—to open the city up to different conceptions of beauty, to different processes of *aisthesis* and different acts of imagination (Jackson 2016).

Before concluding, I would like to mention briefly two forms of intervention that, I suggest, respond to the principles I have proposed as the basis of postsecular urbanism, that of apophatic pluralism and of imagination. The first is provided by the work of the architecture historian and artist Azra Aksamija (2015), a collection of ten artistic projects for ten ways of conceiving the mosque. Playfully decolonial, Aksamija's work interrogates its readers and spectators on tradition and agency, on beauty and imagination, dogma and becoming, and on the potential for participatory art to induce social change: 'What can art actually do to counter the politics of xenophobia without being utilized solely as a vehicle for activism?' (342). Her response opens new venues for rethinking the mosque within and beyond tradition, while challenging Islamophobia through the renegotiation of *aisthesis* and spirituality. The second is a participatory workshop led by Shagufta K Iqbal within the framework *Una moschea per Roma?* held in Rome in May 2017. Drawing on the works of Muslim theologian and poet Jalāl al-Dīn Rumi, it included an improvisational poetry course, run in the space of a *musalla* sited in the neighbourhood of Torpignattara. The initiative prompted an intergenerational, multilingual, and multireligious context. The course used a variety of individual and group writing exercises centred on spatial memory and recollection, making use of sensual dispositions for poetically representing the sanctuary of sacred space unfolded in imagination. The conclusive event has been framed as a promenade performance, were participants from the course have toured audiences around the neighbourhood while performing their poetry. The rationale was that of disclosing through poetry the sacredness of a neighbourhood which was usually negatively depicted by the media precisely by reason of its Muslim population and *musallayat*—assumed as deviant, alien to the cityscape, and indifferent or even hostile towards Italian culture. The performance was co-created with the audience using the creative writing exercises employed during the workshops. Pieces of collaborative poetry have been unravelled, shared, and read at the end of our journey, at the Muslim Centre.

What these experiences testify is the importance of conceiving new modalities of cohabitation, opening up the city to different polities and different images. Altogether, they state that the sensuous experience of the world cannot be separated from the spiritual, and contextually, that spirituality can be elicited by the senses. This can be in turn applied on two levels: the first is that beauty disposes of a spiritual dimension that the postsecular needs to discover and appreciate; second, that groups and individuals must be able to explore their own lived sense of beauty and to give it spatial expression. This principle should address postsecular planning and suggest modes of engagement capable of pluralizing and decolonizing *aisthesis*, imagination, and beauty.

Further reading

Burdett, C. (2016) *Italy, Islam and Islamic world: representations and reflections, from 9/11 to the Arab uprising*, Oxford: Peter Lang.

This work is a detailed and deep reflection on the role of Islam and Islamophobia across the Italian political debate.

Corbin, H. (1989) *Spiritual Body and Celestial Earth: from Mazdean Iran to Shi-ite Iran*, Princeton, NJ: Princeton University Press.

This book offers a dense and stimulating overview of the themes of apophatic theology and creative imagination in Islamic philosophy. The work is complemented with a rich anthology of original works, from Suhrawardi to Abd-Al-Reza Khan Ebrahimi.

References

Abeysekara, A. (2006) 'Desecularizing secularism', *Culture and Religion*, 7: 205–43.
Aksamija, A. (2015) *Mosque Manifesto: propositions for spaces of coexistence*, Berlin: Revolver.
Allievi, S. (2010) *Mosques in Europe: why a solution has become a problem*, London: Alliance Publishing Trust.
AlSayyad, N. and M. Massoumi (2010) *The Fundamentalist City?: religiosity and the remaking of urban space*, London: Routledge.
Asad, T. (2003) *Formations of the Secular: Christianity, Islam, modernity*, Palo Alto, CA: Stanford University Press.
Beaumont, J. and C. Baker (2011) *Postsecular Cities: space, theory and practice*, London: Continuum.
Berking, H, Steets, S. and J. Schwenk (2018) *Religious Pluralism and the City: inquiries into Postsecular Urbanism*, London: Bloomsbury Academic.
Chiodelli, F. and S. Moroni (2017) 'Planning, pluralism and religious diversity: critically reconsidering the spatial regulation of mosques in Italy starting from a much debated law in the Lombardy region', *Cities*, 62: 62–70.
Ciocca, F. (2018) *Musulmani in Italia: impatti urbani e sociali delle comunità islamiche a Roma*, Roma: Melthemi.
Cloke, P. and J. Beaumont (2013) 'Geographies of postsecular rapprochement in the city', *Progress in Human Geography*, 37(1): 27–51.
Cloke, P., Baker, C., Sutherland, C. and A. Williams (2019) *Geographies of Postsecularity: re-envisioning politics, subjectivity and ethics*, London/New York: Routledge.
Conti, B. (2016) 'Islam as a new social actor in Italian cities: mosque controversies as sites of inclusion and separation', *Religion, State and Society*, 44: 238–57.
Degiorgis, N. (2015) *Hidden Islam: Islamic makeshift places of worship in north east Italy, 2009–2013*, 3rd edition, Bolzano: Rorhof.
D'Costa, G. (1990) *The Myth of a Pluralistic Theology of Religions*, Maryknoll: Orbis Books.
della Dora, V. (2018) 'Infrasecular geographies: making, unmaking and remaking sacred space', *Progress in Human Geography*, 42(1): 44–71.
Fincher, R., Iveson, K., Leitner, H. and V. Preston (2014) 'Planning in the multicultural city: celebrating diversity or reinforcing difference?', *Progress in Planning*, 92: 1–55.
Franke, W. (2014) *A Philosophy of the Unsayable*, Notre Dame, IN: University of Notre Dame Press.
Gökarıksel, B. and A. Secor (2015) 'Post-secular geographies and the problem of pluralism: religion and everyday life in Instanbul, Turkey', *Political Geography*, 46: 21–30.
Greed, C. (2011) 'A feminist critique of the postsecular city: God and gender'. In Beaumont J. and C. Baker (eds.) *Postsecular Cities. Space, theory and practice*, London: Continuum, pp. 104–19.
Guolo, R. (2011) *Chi Impugna la Croce: lega e chiesa*, Bari: Laterza.
Hick, J. (1980) *God Has Many Names*, London: Macmillan Press.
——— (1997) 'The possibility of religious pluralism: a reply to Gavin D'Costa', *Religious Studies*, 33: 161–66.
Howard-Snyder, D. (2017) 'Who or what is God, according to John Hick?', *Topoi*, 26: 571–86.
Jackson, M. (2016) 'Aesthetics, politics, and attunement: on some questions brought by alterity and ontology', *GeoHumanities*, 2: 8–23.
Lancione, M. (2014) 'Entanglements of faith: discourses, practices of care and homeless people in an Italian City of Saints', *Urban Studies*, 51: 3062–78.

Lloyd, V. W. and L. Viefhues-Bailey (2015) 'Introduction: Is the postcolonial postsecular?', *Critical Research on Religion*, 3(1): 13–24.
Mahmood, S. (2016) *Religious Difference in a Secular Age: a minority report*, Princeton, NJ: Princeton University Press.
McClymont, K. (2015) 'Postsecular planning? The idea of municipal spirituality', *Planning Theory & Practice*, 16: 535–54.
Milbank, J. (1990) 'The end of dialogue'. In D'Costa, G. (ed.) *Christian Uniqueness Reconsidered: the myth of a pluralistic theology of religions*, Ossining, NY: Orbis Books, pp. 174–91.
——— (2006) *Theology & Social Theory: beyond secular reason*, 2nd edition, Oxford: Wiley-Blackwell.
Modood, T. (2013) *Multiculturalism*, Cambridge, MA: Polity Press.
Morgan, G. (2016) *Global Islamophobia: Muslims and moral panic in the west*, London: Routledge.
Mouffe, C. (2013) *Agonistics: thinking the world politically*, London and New York, NY: Verso.
Nasr, S. H. (1987) *Islamic Art and Spirituality*, Albany: State University of New York Press.
——— (1989) *Knowledge and the Sacred*, Albany: State University of New York Press.
Olson, E., Hopkins, P., Pain, R. and G. Vincett (2013) 'Retheorizing the postsecular present: embodiment, spatial transcendence, and challenges to authenticity among young Christians in Glasgow, Scotland', *Annals of the Association of American Geographers*, 103: 1421–36.
Oosterbaan, M. (2014) 'Public religion and urban space in Europe', *Social & Cultural Geography*, 15(6): 591–602.
Rose, K. (2013) *Pluralism: the future of religion*, New York, NY: Bloomsbury.
Saint-Blancat, C. and O. Schmidt di Friedberg (2005) 'Why are mosques a problem? Local politics and fear of Islam in northern Italy', *Journal of Ethnic and Migration Studies*, 31: 1083–104.
Sandercock, L. and M. Senbel (2011) 'Spirituality, urban life and urban professions'. In Beaumont J. and C. Baker (eds.) *Postsecular Cities. Space, theory and practice*, London: Continuum, pp. 87–103.
Schmidt-Leukel, P. (2005) 'Exclusivism, inclusivism, pluralism: the tripolar typology—clarified and reaffirmed'. In Knitter, P. (ed.) *The Myth of Religious Superiority: multifaith exploration of religious pluralism*, Ossining, NY: Orbis Books, pp. 13–27.
Tse, J. K. H. (2014) 'Grounded theologies: "religion" and the "secular" in human geography', *Progress in Human Geography*, 38: 201–20.
Woodhead, L. (2016) 'Intensified religious pluralism and de-differention: the British example', *Society*, 53: 41–6.

28

Some critical remarks on religious identity

Peter Nynäs

Introduction

Jürgen Habermas was not the first scholar to address the idea of a postsecular condition. I remember that I myself stumbled upon this concept in the title of a book that was published in 1990 (Franck et al. 1990). This volume did not primarily follow from recognition of an increasing coupling of religion with global or other societal public conflicts in a Habermasian sense (Habermas 2008). Instead, many of the authors seemed to envisage a new era that meant a break with the earlier dichotomy between 'knowledge and belief' that was understood to be problematic. The concept of the postsecular is here comprehended in a more epistemological way against the backdrop of secularism and how reason, in contrast to religion, had become central to the Enlightenment narrative and thus a vehicle of progression and emancipation. This resembles Richard John Neuhaus claim from 1982 where he states 'We are witnessing the collapse of the 200-year old hegemony of the secular Enlightenment over public discourse' (Neuhaus 1982: 309).

Later Ursula King (2009: 11–2), for instance, defined the postsecular as a contrast to secularism when she in a more modest but similar manner wrote that postsecularism means 'a renewed openness to questions of the spirit, but one that retains the habits of critical thought which partially defines secularism'. Her emphasis on the need to engage with religion with an open mind involves a potential coexistence of secularism and religion, and the multidirectional possibilities of their relationship. It does not necessarily mean a simple turning back to premodern forms of religion. Rather, as Sarah Bracke (2008: 59) puts it: 'The modern becomes envisionable without the secular'. These views exemplify, on the one hand, that central to many received perspectives on secularization was also the assumption that there is an inherent 'incompatibility between some features of "modernity" and religious belief' (Taylor 2007: 543), and on the other hand alternative ways of addressing the postsecular in comparison with Habermas' observations.

It comes as no surprise that the discussion about postsecularity has not been constituted by a well-defined or well-articulated understanding of the concept. On the contrary, in a critical manner James Beckford points to several clusters of ideas associated with the concept and concludes that 'it goes without saying that the sheer variety of meanings is a notable feature of discourse about postsecularity [and] there are tensions between some of the meanings' (Beckford 2012: 13). He claims that '[t]here is therefore a danger that talking about the postsecular will be like waving a magic wand over all the intricacies, contradictions, and problems of what counts as religion to reduce them to a single, bland category' (Beckford 2012: 17). Nevertheless, this lack of consensus about definition does not necessarily mean a dead end. We might instead see this discussion as a reflection of how the postsecular is not limited to a resurgence of religion, but with the words of Justin Beaumont (2010: 6) point to an interplay of 'religious, humanist, and secularist positionalities' that is furthermore characteristic to many contemporary societies in the West. This focus on interplays and configurations allows us to broaden and nuance the perspective in contrast to Habermas' more simplified emphases of public-consciousness as a differing contrast to sociological indicators of growing secularity in terms of behaviour and convictions. However, in terms of how we view religious identities within such a framework, we might need to put Habermas thinking aside. Habermas' view of religion has throughout the decades changed towards a more positive appreciation of religion as a source in its own right, but his view of religious identities or voices is nevertheless rather abstract in contrast to the lived realities current research bears witness about.

Already more than two decades ago José Casanova (1994) concluded that although many modern societies are clearly marked by a general decline in institutional forms of religion, it does not automatically follow from this that religion as such loses influence and relevance either in the political arena, in the culture of a society, or in the everyday lives of individuals. A brief observation from a Finnish context shows that, on the one hand, we can clearly see a decline in religious membership, participation, and practice, especially during the last few decades (Nynäs et al. 2015b). Public discussion about value-laden issues such as the legalizing of gay marriage and immigration have lately contributed to this decline; and for a variety of reasons people have left the main religious institution in Finland, the Evangelical Lutheran Church of Finland. On the other hand, the public debates around these issues have simultaneously turned religion into an important subject matter, primarily in the media, and engaged a wide range of both religious and secular actors (cf. Moberg and Sjö 2012). This condition reflects, to a large extent, what Habermas referred to when writing about the postsecular. Still, it is not recognized in his writings how conditions like these are constituted by—and themselves form—spaces for negotiations and constructions of identities. The increasing diversification of people's lives addressed in many current discussions of religious change calls for some discussions regarding how we perceive religious identity. When we deal with postsecularity, we might need to critically consider how we conceptualize religious identity and subjectivity (see also Cloke et al. 2019).

Religious identity: a critical concept

In order to address the relevance of identities as part of the complex interplay and configurations at work, we need to conceptually bridge the 'the abstract level of politics and society and the deeply personal, inner landscapes of pluralism' (Illman et al. 2015: 215). In my view, this is of utmost importance when we discuss Habermas' notion of postsecularity, Habermas' view risks reproducing some problematic aspects rooted in an intellectualised and rationalized view of religion shared to a varying degree by many subdisciplines in the field. First, the subject, be it religious or secular, is defined by autonomy, rationality, and individuation, and the social is a matter of recognition. Second, the secular and the religious are essentialized and viewed as separate from each other, and consequently the reconciliation of these two lies at the heart of some of his ideas. (see, e.g., Habermas 1988, 1993, 2006a, 2006b, 2008). Turner (2010) is quite right in stating that arguments about the postsecular character of contemporary Western societies, as well as other closely associated broader arguments such as those having been presented by influential commentators such as Charles Taylor (2007), have difficulties in moving beyond the terrain of abstract speculation. The categories that we impose on people are not necessarily in correspondence with how they live and experience their lives and how they further act and relate based on this within a public sphere.

This is not the place to sketch the main theories pertaining to understanding identity. Still, we need to underline the fact that identity is not easily captured. Identities are not isolated, but connected to different social circumstances within which they operate. Identity may vary from context to context and situation to situation. There are great variations in terms of how we acquire identities, and family relations and ritual processes might be involved. The extent to which we can create and express identities also involves great variations and different forms of symbolic expressions (see, e.g., Bowie 2006: 62–78) Further, we are all an assemblage of multiple or hybrid identities, and we may foreground a specific identity in a given context, contingent upon a host of factors such as expediency and the desire to be included. Our everyday lives take place within a power-infused interactional web which requires us to function as an individual with multiple identities, or at least context-specific identifications (Taylor et al. 2010; Richardson and Monro 2012; Yip and Nynäs 2012; Woodhead and Catto 2012; Nynäs et al. 2015a). With specific reference to ethnicity, race, and nationalism, Banton (2011) offers important insights about multiple belongings and intersectionality:

> [E]very human is assigned to, and identifies with, many social categories; each identification entails costs and benefits. The interrelation between categories presents a generalization of what is currently known as intersectionality. The relative importance of categories changes as individuals trade off advantages associated with one form of social alignment for that of another.
>
> *(2011: 199)*

Kimberlé W. Crenshaw (1991, 1989) coined the concept to make sense of the so-called discrimination on multiple grounds, and it has come to, more in general, refer to the 'mutually constitutive relations among social identities' (Shields 2008: 301). The term draws our attention towards how social identities affect each other and why cannot identities be studied in isolation. The idea of hybrid and multiple identities in contrast acknowledges how people live in-between and combine cultures or aspects of them (cf. Pauha and Jasinskaja-Lahti 2003). In addition, we need to address the everyday aspects of identities. The publicly relevant socio-political aspects of identities are strongly dependent on the more 'mundane' and private aspects of identity. It is self-evident to state that everyday life and public social life are connected, but the implications of the messiness, fluidity, and 'taken-for-grantedness' of everyday life is often overlooked. On the one hand it is constituted by subjectivities, experiences, emotions, bodies, and desires that are lived out on individual and collective levels of spaces and politics, and on the other hand affected by the ways in which individuals—as social actors—appropriate, negotiate, reinterpret, transgress, invert, and challenge various normative practices and spaces (e.g., Miller 2008; Pink 2012). Research on religion, gender, and sexuality has shed light on many aspects related to what multifaceted construction religious identities mean in relation to discussions about postsecularity (Yip and Nynäs 2012).

These brief remarks on identity are relevant for the discussion of the postsecular. To start with, Habermas views are clearly influenced by his broader social-philosophical thinking where the focus primarily has been on the importance of the 'public sphere', and inclusive, participatory democracy and recognition where what is outside the public sphere easily is overlooked (Harrington 2007; Dillon 2010; Martin 2010). This point can be exemplified when Habermas writes that 'in a constitutional state, all norms that can be legally implemented must be formulated and *publicly justified* in a language that all the citizens understand' (Habermas 2008: 28). In this respect, it can be argued that Habermas is maintaining a normative secularist position that requires religious individuals and groups to be able to translate 'their religious norms into a secular idiom' when engaging in public civic and political debate (e.g., Habermas 2006b). This reflects a basic dichotomy in Habermas thinking. It maintains the idea of 'ongoing tensions between religious cultures and civic political life' that is constituted by the presence of religious otherness (Dillon 2010, 141f; 146). In Habermas' thinking, religious voices and identities stand out as the foreign other, the one to accommodate, integrate, and, at the best, also learn and be inspired from through translation (cf. Habermas 2006b; Cooke 2014).

This observation addresses a general risk involved in assuming a simplified notion of religious identity and projecting this onto the postsecular scene. In addition, this problem further risks erasing or neglecting some of the contemporary complexity of religious change. An obvious and problematic juxtaposition of the secular and the religious resides at the core of the discussion around the postsecular. For example, when examining the wider public debates surrounding some contested issue that involves both 'religious' and 'secular' parties, it is not always clear what constitutes a 'religious' idea or argument as opposed to a 'secular' one. It has become increasingly evident that religious individuals not necessarily conceive of themselves as either religious *or* secular. In contrast, contemporary religious changes in the West require us to rethink the dichotomous relation between religious and secular, and they are not necessarily opposites when people construct their identities: they combine spiritual and religious positions with secular values into authentic and unified outlooks on life (Nynäs et al. 2015b).

The relevance of accounting for identities in a nuanced way has recently become increasingly evident. Many researchers in the study of religion emphasize today how people mix ideas, practices, and identities in ways that challenge given scholarly categories of religion and previous expectations (cf. Luhrman 2012; Bender and Taves 2012; Frisk and Åkerbäck 2013; Gilhus and Sutcliffe 2013; Day et al. 2013; Nynäs et al. 2015c). We face, in other words, a situation, where parameters of religion and religious subjectivities seem to have been altered, and consequently we need to broaden the conceptual horizon regarding how we understand these factors. In the real lives of people, this is not necessarily a problem or a contradiction. Following Ann af Burén (2015), we might have to consider current religious subjectivities in terms of simultaneities. Simultaneity is not necessarily the opposite of consistency or coherence, rather a complex and situated *dialogical* interpretation of the boundaries of the subject and his or her surroundings. af Burén (2015: 212) writes that this observation recognizes:

> the cultural expectation of being able to relate to the many meanings of the concept of religion selectively and being able to appreciate and appropriate religious aspects from a variety of contexts. That, however, is not to say that there are no boundaries. As I have shown [---] the local discourses on religion [---] are of crucial importance when it comes to establishing which 'elements' of religion they consider interesting and of value, and what they perceive they are 'allowed' to disclose in different settings and what not.

af Burén's observation adds a relevant concept to the emerging discussion about multiple or hybrid belongings that have recently been accounted for more seriously compared to before.

André van der Braak and Manuela Kalsky (2017) underline that current religious subjects construct their religious identities by combining elements not only from various religious traditions but also through different forms of multiple religious belongings. Empirically oriented research has fostered many similar observations. van der Braak and Kalsky (2017: 662) point to different forms of multiple identity, ranging from the self-conscious combination of two or more traditions to cases where the individuals lack a sense of belonging to any specific tradition. In addition, some people might 'claim to embrace all religious traditions, rather than multiple individual religious traditions' (*ibid.*: 662). This raises the question about the category of religion. These kinds of reflections about religious identities also follow from 'the world religions paradigm', the Western protestant idea about religious traditions as mutually exclusive entities separated by borders (e.g., Hedges 2017).

The fluidity and hybridity that may be involved in the construction of religious identities has implications for many notions about postsecularity including the one proposed by Habermas. It should be underlined that the observations about intersectionality and everyday lives, fluidity and hybridity, do not exclude in any way the fact that a religious movement, community, or tradition can be a strong external source providing material, symbols, roles, means, tools, and so forth for religious identities. Still, the observations do address the relevance of accounting for a growing plurality and diversity on all levels. Hovi et al. (2015) point to how forms of individualism underlie current identity constructions in terms of making a choice or representing a form of distinctiveness in relation to others: current collectivities involve a striving for authenticity across community boundaries where individual agency is more on the surface. Religious subjects are not passive recipients. Rather, they express a high degree of autonomy in relation to external religious influences (see Hunt 2015).

The aforementioned exemplifies that current religious change may have a far-reaching influence. Ulrich Beck (2010) claims that in contrast to previous perceptions:

> we see the formation of a new, religiously determined, global sociality in which increased significance is attached to transnational, religious imagined communities which complement, and enter into competition and conflict with the institutionalized forms of national societies and national institutions.
>
> *(Beck 2010: 42)*

His point is that in times of individualization, contemporary subjects are increasingly forced to reflect on their own subjectivity in the light of religious 'others', and this process 'gains enhanced significance both inside and outside the churches' (Beck 2010: 48). Religious subjects are constantly exposed to both internal and external negotiations. As a result, assumptions about distinct religious voices or agencies that are clearly set apart from secular voices are not in correspondence with the everyday reality research on religion in the West increasingly pointed to.

A matter of socialization and location

From the aforementioned I would like to conclude that investigations of postsecularity cannot neglect a wider discussion about major and recent transformations of modern religiosity such as the 'growth of 'post-institutional spirituality', the development of all forms of popular religion' (Turner 2010: 650) and the changing implications for our comprehension of religious subjects and identities (Nynäs 2017). In terms of identities, the idea of the postsecular may be limited to the extent it is rooted in the juxtaposition of religious and the secular and due to an insufficient account of the complexity embedded in the contemporary religious change in the West. Many observations from current research that religious traditions and institutions, on the one hand, and society and culture on the other hand, converge and develop in novel ways. In other words, religion today develops and is shaped in conjunction with other social phenomena than traditional religious institutions. Some further observations from empirically based research might foster greater attentiveness to the diverse nature of religion in contemporary societies and are more sensitive to its inherent complexity beyond a dichotomous separation of the religious and the secular.

The emerging body of research not only indicates that contemporary religious subjects are situated and constituted differently in comparison to what has previously been assumed. Current religious change has implications for religious socialization, that is, the process through which subjects internalize religion and/or secular worldviews in all their forms and identities are formed. Several accounts of contemporary religious change direct particular focus at the ways in which current developments have profound implications for traditional understandings of religious authority (cf. Brown and Lynch 2012; Hunt 2015). Both psychologists and sociologists of religion have traditionally viewed the family and the locally embedded religious community (e.g., in the form of the 'congregation') as primary loci of religious socialization processes. Even though this still holds true in many respects, some researchers direct our focus to the 'crisis in religious socialisation', referring to the declining capacity of religious institutions to transmit religious practices, views, and beliefs across generations (Hunt 2015: 13). In order to reach an adequate understanding of the place of religion in people's lives today, we need to look beyond the conventional spheres of religious expression and socialization and into the wider contemporary cultural realm (cf. Beckford 2010; Lövheim 2012; Klingenberg 2014; Moberg and Sjö 2015).

Looking briefly in the direction of the role of new media can help us exemplify this role. Socialization has often been seen as key to understanding religious identity, but the social and cultural condition in which later generations have been raised is different in several ways. Young people born after 1990 are not characterized by having been gradually accustomed to consumer culture and digital media during their lifetime. Instead, these social phenomena have constituted an inherent and unquestioned part of their childhood and youth (cf. Possamai 2009). Palfrey and Gasser (2008) refer to this generation as 'the born digital' and the 'digital natives'. The development of modern communications technologies and digital technologies in particular has fundamentally altered not only our means of communication and interaction but also our sense of time and space. This has proved to be relevant for understanding current religious change.

Media is today the main provider of access to the global sociality referred to by Beck (2010). Developments in media during the past decades have had great significance for the new public presence of religion, but it has also altered the character of the religious landscape (Hjarvard and Lövheim 2012). Contemporary forms of media do, on the one hand, provide new tools for the communication of current forms of religion. This makes it possible for religious ideas, attitudes, and practices of ideas to travel across and beyond the boundaries of nations and religious institutions. In addition, an increasing number of researchers have emphasized the importance of popular culture with regard to circulation of current religious notions, ideas, and practices, among both individuals and groups (Partridge 2005, 2006). Consequently, new technologies have contributed to the deregulation of religious ideas and symbols, to the transformation of previous power relations, and to novel individual and collective agency (Lynch et al. 2011; Moberg and Granholm 2012; Granholm et al. 2015).

The relevance of media can be exemplified from a study of LGBT-activists affiliated with the Evangelical Lutheran Church of Finland (Nynäs and Lassander 2015). For this comparatively small group of activists, different forms of media meant a resource in many ways. Social media provided, to start with, a forum and a platform for bringing people together from across the country and sharing experiences, values, and goals, and a tool in the negotiation of internal diversities and boundaries, but also in relation to other institutional, organizational, and societal actors. The shared platform also meant a window to additional media resources that provided the network with both theological resources (e.g., reinterpretations of religious narratives) and secular resources (e.g., LGBT-activism in a human rights perspective) of both national and international character. The nature of the group was clearly marked by a great diversity in terms of both religious and/or secular positions and motivations. In this respect, media played a crucial role in making space for a new identity within the Evangelical Lutheran Church of Finland, rooted in both individual and organizational diversity. In addition, it provided a space for a unique religious LGBT identity in Finland by bringing people in the margins together regardless of previous religious convictions and belongings. In many cases, the blurring of the secular and the religious was significant: both forms of religious identities could be fuelled by secular motivations of convictions.

This shows how parameters of religious identity are changing today. Authority, resources and social patterns, and networks are altered in the contemporary societal and cultural context. In particular, the role of media (e.g., Moberg et al. 2014), consumer-culture (e.g., Gauthier and Martikainen 2013), and social movements (e.g., Nynäs et al. 2013; Nynäs and Lassander 2015) is vital to these changes.

Where does this take us? Ongoing processes of religious change have been described through a range of interrelated conceptual frameworks and with different emphases, such as desecularization (Berger 1999), resacralization (Davie 2010), re-enchantment (Partridge 2005, 2006), postsecularity (Habermas 2008; Nynäs et al. 2012), un-churching (Fuller 2001), and de-Christianization (Brown and Lynch 2012), to name just a few. Some researchers approach religious change against the background of a general 'subjective or expressive turn' (Heelas and Woodhead 2005), while others speak of an 'Easternization of the West' (Campbell 2007). However, in terms of a summary of some relevant findings, Linda Woodhead's outline provides an overview from a sociological perspective. Her outline mirrors the main observations from the wide-ranging AHRC/ ESRC *Religion and Society Programme* (Woodhead 2012). Woodhead refers to the emergence of a 'new style religion' that is replacing 'reformation style religion' (see Table 28.1).

Table 28.1 Reformation and new style religions

Associational characteristics	*Reformation style religion*	*New style religion*
Type	Local congregation, part of a national body	Local or virtual gathering, part of an international network
Membership	Bonding, for life	Voluntary, for need
Size	Medium	Very small or very large
Culture	Authoritarian and uniform (one-size-fits-all)	Egalitarian and fluid (customized)
Focus	Transcendence (salvation religion)	Immanence (life-path religion)
Identity	Unquestioned	A contested (individual) achievement
Members' agency	Little voice or choice	Voice and choice

Source: Adapted from Woodhead (2012).

This means a simplification in light of the many aspects involved as well as in light of many contextual variations. However, this clearly accounts for the multifaceted change and locates the construction of identities and agency in relation to this change. The current reconfiguration involves a shift not only in the social organization of religion but also in how contemporary identities and agencies evolve in terms of 'beliefs', the nature of these, and what they are directed towards. Warner et al. (2010) make an important point when they note in their discussion of Charles Taylor's *A Secular Age* (2007) that secularization is not merely characterized by a decline in religious belief and practice, but it also comprises a change in the conditions for religious belief. A change in the conditions for religious belief has consequences for the ways these beliefs can be expressed (Taylor 2007). In connection with a discussion concerning the transformation of contemporary religiosities, Liselotte Frisk has drawn on the many discussions regarding religion today and provided a lucid and nuanced overview of some of the issues at hand. Although different researchers certainly have their own points of emphasis, there appears to be a number of aspects upon which the majority focus. According to Frisk's reading, one can make the claim that today's religious landscape is characterized by an emphasis on: (a) eclecticism and syncretism, (b) personal experience rather than ideology, (c) radical egalitarianism and/or every individual's own authority in relation to her or his spirituality and religiosity, (d) the self and that which is human over divine beings, and (e) the worldly at the expense of a life after death (Frisk 2011; Frisk and Nynäs 2012).

Many of the characteristics pointed to by both Woodhead and Frisk challenge our received understanding of contemporary religious identities, for example, in terms of a shift away from the institutional and away from a focus on belief or faith. This means that at the individual level, the distinction between the 'religious' and the 'secular' do not define the limits of identity construction. Rather they present subjects with meaningful resources in a dynamic space constituted by intersectionality, fluidity, and a range of individual and social needs and motivations. This may be far from the idea of identities shaped by a perceived and projected dichotomy between the secular and the religious. The location and socialization of religion are interdependent and both have changed in a decisive way, and it challenges our received notion of religious identity today.

Concluding remarks: towards a dialogical view

The increasing diversification of people's lives is a relevant aspect of the postsecular calls for some careful considerations regarding how we perceive religious identity. When we deal with postsecularity, we might need to critically rethink how we conceptualize religious identity and subjectivity. Current relocation of religion is intertwined with a shift in patterns of socialization. This affects identities in the sense that aspects such as lived religion, intersectionality, and fluidity become more evident. In particular, we need to question how Habermas projects an idea of contemporary subjects as being either religious or secular, and consequently scrutinize the sharp distinction between the secular and the religious. A more dialogical notion is called for. This provides a better tool for comprehending current social and cultural reconfigurations of religion and in particular how they are simultaneously played out as a diversity of religiosities and identities that resist and challenge our previous categories such as the religion and the secular. This needs to involve recognition of multiple and hybrid identities and the intersectional configurations that follow.

In light of the aforementioned, I suggest that the term identification rather than identity should direct our theoretical perspectives on postsecularity. A focus on identification accounts better for the relational connotations and the shifting situatedness within which the relational is formed. Identification is a matter of ongoing processes or agencies that involve intersectionality. This has also been better accounted for from an anthropological perspective. Jon Mitchell (2009) draws our attention to a more dialogical understanding of subjects that meets well the observations I have stressed earlier. Mitchell stresses the need to observe the multifaceted complexity of self that is inherently intersubjective (see also Herman et al. 2012). He supports this view by referring to discussions in anthropology which have been influenced by more recent phenomenological reflection (Csordas 1994; Battaglia 1995; Jackson 1996; van Wolputte 2004). The self is viewed here as constituted by being located 'within and emerging through ongoing social process, and conceived as an unfolding condition of being' (Mitchell 2009: 54). This means nuancing how we perceive socially situated subjects as being parts of their environment and its symbolic qualities (Holm 2014). Mitchell (2009) refers to Csordas' (1994: 5) understanding of the self:

> Self is neither substance nor entity, but an indeterminate capacity to engage or become oriented in the world, characterized by effort and reflexivity. In this sense self occurs as a conjunction of prereflexive bodily experience, culturally constituted world or milieu, and situational specificity or habitus. Self processes are orientational processes in which aspects of the world are thematized, with the result that the self is objectified, most often as a "person" with a cultural identity or set of identities.

Despite the fact that Habermas' view of religion has changed towards a positive appreciation, there is throughout his writing an objectification of the religious that renders religious positionalities foreign and simplified in contrast to a lived reality. Habermas' view of the postsecular is rooted in othering subjects, and it does not reflect important observations from current research. The latter confirms how contemporary societies involve an interplay between religious, humanist, and secularist positionalities (e.g., Beaumont 2010; Nynäs et al. 2015c). How such complex interplays are constituted and reconfigured need to be taken more methodologically serious in studies of postsecularity in order to avoid future simplifications. Furthermore, a recognition that the distinction between the religious and the secular is constructed and abstract might help us downplay societal and cultural tensions and conflicts and address other relevant aspects involved. Is the translation that Habermas calls for not only a future task for academics, but an ongoing cultural and societal process that already happens and produces new diversities? How we conceive of the 'religious' subject is a matter that cuts through epistemology, methodology, and theory, and easily causes implicit biases in our studies.

As a final note, I need to underline that a dialogical understanding of religious subjects should not, of course, be equated with a dialogical position; it should not be mistaken for suggesting an open or inclusive attitude on behalf of the subject. Such openness has perhaps implicitly been assumed on behalf of religious subjects in my aforementioned discussions and examples, and it remains important to address also the significant role of radical or fundamentalist positions. Radicalizing tendencies have always been a normative part of history (cf. Hood et al. 1996). Alongside the changes depicted earlier where current diversity and complexity can be played out as pluralist positions or eclecticism, we have lately also witnessed forms of what Almond et al. (2003) call the emergence of 'strong' religion. These tendencies are not reminiscent of 'an earlier form of religion' but a prominent aspect of ongoing negotiations within postsecular spaces. They can be what Manuel Castells (2004) refers to as defensive or resistant reactions to recent developments or, for instance, part of what Peter Beyer (1994) refers to as particularistic religious responses to globalization, such as various expressions of religious fundamentalisms or nationalisms. Due to recent societal and cultural changes, a strict focus on organizational or institutional settings is problematic in the study of radicalized religion; both social and religious movements tend today to develop more ephemeral forms of organization in the nexus of liquid networks and social media (Castells 2012). More radicalized religious expressions and values are also globally transmitted through, for example, the use of media technology (cf. Rajagopal 2001). Despite the fact that strong religion often is impaired by dissociation from dialogue (in the meaning of openness), such forms of religiosity are nevertheless the results of both internal and external negotiations and dialogues (cf. Herriot 2007; Lassander and Nynäs 2016).

Further reading

Cooke, M. (1992) 'Habermas, autonomy, and the identity of the self', *Philosophy and Social Criticism*, 18(3/4): 269–91.

This article by Maeve Cooke discusses critically and in-depth Habermas' conception of self-identity with a particular focus on the relevance of autonomy and self-realization in contrast to, e.g., Foucault's perspective.

Day, A. (2011) *Believing in Belonging*, Oxford: Oxford University Press.

Abby Day's book is important in the sense that it empirically directs our attention to how social relationships provide sites for religion 'belief' and questions the propositional and rational view of what religious identity is constituted by.

Jackson, M. (1998) *Minima Ethnographica: intersubjectivity and the anthropological project*, Chicago, IL: Chicago University Press.

Michael Jackson's book addresses the importance of particular social events and intersubjective encounters as critical lenses to whole historical and contemporary situations, and this asserts the relevance of the links between academic study and lived experience.

McGuire, M. (2008) *Lived Religion: faith and practice in everyday life*, Oxford: Oxford University Press.

Meredith McGuire's book has meant a decisive methodological turn towards an emphasis of lived and experienced religion in peoples' everyday lives. It provides tools and perspectives for comprehending fuller the recurring vitality in Western societies.

Rebughini, P. (2014) 'Subject, subjectivity, subjectivation', *sociopedia.isa*. 2nd colloquium, accessed online 11-06-2018.

Paola Rebughini turns our focus towards problematic terms in social sciences such as subject, individual, agent, person, or social actor that are often used interchangeably. She analyzes these and associated notions of subjectivity including Habermas' view.

References

Almond, G. A., Appleby, R. S. and E. Sivan (2003) *Strong Religion: the rise of fundamentalism around the world*, Chicago, IL: University of Chicago Press.

Banton, M. (2011) 'A theory of social categorie', *Sociology*, 45(2): 187–201.

Battaglia, D. (ed.) (1995) *Rhetorics of Self-Making*, Berkeley: University of California Press.

Beaumont, J. (2010) 'Transcending the particular in postsecular cities'. In Molendijk, A. L., Beaumont, J. and C. Jedan (eds.) *Exploring the Postsecular: the religious, the political and the urban*, Leiden: Brill, pp. 3–18.

Beck, U. (2010) *A God of One's Own: religion's capacity for peace and potential for violence*, Cambridge, MA: Polity Press.

Beckford, J. (2010) 'Foreword'. In Mayo-Collins, S. and P. Dandelion (eds.) *Religion and Youth*, Farnham: Ashgate, pp. xxiii–xxiv.

——— (2012) Public religions and the postsecular: critical reflections', *Journal for the Scientific Study of Religion*, 51(1): 1–19.

Bender, C. and A. Taves (eds.) (2012) *What matters? ethnographies of value in a not so secular age*, New York, NY: Columbia University Press.
Berger, P. L. (1999) *The Desecularization of the World: resurgent religion and world politics*, Washington, DC: Ethics and Public Policy Center.
Beyer, P. (1994) *Religion and Globalization*, New York, NY: Sage Publications.
Bowie, F. (2006) *The Anthropology of Religion: an introduction*, 2nd edition, Oxford: Blackwell.
Braak, A. van der and M. Kalsky (2017) 'Introduction to the topical issue "multiple religious belonging"', *Open Theology*, 3: 662–64.
Bracke, S. (2008) 'Conjugating the modern/religious, conceptualizing female religious agency: contours of a "post-secular" conjuncture', *Theory, Culture & Society*, 25(6): 51–67.
Brown, C. and G. Lynch (2012) 'Cultural perspectives'. In Woodhead, L. and R. Catto (eds.) *Op cit.*, pp. 329–351.
af Burén, A. (2015) *Living Simultaneity: on religion among semi-secular Swedes*, Huddinge: Södertörns Högskola.
Campbell, C. (2007) *The Easternization of the West: a thematic account of cultural change in the modern era*, Boulder, CO: Paradigm Publishers.
Casanova, J. (1994) *Public Religions in the Modern World*, London: University of Chicago Press.
Castells, M. (2004) *The Power of Identity*, Malden, MA: Blackwell.
——— (2012) *Networks of Outrage and Hope: social movements in the internet age*, Cambridge, MA: Polity Press.
Cloke, P., Baker, C., Sutherland, C. and A. Williams (2019) *Geographies of Postsecularity: re-envisioning politics, subjectivity and ethics*, London/New York: Routledge.
Cooke, M. (2014) 'The limits of learning: Habermas social theory and religion', *European Journal of Philosophy*, 24(3): 694–711.
Crenshaw, K. W. (1989) 'Demarginalizing the intersection of race and sex: a black feminist critique of antidiscrimination doctrine, feminist theory, and antiracist politics', *University of Chicago Legal Forum*, 1989(1): 139–67.
——— (1991) 'Mapping the margins: intersectionality, identity politics, and violence against women of color', *Stanford Law Review*, 43: 1241–99.
Csordas, T. (1994) *The Sacred Self: a cultural phenomenology of charismatic healing*, Berkeley: University of California Press.
Davie, G. (2010) 'Resacralisation'. In Turner, B. S. (ed.) *The New Blackwell Companion to the Sociology of Religion*, Chichester: Wiley-Blackwell, pp. 160–78.
Day, A., Vincett. G. and C. Cotter (eds.) (2013) *Social Identities between the Sacred and the Secular*, Farnham: Ashgate.
Dillon, M. (2010) 'Can post-secular society tolerate religious differences?' *Sociology of Religion*, 71(2): 139–56.
Franck, O., Thalén, P. and B. Sahlin (eds.) (1990) *Postsekulariserat Interregnum? från tro och vetande till vetenskap och mystic*, Delsbo: Åsak.
Frisk, L. (2011) 'Globalization: a key factor in contemporary religious change', *Journal of Alternative Spiritualities and New Age Studies*, 5: i–xiv, accessed online 09-02-2015.
Frisk, L. and P. Åkerbäck (2013) *Den Mediterande Dalahästen: religion på nya arenor i samtidens Sverige*, Stockholm: Dialogos.
Frisk, L. and P. Nynäs (2012) 'Characteristics of contemporary religious change: globalization, neoliberalism, and interpretative tendencies'. In Nynäs, P., Lassander. M. and T. Utriainen (eds.) *Op cit.*, pp. 47–70.
Fuller, R. (2001) *Spiritual, But Not Religious: understanding unchurched America*, Oxford: Oxford University Press.
Gauthier, F. and T. Martikainen (eds.) (2013) *Religion in Consumer Society: brands, consumers and markets*, Farnham: Ashgate.
Gilhus, I. S. and S. J. Sutcliffe (2013) 'Conclusion: new age spiritualities—"good to think" in the study of religion'. In Sutcliffe, S. J. and I. Gilhus (eds.) *New Age Spirituality: rethinking religion*, Durham: Acumen, pp. 256–62.
Granholm, K., Moberg, M. and S. Sjö (eds.) (2015) *Religion, Media, and Social Change*, London: Routledge.
Habermas, J. (1988) 'Individuierung durch vergesellschaftung: zu G.H. Meads theorie der subjektivität'. In Habermas, J. *Nachmetaphysisches Denken: philosophische aufsätze*, 2nd edition, Frankfurt/Main: Suhrkamp Verlag, pp. 184–241.
——— (1993) *Justification and Application: remarks on discourse ethics*. Cambridge, MA: The MIT Press.

——— (2006a) 'Religion in the public sphere', *European Journal of Philosophy,* 14: 1–25.
——— (2006b) *Dialectics of Secularization: on reason and religion,* edited by F. Schuller and B. McNeil, San Francisco, CA: Ignatius Press.
——— (2008) 'Notes on post-secular society', *New Perspectives Quarterly*, 25(4): 17–29.
Harrington, A. (2007) 'Habermas and the "post-secular society"', *European Journal of Social Theory,* 10(4): 543–60.
Hedges, P. (2017) 'Multiple religious belonging after religion: theorising strategic religious participation in a shared religious landscape as a Chinese model', *Open Theology*, 3: 48–72.
Heelas, P. and L. Woodhead (2005) *The Spiritual Revolution: why religion is giving way to spirituality,* Oxford: Blackwell Publishing.
Herman, A., Beaumont, J., Cloke, P. and A. Walliser (2012) 'Spaces of engagement in postsecular cities'. In Beaumont, J. and P. Cloke (eds.) *Faith-Based Organizations and Exclusion in European Cities,* Bristol: Polity Press, pp. 59–80.
Herriot, P. (2007) *Religious Fundamentalism and Social Identity*, New York, NY: Routledge.
Hjarvard, S. and M. Lövheim (eds.) (2012) *Mediatization and Religion: Nordic perspectives,* Göteborg: Nordicom.
Holm, N. G. (2014) *The Human Symbolic Construction of Reality: a psycho-phenomenological study,* Zürich: LIT Verlag.
Hood, R. Spilka, B., Hunsberger, B. and R. Gorsuch (1996) *The Psychology of Religion: an empirical approach,* New York, NY/London: The Guilford Press.
Hovi, T., Illman, R. and P. Ingman (2015) 'Interlacing identities, agencies and resources'. In Nynäs, P., Illman, R. and T. Martikainen (eds.) *Op cit.*, pp. 89–110.
Hunt, S. J. (2015) 'Believing vaguely: religious socialization and Christian beliefs in Britain', *Italian Journal of Sociology of Education,* 7(3): 10–46.
Illman, R. H., Martikainen, T., Nynäs, P., Sjö, S. and L. Wickstrom (2015) 'Reframing pluralism'. In Nynäs, P., Illman, R. and T. Martikainen (eds.) *Op cit.*, pp. 11–28.
Jackson, M. (ed.) (1996) *Things as They Are: new directions in phenomenological anthropology,* Bloomington, IN: Indiana University Press.
King, M. (2009) *Postsecularism: the hidden challenge to extremism,* London: James Clarke & Co Ltd.
Klingenberg, M. (2014) *Conformity and Contrast: religious affiliation in a Finland-Swede youth context,* Helsinki: University of Helsinki.
Lassander, M. and P. Nynäs (2016) 'Contemporary fundamentalist Christianity in Finland: the variety of religious subjectivities and their association with values', *Interdisciplinary Journal for Religion and Transformation in Contemporary Society,* 2(3): 154–84.
Lövheim, M. (2012) 'Religious socialization in a media age', *Nordic Journal of Religion and Society,* 25(2): 151–68.
Luhrmann, T. M. (2012) *When God Talks Back: understanding the American evangelical relationship with God,* New York, NY: Alfred A. Knopf Inc.
Lynch, G, Mitchell, J. P. and A. Strhan (eds.) (2011) *Religion, Media and Culture: a reader,* London: Routledge.
Martin, B. (2010) 'Contrasting modernities: postsecular Europe and enspirited Latin America'. In Molendijk, A. L. M., Beaumont, J. and C. Jedan (eds.) *Op cit.*, pp. 63–90.
Miller, D. (2008) *The Comfort of Things*, Cambridge, MA: Polity Press.
Mitchell, J. (2009) 'Ritual transformation and the existential grounds of selfhood', *Journal of Ritual Studies,* 23(2): 53–66.
Moberg, M. and K. Granholm (2012) 'The concept of the post-secular and the contemporary nexus of religion'. In Nynäs, P., Lassander, M. and T. Utriainen (eds.) *Op cit.*, pp. 95–128.
Moberg, M. and S. Sjö (2012) 'The evangelical Lutheran Church and the media in post-secular Finland'. In Hjarvard, S. and M. Lövheim (eds.) *Op cit.*, pp. 79–91.
——— (2015) 'Mass-mediated popular culture and religious socialisation.' In Granholm, K., Moberg, M. and S. Sjö (eds.) *Op cit.*, pp. 91–109.
Moberg, M., Sjö, S. and K. Granholm (2014) 'Introduction'. In Granholm, K., Moberg, M. and S. Sjö (eds.) *Op cit.*, pp. 1–15.
Neuhaus, R. J. (1982) 'Educational diversity in post-secular America', *Religious Education,* 77(3): 309–20.
Nynäs, P. (2017) 'Making space for a dialogical notion of religious subjects: a critical discussion from the perspective of postsecularity and religious change in the West', *Journal of Constructivist Psychology*, 31: 54–71, accessed online 15-03-2018.

Nynäs, P., Illman, R. and T. Martikainen (eds.) (2015a) *On the outskirts of 'the Church': diversities, fluidities and new spaces in contemporary religion*, Zürich: LIT-Verlag.
——— (2015b) 'Rethinking the place of religion in Finland'. In Nynäs, P. Illman, R. and T. Martikainen (eds.) *Op cit.*, pp. 11–28.
——— (2015c) Emerging trajectories of religious change in Finland. In Nynäs, P. Illman, R. and T. Martikainen (eds.) *Op cit.*, pp. 217–26.
Nynäs, P. and M. Lassander (2015) 'LGBT activism and reflexive religion: a case study from Finland in the light of social movements theory', *Journal of Contemporary Religion*, 30(3): 435–71.
Nynäs, P., Lassander. M. and J. Kontala (2013) 'Sekularism och religiös HBT-aktivism inom ramen för religiös förändring: exempel på värdet av perspektiv från studiet av samhällsrörelser, *Teologisk tidskrift*, 5(6): 459–75.
Nynäs, P., Lassander, M. and T. Utriainen (eds.) (2012) *Post-Secular Society*, New Brunswick, NJ: Transaction Publishers.
Palfrey, J. and U. Gasser (2008) *Born Digital: understanding the first generation of digital native,* New York, NY: Basic Books.
Partridge, C. (2005) *The Re-Enchantment of the West: alternative spiritualities, sacralization, popular culture, and occulture, Vol. 1*, London: Continuum.
——— (2006) *The Re-enchantment of the West: studies in sacralization and occulture, Vol. 2*, London: Continuum.
Pauha, T. and I. Jasinskaja-Lahti (2003) '"Don't ever convert to a Finn": young Muslims writing about Finnishness', *Diaconia: The Journal for the Study of Christian Social Practice*, 4(2): 172–93.
Pink, S. (2012) *Situating Everyday Life: practices and places*, London: Sage.
Possamai, A. (2009) *Sociology of Religion for Generations X and Y*, London: Equinox.
Rajagopal, A. (2001) *Politics after Television: Hindu nationalism and the re-shaping of the public in India*, Delhi: Cambridge University Press.
Richardson, D. and S. Monro (2012) *Sexuality, Equality and Diversity,* Basingstoke: Palgrave Macmillan.
Shields, S. A. (2008) 'Gender: an intersectionality perspective', *Sex Roles,* 59(5): 301–11.
Taylor, C. (2007) *A Secular Age*, Cambridge, MA: The Belknap Press of Harvard University Press.
Taylor, Y., Hines, S. and M. E. Casey (2010) *Theorizing Intersectionality and Sexuality*, Basingstoke: Palgrave Macmillan.
Turner, B. S. (2010) 'Religion in a post-secular society'. In Turner, B. S. (ed.) *The New Blackwell Companion to the Sociology of Religion*. Chichester: Blackwell, pp. 649–67.
Warner, M., VanAntwerpen, J. and C. Calhoun (2010) 'Editors' introduction'. In Warner, M., VanAntwerpen, J. and C. Calhoun (eds.) *Varieties of Secularism in a Secular Age*, Cambride, MA/ London: Harvard University Press, pp. 1–31.
Wolputte, S. van (2004) 'Hang on to your self: of bodies, embodiment, and selves', *Annual Review of Anthropology*, 33: 251–69.
Woodhead, L. (2012) 'Introduction'. In Woodhead, L. and R. Catto (eds.) *Op cit.*, pp. 1–33.
Woodhead, L. and R. Catto (eds.) (2012) *Religion and Change in Modern Britain*, London: Routledge.
Yip, A. K.-T. and Nynäs, P. (2012) 'Re-framing the intersection between religion, gender and sexuality in everyday Life'. In Nynäs, P. and A. K.-T.Yip (eds.) *Religion, Gender and Sexuality in Everyday Life,* Aldershot: Ashgate, pp. 1–16.

29

After or against secularism

Muslims in Europe

Kasia Narkowicz and Richard Phillips

Introduction

The postsecular has been conceptualized, both in general terms and also in relation to particular countries and religious groups, as a historical shift in which religion is returning from the margins and the shadows to the centre ground of public life (Habermas 2008). From the perspective of religious groups, minorities in particular, narratives of the return of religion are problematic. We develop this argument with reference to British Muslims of Pakistani heritage, and Polish Muslims with mixed heritage.

Though the UK and Poland differ religiously, politically and social-economically, both countries have experienced a decline in individual adherence to the majority Christian faith. Consequently, both fit within Jürgen Habermas' conception of the postsecular as the historical successor of secularism:

> A "post-secular" society must at some point have been in a "secular" state. The controversial term can therefore only be applied to the affluent societies of Europe or countries such as Canada, Australia and New Zealand, where people's religious ties have steadily or rather quite dramatically lapsed in the post-War period.
>
> *(Habermas 2008: 1)*

For Habermas, the postsecular emerges from and signals a change in consciousness whereby the 'hushed up voice of religion' is invited back into the public sphere (Cloke and Beaumont 2013: 44). The precondition for this postsecularity is conformity, on the part of the religious, to the norms of 'secular discourse' (Habermas 2008: 1). From this perspective, religion is regarded as a moral supplement to secular politics (Asad 2003). The postsecular follows—but fits into the space vacated by—secularism, just as secularism is said to have occupied the space once vacated by religion.

One reason why this understanding of the postsecular is problematic is its exclusivity. Cloke and Beaumont (2013) welcome the inclusion of progressive elements of religion into the public sphere. This claim, though important in itself, raises questions about how we might regard the contributions of religious individuals and groups who identify or do not identify as progressive, and who may identify or be identified as conservative. Distinctions between progressive and regressive religiosity risk ignoring the contributions of conservative religious groups (Fabian and Korolczuk 2017) and fail to capture the messy and complex ways in which religion influences the everyday (Mahmood 2005).

We challenge the 'historical shift' vision of postsecularism, not by advancing another grand history, but rather through a series of smaller stories, which highlight ways in which religious minorities have experienced the postsecular. This approach brings a postcolonial sensibility—one that means listening to subaltern groups rather than making assumptions about them or entertaining dominant representations of them (Spivak 1988)—to understandings of postsecularism. For our purposes, we 'listen' to Polish and British Pakistani Muslims in two ways: firstly by drawing upon published research by others, which offers insights into the ways in which religious experience and identification have been configured at different points in time, with particular reference to the past few decades; secondly, through empirical research of our own. The case study of Muslims in Poland draws upon interviews, conducted by Kasia Narkowicz in 2011–2012, which examined contemporary conflicts around religion and secularism in that country.[1] The second case study touches upon interviews with British Muslims of Pakistani heritage, which was conducted as part of a larger project led by Richard Phillips, on relationship practices and choices in a diverse society. All the participants' names have been anonymized.

Listening to Polish and British Pakistani Muslims, we find that their religion never went away, not, at least, within the historical frame of these case studies, which include periods in which religion was seen to have been banished to the private sphere (in Communist Poland) or overshadowed by national and ethnic identities (in the case of first-generation British Pakistanis). Polish Muslims and Catholics did not simply wait for a non-secular present; they positioned themselves against secularism. Even when it was apparently subsumed within the private sphere by mainstream secularism and official atheism, and when it seemed to play second fiddle to other forms of identification including ethnicity and race, Islam has been present in the day-to-day lives of Polish and British Pakistani Muslims and has not only been limited to the private sphere.

Though these groups have not always been seen as religious minorities, from the perspective of the mainstream, and though they have not always primarily emphasized the religious in their self-identification, it would be wrong to suggest that their religion was forgotten or unimportant, whether privately or publicly (a problematic dualism for many). Rather than rediscovering religion, these groups have always been religious—have always been Muslims—and have articulated this differently, through intersectional identities in which religion, culture, ethnicity, and race are all important, but are advanced differently and strategically in different contexts.

These two case studies therefore contest conceptions of postsecularism as *after secularism*, presenting it instead as *against secularism*. These conceptions are encapsulated in some prominent representations of British Muslims of Pakistani heritage and Polish Muslims of mixed heritage, representations that do not always accord with the ways in which members of these communities represent themselves, doing so for strategic reasons, rather than simply to describe themselves.

Practising Muslims in Poland

The chronological understanding of postsecularism—as after secularism—appears to closely match the experience of countries that were until recently officially secular. Secularism has been abandoned or defeated in different places for different reasons, but a large swathe of European countries has similar experiences, brought about by the demise of Communism and the fall of the Soviet Union. Postsecularism has been most pronounced where, as in Poland, a pronounced religious culture preceded Communism and was repressed, but not eradicated, by the atheism of that period. In such settings, religion was waiting in the wings, occasionally surfacing in public life, and it has resumed its central place in society. Without wishing to deny that there are fewer obstacles to the public expression of religion today than there were before 1989, we will contest the assumption that religion was effectively banished to the private sphere under the Communist system.

Poland was not only non-secular then: it was actively postsecular in the twofold sense that some religious people were contesting the official suppression of religion by the ruling Communist Party and others were practising their faith in ways and spaces that transcended the private sphere. First, the Catholic Church played a crucial role in the mobilization of the Solidarity pro-democracy movement. Danuta Walęsa, the wife of opposition leader, and later President Lech Wałęsa, remembered Sunday mass as a chance to meet, organize, and strategize with fellow opposition movement colleagues in a safe space (Wałęsa 2011). The interventions of religion in the public sphere during key historical moments was similarly evident in the USA where black churches provided spaces for the mobilisation of the civil rights movement (Calhoun et al. 2011). The Polish Church was also active in formal process. It initiated dialogue with the Communist leadership when the regime realized it could no longer ignore dissent (Romanowski 2012). The role of the church has grown and changed since 1989; it is now an important ally of the populist right-wing government, with which it is pursuing the long-held project of Catholic nation-building (Jaskulowski et al. 2018), and since 1989, intervening on issues such as abortion (Fuszara 1993; Guerra 2016) and migration (Narkowicz 2018). And yet, though Poland is now overtly non-secular, the part played by the Church in the pro-democracy movement prior to the fall of Communism reminds us that the nation was not entirely secular, even during that time. Even then, the admittance or imposition of religious voices in the political public sphere, where they articulated ideas of justice, rights, and morality as part of rational deliberation, was consistent with the postsecular, as Habermas (2008) defines the term.

Second, also under Communism, religious minorities were practising their faith in ways and spaces that transcended the private sphere. Muslims, who had been part of the religious Polish landscape since the tenth century when Tatars established a presence in what was then the Grand Duchy of Lithuania (Antonowicz-Bauer 1984: 346; Pedziwiatr 2011), grew in number in the 1970s and 1980s. Despite the ‘closed borders’ and isolation from much of the world during the Communist years, a significant Muslim migration took place during Communism. Arab students from ‘befriended’ countries such as Palestine, Kuwait, Syria, Lebanon, Egypt, and Iraq arrived to study at Polish universities (Narkowicz and Pędziwiatr 2017).

Muslims in Communist Poland might be assumed to have confined their religious expression to the private sphere. But this narrative is challenged by stories that are told by Muslims in Poland about their religious identification and expression at this time which signal a growth and diversification in the Muslim community. Despite a homogenous conception of the Polish nation-state that was exclusive of minorities already during Communism (Zaremba 2001), Muslims were able to actively practice their faith. Upon arrival, during the 1970s, the Muslim students quickly started organizing places of worship and established a vibrant religious community, separate from the Tatar Muslim community. Muslim prayer halls were established in various locations; at universities, in people's homes, and in rented accommodation (Narkowicz and Pędziwiatr 2017). An imam of a Warsaw mosque, interviewed by Kasia in 2011, recalls how he and his fellow Muslim students used the university space to organize and meet their religious needs:

> The brothers arranged to have a room on the top floor of the student halls. It was not a room for other students but like an attic, and the director of the student hall agreed for that room to be a prayer space… And it was similar in other cities like Lublin, in Poznań, in Katowice, in Wrocław... it started with someone offering up their room for Friday prayers or the directors of those student halls designated a room in the building for the Muslim students to pray in.
>
> … And already then Muslims started thinking that they wanted to have mosques and the activity started going in this direction.

As the imam recalled, despite an ostensibly anti-religious landscape in Poland during Communism, the arrival of the Arab students during that time contributed to a further diversification of non-secular spaces within secular institutions such as the university. With time, many smaller prayer halls were established, and some key Muslim centres were formed in the cities of Warsaw, Łódź, Wrocław, Katowice, Kraków, Poznań, Lublin, and Białystok.

In an interview reported in the Polish media, a Muslim convert argued that 'it is not possible for a Muslim to lead a completely secular life' (Gasior 2014: 1). Some Muslim women, interviewed by Kasia, made a similar point, refusing secularist pressure to leave their religions at the door step before entering the public sphere. Oliwia put it this way:

> What frightens non-Muslims, or what they can't understand, is that religion in a Western sense is something private, like you go to your temple, you have something at home, you will pray when you feel like it, and so on. Islam is not like that, of course it depends on how every person understands it, but really… many people accuse Islam of not being a religion but an ideology and true, it is. I always say 'yes you are right'!

She explained further that being a Muslim 'relates to all things in society, what to do, what to eat, what to wear and so on'. From this perspective, secularism is not only not desirable, it is impossible. This impossibility challenges the idea of a rational, religiously neutral public sphere (Calhoun et al. 2011) and of a Habermasian postsecular space of translation, in which religion is welcome but on a limited, conditional basis, particularly one that favours unthreatening forms of Islam and those that are seen by others as 'progressive' and/or 'moderate'. Here, as elsewhere in Europe, it is argued that in order to be accepted into public space, Muslims are required to 'translate' themselves, diluting their religious presence (Asad 2003: 159, 169).

Expression of both Catholicism and Islam in Polish public life runs from the officially secular Communist era to the overtly non-secular present. These and other religions have in some cases been mutually supportive. Some Polish Muslims say that they find it easier to be overtly religious in a society in which religion is accepted and commonplace. Oliwia commented that 'Poland is a religious country' and as such 'it is still easier for us to keep our faith and culture here; as Muslims'. A similar point was made in an interview with a man called Marek, also interviewed in 2011, who told Kasia that '[Catholics] understand that religious convictions might entail practices that might seem weird to some people. I don't know, Catholics in Poland still have got an understanding for fasting, right?' In these interviews, Polish Muslims felt that they did not need to translate themselves in order to be understood.

The continued presence of religion in Polish public life, not only after the fall of Communism but also before, questions simple historical claims that might be made about the secular past and the postsecular present. Catholics and Muslims, in their different ways, have not simply waited for a non-secular present, but they have consistently positioned themselves against secularism. The next section, focussing upon another European Muslim minority, interrogates the ostensibly postsecular present from another angle. It contests simplistic but commonplace narratives about the rise of religiosity and the concomitant eclipse of ethnic and national identification among Muslims in Britain.

Intersectional identities: Pakistani *and* Muslim

Histories of British Pakistani Muslims in the UK typically portray a historical shift from Pakistani and Bangladeshi to Muslim identities. This narrative, describing a recent religious turn, is in keeping with wider histories of race and difference which trace a shift from nationality to race to religion, as primary means through which minorities are identified both in mainstream discourse such as government policies on multiculturalism, and also in claims about the ways in which minorities self-identify. Yet, this periodization is complicated, if not contradicted, by evidence of how members of this group identify and position themselves.

In his Foreword to Tariq Modood's *Multicultural Politics*, Craig Calhoun observes that Britons with Pakistani heritage, many of whom migrated in the postwar period or descend from migrant communities established then, were 'initially mostly termed "Pakistanis" (or more abusively, "Pakis")', but have been 'recategorized first as Asians and then as Muslims' (Calhoun in Modood, 2005: x). Simultaneously, many of those who were previously categorized as Pakistanis have come to distance themselves from Pakistan and disavow Pakistani identities, both of which have assumed 'a burdensome set of undertones in the imaginary of both British Pakistanis and Britons' alike', according to Marta Bolognani (2014: 114). The move away from this (and other) national or ethnic identities and categories has taken place alongside another shift, in this case away from a singular, inclusive, strategically essentialist 'black' identity politics (Alexander 2000: 231). In the place of Pakistani, Bangladeshi, and some other national and ethnic identities, and in place or instead of 'black' or other racial identities, many people have either positioned themselves or been positioned as Muslims (Modood 1992); or so it is said.

This thumbnail sketch of communities and identities is simplistic, and it is at the very least complicated and sometimes contradicted outright by the ways in which members of minority groups speak and say they think about themselves. The shifts in how this minority are variously categorized by others and in how they tend to see or position themselves, which are not always the same thing and frequently diverge, should not be misunderstood as simply a shift in how people identify: moving away from culture and/or race and then to religion. First, the emergent Islamic identification does not banish the racial and political dimensions of the terms it has displaced. Rather, contemporary Islam encompasses and supersedes constructions of race. 'Modood shows clearly that race—and racism—remained inextricably intertwined in the way British Muslims were' and arguably still are 'understood' (Calhoun in Modood 2005: x). Since Islam has come to encompass political and racial dimensions, and is always about more than religion, the increasing propensity of individuals to identify or be identified with Islam should not be seen as a narrowly religious turn, nor as the exchange of one identity for another. It can be understood as a 'positioning' rather than 'essence' (Hall 1990: 226; quoted by Alexander 2000: 231).

It would be wrong to see the identities left behind by the shift towards Islam as irreligious or secular in the sense that religion would have been confined to the private sphere. Among those who identified primarily as Pakistani and those who still do, nationality was and still is widely assumed to define religion. According to Haleh Afshar, those who identify or identified as Pakistani 'automatically assumed that this implied that they were Muslims' (Afshar et al. 2006: 172; see also Modood et al. 1997; Samad 1998). This assumption is reasonable given that most Pakistanis in the UK are indeed Muslims. Among the 625,000 or so Britons with Pakistani heritage who were living in England at the time of the 2001 Census, 92% were Muslims (Change Institute 2009: 30).

The shift from primarily Pakistani to Islamic identification has not really been a religious turn because those describing themselves in these shifting terms have not switched from national to religious identities, but from national identities in which religious identification was implicit, to religious identities in which nationality and/or race are implicit, important, and politicized. This shift is multifaceted, as are its drivers. Important, among these, are the possibilities that minorities find in focussing upon the things they hold in common (including faith) rather than those that have kept them apart (such as national lines between Pakistanis and Bangladeshis). These commonalities bring the promise of solidarity in relation to external pressures. Meanwhile, identification with religion rather than culture can be empowering to particular members of these communities in their relationships with each other. Young British men and women with Pakistani heritage are among those who have found agency in speaking as Muslims. Doing so, they are able to make and justify life choices, some of which go against cultural traditions and family expectations (Lewis 2007; Mondal 2008).

Before focussing upon contemporary articulations of faith, it is important to acknowledge that the sometimes implicit Islam of earlier generations of British Pakistani Muslims was not just a private matter. Then, as now, their faith found public expression, for example, in the construction and attendance of mosques, and the production, sale, and consumption of halal food.[2] First- and second-generation British Pakistanis did not conceal their faith; 'they automatically assumed that [their country of origin] implied that they were Muslims' (Modood et al. 1997; Afshar et al. 2006: 172).

Rather than a narrowly religious turn, the shift from Pakistani to Islamic identification has been a vehicle for some members of overlapping ethnic, racial, and religious communities, particularly younger women, to reject certain practices and relations and embrace others. Afshar et al. (2006: 171) generalise that 'nationality is a poor identifier for women' and that 'the historical construction of nationhood and nationalism is masculine in terms of its character and demands'. It is 'not surprising', they conclude, that while 'the primary identification of many older-generation British Pakistanis is with country … rather than religion' (*ibid.*: 171), this is rejected by 'second-generation migrant women' (*ibid.*: 172). This is not to say that increasingly religiosity is simply a strategic move, motivated by the desire of young women to gain greater control over their lives, though there may be some element of that. In any case, the shift towards greater religious identification reflects changes in self-presentation, with shifting emphases rather than necessarily fundamental shifts in subjectivity.

These observations about the shifts in emphasis from national to religious identification are borne out in some of our own research, which includes interviews with young women of Pakistani heritage.[3] Some of these young women distanced themselves from Pakistan and from Pakistani identities, identifying instead with Islam and as Muslims. Saamiya, who is 18 and lives in Glasgow, explained that she identified primarily as a British Muslim. This, she explained, is 'like not being the same as like a Pakistani Muslim where the girls would kind of stay home and do as they are told'. Pnina Werbner explains the particular attraction of Islam for these young women:

> The girls argue that Islam accords equality to men and women, that it requires young people's consent to a marriage and allows them to choose their own partner, and even to associate with their fiancés before marriage. Islam also opens up a much wider marriage market for young people.
>
> *(Werbner 2007: 161)*

Saamiya, like some others we spoke to as part of this research, acknowledged that Muslims sometimes fall short of ideals such as these, in how they live their lives, and in the ways in which their families, communities, and nations put their religion into practice. Their shift from Pakistani to Islamic (or British Islamic) identities was not a shift from national to religious identity, but from one way of being religious to another, where the latter is conceived as deeper and more committed, and also more assertive and liberated. Moving away from Pakistani identification, another young woman explained to us that she was able to reject some cultural practices that are identified as Pakistani, doing so in favour of others, which may be presented as Islamic and therefore harder for relatives to criticize. Laila, who is 30 and single, distanced herself from Pakistan, explaining that those born in Pakistan are told 'this is how you get married'; they 'weren't allowed to go out dating'. Laila identified instead as British Asian; others present primarily as Muslims; in each case, they find scope to be more assertive about how they approach relationships and marriage, and other aspects of their lives.

These findings resonate with those of other researchers, who have found that increased religious identification can be empowering for women, particularly in relation to their ability to determine and perhaps to reject relationships and marriages. Afshar illustrates this point in relation to 'mohajabehs': women who have adopted strong forms of Islam. She argues that:

> The assertive positions and ideas of the *mohajabehs* have, in practice, made them undesirable brides for the kinship network. They know too much about their rights and have too little respect for the 'traditions'. Often these women are more learned in terms of *Koranic* teaching than their parents. It has become difficult to marry them off in arranged marriages and even more difficult to argue with them, since their arguments are always presented in terms of 'the true Islam'. An interviewee for the Muslim diaspora research, an art student who wears hijab, told Franks, 'The more Islamic I become, the less likely it is that I will be pushed into an unwanted marriage'. This is because her parents are unable to criticize her if she is following *Koranic* teaching. Young women often use their textual understanding of Islam to contest the traditions and restrictions imposed on them by their parents....
>
> *(Afshar et al. 2006: 181; drawing upon Samad 1998; Jacobsen 1988; Shaw 1988: 435)*

Here, Afshar (2006: 181) argues, 'faith supersedes nationality', but nationality persists, along with regional and kinship identity and allegiance. 'Muslim women who talked to' Haleh Afshar 'generally accepted cultural, ethnic and national identities that defined them in different ways in different circumstances' (Afshar et al. 2006: 176; drawing upon Afshar 1989; Vertovec 1998). In this context, faith is never reduced to a singular identity, and it is always more than abstract religiosity; for some, it is a form of self-assertion and empowerment. This assertiveness and self-possession comes across in many of the interviews we conducted with young British Pakistani Muslims. Ifrah, who has transitioned from wearing the hijab to the abaya, stresses the importance of her faith, and the fact that she has not simply gone along with peers or family, but has actively chosen how to express it in her clothing. She told us: 'I am actually like properly into the religion rather than, I am just wearing the scarf because everyone else wears it, because my parents told me to wear it' (Ifrah, 21, Female).

Crucially, though they relate primarily to personal relationships, and ultimately to marriage and family, these expressions of religious identity are not simply private, as would be the case in an entirely secular context; they also reach into areas of public life. To the feminist mantra that 'the personal is political', we must add another: 'the religious is public'. Indeed, many of those whom we interviewed as part of this project stressed that their religion could not be compartmentalized; it informs and is expressed in all areas of their lives. As one 29-year-old woman put it, 'I'm Muslim and that kind of influences my morals and my decisions and the way I interact with people' (Janan). A similar point was made, more specifically and concretely, by Noor, a 27-year-old woman, who also lives in Yorkshire. She contested the commonplace assumption that to be modern or liberal—to participate in mainstream culture and politics—one must leave religion at the door, particularly if one is a Muslim:

> I view myself as someone liberal, but that is in my cultural attitudes. It might be that you shouldn't enforce things on other people, but as a Muslim I would view myself as quite kind of, I wouldn't say strict because it is another word I don't like, but I don't think that there is anything that I am not doing that kind of makes me...liberal or modern, but at the same time I don't think I am necessarily backwards. So I just think the need or the suggestion people to clarify what camp they are born into, is more damaging than useful.

Noor refused to separate her liberal convictions and Islamic faith, and as a result she brought a new dimension to the former, challenging widespread assumption that liberalism and Islam are incompatible, and at odds with each other (Massad 2015; Phillips 2016).

The tendency for women with similar backgrounds to identify in different ways—as British Asians, Asians, British Muslims, Muslims, and so on—underlines the complexity of histories of difference, and warns against a periodization that might suggest one form of identification has been supplanted by another, or even the more modest claim that one has assumed greater significance over time. In other words, this cautions against understanding the postsecular as after secularism—the period in which religion has resurfaced and been restored to prominence—and argues instead for an understanding of the postsecular as religious expression that is important not only in its own right, but also as a politics, not just after but actively against something: in this case, against the constraints of nationalism and national identification, and for the rights and freedoms that may come with the reduction of such constraints.

Conclusion

Contesting a simply linear, chronologically historical conception of postsecularism, in which this term is understood as *after secularism*, we have advanced an understanding of the postsecular as *against secularism*.

This distinction has counterparts in postcolonial theory, where the postcolonial has variously been understood as after *and against* colonialism. As in postcolonial criticism, this understanding of the post as more than chronology is important because it politicizes the term it qualifies, in this case postsecularism, and it recognizes the agency of those who identify with and mobilize it and, with it, their identification as religious. It also recognizes that some groups have been entirely excluded from the concept of the postsecular (see Habermas 2008) or included on the condition that they translate themselves into a liberal public sphere (see Cloke 2012). This is particularly relevant for Muslim minorities, who have been constructed as 'external to the essence of Europe' (Asad 2003: 165). While the Habermasian (2006) concept of postsecularism offers a place for religion in the public sphere, where its role was previously limited, the motivation still is to enrich the secular society with 'key resources for the creation of meaning and identity' (Habermas 2006: 10).

For the Muslim minorities in two national contexts, post-Communist Poland and post-colonial Britain, leaving some parts of one's faith in the private sphere was not possible because, as our interviewees stressed, 'Islam is concerned with every aspect of life and you don't divide the sacred and profane' (Dagmara). 'It relates to all things in society, what to do, what to eat, what to wear and so on, every sphere of life. Relations within your family, relations with your neighbour' (Oliwia). Similarly, in the British case study, religion impacted on the ways people behaved in the private and the public sphere. One interviewee said: 'I'm Muslim and that kind of influences my morals and my decisions and the way I interact with people' (Maryam). Another emphasized how the abaya she chooses to wear outside her house represents her beliefs in public (Ifrah).

The distinction between *after* and *against* secularism, which we have stressed, is not simply a matter of definition. When the postsecular is conceived as *against* secularism, it assumes a more active—one might even say implicitly political—dimension. Here, religious minorities refuse to apologize for themselves, refuse to be self-effacing, and demand to be recognized. Through their overt religiosity, they also bring new dimensions to public life.

Notes

1 The PhD, funded by the European Research Council as part of the LiveDifference Research Programme led by Gill Valentine at the University of Sheffield, focused on spaces of conflict in the Polish public sphere. Fieldwork was conducted by Kasia Narkowicz in Warsaw between 2011 and 2012.

2 This project was led by Richard Phillips, with the involvement of Nafhesa Ali, Claire Chambers, Kasia Narkowicz, Raksha Pande, and Peter Hopkins. Interviews were conducted in 2016–2017. Interviewees discussed their relationship choices with reference to their Pakistani heritage and Islamic faith. Doing so, they revealed dynamics between ethnicity and religion, which inform this discussion of postsecularism.

Further reading

Górak-Sosnowska, K. (2011) *Muslims in Poland and Eastern Europe: widening the European discourse on Islam*, Warsaw: University of Warsaw.

This edited book provides a range of perspectives on Muslims and Islam in Poland and other Eastern European contexts, historically and today.

Lewis, P. (2007) *Young, British and Muslim*, London: Continuum.

This book tracks the rise—we would say the resurgence—of identification with Islam among young people in Britain. It shows how this form of identification was empowering for those involved.

Mondal, A. (2008) *Young British Muslim Voices*, Oxford: Greenwood.

Similarly, this book demonstrates identification of young people in Britain with Islam, with an emphasis on questions about Britishness and the place of Muslims in British society.

References

Afshar, H. (1989) 'Gender roles and the "moral economy of kin" among Pakistani women in West Yorkshire', *New Community*, 15(2): 211–25.
Afshar, H., Aitken, R. and M. Franks (2006) 'Islamophobia and women of Pakistani sescent in Bradford: the crisis of ascribed and adopted identities'. In Moghissi, H. (ed.) *Muslim Diaspora*, London: Routledge.
Alexander, C. E. (2000) *The Asian Gang: ethnicity, identity, masculinity*, Oxford: Berg.
Antonowicz-Bauer, L. (1984) 'The Tatars in Poland', *Journal Institute of Muslim Minority Affairs*, 5(2): 345–59.
Asad, T. (2003) *Formations of the Secular: Christianity, Islam, modernity*, Stanford, CA: Stanford University Press.
Calhoun, C., Juergensmeyer, M. and J. VanAntwerpen (2011) 'Introduction'. In Calhoun, C., Juergensmeyer, M. and J. VanAntwerpen (eds.) *Rethinking Secularism*, New York, NY: Oxford University Press.
Change Institute (2009) *Pakistani Muslim Community in England*, Report commissioned by DCLG (Department for Communities and Local Government, Home Office), London: HM Government Office of Public Sector Information.
Cloke, P. (2012) 'Post-secular stirrings? Geographies of hope in amongst neoliberalism', Queen Mary's 9th Annual David Smith Lecture.
Cloke, P. and J. Beaumont (2013) 'Geographies of postsecular rapprochement in the city', *Progress in Human Geography*, 37(1): 27–51.
Fabian, K. and E. Korolczuk (2017) *Rebellious Parents*, Bloomington: Indiana University Press.
Fuszara, M. (1993) Abortion and the formation of the public sphere in Poland'. In Funk, N. and M. Mueller (eds.) *Gender Politics and Postcommunism: reflections from Eastern Europe and former Soviet Union*, London: Routledge.
Gasior, M. (2014) 'Polska Muzułmanka Agnieszka Amatullah Wasilewska: Wyznawca Islamu nie może prowadzić świeckiego życia' ("Polish Muslim Agnieszka Amatullah Wasilewska: a Muslim cannot lead a secular life"], NaTemat.pl.
Guerra, S. (2016) 'The Polish Catholic Church has become intertwined with Euroscepticism and the promotion of conservative "national values"', *Democratic Audit UK*. http://www.democraticaudit.com/2016/04/28/the-polish-catholic-church-has-become-intertwined-with-euroscepticism-and-the-promotion-of-conservative-national-values/
Habermas, J. (2008) 'Notes on a post-secular society', *New Perspectives Quarterly*, 25: 17–29.
Jacobsen, J. (1998) *Islam in Transition: religion and identity among British Pakistani youth*, London: Routledge.
Jaskulowski, K., Majewski, P. and A. Surmiak (2018) 'Teaching the nation: history and nationalism in the Polish school history education', *British Journal of Sociology of Education*, 39(1): 77–91.
Massad, J. (2015) *Islam in Liberalism*, Chicago, IL: University of Chicago Press.
Modood, T. (1992) *Not Easy Being British: colour, culture and citizenship*, Stoke-on-Trent: Runnymede Trust and Trentham.
Modood, T., Berthoud, R., Lakey, J., Nazroo, J., Smith, P., Virdee, S. and S. Beishon (eds.) (1997) *Ethnic Minorities in Britain: diversity and disadvantage*, London: Policy Studies Institute.
Narkowicz, K. (2018) 'Refugees not welcome here: state, church and grassroots responses to the refugee crisis in Poland', *International Journal of Politics, Culture, and Society*, doi: 10.1007/s10767-018-9287-9, accessed online 15-06-2018.
Narkowicz, K. and K. Pędziwiatr (2017) 'From unproblematic to contentious: mosques in Poland', *Journal of Ethnic and Migration Studies*, 43(3): 441–57.
Pędziwiatr, K. (2011) 'Muslims in contemporary Poland'. In Bures, J. (ed.) *Muslims in Visegrad Countries*, Prague: Anna Lindh Foundation and Visegrad Fund, pp. 10–24.
Phillips, R. (2016) 'Islam in sexuality', *Dialogues in Human Geography*, 6(2): 229–44.
Romanowski, A. (2012) 'Kościół, lewica, wojna' ("Church, the left, war"), *Znak*, nr: 681.
Samad, Y. (1998) 'Media and Muslim identity: intersections of generation and gender', *Innovation*, 11(4): 425–38.
Shaw, A. (1988) *A Pakistani Community in Britain*, Oxford: Basil Blackwell.
Spivak, G. C. (1988) 'Can the subaltern speak? In Nelson, C. and L. Grossberg (eds.) *Marxism and the Interpretation of Culture*, Basingstoke: Macmillan Education, pp. 21–78.

Vertovec, S. (1998) 'Young Muslims in Keithley, West Yorkshire: cultural identity, context and "community"'. In Vertovec, S. and A. Rogers (ed.) *Muslim European Youth: reproducing ethnicity, religion, culture*, Aldershot: Ashgate, pp. 87–107.

Wałęsa, D. (2011) *Marzenia I Tajemnice* ("Dreams and Secrets"), Krakow: Wydawnictwo Literackie.

Werbner, P. (2007) 'Veiled interventions in pure space: honour, shame and embodied struggles among Muslims in Britain and France', *Theory, Culture & Society*, 24(2): 161–86.

Zaremba, M. (2001) *Nacjonalizm, Komunizm, Legitymizacja: nacjonalistyczna legitymizacja władzy w komunistycznej Polsce* ("Nationalism, Communism, Legitimisation: nationalist legitimisation of power in communist Poland"), Warsaw: Trio.

30

Postsecularity in twenty questions

A case study in Buddhist teens[1]

Phra Nicholas Thanissaro

Introduction

Postsecularism is a scholarly position that observes, that despite steady decline in religious ties in the postwar period, there has been a manifest resurgence in conspicuousness of religion and religion-related issues throughout the affluent societies of Europe, Canada, Australia, and New Zealand.[2] Habermas (2006) reflected that with postsecularity, the nature of religious behaviour and convictions had not changed particularly, but *awareness* of it has.

To move beyond the usually limited potential of the postsecularism concept merely to evoke reaction, Moberg and Grunholm (2012) recommended that if research data could be gathered, it would help postsecularity gain greater analytical value—bringing theorists closer to the subjects, politics, and movements involved in the reconfiguring of the religious and the secular.[3] For postsecularity to become the subject of empirical sociological research, it needs to be operationalized. If postsecularity were a unidimensional phenomenon, one might conceive a reliable scale that could measure the extent of postsecularity and perhaps correlate it with other social forces. From what sociologists have written, however, it would appear that postsecularity is *multi-faceted* rather than unidimensional, meaning the best starting point for its operationalization would be to describe the cluster of different dimensions included within the concept. Dimensions one would expect to find in a multidimensional hermeneutic of postsecularity (and this forms my tentative definition of postsecularity) might include diminished secularity and modernity, increased visibility of religious practice in the public sphere, and an increase in what de Groot (2008) calls 'liquid religion'. Postsecularism's near-neighbours of projection, religious revivalism and resistance identity, might feature in postsecularity to an extent that remains to be seen. It is these dimensions that underpin the 20 questions used as the empirical basis for this chapter.[4]

To offer some background context to this study, it should be mentioned that Buddhism is a minority religion in Britain for which Thanissaro (2014a) has previously invoked postsecularity to explain growth when secularization or melting-pot theories would lead us to expect decline. Buddhists have some special features with regard to postsecularity which should be indicated at the outset. Firstly, not all practising Buddhism self-identify as 'Buddhist'. Secondly, Buddhism is one of the few world religions where followers do not believe in some form of God. Thirdly, for Buddhism, as with many of the other Dharmic religions, there is difficulty generally in distinguishing between public and private expressions of religion. Finally, for Buddhism in the West, it is important to distinguish between the styles of religious practice found in heritage Buddhists (those who have ethnic roots in the countries of Asia) and convert Buddhists (those who have converted to Buddhism independent of their family's influence)—a dichotomy that will help shed light on boundary marking's impact on certain aspects of liquid religion.

Thanissaro's (2014a) preliminary foray into the postsecular nature of Buddhist faith in Britain was limited by the qualitative nature (focus groups) of the research. For that study it was not possible to tell whether Buddhist opinions expressed were statistically significant, and since research was restricted to heritage Buddhists, conclusions about Buddhism could have been confounded with 'Asianness'. This chapter aims to unpack the question of the sort of religiosity for which postsecularity reflects a resurgence. Most sociological literature portrays postsecularity as a quality of an age or a culture. Here, I study postsecularity additionally as a quality of peoples' religiosity, while going on to explore its relationship with boundary marking. Furthermore, I aim to apply quantitative analysis to postsecularism and to compile a set of questions identifying postsecular attitudes—looking specifically at: (1) facets of modernism and secularity, (2) public and private spheres of religion, (3) liquid religion, (4) projection, and (5) boundary marking in a case study of teen Buddhists in Britain.

Modernism and secularity

The first telltale sign of postsecularity should be erosion of the values of secularity and modernity. I have combined modernity and secularity under the same heading because Taylor (2007) observed both are closely related in the way they cast doubt on non-scientific aspects of religious belief. Secularity tends to show the characteristics of pluralization and relaxation of the sacred, a breakdown of boundaries between the sacred and the profane (Connolly 2005), being seen in both political and social spheres (Turner 2010), not being merely a net reduction in religious belief and practice but a change in the very conditions of belief (Warner et al. 2010). These elements, taken together, herald a redefinition of what is considered a 'neutral' perspective to reduce the influence of religion and to emphasize secular humanism as the common creed even in a diverse society. The presence of Buddhism in the West might be considered a symptom of secularism by mainstream Christians.

Modernism anywhere in the world represents the institutional or 'solid' manifestations of religion. Buddhist modernism throughout the world shares the discourse of rational and scientific Buddhism that downplays devotional acts, merit-making, miracles and pre-scientific cosmologies. In Asia, Buddhist modernism picks up additional meaning in terms of Buddhists essentializing their own identity and institutionalizing themselves around temple life and spiritual leaders (McMahan 2008) and possibly agitating for Buddhism as a state religion. In the West, however, Buddhist modernism has taken on a slightly different set of features from those seen in Asia; it tends to include the expectation of *proof* based on scriptural research or individual experience, a sense of universalism and downplay of moral injunctions (Baumann 2001; Braun 2015).

In England, cross-sectional surveys conducted since 1974 at four-yearly intervals have charted the decline in adolescent religion in terms of attitude towards the tenets of (Christian) faith and church attendance (Kay and Francis 1996) that are indicative of secularism and modernism. In terms of choice of indicative questions, one of the first facets of postsecularity would be a diminishing of modernity and secularism as reflected by *increased* belief in the value of religion relative to science, increased religious self-labelling, increased perception of the relevance of religious teachings, and value of religious clergy. In this study, Buddhists were found to be significantly less likely than religiously undifferentiated adolescents (hereafter RUA) to believe that religion had been replaced by science. Furthermore, they were significantly more likely to identify themselves as religious. Buddhists were more likely to think clergy did a good job, and less likely to think the religious teachings had lost their relevance to contemporary life.

Public and private religion

A second telltale sign of postsecularity derives from religion appearing more in the public sphere rather than being hidden away in the private sphere *as per* modern times (Casanova 1994), as the result of postsecularity (Utriainen et al. 2012). A previous small-scale study of heritage Buddhists in Britain found temples were attended on a weekly or more occasional basis to present meals to the monastic community, for festivals in the Buddhist calendar, or for the anniversaries of the passing of relatives (Thanissaro 2011). Temple attendance in previous research was cited as one of the hallmarks of being a 'proper Buddhist' and considered advantageous because teens could meet experts face-to-face to receive instruction in Buddhism (Thanissaro 2014a) in a conducive peaceful setting (Thanissaro 2014b). While at the temple, teens were more focussed on learning the Buddhist message as illustrated by the words of Vari, a 20-year-old Thai Buddhist: 'There is more motivation ... if you come to the temple; you *have* to listen' (Thanissaro 2014b: 320). It should be borne in mind that having a local temple to visit in the UK is a fairly recent phenomenon and has only featured in UK Buddhist life since the 1990s (Thanissaro 2013).

Empirical measures of public religion are assessed in terms of the frequency of attendance at a place of worship. The presence of postsecularism would lead us to expect increased frequency of public aspects of religious participation. Temple attendance was found to be the most typical religious involvement for Buddhist teens. Religious participation in the public sphere in our Buddhist case study is represented by frequency of attendance at a temple or meditation centre—82% of Buddhists attended at least on a weekly basis which is much more than the 5%–6% of the general population who attend a place of worship weekly for the same age-group (Brierley 2006). Religion in the public sphere thus seems to remain at a high level for Buddhists.

Liquid religion

A third set of indicators relate to 'liquid religion' which serves to discern ways in which non-religious life has been resacralized, revitalized, and re-enchanted by religious values (Utriainen et al. 2012). When discussing the shift from modernity to 'liquid modernity', specifically in reference to religion, Bauman's (2001) expectation was that religion would disappear along with other 'solid' establishments. Ironically, Kees de Groot was able to apply Bauman's (2000) concept of liquid modernity to religion, identifying three places where liquid religion could be observed, namely: (1) in the religious sphere, (2) at the boundaries between the religious and the secular spheres, and (3) outside the religious sphere (de Groot 2008). To these three I have added 'lived religion' and will examine all four in turn, allocating indicative questions.

Thanissaro (2013) saw the postsecular aspect of 'liquid religion' as accounting for the seemingly porous interface between Asian cultural identity and the Western cultural mainstream, a compromise between pleasing parents and the Buddhist community, fitting in enough with non-Buddhist peers to avoid being ostracized while succeeding in education. Using these four aspects of liquid religion, our understanding of the relationship between Buddhism and liquid religion becomes more nuanced.

Liquid religion in a religious context

Liquid religion in a religious context can mean both empowerment of the temple congregation and the deregulation of religious ideas and symbols, allowing them to circulate in society in ways increasingly beyond the control of religious institutions, empowering believers within their community to feel they could do more to solve the problems of the world. In previous research, being part of a Buddhist temple community was cited as giving a sense of *belonging* also conducive to learning. In the words of Maya, a 15-year-old Sri Lankan Buddhist, there:

> is a nice community that we have in the temple. It is like something we can always depend on. It's by coming to the temple, like with any religious place, you do kind of become a part of the community and it is your second family. It is your 'family away from home'.
>
> *(Thanissaro 2014b: 320)*

Indicative questions pointing to increased liquid phenomena in the religious milieu include disagreement with the statements 'there is nothing I can do to help solve the world's problems' and 'the temple community seems irrelevant to life today'. The present study found Buddhists were significantly *less* likely to feel powerless in the face of the world's problems and also less likely to think the religious community was irrelevant to contemporary life.

Liquid religion on the secular borderline

Liquid religion may be manifested by religious-type experiences being sought on the borderline with secular activities. Given that modern cosmopolitan culture would expect people to emphasize their individuality and independence, it would be deemed a particular sacrifice to go *against* this trend and express attitudes where they make long-term commitments to something larger than themselves, whether it be ideals or communities, secular or otherwise. With this aspect of liquid religion, religiosity becomes orientated away from traditionally 'solid' religious values such as security, conformity, and tradition, and towards self-expression and individuality (Lassander 2012). De Groot (2008) gives the example of those who take part in festivals where a religion organizes but does not dominate the event, and where celebrities may be deployed as its public face. Postsecular participants might simply want to feel part of something larger than themselves *without* pressure to make a long-term commitment to shared values—their connection more aesthetic. For such manifestations of 'instant community', any appearance of a mass movement is illusory, since like a 'flash mob' demonstration, any allegiances apparent during the event rapidly fade the moment it ends. To this category Wickström and Illman (2012) add the example of treating the environment as sacred. Indicative questions for liquid religion on the secular borderline included: 'I am a spiritual person', 'I am influenced by celebrities', the filial piety question 'we should keep our aging parents with us at home' and the environmental question 'I am concerned about the risk of pollution to the environment'.

This study showed that disconnection from religious roots was not found in the UK sample, since Buddhists were *less* likely to estimate that they were influenced by celebrities and *more* likely to believe they should care for parents in old age. Buddhists, however, were found to be more likely to identify as a spiritual person—but this observation should be put in context by comparison with Buddhist self-identification as 'religious', since converts were significantly more likely to say they were spiritual rather than religious, and for heritage Buddhists it was the other way around. There was no evidence to suggest environmental concern was linked with being a Buddhist in a way corresponding to liquid religion on the secular borderline. The statistics fail to reflect the 'instant community' phenomenon being present in Buddhists—especially heritage Buddhists—and therefore the aspect of liquid religion on the secular borderline would seem to break down, at least for Buddhist self-identifiers—although this will be discussed as interference by boundary marking.

Liquid religion outside the religious sphere

The third facet identified by de Groot is where liquid religion is found completely outside the religious sphere. It may manifest as religion being treated as a consumer commodity (Frisk and Nynäs 2012) fuelling a 'smorgasbord' approach to spirituality following free market principles (Carrette and King 2012). Included in this aspect are tensions and contradictions on the part of subjects where identity boundaries have become porous. Individuals in Western society consider themselves often erroneously as *either* secular or religious—however, all combinations of believing in and belonging to religion have latterly been observed, including the permutations in between (Beaumont 2010). Authority may be plural and increasingly complex (Utriainen et al. 2012). The sort of questions asked need to be those that represent values *conflicting* with those of the respondent's own religion. Postsecularity would be demonstrated by the simultaneous presence of seemingly conflicting sources of authority. Although for other religions, suitably conflicting attitudes would need to be chosen accordingly, for Buddhists, making use of the assumption of atheism, the question asked concerns belief in God. The presence of postsecularism would predict evidence of competing sources of authority coexisting. Buddhist adolescents were found, nonetheless, to be significantly less likely to say they believed in God.

Lived religion

A possible fourth facet of liquid religion *not* specifically mentioned by de Groot but included as a postsecular religious phenomenon by authors such as Utriainen et al. (2012) is that of 'lived religion'. Lived religion entails well-being practices such as yoga, charismatic healing, and angel therapy engaged in and made use of and by individuals in their everyday lives. The indicative question chosen for lived religion was agreement that life had a sense of purpose. The presence of postsecularism would predict the religious to have more of a sense of purpose in life. It was found that Buddhists *were* significantly more likely to have lived religion in that they were more likely to say that they felt their life had a sense of purpose.

Projection

Projections are a phenomenon described by Berger (1967) where 'apparitions' are perceived by those who have been raised in a certain religious culture to anticipate and have a vocabulary to deal with certain aspects of the supernatural. Experiences reported in social science that would seem to conform to this category include Bauman's (1997) reference to Maslow peak experiences and Abby Day's (2013) description of normally unreligious young people having supernatural encounters with apparitions of recently deceased relatives. For those assuming the contemporary age to have normalized towards Dawkins-style hyper-rationalism, to see angels might suggest a throw-back to the beliefs of preliterate cultures. Many would expect such beliefs to have been displaced by modernity in the way Pirsig (1974) argued that 'ghosts' had been banished by 'rational belief' in his novel *Zen and the Art of Motorcycle Maintenance.* Indicative questions pertaining to projection and which would indicate the presence of postsecularism relate to belief in angels and life after death because they push back against values supposed to have been displaced by modernity. Buddhists were found less likely to believe in angels but were *more* likely to believe in life after death.

Boundary marking

Differences in boundary marking may relate to what people *do* with their religion. Where at first sight 'revivalism' or 'resistance identity' might seem to indicate reversion to modern or premodern identity, it may indicate postsecularism if it occurs as a defence against secularization or the melting pot of an 'other' mainstream identity. For the purpose of interpreting the features of postsecularity, in this section data concentrate on the differences of attitudes between heritage and convert Buddhists.

Religious revivalism

Religious revivalism means entrenching of identity in defensive reaction to a perceived 'other' and might be perceived as a resolidifying of religion towards 'modern' forms. Although Habermas (2006) referred specifically to fundamentalism, Bauman only alluded to religious revivalism in the face of secularism as 'ghettoization' (Bauman 2001). Indicative questions for religious revivalism include considering yourself to be a 'proper' example of your own religion, considering your religion to be the only true religion, and having a sense of national pride. Heritage Buddhists were found to be more likely than convert Buddhists to say that they had a strong sense of national pride, to consider themselves a proper Buddhist, and to think Buddhism to be the only true religion. Heritage Buddhists are therefore more likely to use religious identity to mark boundaries against 'the other' than converts by entrenching themselves in the attitudes of religious revivalism.

Resistance identity

Unlike revivalism which resists 'the other', resistance identity counters individualism by emphasizing aspects of religion that bind the community together as a collective. In the context of this discussion of postsecularity, resistance identity works against mainstream secular forces (for example, as we shall see later in the chapter, 'the Establishment' for convert Buddhist teens or dissolution of the 'in-group' for heritage Buddhist teens) that seek to undermine collective identity. Resistance identity was exemplified in Thanissaro's (2014a) description of Buddhist teen reaction against stereotyping and special effort to ensure that funerary practices would not die with their parents. It would also include efforts to counter relativism concerning religious truth claims. Resistance identity is implicit to the extent the religion serves to bring people together in a Deleuzian (Deleuze 1993) sense of religion serving merely as an arbitrary set of shared values that cement a group of people together—thus putting community before themselves in a way I would argue could be a form of resistance identity that indicates another aspect of postsecularity. Following Singelis et al. (1995) to ascertain the degree of collectivism within a hierarchy (vertical collectivism), the following question was chosen: 'I would do what pleases my family even if I detest that activity'; and to determine the degree of collectivism on the basis of equality (horizontal collectivism), the following question was selected: 'the well-being of my fellow students or workers is important to me'. The presence of postsecularism would predict a resurgence of positive attitudes towards submission to community values at the expense of individualism—so as resistance identity postsecularism may manifest as community values subduing the modernist tendency to individualism. Collectivism was found to be stronger for heritage than convert Buddhists on both the horizontal and vertical dimensions. Heritage Buddhists were significantly more likely than convert Buddhists to agree that they would do what pleased their family even if they detested the activity and to agree that the well-being of their fellow students or workers was important to them.

Conclusion

It would appear that the quantitative approach to postsecularism employed in this chapter has succeeded in bringing research closer to the subjects, politics, and movements involved in the reconfiguring of the religious and secular. Although philosophical, theological, theoretical, or normative approaches have some analytic power in their description of postsecularity, the more empiricist epistemology of the approach in this chapter provides an innovative potential for *predictive* power. It has revealed the weakness of omitting to use self-identifiers, and it is emancipatory to the extent it has revealed the diversity within the Buddhist category.

Assuming that Buddhist teens in Britain are a valid testing ground for the theory of postsecularity and given that postsecularism seems to be reflected generally in the way Buddhists experience resurgence in their religion, to help assess whether we have captured all the relevant facets of postsecularism in our questions, we have to see whether postsecularity's sub-components follow the same general trend. Despite the multifaceted characteristics of postsecularism, data from this study of British Buddhists would indicate that diminished modernism and secularity correspond with postsecularism.

Increase in public expression of religion also conforms to postsecularism, but it is pertinent to add that for Buddhists there is a high frequency of involvement with private expressions of religion (meditation, prayer, chanting, bowing to parents, and scripture reading) alongside public expressions of religion—so it cannot be said that public expressions have *replaced* private expressions. Increased prevalence of liquid religion in the religious context, lived religion, and some aspects of projection also bear witness to these aspects being part of the postsecular.

Results were less clear, however, in the case of liquid religion on the secular borderline. Buddhists were more likely to say they were spiritual, but contrary to expectations, less likely to show signs of instant community. Buddhists were also less likely to show the expected signs of liquid religion outside the religious context where mixing and matching of conflicting authorities would have been required. This disparity, at least in the case of Buddhists, would seem to correspond with efforts to mark boundaries through resistance identity and religious revivalism for Buddhists where certain aspects of modern religion are clung to or reinstated—not in reaction to secularity, but in response to individualism or competing expressions of religious faith.

There are many challenges to this exercise in finding empirical indicators for postsecularism. For example, it is a sort of tautology to assume the religious in a postsecular age will show more signs of postsecularism than the secular, and then use their individual differences as the definition for postsecularity. Furthermore, the questions chosen to operationalize postsecularism in this study may not have been ideal since use was made of questions on a survey designed for more general purposes. Nonetheless, there is huge potential for applying this methodology to map postsecularity in different sectors of society. As these 20 questions are worded neutrally and show statistical significance for content validity matching postsecularity, they are commended for wider use in postsecular research in non-Buddhist religions.

There are several aspects of religiosity that are measurable (Francis 2009)—religious affiliation, religious participation, religious attitudes and beliefs—and many other aspects of religion which are not so easy to quantify, whether it be doctrine, mythology, ethics, ritual, experience, institution, or material (Smart 1992). It would appear that of all these aspects of religion, it is *attitudes* that best reflect the religious resurgence described by postsecularity. The findings of this study with Buddhists correspond with Habermas' observation that awareness is the aspect of religion that has changed with postsecularity. This study adds weight to Habermas' argument for postsecularity having transformed awareness of religion, by showing that postsecularism can be considered a quality of the people as well as being a feature of the age or the culture, because it is reflected and measurable in participants' attitudes.

Although postsecularism appears useful as more than just a sociological concept—as shown by the wide range of essays in this book—it needs to be distinguished from border marking, otherwise the instant community aspect will be obscured. De Groot's three aspects of liquid religion appear to be borne out by empirical data (within the boundary marking caveat) but could usefully be strengthened by explicit inclusion of the category of 'lived religion'. Bauman's general notion of secularization may require revision since there is not always a clear opposition between the 'modern' and the postsecular. It may be that in describing liquid modernity in relation to religion, Bauman has assumed that religion must be *either* solid or liquid—either with an objective fixed structure or a socially constructed one.

It seems likely that religion has always been of a more hybrid nature than academics have acknowledged—but with new paradigms of social research, the liquid aspects have become more apparent. However recent religious hybridity might be, we are still left with the puzzle of whether revivalism is reversion to modernist religion or whether it has somehow become further transformed in a way that would constitute a new postsecular form caused by the process of liquefaction and recongelation. To extend Bauman's metaphor, there are two scenarios for materials that melt and recongeal—some resolidify in mixed form (e.g., neapolitan ice cream which forms an interesting pattern of marbled variegations) forming solid forms in new combinations—possibly illustrated by the pan-Asianism phenomenon seen in the USA; whereas some resolidify by separating out more completely than previous to melting (e.g., butter). The case for Buddhists would seem to follow the butter metaphor—since as a result of the porous boundaries between Western and eastern cultures, Buddhists have artificially shored up their boundaries in a way more complete than is seen in more religiously homogenous cultures.

This study reaffirms Thanissaro's (2014a) findings with the heritage Buddhist community that postsecularism is displayed in the characteristics of religious resurgence for British Buddhists. This chapter has shown additionally that postsecularity can be observed in the *convert* Buddhist community. The Buddhist community seems to display nearly all the facets of postsecularity, whether it be rejection of modernism and secularity, liquid religion in the religious context, lived religion or projection. Aspects of liquid religion on the borderline of the secular or in non-religious contexts are interfered with by Buddhist boundary marking. Since the heritage and convert Buddhists are using religion to mark boundaries in different ways, the aspect particularly of 'mixing and matching' of religious authority and 'instant community' are obscured. Heritage Buddhists are aligning themselves relative to pressure to maintain and perpetuate the identity of their in-group. By contrast, convert Buddhists seem to be aligning themselves relative to rejection of establishment values and social status while advocating alternative spirituality (Thanissaro 2016).

Convert Buddhists have thus demonstrated cultural agency by reinterpreting and mobilizing an array of resources in a way that would not have been possible in the modern era. These differences may be due to alternative ways of interpreting the term 'modernity' depending on the boundaries/identities being promoted or defended—which for Buddhists would differ in Asian and Western contexts (McMahan 2008)—although, it could be argued that self-identification as belonging to a religion is, in itself, a vestige of religious modernism.

Many of the characteristics of postsecularism are useful for explaining the tenacity with which Buddhist identity features, very different to the values of a mainstream culture, are being passed down to a second generation of young Buddhists. The melting pot theory of acculturation would predict this minority religion would gradually lose its identity to that of a majority secular mainstream. But this prediction fails to correspond with the observed reality of Buddhist teens living in Britain receiving nurture into their family's tradition. Postsecularism allows for the way modern forms of religion may recombine with postsecular forms. Nonetheless, postsecularism-like examples of 'Buddhist' mixing and matching of spirituality such as the mindfulness movement or juxtaposition of mindfulness with contemporary art, literature, and Judaism (Mitchell and Quli 2015), are often applicable only to those who have an aesthetic interest in Buddhism rather than an affiliation to Buddhism. It should be pointed out that failure to define the level of self-identification with Buddhism in research participants weakens any conclusions to be drawn in research about postsecularity. I have dealt with postsecularity in Buddhist self-identifiers, but different conclusions might be reached if the same research questions had been fielded in the quasi-community of Buddhist sympathizers.

Notes

1 The author would like to thank Revd. Canon, Prof. Leslie J. Francis (University of Warwick), and Prof. Mandy Robbins (Glyndŵr University) for supervision of this research, and to St. Mary's RE Centre for hosting the online version of the survey used in this study. Special tribute is also extended in the year of writing (2017) to three social commentators cited in this chapter—Zygmunt Bauman, Robert M. Pirsig, and Peter L. Berger—who passed away on 9 January, 24 April, and 27 June respectively.

2 But not the USA—with the possible exception of 'white guilt' which shows aspects of boundary marking (Steele 2006).

3 The case study on postsecularity described in this chapter is based on data derived from a quantitative survey including the 20 questions described and general demographic questions fielded in the period 2013–2014 as a paper and online survey to self-identifying Buddhists aged between 13 and 20 and resident in Britain, as part of a University of Warwick-funded doctoral research project. Where comparisons are made with non-Buddhists, Buddhist adolescent data case has been compared for significant differences against a dataset of answers to identical questions published by Francis (2001).

4 A summary of the 20 questions and the statistical analysis behind the conclusions can be found at doi: 10.13140/RG.2.2.11696.46085.

Further reading

Batchelor, S. (2017) *Secular Buddhism: imagining the dharma in an uncertain world*, New Haven, CT: Yale University Press.

Postsecularity has not caught up with Buddhism in the academic literature. Closest to the topic under examination has been limited discussion of Buddhism and postmodernity. Although running counter to Asian Buddhist views, this chapter argues that postmodernity and Buddhism (in the West) have much in common because of Buddhism's inherent fluidity.

Page, S.-J. and A. K.-T. Yip (2017) *Understanding Young Buddhists: living out ethical journeys*, Leiden/Boston, MA: Brill.

A recent example of mixing and matching of religion, sexuality, and other aspects of identity in young British convert Buddhists—which the authors specifically identify as lived religion.

Park, J. Y. (2008) *Buddhism and Postmodernity: Zen, Huayan, and the possibility of Buddhist postmodern ethics*, Lanham, MD: Lexington Books.

The third section of this book approaches the argument that both Buddhism and postmodern philosophy are perceived to have a shared problem in ethics—the non-identity of identity, interconnectedness of opposites, and the lack of a transcendental foundation of an entity—comparing Nagarjuna amongst others with Kristeva, Lyotard, and Derrida.

References

Bauman, Z. (1997) *Postmodernity and its Discontents*, Cambridge, MA: Polity Press.

——— (2000) *Liquid Modernity*, Cambridge, MA: Polity Press.

——— (2001) *Community: seeking safety in an insecure world*, Cambridge, MA: Polity Press.

Baumann, M. (2001) 'Global Buddhism: developmental periods, regional histories, and a new analytical perspective', *Journal of Global Buddhism*, 2: 1–43.

Beaumont, J. (2010) 'Transcending the particular in postsecular cities'. In Molendijk, A. L. M., Beaumont, J. and C. Jedan (eds.) *Exploring the Postsecular: the religious, the political and the urban*, Leiden: Brill, pp. 3–17.

Berger, P. L. (1967) *The Sacred Canopy: elements of a sociological theory of religion*, New York, NY: Doubleday.

Braun, E. (2015) 'The United States of Jhāna: varieties of modern Buddhism in America'. In Mitchell, S. A. and N. E. F. Quli (eds.) *Buddhism beyond Borders: new perspectives on Buddhism in the United States*, Albany, NY: SUNY Press, pp. 163–80.

Brierley, P. (2006) *Pulling out of the Nosedive: a contemporary picture of churchgoing: what the 2005 English Church census reveals*, London: Christian Research.

Carrette, J. and R. King (2012) 'Spirituality and the re-branding of religion'. In Lynch, G., Jolyon, M. and A. Strhan (eds.) *Religion, Media and Culture: a reader*, London: Routledge, pp. 60–9.

Casanova, J. (1994) *Public Religions in the Modern World*, Chicago, IL: University of Chicago Press.

Connolly, W. E. (2005) *Pluralism*, Durham, NC: Duke University Press.

Day, A. (2013) *Believing in Belonging*, Oxford: Oxford University Press.

Deleuze, G. (1993) 'Three aspects of culture'. In Boundas, C. V. (ed.) *The Deleuze Reader*, New York, NY: Columbia University Press, pp. 245–52.

Francis, L. J. (2001) *The Values Debate: a voice from the pupils*, London: Woburn Press.

——— (2009) 'Comparative empirical research in religion: conceptual and operational challenges within empirical theology'. In Francis, L. J., Robbins, M. and J. Astley (eds.) *Empirical Theology in Texts and Tables: qualitative, quantitative and comparative perspectives*, Leiden: Brill, pp. 127–52.

Frisk, L. and P. Nynäs (2012) 'Characteristics of contemporary religious change: globalization, neoliberalism, and interpretative tendencies'. In Nynäs, P., Lassander, M. and T. Utriainen (eds.) *Post-Secular Society*, New Brunswick, NJ: Transaction Publishers, pp. 47–70.

Groot, K. de (2008) 'Three types of liquid religion', *Implicit Religion*, 11(3): 277–96.

Habermas, J. (2006) 'Religion in the public sphere', *European Journal of Philosophy*, 14(1): 1–25.

Kay, W. K. and L. J. Francis (1996) *Drift from the Churches: attitude toward Christianity during childhood and adolescence*, Cardiff: University of Wales Press.

Lassander, M. (2012) 'Grappling with liquid modernity: investigating post-secular religion'. In Nynäs, P., Lassander, M. and T. Utriainen (eds.) *Op cit.*, pp. 239–67.

McMahan, D. L. (2008) *The Making of Buddhist Modernism*, Oxford: Oxford University Press.

Mitchell, S. A. and N. E. F. Quli (2015) 'Preface'. In Mitchell, S. A. and N. E. F. Quli (eds.) *Op cit.*, pp. vii–xiii.
Moberg, M. and K. Granholm (2012) 'The concept of the post-secular and the contemporary nexus of religion, media, popular culture, and consumer culture'. In Nynäs, P., Lassander, M. and T. Utriainen (eds.) *Op cit.*, pp. 95–127.
Pirsig, R. M. (1974) *Zen and the Art of Motorcycle Maintenance*, New York, NY: HarperCollins.
Singelis, T. M., Triandis, H. C., Bhawuk, D. P. S. and M. J. Gelfand (1995) 'Horizontal and vertical dimensions of individualism and collectivism: a theoretical and measurement refinement', *Cross-Cultural Research*, 29(3): 240–75.
Smart, N. (1992) *The World's Religions*, Cambridge: Cambridge University Press.
Steele, S. (2006) *White Guilt: how blacks & whites together destroyed the promise of the civil rights era*, New York, NY: HarperCollins.
Taylor, C. (2007) *A Secular Age*, Cambridge, MA: Harvard University Press.
Thanissaro, P. N. (2011) 'A preliminary assessment of Buddhism's contextualisation to the English R.E. classroom', *British Journal of Religious Education*, 33(1): 61–74.
——— (2013). 'Temple-going teens: religiosity and identity of heritage Buddhists growing up in Britain', *Religious Education Journal of Australia*, 29(1): 9–15.
——— (2014a) 'Almost a proper Buddhist: the post-secular complexity of Heritage Buddhist teen identity in Britain', *Journal of Global Buddhism*, 15: 1–14.
——— (2014b) 'Internal diversity in Buddhism: comparing the values of Buddhist teens raised by heritage & convert parents', *Journal of Contemporary Religion*, 29(2): 315–30.
——— (2016) *Temple-Going Teens: religiosity and identity of Buddhists growing up in Britain*, unpublished Ph.D. thesis, University of Warwick.
Turner, B. S. (2010) 'Religion in a post-secular society'. In Turner, B. S. (ed.) *The New Blackwell Companion to the Sociology of Religion*, Chichester: Blackwell, pp. 649–67.
Utriainen, T., Hovi, T. and M. Broo (2012) 'Combining choice and destiny: identity and agency within post-secular well-being practices'. In Nynäs, P., Lassander, M. and T. Utriainen (eds.) *Op cit.*, pp. 187–216.
Warner, M., VanAntwerpen, J. and C. Calhoun (2010) 'Editors' introduction'. In Warner, M., VanAntwerpen, J. and C. Calhoun (eds.) *Varieties of Secularism in a Secular age*, Cambridge, MA: Harvard University Press, pp. 1–31.
Wickström, L. and R. Illman (2012) 'Environmentalism as a trend in post-secular society'. In Nynäs, P., Lassander, M. and T. Utriainen (eds.) *Op cit.*, pp. 217–38.

31

Unofficial geographies of religion and spirituality as postsecular spaces

Edward Wigley

Introduction

Since the early 2000s, geographers have been encouraged to investigate religious experience and spirituality 'beyond the official' spaces of institutionalized religion (Kong 2001). What is 'official' is often constituted by the spaces sanctioned for religious purpose, practice, symbolism, and expression by governmental or specific organizations perceived to have legal or social authority. To be 'unofficial' is considered that which is not sanctioned by, and in some cases even resisted by, local authorities or religious organizations and groups. Western societies are increasingly resisting official narratives, and schemes of social organization are moving towards unofficial actions and praxis as part of the 'massive subjective turn of society' (Taylor 1991, cited in Heelas and Woodhead 2005).

Meanwhile the postsecular agenda has suggested a plurality of different channels in which religion has infiltrated popular, and 'secular', appropriations of discourses and spaces (Molendjik et al. 2010; Beaumont and Baker 2011; Beckford 2012). These different channels can be seemingly dramatic and dominant, such as the return of religion via fundamentalist groups (Wilson and Steger 2013), or less visible through faith-based organizations and groups providing social or public services without emphasizing their religious identities through symbolism or codification (Dalferth 2010; Beckford 2012; Cloke and Beaumont 2013; Reynolds 2015) as well as the continuing and affecting prevalence of religious identities of people and sites (Knott 2010; Stevenson et al. 2010; Havlicek and Klingorova 2017).

This contribution considers the postsecular in the light of the subjectivization of religion and emergence of spirituality. More specifically it draws on empirical research to focus on the subjective experience and construction of geographies that maybe framed as postsecular: the non-institutional and non-religiously coded everyday urban spaces that can accommodate, through the acts of contemplation, practice, and mobility, sacred spaces within the profane. Such experiences often draw from the official institutions and narratives of religion, but blend these sources into unofficial spaces of home, employment, recreation, and journeys. The next section will first consider the literatures investigating geographies of religion and spirituality and the mobilities paradigm before relating these to contemporary discussions of the postsecular. Following this will be a discussion drawing on the author's empirical research. This work demonstrates the postsecular within the context of the individual's everyday life and activities, and how this is understood on a practical and aesthetic level. The entry then concludes that such practices as they merge into everyday life, 'spiritually re-appropriate' (Beaumont and Baker 2011) secular landscapes on a personal, rather than institutional or organizational, scale.

The postsecular

Postsecularity emerged during the early twenty-first century, employing a range of definitions and being deployed as a concept in a range of different discursive contexts (Molendjik et al. 2010; Beckford 2012; Olson et al. 2013). Definitions have been diverse; however, there are common themes to the different interpretations of the postsecular. While some commentators have suggested that the term implies the decline of the secularization narrative and resurgence of the religion in Western societies where secularization is understood to have taken place in the twentieth century (Bruce 2013), this does not represent the majority of views. The 'established' churches of Western Europe continue to decline. In the UK, the Church of England as well as the Methodist Churches continues to dramatically reduce in numbers decade after decade. Religions that have seen growth are those that became an established presence in the last half of the twentieth century, with a clear (often conservative) theological message and/or a core base within migrant communities and their children rather than a resurgence among those communities who had previously abandoned their churches (Parsons 1993; Davie 2006; Guest et al. 2012).

Religion never really disappeared in the way that the sociologists of the twentieth century predicted (Woodhead 2012; Davie 2015). Beliefs and practices became fragmented and disassembled from larger structures of religious institutions, instead finding more individually orientated models of 'spirituality' (Davie 1994; Heelas and Woodhead 2005). The late twentieth century saw the growth of participation in 'New Age' or alternative spiritualities or activities and their wider diffusion across different and diverse segments of society, from business offices that accommodate massage practitioners to aromatherapy candles available in supermarkets (Heelas and Woodhead 2005; Harvey and Vincent 2012). Alongside this prevalence of 'spirituality' has been the political presence of religion in institutional and popular forms identified by many postsecular studies. Recent social and political developments have drawn religious organizations into the public sphere in more 'official' capacities. In the UK, for example, the economic measures and political drive through the short-lived 'Big Society' ideology, has seen increased visibility of Christian and non-Christian faith-based organizations delivering public and social services such as adoption agencies and food banks (Cloke 2011). Such shifts are driven by a lack of concern regarding, or downplaying of, the faith- or non-faith-based motivation of an organization or group, provided that the outcomes are delivered and accessible to all faiths and none (Dalferth 2010; Cloke and Beaumont 2013). For the organizations and groups delivering such services, their new-found role represents an opportunity to reclaim a position of importance in the public sphere—a reterritorialization of the neoliberal and urban society (Beaumont and Baker 2011).

Cities, as seen earlier, have been critical spaces for the emergence of postsecular communities, whether in the delivery of public services, the co-existence and contestation between different faiths and none (Fenster 2011; Olson et al. 2013), and the complexities of decline and growth within Christian and non-Christian religions (Stevenson et al. 2010). Reterritorialization of political and social spaces is a key concern in this debate; however, it is the transformation of religious praxis and emergence of 'spirituality' that people perform consciously and unconsciously that is examined in this essay. Knott (2005) presents a possible pathway when considering the multiple spatial properties of objects and their relationships between the religion, the secular, and the postsecular. Meanings are generated by the juxtapositions of particular symbols, objects, or people in different spaces, and the recombination of these things create, reinforce, or challenge values associated with places.

The manifestation of religion or spirituality in spaces outside of official sites therefore generates new meanings or challenges or confirms other values. This concern for understanding the individual at the centre of processes of the postsecular coincides with Zock's (2010) call for more research of the individual in hybrid and culturally diverse cities from which the postsecular emerges. Havlicek and Klingorova's (2017) survey of public 'sacred sites' found within postcommunity Prague illustrates the prevalence of religion and its materialities in otherwise (and historically reinforced) secular urban space, and in the process accruing new function and meaning for the city, and its residents and visitors. In Jerusalem, residents of conservative neighbourhoods impose religious values relating to modest clothing in the area with (illegal) signs and regulation, reclaiming and territorializing secular public space (Fenster 2011). In a recent article, della Dora argues for 'infrasecular geographies' of sacred spaces transformed, modified, or 'unmade' into secular space, yet retain the visual expressions of religion, contending her term is effective 'in capturing the complexity of a society in which the secular and the religious coexist, overlap and compete' (2018: 48). In the context of the city, the postsecular has 'crossed the boundary, made incursions and inroads, and arguably begun to transform its secular host' (Knott 2010: 28). Beaumont and Baker (2011) discuss the 'spiritual re-appropriation' of the secular landscape in the context of organizations that resist secular neoliberal claiming of the city. Following on from Havlicek and Klingorova's survey of publicly visible 'sacred sites', the territorializing of urban neighbourhoods (Fenster 2011), and an embodiment perspective, the research presented here focuses on the postsecular space and spatialities of the body.

Religion and mobility

Geographers of religion and spirituality have also engaged with the new mobilities paradigm and the call by Kong for research beyond the official borders and sites of faith to understand '[t]he ways in which religion is experienced and negotiated are also multifaceted and multi-scaled, from the body to the neighbourhood, city, nation, and across nations' (2001: 769). The new mobilities paradigm recognizes the centrality of mobility to contemporary social life (Urry 2007): who or what is moving, what speed are they moving, how are they moving, where are they moving. Everyday movement encompasses multitudes of social life and can bring travellers into encounters with a diverse range other people, objects, and places. Additionally, mobilities research has uncovered a range of activities and strategies that drivers, passengers, pedestrians, and other travellers employ to increase the enjoyment or utility of their travel time.

While geographers of religion have engaged with mobility, traditionally this has been focussed on the area of pilgrimage: extraordinary journeys perceived to be difficult, expensive, and arduous. More recent literature has challenged this perception and investigated the socio-economic continuities with everyday life within contemporary pilgrimage (for example, Eade and Sallnow 1991; Coleman and Eade 2004). Maddrell et al. (2015a)-edited collection *Sacred Mobilities* potentially introduces a new dimension to this research that understands the significance that movement can enable to a diverse range of secular-sacred spaces. The editors' own contribution (2015b) to the collection, a chapter regarding the annual TT motorbike races in the Isle of Man, underlines their interpretation of sacred spaces as inclusive of secular events. This research reflects innovative approaches to understanding the space between religion and the secular dimensions of social life.

Mobility has also featured, if less explicitly, in other studies since the turn of the twenty-first century. The practice of spirituality and activities, such as yoga and meditation, originate in spiritual or religious contexts, in and amongst everyday life and habits (Holloway 2003; MacKian 2012; Philo et al. 2011, 2015). Such studies have addressed the shift of religion and spirituality from institutionally focussed to individually centred approaches to developing a set of practices and beliefs that are relevant to and draw from personal experience (Heelas and Woodhead 2005; Giordan and Pace 2012). In the process, religion and spirituality have continued significance yet simultaneously reduced visibility in the public sphere. Instead, religion and spirituality have moved beyond traditional boundaries into everyday spaces from urban parks to driving spaces, merging the secular and the religious. While these studies have been very fruitful, engagement with mobilities is still a relatively growing area, and, in addition, the focus on non-institutional religion has unduly emphasized alternative spiritual and holistic practices rather than subjectively constructed experiences of mainstream religions such as Christianity. These practices demonstrate an emerging approach of combining official and unofficial practices and beliefs and suggest new approaches to understanding the postsecular.

This research addresses these issues and explores how spiritual and religious ideas and practices are realized in the daily lives of participants—especially in those time-spaces considered as secular and outside of the coding, regulation, and maintenance of the 'official spaces' of religion, in this case the Baptist church or Buddhist meditation centre of Bristol. This work understands the postsecular set of relations on the scale of the individuals, as they practice and perform actions that enable their spirituality to emerge in otherwise secular spaces.

Methods

The empirical work presented in this essay investigates the subjective experience of combining the spiritual with the secular in time-spaces of mobility. The sample was therefore drawn from two groups that represent the 'congregational domain' and the 'holistic milieu' as mapped by Heelas and Woodhead (2005) to explore how both a traditionally embedded congregational sample and a holistic sample adapt to time-spaces of mobility. The Baptist Christian denomination was selected as an example of the congregational domain, whilst Buddhist meditation centres were selected from the holistic milieu. Many of the participants did identify themselves as 'Baptist Christians' and 'Buddhists'; however, this was not a requirement of the research. The participants were selected as people who attended sites of Christian worship or Buddhist-led meditation whether or not they affiliated with the broader religious movement. Participants, recruited via a survey earlier in the research, were asked to keep a diary of their activities for one week and then discuss this in the context of an interview.

Similar to Cadman et al. (2013), the project sought out the time-spaces in which religion and spirituality 'structure, frame and bleed into everyday practices', necessitating ethnographic research to be undertaken. A diary-interview method was employed to allow access to the personal experiences of the participant without the disruption of a researcher. Diaries enabled an intimate portrait of the participant's to be developed; the interweaving of their spiritual lives and their secular lives while the interviews that accompanied the diaries in this research were semi-structured, usually lasting around one hour in duration. The interview provided an opportunity to clarify activities and to increase the resolution of data from generic terms of 'walking' to 'walking to work, thinking about *x*, acting in this way because of *y*', deepening the diary data (Phipps and Vernon 2009). Along with the diaries, thorough thematic analysis was carried out on the participant's corresponding interview transcript. In total, 16 sets of diaries and interview transcripts were collected, coded, and analyzed for themes emerging from their everyday journeys and spiritual practices. All participants have been assigned pseudonyms to maintain their confidentiality whilst survey respondents are left unnamed. Church and meditation centre names have also been changed for this reason.

Postsecular spaces of the everyday

This section examines the participant's use of particular places and times for spiritual or religious purposes within the socio-economic schedule of everyday life. It explores the participant's practice of religion and spirituality within environments considered to be public and secular in identity. In doing so, the diminishing distinctions between religion and the secular, the sacred and profane, are examined. While recognizing that these modes are not polarized in everyday life but blended into each other, this research asserts that there is an affinity between the two modes of space that allows participants to negotiate this pairing within the modern environments in which they are positioned. Within these spaces, the engagement of socio-economic responsibilities of everyday life synchronises and supports the rhythmic, behavioural, and mental manifestations of religious and spiritual activity among participants. Such synergies act to generate additional meaning and significance for the participant as the two rhythmic sources interact and adjust to each other.

The temporary time-spaces that participants construct or manipulate in otherwise transient spaces, those of employment and leisure are identified and explored here. These time-spaces do not 'belong' to, nor are specially constructed for, the participant, yet they are able to use these spaces for prayer and meditation activities before they are returned to normal use. As they create space-time paths throughout the day, from waking up in the private space of their bedroom in the morning to returning to the same room at night, the individuals thread together a range of time-spaces within their everyday spatial range and activities. For participants in this research, these spaces can also accrue meaning through the practices and performances enacted. These public spaces are usually considered as 'secular', such as the workplace, the mobility-centred spaces of a journey, and the recreational spaces of parks, amongst other places.

These spaces contain specific circumstances that can influence the form of religious or spiritual activity practised or drawn upon. Pockets within employment time-spaces could be transformed, temporarily, into spaces of renewal and spiritual practice, and this occurred with several participants. Paul and Richard of the meditation centre sample, for example, developed strategies and networks to accommodate brief moments of meditation practice throughout their working day. Paul made time for a series of two-minute meditations at certain points throughout the day, while Richard had found a quiet room at his workplace for brief moments of meditation. Throughout his participation in this research, it was clear his job was a significant source of stress and anxiety to Richard. When he could anticipate heightened stress levels or a greater level of commitment to his meditation, Richard would also carry with him an MP3 player to aid his practice, containing audio files for meditation practice including a recording by Jack Cornfield, a well-known teacher within Theravada Buddhism. As well as this room at work, Richard also sought to use opportunities presented within his working and travel schedule for brief pockets of meditation practice:

> I'm pretty good in traffic jams. [...] and sometimes actually it's an opportunity [...] I think well here's an opportunity [for breathing exercises]. If I get somewhere early think well

> I can get out of the car and go and sit in there and scrounge a cup of tea or...I can sit here for ten minutes and shut my eyes and do a sort of breathing, slow myself down.

Here, Richard has identified time-spaces that are often negatively perceived—sitting in a traffic jam and the tedium or anxiety of waiting for a meeting—and transformed these for positive effect using meditative practice.

Unlike Richard and Paul, who removed themselves from other people and sources of distraction for spiritual practices, Michelle found a benefit in remaining at her desk and using podcasts on Buddhist meditative instruction and teachings in the background while working, finding that the occasional comment will disrupt the working body:

> And then you pick up snippets. You think oh yeah, god they've got a point there, let me try it. [...] But you always, you kind of tend to go back to your old habits but that's the whole point isn't it? Eventually you won't.

Michelle's final comment here makes visible an underlying theme also present within Richard's and Paul's accounts: the attempt to alter pre-established and continually reproduced habits. Paul has embedded a new habitual routine within his working schedule to pre-empt employment-related stresses and anxieties. Richard aims to transform time-spaces that are likely to lead to frustration and tedium, such as traffic jams and waiting around, into positive opportunities for renewal. Michelle acknowledges the challenging nature of changing habits, but uses podcasts and practice to try to affect change. Diverging from Philo et al.'s (2015) participants, Michelle, Richard, and Paul, along with other participants in this sample, embed their individual spiritual practices within the time-spaces that are likely sources of stress and anxiety in an attempt to control their response to the stimuli accommodated by these environments.

Spiritual practices during work hours amongst the Baptist church sample were also observed during this research. However, these were manifested with a greater degree of spontaneity and in many respects greater level of embedding into the events and surroundings of the day. Spiritual activities manifested themselves in these environments in the forms of brief prayers that could be uttered or contemplated at the desk in between workloads and tasks.

> Max: [...] It tends to be when there's particular issues coming up and then they crop up in your mind in the day so yeah. Prayers are a kind of amorphous thing.

Max's comments and similar remarks by other participants from the Baptist Church sample imply a wave-like quality to prayer and spiritual practice, catalyzed by internal and external stimuli throughout the day. More overtly communicated here than with the earlier examples from meditation centre participants is the sense that prayer emerges and dissolves into the fabric of daily activities seamlessly as a stream of consciousness. That Max (and other participants from the Baptist church sample) recognizes moments of prayer implies the embeddedness of this practice even though it does not require specifically codified spaces, behaviours, or gestures.

Prayer, moments of meditations, or listening to Buddhist teachings or instructions are practices which originate and are imported from the participant's involvement with Buddhist meditation centres—a transfer from 'official' spaces of the meditation centre to the 'unofficial' spaces that the participant inhabits in everyday life. The practice contains the original intention of bringing contemplation, mindfulness, or calmness, but is modified in format to fit in with the participant's secular socio-economically shaped requirements. Such seamless integration of religion and secular activities resonates with Beaumont and Baker's (2011) assertion that religion mimics the designs and flows of neoliberal urban spaces. In turn, the manifestation of the religious or the spiritual within the secular everyday creates new meaning as the practice helps the participant to find the relevance of the sacred within the profane. The religious and the spiritual emerge and transform the secular (Knott 2010) space of the workplace, allowing a postsecular identity to emerge in the thoughts and practices of the individual through their prayers, meditations, and contemplations.

Experiences of participants utilizing everyday spaces for religious or spiritual purposes were not restricted to places of work. Across both samples, there were numerous reports of everyday spaces of leisure being employed by participants for a form of renewal. These most often consisted of public green spaces across the city. The theme of renewal for the majority of accounts was centred upon a communion with nature and natural environmental processes. Michelle, for example, spoke at length about the renewing effects of green spaces where natural cycles of birth, growth, decay, and death could be observed.

> If I'm feeling that I need to be resourced…so if I'm feeling vulnerable or tired…erm…or I need to be in touch with something that's greater than myself…something…if I, yeah, it's usually when I'm feeling…that I need resourcing then I, then I can feel when I walk through the park it gives me something back.

Michelle speaks here of seeking out specific sites that can 'resource' her, echoing the need of participants for pockets from which to draw 'energy' within everyday public space (Philo et al. 2011, 2015). Aesthetic value could also be identified in the human-constructed landscape; however, for Michelle, there was a distinct difference in the 'resourcing' effects of the natural over the human landscape:

> Walking down the street can be spiritual in that sense cos you'll be looking at all of the textures of the brick or the concrete or…erm…reflections of oil. You know, light refracted on oil. And you can get a sense of spirituality but you're doing it. You're, you've got to practice mindfulness to get that and you'll see it but it doesn't resource me in the way…that I could just be in nature and actually not have to try and it'll resource me. That I think is, feels like the difference, for me…erm…the human environment doesn't give back, you've got to try, you've got to look for those things, you've got to look for the beauty and there's some-, and when you have found it can, it really does open it up. But there's something about, if you're in nature it just seems to be naturally there. That's the difference for me.

Rhythms of nature that are allowed into the urban environment command a restorative effect and energy that can be drawn upon by the individual. The cycles of death and rebirth that are present within nature are present within green spaces rather than the impermeable surfaces of the concrete streets. Rhythms of nature intersect with the rhythms of daily life for this participant, slowing down the pace of her otherwise busy life (Wunderlich 2013) to enable moments of rejuvenation.

Modern surfaces, in Michelle's account, are 'impermeable' and non-reactive compared with 'natural' surfaces, although not entirely absent of 'resourcing' capacity. The overall theme here of the restorative effects of natural rather than human spaces echoes similar conclusions from environmental psychologists (Maddrell and della Dora 2013)—a sense of enchantment that recharges the recipient (Bennett 2001; Philo et al. 2015). Despite the impermeable appearance of modern environments, Michelle was able to identify natural processes of erosion and decay within the urban fabric. Landscapes that are perceived as durable and stable are also subject to natural processes of degeneration and regeneration. Such observations here echo Beckford's (2012) suggestion that a revival of enchantment in the world and shift away from the twentieth century's singular focus on technology and rationality (Beaumont and Baker 2011) are part of the emergence of postsecular landscapes.

Another participant from the meditation centre sample, Sean, talked of 'awareness' when outside in physical movement or in the pauses between physical movements. This awareness could often be linked to the mobility he was directly engaged in, such as cycling along motorized traffic-free routes, for example, a former railway path that had been converted to a dedicated pedestrian-cycle path:

> on the cycle path that's specified single u[se], well pedestrians as well [as cyclists] [...] you don't have to be too worried about traffic coming and going...erm...and you just kind of yeah, trees and nature all around...which does allow for a bit more of a kind of...dreamy sort or, imagin[ation], allows the imagination to wander a bit more [...]

The separation of motorised and non-motorized traffic in this park provides the protection from the dangers of the road. Similarly, the separation of churches and meditation spaces from the outside world enables participants to feel comfortable and secure enough to engage in activities where they may otherwise feel vulnerable. In this case Sean is able to let the imagination wander. Sean, like Michelle, could also see a form of aesthetic beauty in the built environment to a certain extent:

> in town there's still room for the imagination to wander. Sort of into shop windows or into people passing you or I don't know whatever you're doing. Yeah I think that's sort of...there's something about the, pedestrianism that...erm...I guess lends itself a bit to... just to a bit more sort of...erm...an open awareness?

Within these quotes and their interviews, Michelle and Sean interpret the urban environment through their own spiritual framework. The use of terms such as 'mindfulness', 'resourcing', and 'awareness' can be traced to the discourse of the holistic milieu. Participants here are able to find pockets within everyday public space which fulfils a spiritual purpose, re-energising their day.

This use of space, particularly of urban parks, was not limited to the Meditation Centre sample but also present within the Baptist Church participants' experience as well. Rita had developed a specific form of practice that encompassed her cycling through a park on the way to work in the mornings she called a 'cycling liturgy' (see Wigley 2018). Through the reciting of prayer and scripture at specific marker points in the park, her experience of the place had changed as these words had become embedded within the landscape. On another morning during her diary week and in a different park, on her way to the local supermarket, Rita described an awareness of being 'alone with God'. The interview probed this comment further and found this feeling was partially stimulated at least by the temporary cessation of socio-economic duties, and the language which she employed was spirituality-centred. Protestant denominations such as the Baptists traditionally reject the idea of certain spaces allowing for increased access to God based on the eternal and all-pervading presence.

Rita frames this connection to God in this park as 'being alone' rather than necessarily being closer; hence, it may be that as distractions fade into the background and the sensory field changes from one of a human-constructed to a natural (if ultimately humanly engineered) environment, that God occupies more of the believer's perceptive field.

Urban space is often associated with the concrete greyness of roads and buildings as part of a secular, neoliberal, and mundane environment. Green spaces, usually parks but also cycle paths, appear to be spaces where the participants could find a connection with a higher, underlying power. These spaces open up the urban greyness and the impermeable character of concrete and private-owned properties that alienate the individual. Nature imposes its rhythms through the sense channels of sound, vision, smell, and touch, reminding the individual of the transcendent cosmic processes that underlie artificial surfaces, enabling a postsecular landscape to emerge, recognizing enchantment in everyday life (Beckford 2012, even if he objects to the term). Distinctions between religion and secular become blurred and challenged (Knott 2010) as 'secular' public spaces are appropriated for spiritual practices rooted in Christian and Buddhist traditions. In such spaces, a sense of a transcendent reality accessible within the mundane secular realm leads to an emergence of the postsecular as bringing together the spiritual and the secular in coexistence.

Conclusion

This contribution has drawn upon empirical material to illustrate the potential for modern urban and economic landscapes to afford opportunities for spirituality and religion, as well as be transformed in the perceptions of those actors who utilize the environment. Spaces and places of work and of leisure have been explored in the experiences of the diary participants, and their uses of these spaces have revealed a spiritual depth that has been constructed in spaces built on secular values. It is not simply that participants used their places of work to engage in meditation, for example, as a means of supporting their economic or productive activity but that these spaces are used as opportunities to further spiritual development, to disassemble habits they consider to be impediments for well-being. Similarly, the parks or the cycle paths provide breaks within modern rhythms of movement, mobility, and activity, enabling the participants here to reconnect with a spiritual awareness.

Activities and practices of these participants are relatively silent or not-outwardly obvious to external audiences. Their diaries do not reveal any confrontations or challenges despite these practices often taking place in public or shared spaces and visible to friends, family, colleagues, or strangers. The practices have escaped from the 'official' spaces of the meditation centre or the church, but are equally not confined to the private spaces of the home as secularist and secularisation theory would imagine. These practices blend or complement otherwise secular actions such as walking, cycling through parks, or sitting in traffic jams. They merge into and 'spiritually appropriate' (Beaumont and Baker 2011) everyday routines as the gap between the religious/spiritual and the secular enables a postsecular experience at the heart of a subjective spiritual geography of urban time-spaces.

Further reading

della Dora, V. (2018) 'Infrasecular geographies: making, unmaking and remaking sacred space', *Progress in Human Geography*, 42(1): 44–71.

Provides an up-to-date review of the postsecular literature while also offering a new concept of 'infrasecular' geographies that capture the rich layering of sacred and secular spaces in Western societies.

Gökarıksel, B. and A. Secor (2015) 'Post-secular geographies and the problem of pluralism: Religion and everyday life in Istanbul', *Political Geography*, 46: 21–30.

Study of Muslim women in Istanbul as they encounter and coexist with secular and Alevi neighbours, investigating the lived experience of pluralities on the scale of the body.

References

Beaumont, J. and C. Baker (eds.) (2011) *Postsecular Cities: space, theory and practice*, London: Contiuum.

Beckford, J. A. (2012) 'SSSR presidential address: public religions and the postsecular: critical reflections', *Journal for the Scientific Study of Religion*, 51(1): 1–19.

Bennett, J. (2001) *The Enchantment of Modern Life: attachments, crossings, and ethics*, Princeton, NJ: Princeton University Press.

Bruce, S. (2013) 'Post-secularity and religion in Britain: an empirical assessment', *Journal of Contemporary Religion*, 28(3): 369–84.

Cadman, L., Philo, C. and J. Lea (2013) 'Using time-space diaries and interviews to research spiritualities in an everyday context'. In Woodhead, L. (ed.) *How to Research Religion: putting methods into practice*, Oxford: Oxford University Press.

Cloke, P. (2011) 'Emerging postsecular rapprochement in the contemporary city'. In Beaumont, J. and C. Baker (eds.) *Postsecular Cities: space, theory and practice,* London: Continuum.

Cloke, P. and J. Beaumont (2013) 'Geographies of postsecular rapprochement in the city', *Progress in Human Geography*, 37(1): 27–51.

Coleman, S. and J. Eade (2004) 'Introduction: reframing pilgrimage'. In Coleman, S. and J. Eade (eds.) *Reframing Pilgrimage: cultures in motion*, London/New York, NY: Routledge.

Dalferth, I. U. (2010) 'Post-secular society: Christianity and the dialectics of the secular', *Journal of the American Academy of Religion*, 78(2): 317–45.

Davie, G. (1994) *Religion in Britain since 1945: believing without belonging*, Oxford: Blackwell.

——— (2006) 'Religion in Europe in the 21st century: the factors to take into account', *European Journal of Sociology*, 47(2): 271–96.

——— (2015) *Religion in Britain: a persistent paradox*, Malden MA: Wiley-Blackwell.

Eade, J. and M. J. Sallnow (1991) 'Introduction'. In Eade, J. and M. J. Sallnow (eds.) *Contesting the Sacred: the anthropology of Christian pilgrimage*, London: Routledge.

Fenster, T. (2011) 'Non-secular cities? Visual and sound representations of the religious-secular right to the city in Jerusalem'. In Beaumont, J. and C. Baker (eds.) *Postsecular Cities: space, theory and practice,* London: Continuum.

Giordan, G. and E. Pace (2012) 'Introduction: mapping religion and spirituality in a postsecular world'. In Giordan, G. and E. Pace (eds.) *Mapping Religion and Spirituality in a Postsecular World*, Leiden/Boston, MA: Brill.

Guest, M., Olson, E. and J. Wolffe (2012) 'Christianity: loss of monopoly'. In Woodhead, L. and R. Catto (eds.) *Religion and Change in Modern Britain*, London: Routledge.

Harvey, G. and G. Vincent (2012) 'Alternative spiritualities: marginal and mainstream'. In Woodhead, L. and R. Catto (eds.) *Religion and Change in Modern Britain*, London: Routledge.

Havlicek, T. and K. Klingorova (2017) 'City with or without God? Features of post-secularism in religious landscape of post-communist Prague', *Social & Cultural Geography*, doi: 10.1080/14649365.2017.1312696, accessed online 16-06-2018.

Heelas, P. and L. Woodhead (2005) *The Spiritual Revolution: why religion is giving way to spirituality*, Oxford: Blackwell.

Holloway, J. (2003) 'Make-believe: spiritual practice, embodiment, and sacred space', *Environment and Planning A*, 35(11): 1961–74.

Knott, K. (2005) *The Location of Religion: a spatial analysis*, London/Oakville: Equinox Publishing Ltd.

——— (2010) 'Cutting through the postsecular city: a spatial interrogation'. In Molendjik, A. L., Beaumont, J. and C. Jedan (eds.) *Exploring the Postsecular: the religious, the political and the urban*, Leiden/Boston, MA: Brill.

Kong, L. (2001) 'Mapping "new" geographies of religion: politics and poetics in modernity', *Progress in Human Geography*, 25: 211–33.

MacKian, S. (2012) *Everyday Spirituality: social and spatial worlds of enchantment*, Basingstoke: Palgrave Macmillan.

Maddrell, A. and V. della Dora (2013) 'Editorial: spaces of renewal', *Culture and Religion*, 14(1): 1–7.
Maddrell, A., Terry, A. and T. Gale (2015a) *Sacred Mobilities: journeys of belief and belonging*, Farnham: Ashgate.
Maddrell, A., Terry, A., Gale, T. and S. Arlidge (2015b) 'At least once in a lifetime: sports pilgrimage and constructions of the TT races as "sacred" journey'. In Maddrell, Terry, A. and T. Gale (eds.) *Sacred Mobilities: journeys of belief and belonging*, Farnham: Ashgate.
Molendjik, A. L., Beaumont, J. and C. Jedan (eds.) (2010) *Exploring the Postsecular: the religious, the political and the urban*, Leiden/Boston, MA: Brill.
Olson, E., Hopkins, P., Pain, R. and G. Vincett (2013) 'Retheorizing the postsecular present: embodiment, spatial transcendence, and challenges to authenticity among young Christians in Glasgow, Scotland', *Annals of the Association of American Geographers*, 103(6): 1421–36.
Parsons, G. (1993) 'Contrasts and continuities: the traditional Christian churches in Britain since 1945'. In Parsons, G. (ed.) *The Growth of Religious Diversity: Britain from 1945*, London: Routledge, pp. 23–94.
Philo, C., Cadman, L. and J. Lea (2011) *The Everyday Urban Spiritual: placing spiritual practices in context* (Working Paper). University of Glasgow: Religion and Society.
——— (2015) 'New energy geographies: a case study of yoga, meditation and healthfulness', *Journal of Medical Humanities*, 36: 35–46.
Phipps, P. A. and M. K. Vernon (2009) 'Twenty-four hours: an overview of the recall diary method and data quality in the American Time Use Survey'. In Belli, R. F., Stafford, F. P. and D. F. Alwin (eds.) *Calendar and Time Diary*, London: Sage.
Reynolds, N. (2015) 'Discourses of love, compassion, and belonging: reframing Christianity for a secular audience', *Journal of Contemporary Religion*, 30(1): 39–54.
Stevenson, D., Dunn, K., Possamai, A. and A. Piracha (2010) 'Religious belief across "post-secular" Sydney: the multiple trends in (de)secularisation', *Australian Geographer*, 41(3): 323–50.
Urry, J. (2007) *Mobilities*, Cambridge, MA: Polity Press.
Wigley, E. (2018) 'Constructing subjective spiritual geographies in everyday mobilities: the practice of prayer and meditation in corporeal travel', *Social & Cultural Geography*, 13(3): 411–25.
Wilson, E. K. and M. B. Steger (2013) 'Religious globalism in the post-secular age', *Globalizations*, 10(3): 481–95.
Woodhead, L. (2012) 'Introduction'. In Woodhead, L. and R. Catto (eds.) *Religion and Change in Modern Britain*, London: Routledge.
Wunderlich, F. M. (2013) 'Place-temporality and urban place-rhythms in urban analysis and design: an aesthetic akin to music', *Journal of Urban Design*, 18(3): 1–26.
Zock, H. (2010) 'Voicing the self in postsecular society: a psychological perspective on meaning-making and collective identities'. In Molendjik, A. L., Beaumont, J. and C. Jedan (eds.) *Exploring the Postsecular: the religious, the political and the urban*, Leiden/Boston, MA: Brill.

Part IV

Political and social engagement

32

(Re)enchanting secular people and politics

Timothy Stacey

Introduction

When we talk about the secular, we're really also talking about the modern. Ironically, the denial of created order begat beget an era defined by the need to create order: rational order (Taylor 2007). When we talk about the postsecular, though, we are not only talking about the postmodern. We're also talking about a response to the situation of postmodernity. We are reflecting, of course, on the rejection of rational order. But we're also reflecting on a particular response to the apparent absence of any rational order, namely, a reawakening to the value of religion. There are two ways of interpreting this reawakening: as a fundamental 'no' to nihilism, a recognition of the emptiness of rationality, a leap of faith; or as an enlightened awakening to the use of poetic language and ritual in conveying truths that elude rational description.

Now before I go on, I have to acknowledge that some might intend by the postsecular a mere renewed awareness of religion: both as a security threat and as a resource in a mixed economy of welfare (Dinham 2009; Casanova 2011; Beckford 2012). It may well be that this is by far the most prominent lens through which people have reawakened to religion. But there is also, among some, an authentic reawakening to religion and religiosity. It is this development that interests me in in this essay. Specifically, within this development, there are two aspects that might be of interest. The first is a 'rapprochement' between religious and secular people (Cloke and Beaumont 2013; Cloke 2015; Williams 2015). The second, on which I will focus, is an exploration of one's own beliefs and values: what might be called postsecular enlightenment.

I should also make clear that I am taking a normative approach. Although I draw on findings from original social scientific research, I am ultimately interested less in interpreting the world than changing it. I do not read the postsecular as merely a naturally occurring phenomenon to be interpreted, and thus am not all that interested in questions as to whether a postsecular era has 'really' arrived or not (cf. Beckford 2012). Instead, I take the postsecular as an opportunity, a discursive intervention that may serve to bring about meaningful change in the world. Specifically, I am interested in the role of the postsecular in developing solidarity.

I take the postsecular as a theoretical concept, the embodiment of which is a state of postsecularity. The two ways of interpreting postsecularity might best be described as the 'leap of faith' and the 'poetic dwelling' models. In the course of my ethnographic research into the beliefs and values of secular people engaged in acts of solidarity in Vancouver, Canada, it is the latter that is most likely to bring about meaningful change. I will introduce these two interpretations theoretically before drawing on them to explain the interactions between the participants and myself. I stress that developing this sense of postsecularity among secular subjects is vital to restoring solidarity in the world. To be clear at the outset, I do not intend that we should impose religion or religious categories on secular people. Instead, my desire is to awaken secular people to the beauty and mystery of their own beliefs and values. I do so because, as I will argue, although people's motivations can be rationally explained, people cannot be motivated by rationality alone.

Almost every one of my participants mentioned how grateful they were for my help in constructing their imaginary. I came to feel like a travelling priest for secular people—although in reality the metaphor is imperfect since my method was the opposite of that of a priest: I was not imputing a theology but helping my participants to bring to the surface their own beliefs, values, and the myths through which these were articulated. Rather than detailing these beliefs and values, I reflect on the relationship between the participants and myself as a researcher.

Perhaps the key aspect that becomes distinctive in highlighting the poetic dwelling interpretation is the notion that belief is primarily performed. While other research (Day 2010, 2013) has suggested that beliefs are performed, my work is distinctive in claiming that performance must be extended beyond those actions we might normally associate with belief, such as prayer, and instead focus on the kinds of actions in which people are sacrificing something, even if only their time. What, I ask, are people willing to stand up for? Indeed, the starting point for my bringing people's beliefs to the surface is to ask them why they are motivated to be involved in acts of solidarity. I call this performative postsecularity. Having thus explained performative postsecularity, I move on to describe the kind of system that could best encourage this way of being. I call this performative postsecularism.

From leaps of faith to poetic dwelling

The leap of faith approach to postsecularity is best exemplified by John Milbank in his *Theology and Social Theory* (Milbank 2013). For Milbank, the modern inevitably leads to the postmodern. A universal rational order is impossible. Allowing the assumption that such an order is possible to become dominant will inevitably lead to disappointment, and thus the conclusion that there is no objectivity, no order, only power, and only the chaotic domination of one idea, one people by another. Any attempt to overcome postmodern discord by the use of reason alone is similarly doomed to fail. For Milbank:

> The only possible response to nihilism is to affirm one's allegiance to a particular tradition, and derive an ontology from the implicit assumptions of its narrative forms.
> *(Milbank 2013: 262)*

From this perspective, described elsewhere as RO (Milbank et al. 2006), reason can only take us so far; the rest is faith. What might the implications of this position be for postsecularity? For Milbank, the greatest narrative remains the Catholic narrative, albeit liberally understood (Milbank 2013: xxiii). From this perspective, the ideal model for postsecularity would presumably be mass conversion to Catholicism.

Yet one might also imagine a slightly alternative reading. It might be, for instance, that people become disillusioned with nihilism and, by way of escaping the void, clasp at alternative narratives. At this point I find it useful to introduce my use of the term myth.[1] Rationality alone, certain people may concede, is insufficient. Rationality has not prevented the slide into ideological and social fragmentation characteristic of postmodernity, and nor does it seem to reverse this slide. To simply speak of narrative may simply be taken to imply an alternative rationality, rather than reaching at something beyond the rational. By myth, then, I intend the stories of great events and characters that point to an ideal of how the world should be. Myths tell of an imagined future, the possibility of which is less significant than the kinds of action they inspire in the present. What I am suggesting is that it might be that people are saying 'no' to nihilism and taking a leap of faith into alternative mythologies—towards a socialist utopia of some kind perhaps.

It is worth noticing that this reading is particularly demanding: faith seems to be something absolute and unconditional. I do not see a leaping to this kind of faith much among my participants. Instead, contrary to Milbank's intentions, it may be that the leap of faith, or plummet into the darkness as one may choose to see it, is best exemplified in the retrenching in divisive myths of geographic, racial, or religious privilege associated with Modi, Brexit, and Trump. In such cases, it may be argued, people long ignored by a liberal establishment, and forlorn in the face of an ever more complex and global political and economic society, are taking a leap of faith into the simple and divisive narratives offered by charismatic leaders (Inglehart and Norris 2016).

What I do tend to find among my participants is a more enthusiastic navigation of a world of chaos and difference, in which religion plays an ambivalent role: sometimes as a warning against what can happen if faith goes too far, sometimes as an alternative way of relating to the world that inspires people to explore their own beliefs and values.

I want to begin by introducing Heidegger's concept of poetic dwelling. For Heidegger (2001: 228), the evidence that people dwell poetically is best drawn out by how unpoetically we experience life; that is, we experience life as lacking in poetry. Thus, human life is defined by an awareness of and a desire for the poetic: the poetic is a lens through which life is interpreted. In Heidegger's words: 'Our dwelling is harassed by the housing shortage…by work, made insecure by the hunt for gain and success, bewitched by the entertainment and recreation industry' (*ibid.*: 213). And yet '[t]hat we dwell unpoetically, and in what way, we can in any case learn only if we know the poetic' (*ibid.*: 228).

Ricoeur extends this concept. For Ricoeur, poetic language, in particular metaphor (Ricoeur 2008) and narrative (Ricoeur 2009), is the means by which we access truths that rational language alone is incapable of capturing. Says Ricoeur: 'poetic discourse brings to language aspects, qualities, and values of reality that lack access to language that is directly descriptive' (*ibid.*: xi). Narrative in particular imputes events and characters with an imagined order. In so doing, it is able to collapse chronology into causality, thus allowing a people or an individual to say, for instance, we began to believe this and thus began to behave like that. These narratives then form a hermeneutic relationship with life as it is actually lived. The story can be rewritten as life changes, but so too do one's interpretations of events, and thus one's ways of acting and reacting, that is, in short, one's lifestyle choices, change as a result of the telling of the story. In the following paragraphs, I will reformulate this conception to include myth: narratives that articulate one's deepest held beliefs and values.

Seligman (2008) offers a way of rendering the notion of poetic dwelling accessible across religious and secular differences. For him, religion often has a subjunctive quality. In particular, 'ritual creates a subjunctive, an "as-if" or "could-be," universe. It is this very creative act that makes our shared social world possible' (*ibid.*: 7). Ritual, and I would add myth, is thus a means of creatively responding to the world.

In a similar vein, Day (2013) suggests that among many self-identified Christians in the UK, belief is performative: it is less a propositional position than a demonstration of one's belonging: to a geography or to a family. In my own research, I have extended this notion of performativity. Though I do so for primarily normative reasons, the decision yields interesting findings for social science. Rather than exploring those performances ordinarily associated with religion, I explore what proverbially 'gets people out of bed in the morning' and, more than this, what gets them out the front door and into the street.

What I find useful in the foregoing for my purposes here is less a means of understanding religion—again, I am not all that interested in understanding just how many people this interpretation can be applied to—than as a point of access for secular people to new creative ways of engaging with the world. By demonstrating the as-if, performative nature of something as seemingly, at least for the outsider, ontologically determining as religion, it may be that we can begin to tease out deep-seated secular beliefs and values. In what follows, I try to share how I have used these insights to awaken secular activists to their beliefs and values: what I call performative postsecularity.

Confessions of a travelling priest

Vancouver is a utilitarian metropolis imposed on a utopic geography. The city is geographically and culturally isolated from the rest of the world by a landscape regularly associated with Eden: besieged and infiltrated by the Pacific Ocean to the West; closed off from the rest of the world by the North Shore Mountains to the north, and the Rockies to the east and south.

All this beauty has drawn swathes of immigrants seeking to build a new utopia. Yet Vancouver is a troubled utopia. The city itself is built on unceded Coast Salish territory, and the quality of life enjoyed by so many settlers has come at the expense of the Musqueam, Squamish, and Tsleil-Waututh people. On top of this, for every utopian, the city has brought a myriad of others, looking for little more than an apartment with a nice view and good access to the sea in the summer and skiing in the winter. The frontier settlers are now being joined by a new wave of high-wealth immigrants. The result is some of the most unaffordable property in the world, stark inequality, rampant homelessness, and communities divided along racial lines. All this beauty, in short, has brought a lot of suffering.

Amid these troubles, Metro Vancouver Alliance (MVA) draws together faith groups, unions, academia, and community groups to 'organize for the common good'. MVA is an Industrial Areas Foundation (IAF) affiliate. The IAF is one of many community organizations founded by Saul Alinsky in the middle of the twentieth century. Conscientized in the slums of Chicago in the post-Depression era, the premise of Saul Alinsksy's work was that civil society could only ever wield effective power if the key stakeholders acted together with one voice: faith groups, unions, community groups, and academia. Since that time, while the most orthodox organizers may always speak with the language of power, at least amongst the organizers I observe, the notion of bringing people together across differences is also an end in itself.

MVA draws the racially, culturally, and economically privileged together with people for whom life may have never seemed so unpoetic. They are the people for whom 'dwelling is harassed by the housing shortage...by work, made insecure by the hunt for gain and success, bewitched by the entertainment and recreation industry' (Heidegger 2001: 213): elderly people who 'surf' from friend's couch, to spare room, to pet-sitting gig in hope of avoiding astronomic rent prices; young students living eight people in a three-bedroom house; people desperately short of time on account of working two, sometimes three jobs; and people competing with the entertainment industry for the attention of their children.

My research is designed in four layers: (1) wide ethnographic background research involving 'hanging around' at the many civic and political events around the city, from lunch clubs to anti-racism rallies (Skeggs 2012: 22); (2) more focused ethnographic research into the many training events organized by MVA, wherein members are inducted into the beliefs and practices associated with organizing; (3) 'Forty stories' of secular individuals working with MVA, involving a formal, semi-structured interview alongside multiple informal conversations, and designed to understand the motivations of a broad array of people across age, race, and gender differences; and (4) three focused ethnographic case studies with secular institutions associated with MVA. As I write, the fourth and final stage is yet to be executed.

In the course of my gathering 40 stories, I have enjoyed travelling the city by bicycle, getting a sense of the topography of a place whose 'majority religion' is an earth-based spirituality (Shibley 2008). I journey across the city, wherever possible taking in the ocean and mountain views that provide the scenic backdrop for those living here, to meet people at their favourite hang-outs—usually a café, beach, park or hike in this health-conscious, nature-obsessed city, though occasionally a pub.

My previous research has suggested that the central difference between religious and secular people is not categorical but reflexive: secular people have few institutions, rituals, and myths through which their beliefs and values are intentionally explored and brought to life. As a result, it is difficult to bring to the surface the most fundamental beliefs and values, those that have a deep and lasting influence on what people are willing to fight for. I have had to be creative in my line of questioning: always getting to know people a little informally before interviewing them; being very open about my personal explorations of my own beliefs. In particular, my exploration revolves around some of the following key questions. How did you end up working here? Did you always know that you wanted to work here? When was the moment you knew you wanted to do something? What are the key moments that inform who you are? What are the public moments that make you angry? What are the public moments that make you cry? What are the stories, individuals, and events that most inspire you? As I ask myself these questions, I try to develop a picture of the way I feel the world should be. As I share my story, I find my participants begin to open up about their own.

In almost every case, my participants have thanked me for helping them to understand the beliefs, values, people, stories, and events that most inspire them. The reason for this is that they feel bringing their ideals to the surface energizes them to act. Contra Kant and more recently Habermas (2006), rationality alone, the mere logical demonstration of the most appropriate way to behave, is insufficient to the task of motivating solidarity. People's motivations can be rationally explained, but people cannot be motivated by rationality alone.

What I had not anticipated before undertaking the research is that the meticulous effort I had put into developing strategies for understanding people socially and scientifically would turn out to be practices for making people feel whole, determined, and energized to act. In a sense, my work to reflect my participants' beliefs and values had become a process of co-constructing those beliefs and values. Put another way, I was not so much mirroring the melodies of a full heart as helping it to feel full. I became, for want of a less culturally domineering word, a travelling priest or, for want of a term with fewer culturally vacuous connotations, a motivation coach. I was helping people to develop a poetic discourse to understand their beliefs and values in a manner that eludes directly descriptive language.

The importance of aspects beyond purely secular rationality is further evidenced in three ways: the first is that many of my participants were openly appreciative of their contact with religious colleagues, which they felt had led to productive conversations and had helped them to explore their own beliefs and values. In particular, they suggested that such conversations were less common with secular colleagues, and together we speculated as to whether this might be related to a lack of resources for reflection on beliefs and values. Second, and related to this is how important it proved for a number of my participants to draw on religious belief in distinguishing their own beliefs and values: 'I wish I could have that sense of comfort', as one participant put it. Religious conviction is constructed as engendering a sense of security that some of my participants experienced as an absence. Like Heidegger's notion of poetic dwelling, religiosity is still tentatively constructed as a quintessentially human way of relating to the world, primarily by means of performing a lamenting awareness of its absence, even where my participants consider themselves incapable of accepting religiosity on account of their own rationality.

The second aspect eluding a purely rational equivalent is that in the course of our discussions, what my participants have found most beneficial is the plotting of a narrative from childhood to present. Such narratives help my participants to understand the ideals that most inspire them. By reading between Ricoeur (2009: 64) and Seligman (2008), we can suggest that there are clear parallels between the as-if world constructed in religious ritual, and the performative plotting of a narrative by secular individuals. My primarily liberal-arts university-educated participants are well aware of the inevitable sociological, anthropological, and psychological deficiencies of the narratives they produce, that is, of the lack of correspondence between their narratives and an ever intangible 'objective' reality, while nonetheless finding those narratives empowering. They are thus confirming the imaginative nature and performative power of the process of plotting their narrative. Moreover, they are constructing a particular notion of the self, as, for example, shaped by certain beliefs, that is equally tentative.

Because my participants are focusing on key events and characters and developing these in a trajectory towards their deepest held beliefs and values, I prefer to use the term myth rather than narrative. Rather than being propositionally 'believed', these myths are primarily performed: both in their telling, and in their impact. As Ricoeur has suggested, these myths exist in hermeneutic relation to how life is lived. My participants are thankful for my help in bringing their stories to the surface, not merely because they provide a moment of epiphany, but because this moment can be drawn on as a resource to energize acts of solidarity. In this way, our discussion itself may become a moment in that individual's evolving myth of personal transformation.

In this tentative development of myths of personal transformation, the notion of a leap of faith is not viable: my participants are not yet always aware of the beliefs and values that lurk below, and the myths through which these are constructed are loose, like a sketch of a story—a draft rather than a fixed text. They are not ready to take a leap. But nor are they willing: a principled distance from their own deepest values, an ability to adapt and be cajoled in another direction is an important value in itself in this postmodern environment. As one participant put it when I asked her 'what do you believe?', 'that kind of flies in the face of postmodernism'. Thus, the point of bringing my participants' beliefs and values to the surface is not to encourage them to hold fast to these beliefs and values at the exclusion of others, but to give them ideational resources from which to draw in developing their performances of solidarity. It is in this twofold sense that I suggest the value of performative postsecularity: my participants are performatively conjuring beliefs and values which then help to shape their performances of solidarity. The performance is *both* in the development of an extra-rational poetic discourse *and* in the employment of this discourse in the shaping of political actions.

One of the advantages of continuously hanging around one's participants is that the interview is not the last contact. Instead, our conversations are iterative. As I began to realize the importance of my travelling priest role, I shared my insights with my participants. We began to discuss the merits of having these kinds of conversations within their institutions. While it was regularly agreed that such conversations would be useful, two anxieties arose: that such conversations might be inherently exclusionary, potentially undermining the image of their institution as an ideologically neutral, safe, and open space; and that in a competitive funding environment, there is a lack of resources, especially time, for such conversations. The first anxiety neatly demonstrates how assumed connections between religion, ideology, and violence lead people to reproduce secular polities, while the second demonstrates how even people whose lives are devoted towards resisting the worst atrocities of capitalism are nonetheless forced to reproduce its assumptions. In the next section, I am sensitive to these anxieties as I ask how political and civic institutions might reform to draw out performances of postsecularity.

Encouraging postsecularity

Bringing beliefs and values to the surface is core to motivating solidarity. So how can institutions help people to surface their beliefs and values in safe and inclusive ways? The MVA answer, and perhaps by extension the answer of the IAF of which MVA is an affiliate, is poetic method. I use the term poetic in a deliberately poetic sense. The term seems descriptively inappropriate given that the methods employed by MVA appear anything but poetic: they are intentionally confrontational, since they see agitation as a means of developing energy. But in reality, everything MVA does is laced with the notion that poetic discourse motivates people in a way that rational discourse cannot.

I have elsewhere discussed extensively how IAF institutions are able to construct what I call 'generalised transcendental frameworks' that simultaneously provide much of the cohesive social function often associated with religion, while doing so across religious and secular differences. The most common means of articulating this framework is 'the common good'. By having participants leave their differences at the door and focus instead on the common good, IAF organizations are able to make people of very different religious and secular backgrounds recognize they share common aims and values, thus facilitating a form of postsecular rapprochement. In this manner, IAF institutions are able to allay anxieties regarding religious and ideological division. Moreover, since they are focused primarily on power and measurable outcomes, it can be assumed that IAF institutions will not seek to develop these rituals unless they are instrumentally effective. We can thus also attenuate capitalist anxieties regarding a lack of resources for exploration of beliefs and values: such rituals are an effective means of developing power. Previously I focused on the idea of performative postsecularity as consisting of three aspects:

1. A reflexive awareness of the ways in which one's deepest held beliefs and values are articulated through myths: stories of great events and characters.
2. Recognition that the truth of a myth is derived less from its discursive rationality than its performative rationality: not in its internal logic or empirical veracity, but in its impact on society.
3. The development of shared spaces and actions in which myths can be inclusively shared and collectively explored.

Here I want to focus on the role of IAF institutions in facilitating performative postsecularity by means of encouraging narrative self-reflection. In this way, I am deepening the first tier, further emphasizing that developing shared spaces in diversely religious and secular societies may also require as a prerequisite the devotion of specific resources, in terms of staff and training material, for enabling secular people to explore their beliefs and values through myths. The practice that most closely mimics my own research method in this regard is one-to-ones. Lasting about 20 minutes, one-to-ones are designed for sharing one's emotional journey from relative political disengagement to activism. One-to-ones are the first and most frequently used tool in recruiting allies. They consist of simple questions such as the following. How did you end up working in this job? What do you want to get out of your work here? Is there any particular cause that is really significant to you?

I tend to treat one-to-ones as key to my development of performative rationality. They are judged, I suggested, less in terms of their truth content than in terms of the kinds of performances they inspire. There I was primarily focused on the performances inspired in the listener: inducing them to reciprocate by sharing their story or pressuring them to act. Here I want to focus on the construction of the myth, and the consequent moral energy inspired in the storyteller.

I suggested earlier that my participants are aware of the rational and empirical shortcomings of the myths they conjure. As I have explained elsewhere, IAF organizers are similarly aware of these shortcomings. Indeed, more experienced organizers often seek to help new organizers to construct their story. I have had numerous conversations with organizers in which we have discussed the relative merits of one story over another. On one level, this might seem inauthentic: a story is being conjured merely to develop performative force. This is perhaps to miss the point. Experience with my participants suggests that a myth is not representative of identity but rather is constitutive of it. Prior to the telling of their story, people do not have a different identity that is waiting to be conveyed; rather, they have less identity: they are not able to convey who they are in narrative form. With each performance of their evolving myth, their identity gets thicker. It is entirely possible that their myth does not accurately represent the rational chain of causality that led them to where they are; but what the myth does is to convey in narrative form the moments of significance, their emotional reactions to these moments, and thus their deepest held beliefs and values. In short, myth is not representative of an empirically falsifiable chain of events but of the storyteller's beliefs.

MVA thus exhibits performative postsecularism in its practice in that it encourages secular people to construct an as-if narrative that allows them to foreground their deepest held beliefs and values. Yet, perhaps more powerfully, and this is perhaps where I need to add a fifth tier to my explication of performative postsecularism, the organization is simultaneously secular and postsecular: it is secular in that it is not imposing or indeed even proposing any particular myth as fundamental to solidarity. Instead, it is proposing a model of drawing myths out of participants, who in turn face outwards to the rest of society and enable each person they meet to construct their own stories. Each participant becomes their own priest and proselytiser of the priesthood. It is a kind of Lutheranism gone radical: people are encouraged not merely to move beyond mediated faith to have a direct relationship with *the* truth, but to develop their own personal relationship with their own personal truth. From this perspective, there is no need to worry that a discussion of values infringes on a secular, liberal ideal of principled distance from ideology. On the other hand, as my previous work makes clear, postliberals need not worry as the cultivation of these very personal stories is ultimately aimed at the development of shared moral responsibility.

There is, moreover, little need to worry about a lack of resources for such discussions in a capitalist society. The IAF approach suggests that these stories ignite the tellers and listeners alike, igniting more energy. This point should not be used to deny that these narratives are being developed in a context that is deeply critical of capitalism. Rather, the point is that in neoliberal societies, in which even those organizations dealing with the worst atrocities of capitalist inequality are forced to make neoliberal 'efficiencies', performative postsecularism should be considered a model for developing resources of resistance, rather than, taking a neo-Marxist lens, as a distraction from 'real' work.

Conclusion

The eighteenth-century Swedish Enlightenment theologian Emanuel Swedenborg once suggested that if the Old Testament represented the infancy of Judeo-Christians' relationship with God, and the New Testament its early teens, then perhaps his was a time of adolescent rebellion and rejection. Cultural hegemony aside, and poetically speaking, I find that extending this metaphor may be useful for understanding the foregoing discussion of performative postsecularity and performative postsecularism. Having achieved the safety, stability, and distance from organized religion necessary for independent thought, secular people living in post-Christian societies are now finding it useful to draw on pseudo-religious language and to interact with religious people in developing their own as-if myths. Secular people are reaching a magnanimous adulthood in which, no longer fearing of the power of religion, they are willing and able to playfully toy with religion-as-metaphor, adopting some of its language to inform the surfacing of their own beliefs and values: they are undergoing a new, postsecular enlightenment.

Emerging work in the study of non-religion and secularity is often vehement about protecting the non-religious from the discourses of religion. There is clearly merit in avoiding this unhelpful imposition of categories which may foreclose imaginative exploration of alternative and beautiful ways of living. Yet if, as Asad (2003) has suggested, the religious and the secular are two sides of a geographically, historically, and culturally specific binary, then the vehement protection of the secular from any supposed interference from supposedly religious categories, potentially contributes to the parasitic expansion of religion as a category, while further reifying this unhelpful binary. In particular, it is important to avoid the mistake of reifying what I have elsewhere called a religious/secular, mythic/rational binary, thus foreclosing ways in which secular people can explore the extra-rational elements of their own identity.

I have suggested that my participants dwell poetically and have particularly emphasized the importance of myth in constructing their identity and understanding their motivations. Perhaps these terms themselves will be perceived as encouraging the domination of religious categories. Yet I do not claim that these categories are fixed, but rather that they have proved useful terms to think with. More importantly, my experience with my participants suggests they are mature enough to toy with various religious and secular concepts in constructing their own identity. A similar propensity among academics studying religion would be an important step in our deeper engagement and understanding.

Note

1 Since this is a deliberately normative piece, I shall refer the reader throughout to my own social scientific research. Of particular relevance are Stacey (2017, 2018) (see further reading for more details).

Further reading

Beaumont, J. and C. R. Baker (2011) *Postsecular Cities: space, theory and practice*, London/New York, NY: Continuum.

Beaumont and Baker offer a comprehensive and accessible introduction to the concept of the postsecular. Numerous authors place the postsecular into dialogue with original research in urban settings.

Stacey, T. (2017) 'Imagining solidarity in the twenty-first century: towards a performative postsecularism', *Religion, State and* Society, 45: 141–58.

I highlight the role of myth in motivating political action among both religious and secular individuals in London, UK.

——— (2018) *Myth and Solidarity in the Modern World: beyond religious and political division*. New York, NY/London: Routledge.

I demonstrate how organizations are able to facilitate the inclusive sharing of myths across religious and secular differences in order to facilitate inclusive political resistance to the hegemony of the state and market.

References

Asad, T. (2003) *Formations of the secular: Christianity, Islam, modernity*, Standford, CA: Stanford University Press.

Beckford, J. A. (2012) 'SSSR presidential address: public religions and the postsecular: critical reflections', *Journal for the Scientific Study of Religion*, 51(1): 1–19.

Casanova, J. (2011) *Public Religions in the Modern World*, Chicago, IL: University of Chicago Press.

Cloke, P. (2015) 'Postsecular Stirrings? Geographies of rapprochement and crossover narratives in the contemporary city'. In Brunn, S. D. (ed.) *The Changing World Religion Map*, Dordrecht: Springer Netherlands, pp. 2251–64.

Cloke, P. and J. Beaumont (2013) 'Geographies of postsecular rapprochement in the city', *Progress in Human Geography*, 37(1): 27–51.

Day, A. (2010) 'Propositions and performativity: relocating belief to the social', *Culture and Religion*, 11(1): 9–30.

——— (2013) *Believing in Belonging: belief and social identity in the modern world*, Oxford: Oxford University Press.

Dinham, A. (2009) *Faith, Public Policy and Civil society: problems, policies, controversies*, Basingtoke: Palgrave Macmillan.

Habermas, J. (2006) 'Religion in the public sphere', *European Journal of Philosophy*, 14: 1–25.

Heidegger, M. (2001) *Poetry, Language, Thought*, New York, NY: Perennial Classics.

Inglehart, R. and P. Norris (2016) 'Trump, Brexit, and the Rise of Populism: economic have-nots and cultural backlash', *HKS Working Paper No. RWP16–026*, available at SSRN, doi: 10.2139/ssrn.2818659.

Kant, I., Gregor, M. J. and J. Timmermann, J. (2012) *Groundwork of the Metaphysics of Morals*, Cambridge: Cambridge University Press.

Milbank, J. (2013) *Theology and Social Theory: beyond secular reason*, Malden, MA: Blackwell.

Milbank, J., Pickstock, C. and G. Ward (2006) *Radical Orthodoxy: a new theology*, London: Routledge.

Ricoeur, P. (2008) *The Rule of Metaphor Multi-Disciplinary Studies of the Creation of Meaning in Language*, Toronto: University of Toronto Press.

——— (2009) *Time and Narrative Vol. 1.*, Chicago, IL: University of Chicago Press.

Seligman, A. B. (2008) *Ritual and its Consequences: an essay on the limits of sincerity*, Oxford/New York, NY: Oxford University Press.

Shibley, M. (2008) 'The promise and limits of secular spirituality in Cascadia'. In Todd, D. (ed.) *Cascadia: the elusive utopia*, Vancouver: Ronsdale Press.

Skeggs, B. (2012) *Formations of Class and Gender: becoming respectable*, London: Sage.

Taylor, C. (2007) *A Secular Age*, Cambridge, MA: Harvard University Press.

Williams, A. (2015) 'Postsecular geographies: theo-ethics, rapprochement and neoliberal governance in a faith-based drug programme', *Transactions of the Institute of British Geographers*, 40(2): 192–208.

33

The Nordic far-right and the use of religious imagery[1]

Øyvind Strømmen

Introduction

In a highly critical essay, sociologist James A. Beckford (2012) points out that the 'sheer variety of meanings is a notable feature of discourse about postsecularity', adding that there are 'tensions between some meanings':

> For example, it is not easy to reconcile the idea that the secular has somehow come to an end with the idea that postsecularity represents a refinement—or a more productive phase—of secularity. Again, there is tension between the claim that postsecularity enables a return to presecular forms of religion and the contrary claim that any forms of religion that emerge in postsecularity must be new and nondogmatic or "spiritual".

He follows up with scathing criticism, opining that the concept 'trades on simplistic notions of the secular', and risks becoming a magic wand, something to wave over 'all the intricacies, contradictions, and problems of what counts as religion to reduce to them to a single, bland category' (Beckford 2012: 16–7).

While I do not share his far-reaching criticism of the term, Beckford has a point when he points to the large variety of meanings ascribed to 'postsecularity'. Some use it in connection with the alleged re-enchantment of culture, others to describe the public resurgence of religion. And—importantly—the term is used both descriptively and normatively. In my view, postsecularity is a largely Eurocentric concept. However, it is useful in describing a very tangible change that has taken place within secular European countries in the past few decades, and particularly in the post-911 era. While one can make a strong argument that secularization itself is still very much ongoing, the *discourse* on religion has changed. In short, we talk *more* about religion and we speak *differently* about religion.

The Scandinavian countries are, as Norwegian scholar of religious studies Egil Asprem points out, amongst the most secular countries in the world. At the same time, he argues, religion has never been taken more seriously as an influence on societal development. While secularization—measured by decreasing participation in organized religious groups, and by decreasing self-reported faith—is still ongoing, religion is increasingly visible in the media and in public debate. Asprem furthermore underscores that the discourse on Islam plays a particular role here, pointing at how right-wing populists describe Islam as a threat against 'secular' and 'democratic' values, and at how Islam is portrayed as alien, as non-European, as an external threat now finding its way into Europe, and as capable of undermining European secular traditions from the inside through a process of stealthy Islamization (Asprem 2016).

As I prepare to wave the wand of postsecularity over the Nordic far-right and their contemporary usage of religious symbolism, these are the images in the background: The Nordic countries are all largely secular countries. However, religion has received an increased amount of media attention in the past few decades and has played an increasing role in public debate. This is particularly true for Islam, a subject of constant debate, a debate that has been fuelled by terrorist attacks carried out by extremist Islamists, as well as by—for instance—the 2005 cartoon controversy beginning after the Danish newspaper *Jyllands-Posten* published 12 cartoons, most of which depicted the Prophet Muhammad. In fact, as a Norwegian study shows, Christianity is getting *less* media attention, while both Islam and new religious movements ('alternative spirituality') are receiving more of it, although the latter has the benefit of largely being positively portrayed (Furseth 2015).

The chaotic landscape of the Nordic far-right

The modern-day extra-parliamentary far-right in the Nordic countries is a varied and somewhat chaotic phenomenon. It encompasses openly neo-Nazi groups such as the Nordic Resistance Movement (Nordiska Motståndsrörelsen, NMR), which is active in Norway, Sweden, and Finland, has attempted to establish itself in Denmark, and seeks to establish an Icelandic branch. It also includes often short-lived groups based entirely on a looser anti-Muslim or anti-immigrant ideology, including—in recent years—groups inspired by the English Defence League and by German Pegida (Patriotische Europäer gegen die Islamisierung des Abendlandes/Patriotic Europeans against the Islamization of the Occident) (Meleagrou-Hitchens and Brun 2013; Berntzen and Weisskircher 2016).

Furthermore, the Nordic far-right includes movements and individuals adapting an 'Identitarian' ideology, a form of neo-fascism or post-fascism drawing heavily on the French *nouvelle droite* (New Right)-movement (for more detail, see Zúquete 2018). One example was the formerly neo-Nazi Swedish organization Nordic League (Nordiska Förbundet, NF), which ceased its activities in 2010. Another example is the 'think-tank' Motpol (i.e., 'polar opposite'), originally connected to the Nordic League but now functioning as an independent Internet magazine, which also arranges an Identitarian conference, Identitarian Idea (Identitär Idé) (Teitelbaum 2017).

Recently, one has also seen attempts at importing the so-called *alt-right*—originally an American phenomenon combining far-right ideology with Internet culture—to Norway. In August 2017, for instance, an 'alt-right' art exhibition took place in an apartment in Oslo. The artist, Lilith Keogh, also ran as a candidate for the minor political party Alliansen (The Alliance) in the parliamentary election that fall. It is a political party whose leader, Hans Jørgen Lysglimt Johansen, has combined criticism of 'mass immigration' and a 'rigged political system' with attacks on 'the Jews in the Israeli lobby' and on 'Negro gangster rap culture'. Other candidates for the party go further, such as Jarle Johansen, whose stated goal is the reintroduction of the anti-Jewish and anti-Jesuit paragraph of Norway's otherwise liberal 1814 constitution (Færseth 2017a, 2017b).

The example of Alliansen is somewhat illustrative of how murky the landscape of the Nordic far-right has become, different phenomena floating together. While Lysglimt Johansen could be seen as a representative of what has elsewhere been described as the alt-light, his namesake Jarle Johansen draws inspiration from older anti-Semitic ideology, while Keogh openly identifies with the alt-right. Lysglimt Johansen himself has also stated that he believes that his political party shares members with the already mentioned neo-Nazi group NMR.

Another similar example can be found in the anti-immigrant vigilante group *Soldiers of Odin*. The group was founded in Kemi, Finland, in October 2015, ostensibly as a response to the increasing amount of asylum seekers arriving in the country amid the European migrant crisis. The group soon spread to other towns in Finland, then to other countries including Sweden, Norway, and Denmark. While their ideology was unclear at best, the various Soldier of Odin groups *appeared* to be based on similar anti-immigrant and anti-Muslim ideas as the EDL and Pegida and copied in the Nordic countries. That the Norwegian branch chose Ronny Alte as their spokesperson is indicative. Alte had previously filled similar roles in both the Norwegian Defence League and in (one of several variants) Pegida Norway (see Figure 33.1) (Støbakk 2016). The Finnish founder of the group, Mike Ranta, was a member of the Finnish Resistance Movement (Suomen vastarintaliike) which merged with its Swedish and Norwegian sister organizations to become the NMR one year later (see YLE 2016 for more information).

In some ways, Soldiers of Odin is an example of how an organizational concept in itself became a *meme* on the far-right, an idea spreading to other countries and being picked up by individuals with varying ideological motivations. The same is true for symbols, where both general concepts (as parts of wider discourses) and specific imagery are being spread via the Internet. This imagery ends up being employed by *different* groups on the far-right, espousing *different* ideologies. This is certainly the case for the most well-known of all alt-right symbols, Pepe the Frog. Originally a comic strip figure, Pepe became 'a classic mutating meme', a part of Internet culture, before ending up being co-opted by outright white supremacists (Lantagne 2017: 4).

Figure 33.1 Norwegian Defence League material
Source: now defunct NDL website.

On infidels and crusaders

As sociologist Joel Busher (2012) points out:

> [t]he language and protest performances of nationalist political parties and social movement organizations» can provide «especially intriguing case studies of how national civil religions are adapting to the pressures associated with globalization.

Busher looks specifically at the English Defence League, whose leadership—he notes—sought to distance itself from both the overt biological racism and the anti-Semitism that have often been an important part of far-right-wing groups in the UK. Instead, the group focused on threats against a culturally defined nation. This approach provided activists 'with a means of articulating exclusivist group boundaries that have not necessarily carried with them the kind of public stigma associated with traditional far-right discourses of race' and allowed them to 'tap into and contribute to powerful discursive currents within the political mainstream' (Busher 2012: 419), most importantly ideas about a clash of civilizations. Busher mentions how many activists within the English Defence League adopted the appellative 'infidel' and how the very emblem of the EDL incorporates not only the cross of St. George but also the motto 'In hoc signo vinces' ('In this sign thou shalt conquer'). He then points to 'the repeated use of images of medieval crusades and crusaders—the symbol *par excellence* of a "clash of civilizations" between a Christian Europe and an Islamic Middle East' (Busher 2012: 421).

The Norwegian, the Swedish, and the Danish variants all adapted similar logos as the EDL, simply replacing the St. George cross with their respective national (Nordic Cross) flags, while retaining the Latin motto. Activists in all three organizations also adapted a similar Crusader imagery as their EDL counterparts, often sharing, spreading, or adapting anti-Muslim memes of British or even American origin. This usage of crusader imagery should be seen as part of a wider subculture of self-declared counter-jihadists (see Figures 33.2 and 33.3).

Figure 33.2 Anti-Muslim meme spread on the Internet
Source: YouTube video, uploaded by the user Whiteboy Rising in July, 2016.

Figure 33.3 'Dealing with Muslims—they got it right the first time'
Source: Free Republic.

Within counter-jihadist circles, their struggle against Islam is often seen within the context of a 'third jihad', as summed up in a post on the blog *Citizen Warrior* in 2008:

> The first jihad started with Mohammed. His armies conquered all of Arabia. In the hundred years after his death, his armies conquered most of the Middle East, North Africa and Spain. [...] The second major jihad started in 1071. Islamic armies toppled Constantinople and spread into Europe, India, and further into Africa. The second jihad began to decline when the Muslim army was stopped on September 11th, 1683, at the gates of Vienna, Austria.[...] Now we are in the third jihad, the third great wave.
>
> *(Citizen Warrior 2008)*

A similar reference can be found on the blog *Gates of Vienna,* which has been a central blog in the counter-jihadist blogosphere (May 2006). Writing on the battle of Vienna in 1683, the blog author, Edward S. May (or 'Baron Bodissey'), concludes with these words:

> On a bright September morning in 1683 the tide of the second wave of the Great Islamic Jihad turned and began to ebb.
>
> On a bright September morning in 2001, the tide of the third wave of the Great Islamic Jihad was at its flood.
>
> But this tide will turn too, Sobieski is out there, waiting to be awakened and ready for the call to raise the siege at the Gates of Vienna.

In other words, both their struggle against Islam and conflicts of the distant past are seen as part of the very same struggle. The crusaders, consequently, are just one of several groups and individuals referenced. Others include, of course, both Sobieski, the Frankish King Charles Martel and the Wallachian Prince Vlad Tepes. In the descriptions of the counter-jihad movement, the everlasting struggle between Islam and the West takes the form of an epic fight between good and evil, more akin to *The Lord of the Rings* than to the complexities of real history. It provides—in other words—a historical framework, a construction of 'self' and 'other' in which Muslims are the everlasting enemy, while there is little room for individuals such as the Hungarian Protestant and anti-Habsburg leader Imre Thököly, an ally of the Ottomans in 1683. In this way, both the grand story of 'the third Jihad' and various historical imagery provide an imagined historical legacy.

The same thing is true for the use of Crusader imagery. While some counter-jihadist writers and some online and street activists of the EDL and copycat groups surely self-identify as Christians, the message of Crusader imagery appears to be more about Christendom – as a metaphor for the West – than about Christianity. Interestingly, in the Nordic Defence League groups, Crusader imagery has been mixed with *Viking* imagery (see Figures 33.4 and 33.5).

Figure 33.4 complains about 'halal meat in our hospitals and no pork in our kindergartens', about firemen having rocks thrown at them, and about 'schools changing Christmas' for Muslims. It ends with the words: 'What the f*ck does it take for the Danes to wake up and say "stop, it's our country"'. At the top glares a Viking king.

Figure 33.5 was shared by the Danish Defence League, with the words below: 'We love battles. Our paradise, Valhalla, therefore, is a place where one is allowed to fight every day, and then to party afterwards. Just come'. Once again, the references to Vikings, including a

Figure 33.4 Danish Defence League poster
Source: Facebook posting November 2013.

Figure 33.5 Danish Defence League
Source: Facebook posting April 2015.

reference to the Valhalla of the Old Norse religion, is less about religion than about setting one's own nationalist, anti-Muslim ideas into a wider historical framework.

The group Soldiers of Odin might be the *oddest* example of this. As mentioned earlier, it was founded in Kemi, Finland, in October 2015, by an activist belonging to a neo-Nazi organization. The vigilante group—named after the chief God of the old Norse pantheon—also has a logo featuring the one-eyed God (mistakenly described as 'a symbol of a Viking and the Finnish flag' by Reuters, see Rosendahl and Forsell 2016). Figure 33.6 shows the logo of the Finnish department of Soldiers of Odin. Similar logos, with differently coloured flags, were used by groups in other countries. A *Finnish* nationalist group, thus, chose to reference *Norse* mythology. The organizational concept, including both Odin's name and the image of the one-eyed god, soon spread to other countries, including Canada and the USA. Here, actual historical legacy seems to play far less of a role than underlying discourses, than—in fact—the idea of the Viking being a strong and masculine warrior, ready to defend his people, rather than—say—a marauding outlaw, raping and pillaging. Odin is a symbol of the *imagined* Viking, complete with the entirely ahistorical horned helmet, rather than the old Norse god of Wisdom.

This idea of the Viking is also employed by other far-right groups, as exemplified by the tweet from Allianseungdommen (Alliance Youth), stating: 'The Alliance are True Vikings. All the other parties are morons' (see Figure 33.7). The tweet tries to assert that

Figure 33.6 Logo of the (original) Finnish branch of Soldiers of Odin
Source: Soldiers of Odin.

Figure 33.7 'Alliansen are True Vikings'
Source: Allianseungdommen on Twitter, May 2017.

Alliansen are 'True Vikings' and shows a viking, featuring a Norwegian flag, chopping down a signpost, the sign reading 'Multiculture'. It was originally posted by the meme-oriented Facebook site 'Frie Nasjonalister—Politisk ukorrekt info', which is run by the minor group Frie Nasjonalister (Independent Nationalists), a group that describes Norway as 'under occupation'. The group also claims that Norway has become led by politicians and intellectuals who have betrayed the people 'with great evil' (Frie Nasjonalister n.d.). Their logo is a variant of the *Elhaz* rune, which Germanic mysticism of the early twentieth century (and consequently also the German Nazi party) saw as a symbol of life and described as the *Lebensrune.*

Even when the use of old Norse symbols and Viking imagery does cross into the realm of Norse mythology or pagan religion, it rarely carries any actual *religious* connotations. Few activists in any of these groups would describe themselves as paganist or as followers of Old Norse religion. 'In most Western and Northern European countries, racism, right-wing extremism and fascism are predominantly secular movements', noted Tore Bjørgo in *Terror from the Extreme Right* in 1995, republished in 2013. He added that this secularity contrasts with militant American racism, 'which is highly influenced by religious movements, in particular "Christian Identity," an anti-Semitic and rather unconventional (to say the least) reinterpretation of the Bible' (Bjørgo 2013: 9). In the Western and Northern European context this evaluation still rings true, while the situation in the USA could possibly be described as more complex today.

Odin and the race war

In the same volume, Jeffrey Kaplan pointed out that Odinism, 'a reconstruction of the Viking-era Norse pantheon', did play a 'vital role in the world of the radical right and in the wider universe of the cultic milieu':

> As denizens of the cultic milieu, Odinists practice an imaginative blend of ritual magic, ceremonial forms of fraternal fellowship, and an ideological flexibility which allows for a remarkable degree of syncretism in adopting elements of other white supremacist appeals – Nazism and, remarkably, Christian Identity in particular. More [...], Odinists tend to subscribe to beliefs which are explicitly Christian.
>
> *(Kaplan 2013: 60)*

A few years later, Mattias Gardell (2003) observed that racist paganism had 'emerged as one of the most dynamic trends of the increasingly radicalized but highly fragmented and schismatic radical-racist milieu in the United States'. He also noted that even 'the most cursory glimpse at white-racist publications, Web pages, and white-power lyrics reveals muscular heathens, pagan gods and goddesses, runes and symbols, magic and esoteric themes in abundance' (Gardell 2003: 1).

While both Gardell and Kaplan focus on North America, it should be underscored, as Kaplan does, that Odinism 'travels well, linking racialist adherents in North America with like-minded groups in Germany, Southern Africa and Scandinavia' (Kaplan 2013: 62). In fact, much to the chagrin of non-racist neo-paganists following Ásatrú, that is, neopaganism with a Scandinavian focus, self-declared Odinism has become a lasting phenomenon within neo-Nazi circles. In Norway, the most well-known example of Odinism is the now largely defunct neo-Nazi group Vigrid. Figure 33.8 shows their logo of Vigrid which is similar to

Figure 33.8 National Alliance logo
Source: National Alliance.

the logo of the American extreme right-wing group National Alliance. It has the 'life rune' as its central symbol.

The group is led by Tore Wilhelm Tvedt, who describes himself as a prophet of Odin and who has also been carrying out various religiously laden rituals and ceremonies (see Tveito 2007). Another example is Varg Vikernes, a former black metal artist and convicted murderer and church arsonist, who has been promoting an ultranationalist ideology he describes as Ôðalism:

> an ideology based on blood [...] and soil [...]; protecting, promoting and if necessary reviving the customs, traditions, world view, values and religion that naturally came from each particular population in their homeland.
>
> *(Vikernes 2013)*

As journalist Will Carless (2017) points out, Odinism has also been mentioned in connection with several terrorism cases in recent years. Carless mentions six examples from the USA where white supremacists claiming an Odinist belief system have been convicted of plotting or carrying out terrorist attacks. He also draws attention to the Norwegian right-wing extremist Anders Behring Breivik.

Shortly before carrying out his twin terrorists attack against a government building in the centre of Oslo and a Labour Party youth camp at Utøya, Breivik set up a Facebook page where he described himself as a 'Christian conservative'. His manifesto, emailed to a large number of people just prior to the attack, was filled with counter-jihadist ideas, including a great deal of material written by others. In the manifesto he described himself as 'extremely proud of my Odinistic/Norse heritage' and called for 'a nationalistic Church which will tolerate and allow [...] native cultures/heritage/thought systems such as Odinism'. He also claimed, however, that Odinism did not have 'the potency to unite us against such a devastating force as Islam, cultural Marxism/multiculturalism and capitalist globalism' (Breivik 2011). From his prison cell, Breivik has started to openly identify as not only a fascist and National Socialist, but also as an Odinist and one attacking Christianity. In a letter sent to several newspapers, he declared: 'Few things in this world are more pathetic than the Jesus figure and his message, and I have always despised the weakness and the internationalism represented by the Church'. While all this can be seen as a 'rebranding' effort from Breivik, it's worth noting that he had named and marked the weapons used in the Utøya attack 'Mjölnir' and 'Gungnir', that is the hammer of Thor and the spear of Odin (Rognsvåg 2015). In his personal cut-and-paste ideology, Breivik has been drawing from various sources and from various nationalist, ultranationalist, and right-wing extremist discourses.

Old news

Today, the most influential neo-Nazi group in the Nordic countries is the so-called Nordic Resistance Movement. This group does not self-identify as Odinist and barely mentions religion in their manifesto. It should be noted, however, that it is difficult to find anti-Christian texts on their site, and that the group draws inspiration from American right-wing extremists such as David Lane and Richard Scutari, both self-declared Odinists. The logo of the NMR also combines the *Tiwaz* rune and a variant of the *Ingwaz* rune, which—according to their activist handbook—represent various values, including boldness, sacrifice, fertility, and creativity (quoted by Holmqvist 2017). Once again, we see how the use of historical symbols is used as a form of framework, providing an extreme right-wing group with an imagined historical legacy. All of this is, of course, old news on the extreme right. As Helene Lööw (2015) notes in *Nazismen i Sverige 2000–2014,* Viking mythology and a focus on Norse heritage have been central to the National Socialist mythology since the 1930s.

The Danish Nazi Party Danmarks Nationalsocialistiske Arbejderparti (DNSAP) can serve as an example. Established in 1930, and officially dissolved in 1945, the party used Norse symbolism from the very beginning, combining the imported Nazi swastika symbol with Huginn and Muninn, Odin's messenger ravens. Later, the party employed various symbols of national identity, including viking ships, runes, and also the Trundholm sun chariot, a Bronze Age artefact (see Figure 33.9).

In addition, the party drew on Viking role models befitting a Nazi ideal of masculinity. The Norwegian equivalent of the DNSAP, Nasjonal Samling (National Union, NS), also heavily referenced the Old Norse heritage, such as Norwegian kings, the old Norse gods, as well as borrowing terminology and using runes and other Old Norse symbols (Thaule 2007). It should be noted that NS largely saw itself as a Christian party. As a party symbol, it used the Solar Cross, which was, however, referred to as the Cross of Olaf. Olaf II was seen as leading the Christianization of Norway. During the period of Romantic Nationalism, Olaf had become a symbol of Norwegian independence and pride. The use of the Solar Cross as party symbol thus carried both Old Norse and Christian and national connotations. The symbol was 'the ancient symbol of the Solar cross', in the 'old Norwegian colours: red and gold', but also the symbol 'used by Olaf the Holy when Christianity was introduced to Norway' (quoted in Thaule 2007: 91). In short, it connected NS with the past.

Figure 33.9 Front page of programme meeting between leaders of Nazi-oriented parties in Denmark, Sweden, and Norway, Copenhagen, November 1939

Source: John T. Lauridsen (1993).

Conclusion

As discussed in this chapter, the imagery used by the far-right today often does the very same thing: it connects the groups and activists of the far-right with the past, or rather with an *imagined* past, often built around ideals of masculinity. If we seek to understand how the far-right in the Nordic countries uses religious imagery, this connection remains essential. Groups and individuals with highly varying ideologies, ranging from neo-Nazis espousing openly anti-democratic ideology to specifically anti-Muslim groups claiming to be champions *of* democracy and freedom of speech, employ symbols taken from wider national and nationalist discourses. Concepts and specific imagery alike are often spread in the form of Internet memes, and cross borders without much difficulty; something which is true for both Crusader imagery and Viking imagery.

Religious belief only rarely plays any real role; with the exception of a few handfuls of self-declared Odinists, there are not many in the contemporary Nordic far-right who integrate religious beliefs with their ideology. In spite of this, religious imagery—including both Christian and Pagan symbolism—is deployed. Anti-Muslim groups gladly *combine* Viking symbolism, including pagan references, with Crusader imagery, contrasting themselves from Muslims by using both the appellative 'infidel' (or even the Arabic term *kafir*) and by painting themselves as the modern-day variant of the Crusaders. Scandinavian neo-Nazi groups, on the other hand, rarely reference Christianity positively, but employ references to both Old Norse culture, the idea of the Viking and pagan religion, in an attempt to give themselves historical legacy.

In 2004, in an article later translated to English, Alexander Kyrlezhev, now a consultant of the Synodal Biblical and Theological Commission of the Russian Orthodox Church, wrote that religion has returned from 'the solitary confinement to which it was banished by the modern' (Kyrlezhev 2008: 27), but that this does not entail a return to the old structure of sacred and profane:

> While secularism chased religion into the religious ghetto, the postsecularism of the postmodern accepts and dissolves religion within itself. Religion may exist everywhere together with the secular, but not in the old forms – not, for example, in the form of a church as a social and cultural institution, claiming universality and a dominating role in culture.
>
> *(Kyrlezhev 2008: 26)*

He added that 'religion in the postsecular age may resurface in symbolic form', as a form of marker of tradition that 'does not necessarily imply belonging to a religious tradition' (Kyrlezhev 2008: 27). This could certainly be said to be the case when it comes to how activists on the Nordic far-right employ religious imagery, *not* as an expression of actual religious faith, but rather as an ethno-nationalist marker, portraying themselves as true Norwegians, true Swedes, or true Europeans. Instead of viewing this as an example of postsecularism, however, I suggest that the use of such imagery is *both* old news on the far-right, *and* a product of the most obvious postsecular turn in Europe: while there is little indication that we are becoming increasingly religious, we are increasingly talking about religion, and especially about Islam.

Note

1 The article builds on a presentation given at the conference 'Radicalized religion: religion as a resource for political theory and practice' at the University of Chichester, UK, June 2017.

Further reading

Bruce, S. (2011) *Secularization: in defence of an unfashionable theory*, Oxford/New York, NY: Oxford University Press.

Bruce elaborates the secularization paradigm and defends it against a variety of recent criticisms, drawing on a range of empirical studies.

Busher, J. (2012) 'From ethnic nationalisms to clashing civilizations: reconfigurations of (un)civil religion in an era of globalization', *Religion Compass*, 6: 414–25.

Busher discusses how EDL members use symbols of national belonging, and how they have tapped into a political and social discourse about a 'clash of civilizations' and the threat posed by Islam to Western national cultures.

Griffin, R. (2012) *Terrorist's Creed: fanatical violence and the human need for meaning*, Basingstoke: Palgrave Macmillan.

This book explores the nature of fear in postmodern society, alongside the 'metapolitical' universe of the terrorist. Widely studied is the terrorist's 'rational' aim to achieve objectives through violent means, but this work highlights the impulsive and passionate drivers of violence.

References

Asprem, E. (2016) 'Islam og Europa: om religionspluralisme, reformasjon, og postsekularitet', *Internasjonal Politikk*, 74(4): 1–8.

Beckford, J. A. (2012) 'SSSR presidential address: public religions and the postsecular: critical reflections', *Journal for the Scientific Study of Religion*, 51: 1–19.

Berntzen, L. E. and M. Weisskircher (2016) 'Anti-Islamic PEGIDA beyond Germany: explaining differences in mobilization', *Journal of Intercultural Studies,* 37(6): 556–73.

Bjørgo, T. (2013) 'Introduction'. In Bjørgo, T. (ed.) *Terror from the Extreme Right,* New York, NY: Routledge.

Breivik, A. B. (2011) *2083—A European Declaration of Independence: de laude novae militiae pauperes commilitones christi templique solomonici,* London: self-published manifesto, accessed online 19-08-2018.

Carless, W. (2017), 'An ancient Nordic religion is inspiring White supremacist terror', *revealnews.org,* 25 May.

Citizen Warrior (2008) 'The Third Jihad', *citizenwarrior.com,* accessed online 23-05-2018.

Frie Nasjonalister (n.d.) 'Om Frie Nasjonalister', *frienasjonalister.wordpress.com,* accessed online 23-05-2018.

Furseth, I. (2015) 'Kristendom taper plass, mens islam og alternativ spiritualitet vinner fram', *Aftenposten,* 24 April, accessed online 23-05-2018.

Færseth, J. (2017a) 'Partilederen unnlater å nevne at toppkandidatene inkluderer holocaustbenektere', *Dagbladet,* 21 June, accessed online 20-03-2018.

——— (2017b) 'Alt-art: når frosken kommer ut av sekken', *Humanist,* 25 August, accessed online 20-03-2018.

Gardell, M (2003) *Gods of the Blood,* Durham, NC: Duke University Press.

Holmqvist, S. (2017) 'Nationaldagsreflektion: Vilken fana ska vi samlas under?', *nordfront.se,* 6 June, accessed online 23-05-2018.

Kaplan, J. (2013) 'Right wing violence in North America'. In: Bjørgo, T. (ed) *Terror from the Extreme Right,* New York, NY: Routledge.

Kyrlezhev, A. (2008) 'The postsecular age: religion and culture today', *Religion, State & Society,* 36(1): 21–31.

Lantagne, S. (2017) 'Famous on the internet: the spectrum of internet memes and the legal challenge of evolving methods of communication', available at SSRN: https://ssrn.com/abstract=2944804.

Lauridsen, J. T. (1993) 'Vikingernes sande efterkommere', *Magasin fra Det kongelige Bibliotek,* 7(4): 14–28.

Lööw, H. (2015) *Nazismen i Sverige, 2000–2014,* Stockholm: Ordfront.

May, E. (2006) 'The Other September 11th', *Gates of Vienna,* 11 September 2006, accessed online 23-05-2018.

Meleagrou-Hitchens, A. and H. Brun (2013) *A Neo-Nationalist Network: the English Defence League and Europe's counter-jihad movement,* London: ICSR.

Rognsvåg, S. (2015) 'Breivik mener Jesus er "patetisk"', *Dagen,* 19 November, accessed online 23-05-2018.

Rosendahl, J. and T. Forsell (2016) 'Anti-immigrant "Soldiers of Odin" raise concern in Finland', *reuters.com,* 13 January, accessed online 23-05-2018.

Støbakk, T. (2016) 'Ronny Alte om maktkampen i Odins Soldater: når de stempler meg som rasist, må jeg ta til motmæle', *Dagbladet,* 29 February, accessed online 23-05-2018.

Teitelbaum, B. R. (2017) *Lions of the North: sounds of the new Nordic nationalism,* Oxford: Oxford University Press.

Thaule, J. (2007) *På Nasjonal Grunn,* Master thesis in History, Oslo: Universitetet i Oslo.

Tveito, L.-H. (2007) *Kampen for den Nordiske Rases Overlevelse: bruken av den norrøne mytologien innenfor Vigrid,* Master thesis, Tromsø: Universitetet i Tromsø.

Vikernes, V. (2013) "Why ôðalism", *Thulean Perspective,* 31 July, accessed online 23-05-2018.

YLE (2016) 'Soldiers of Odin's secret Facebook group: weapons, Nazi symbols and links to MV Lehti', *yle.fi,* 16 March, accessed online 23-05-2018.

Zúquete, J. P. (2018) *The Identitarians: the movement against globalism and Islam in Europe,* Notre Dame: University of Notre Dame Press.

34
Postsecularity prefigured

Roger Speare

Introduction

My essay will probably be the odd-one-out in this Handbook. I am not an academic, and my interest in the subject is that of a practitioner who wants to better understand what is happening and accelerate it. Let it be said I am one of many Christians that are applying gentle but persistent pressure on society and the churches to embrace a postsecular view. I have worked as a volunteer with Emmaus since 1998, and this worldwide solidarity movement together with its inspirational founder, the Abbé Pierre (see Figure 34.1), will provide much of the material for this study.

Despite my practical stance, it has, however, been fascinating and inspiring to read some of the pioneering academic work that has gone into unpacking the idea of postsecularity during the past decade or so (see Chapter 1). These works alongside concepts such as the 'enlightened city', the enigma of capital and the politics of hope (Beaumont 2018) and the wave of ground-breaking inquiries into faith-action on homelessness and foodbanks in the UK (Cloke et al. 2011; May and Cloke 2014; Williams et al. 2016; Cloke et al. 2017; Cloke et al. 2019) have reinforced my efforts to explore, and conceptualize, the genius of Abbé Pierre's oeuvre.

It is my contention that the Abbé Pierre prefigured postsecularity. He anticipated, no later than 1969 and arguably as early as 1949, a social and religious trend that was not mentioned directly in the published literature until the late 1990s with Philip Blond's *Post-Secular Philosophy* (1997) and Jürgen Habermas' notes on postsecular society (2008), although the latter thoughts were arguably sparked by the 2001 assaults on the World Trade Center in New York.

I have little doubt that Abbé Pierre was a prophet and pioneer of the postsecular bringing together people of faith and no-faith to work collectively for humanitarian and social justice causes. He has bequeathed to us a humanitarian vision for our world and shown us a way to realize that it encapsulates the essence of postsecularity. This idea is well supported by the Emmaus literature and will be developed in a brief historical narrative. Abbé Pierre's personal spiritual journey is inseparable from the Emmaus story. This narrative mirrors contemporary developments within Christianity that have fuelled the shifts that one considers the tentative steps to a postsecular society from the faith community side of that process.

Figure 34.1 Abbé Pierre in his prime (1955)
Source: Wikipedia Commons: photo: Nationaal Archief.

I will first set the idea of postsecularity in context before explaining how it particularly appears to me.

Postsecularity

Building on Klaus Elder and Jürgen Habermas, Cloke and Beaumont (2013) embraced what they termed geographies of postsecular rapprochement in the city as those urban spaces of partnership between people of faith and those of no religious faith who come together to offer care, welfare, and justice to socially excluded people (see Chapter 1). Despite growing attention to the postsecular in human geography and the social sciences and humanities more generally, reactions to the notion at least in human geographical circles until recently were somewhat hostile (Kong 2010; Ley 2011; Wilford 2010; see also Joas 2008; Reder 2010).

Taking the cue from Beaumont (2018) I draw attention to a series of misunderstandings about 'postsecularity' from current perspectives in philosophy, theology, sociology, and human geography. I argue that a way out of the deadlock requires a judicious coming together of insights on postsecular ethics, new conceptualizations of radical difference beyond the crisis of multiculturalism, and new theorizations of the urban-social that entangle religion, belief, and faith with selective currents within socio-spatial theory. A postsecular ethics of the city, derived in part from a reading of Albert Camus's *L'Homme Révolté* ('The Rebel'), offers a way to confront creatively critical urban theory and the postsecular by simultaneously emphasizing: (1) motivations for rebellion against prevailing *status quo,* (2) inevitable failure of attempts at human perfection, and (3) ongoing struggles for social justice that do not abandon values such as the intrinsic worth of human life. The 'enlightened city' that might emerge from these deliberations cannot be dismissed as utopian or a normative dream. It exists in numerous spaces of engagement for a more just world. Attention has been drawn to the possibilities but also limitations of these spaces of engagement with reference to my personal experiences of Emmaus, an international solidarity movement founded in Paris in 1949 by the Catholic Priest Abbé Pierre to combat poverty and homelessness.

What do I understand by 'postsecularity'? To me postsecularity is the coexistence, mutual respect, and collaboration of believers and non-believers in pursuit of humanitarian goals. In taking this position I do not wish for a return to an earlier world where religion ruled the roost. Nor do I mean that the starkly different worldviews of atheism and religion should in some way to be watered down in order to arrive at a ceasefire. By postsecular I mean believers and non-believers learning to appreciate what the other brings to the party, agreeing that at an intellectual level, the question about God's existence is unanswerable, embracing each other as human beings and getting on with the job of making our world a better place for everyone.

This sounds like a great scheme, but it is not without its problems. There will be a continuing intellectual tension because humans are obsessive about pigeonholing their ideas. The good news is that in Emmaus we have the makings of a paradigm for a postsecular world. Emmaus actually describes itself officially as a secular movement because it does not practice or propagate any particular ideology or religious belief. It is not, however, anti-religious, and as we will see, its Universal Manifesto aims to bring together believers and non-believers in a common purpose. For this reason, I think postsecular is a more accurate description.

Emmaus is working and has been for nearly 70 years. It has had and continues to have its problems, and we will touch on some of those later, but it has demonstrated its resilience and ability to return to its core values when these are in danger of being lost. This strength comes down to the vision, genius, and inspiration of Henri Grouès, better known as the Abbé Pierre. He was, in my view, a spirit-filled Christian well ahead of the postsecular curve in the country that invented secularism, and the story of his life is the topic to which we will now turn.

Abbé Pierre and the origins of Emmaus

With some justification it could be claimed that the Abbé Pierre was for half a century the conscience of France. Jean Rousseau, former president of Emmaus International, introduced his 2015 collection of Abbé Pierre's previously unpublished works with the words:

> Recognized as a defender of the excluded and the most impoverished, matching words with actions, the Abbé Pierre was above all a free man. All his life he helped the poor by fighting every injustice.
>
> He was in all the important battles of the 20th century, wartime resistance worker, elected as a Deputy (MP) at the time of liberation, third-world person, citizen of the world and advocate for conscientious objectors and migrants. Every action by the Abbé Pierre was guided by the search for justice and dignity for each person, and by the abiding concern to fundamentally transform modern society.
>
> *(Abbé Pierre and Rousseau 2015: Back cover)*

Henri Grouès was the fifth child of eight, born in 1912 into a devoutly Catholic upper-middle class family in Lyon. He was a turbulent and strong-headed but sensitive boy who combined in his teens an absorbing interest in Cartesian thought with a hunger for the Absolute and the mystical.

At the age of 11, his father, director of the Foundries of the Rhone, took him and one of his brothers to a place where he went every Sunday. There Henri saw his father roll up his sleeves and minister to the needs of 50 beggars, shaving them, cutting their hair, ridding them of vermin, and serving them breakfast. Without a hint of condescension, the rich man became the poor man's servant.

Easter 1927 marked the defining moment of his life, an event to which he continually referred for the rest of his days. While on a pilgrimage to Assisi he received a double revelation, two inner events which he never felt able to express fully in words. First, that during adoration there is communion with the whole universe, and, secondly, that it is only through adoration one gains a clear understanding of the intensity of action that one has to undertake. In the Abbé Pierre's own words, 'the boy who came back from Assisi was no longer the same person' (See note 1).

He subsequently spent seven years as a Franciscan friar before fragile health, something that dogged him for his whole life, led him to become a parish priest. This coincided with the outbreak of World War II, and his army service was cut short by a bout of pleurisy. In an effort to improve his health, he had earlier taken a course in mountaineering, and he was able to put this to good use guiding scores of oppressed people from occupied France over the Alps into Switzerland. This together with the forging of documents and his chaplaincy and recruitment work for the Maquis resulted in a public profile that propelled him into becoming a member of parliament when the war came to an end in 1945.

Since 1941 he had been meditating on the story in Luke's Gospel about two disciples of Jesus Christ, devastated by his crucifixion and walking away from Jerusalem, the place of his death, to a town called Emmaus. On the way a stranger joins them and enquires why they are so despondent. He then goes on to explain that these horrifying events are all part of a divine plan. Eventually they recognize him as Jesus, raised from the dead. Abbé Pierre saw in this 'a mystery of invigorating disillusion' (Brodiez-Dolino 2013: 44). When in 1945 he started creating homes for people dispossessed by the war, he named the first hostel 'Emmaus'.[1]

The birth of the Emmaus Movement

In November 1949, Abbé Pierre met Georges Legay. Georges had been serving a life sentence for murdering his father in a fit of anger, but he had been pardoned for heroic conduct during a prison fire and released. His family, however, wanted nothing to do with him and, overcome with despair, he had attempted suicide. Abbé Pierre was asked to help, and this is his account of what happened:

> Without thinking, without premeditation, I did the opposite of charity. Instead of saying, "You are miserable, I will give you a roof over your head, work and money" circumstances made me do exactly the opposite. I said, "You are desperately unhappy but I have nothing to give you. My family fortune is already gone. My parliamentary allowance is spent before I even receive it to repair the house and to lodge those in need. I am tired and I can't respond to all these calls for help. But you, because you want to die there's nothing holding you back. So, please, wouldn't you like to help me to help others?"

> That was the birth of the Emmaus movement. If that moment were ever forgotten, Emmaus would cease to be. Above all else Emmaus is telling someone who feels unwanted or useless, "I have nothing to give you but my friendship and my plea to share in my efforts so that together we can save others".
>
> The only true gift is to help another to be able to become a giver in their turn thus creating a wonderful chain, without limits, of sharing and initiation in the reality of love.[2]

Georges later said of this encounter, 'No matter what he could have given me, I'd have tried to kill myself again, because what I needed wasn't just the means to live but reasons to live'.

And so, the first community of Emmaus came into being. Supported by Abbé Pierre's parliamentary allowance, the Companions (community members) worked at creating homes for homeless people. When this source of income dried up, they were forced to find something else and this turned out to be rag-picking, salvaging stuff found on rubbish dumps and recovering items from neglected cellars and attics. By 1953, there were around 160 people living in a handful of communities; a quarter of them were builders and three quarters were rag pickers, the building programme being funded by the recycling operation. The series of events that would thrust Emmaus into the public consciousness was just about to happen.

The winter of 1953–1954 was the coldest that Paris had suffered in ten years. Two significant events happened on the night of 3rd/4th January: a parliamentary budget council failed to approve spending that would have allowed Emmaus to build several hundred emergency dwellings; and a homeless baby froze to death. Abbé Pierre wrote an open letter to the minister for reconstruction who had presided at the budget council and invited him to the funeral. The invitation was accepted, and the reality of the homelessness situation thus became firmly embedded in the political will. Things began to move.

Then at 3 am, on 1 February, an elderly woman died on the pavement of the Boulevard Sebastopol. Her only identification was an eviction notice from her attic flat two days earlier. This time, Abbé Pierre broadcast live on radio. He appealed for blankets, tents, and stoves. The response was overwhelming. The emergency items poured in, and 1,000 people volunteered to help. These and similar subsequent events that winter became known as 'The Uprising of Kindness' and triggered a sustained period of growth for Emmaus.

The Universal Manifesto

In 1969, a total of 20 years after the birth of Emmaus when the movement had reached every continent, the first Emmaus world assembly was convened in Berne. The purpose of that meeting was to create a Universal Manifesto.[3] Here it is in full.

> Our name, "Emmaus", comes from the name of a village in Palestine where despair was transformed into hope. For all, believers and non-believers alike, this name evokes our shared conviction that only love can unite us and allow us to move forward together.
>
> The Emmaus Movement was created in November 1949 when men who had become aware of their privileged situation and social responsibilities in the face of injustice and men who no longer had any reason to live crossed paths and decided to combine forces and take action together to help each other and come to the aid of those who were suffering, in the belief that it is by saving others that you yourself are saved.

To this end, the Communities were set up, working to live and give. Groups of Friends and Volunteers were also set up to continue the struggle in the private and public arena.

1. OUR LAW applies to all humankind and is that on which depends any life worth living, true peace, and joy for the individual and society: "Serve those who are less fortunate before yourself". "Serve first those who suffer most".

2. OUR CONVICTION is that respect for this law should guide any pursuit of justice and therefore peace among peoples.

3. OUR AIM is to take action to ensure that every person, society and nation can live, have a place and be fulfilled through communication and sharing in equal dignity.

4. OUR METHOD involves creating, supporting and coordinating a system in which everyone, by being free and respected, can meet their own needs and help each other.

5. OUR PRIMARY MEANS wherever possible, is the salvage work that gives new value to any object and increases the potential to provide emergency relief to help those suffering most.

6. ANY OTHER MEANS to raise awareness and meet this challenge should also be used to ensure that those suffering most are served first, by sharing their troubles and struggles – whether public or private – until the cause of each ill is eliminated.

7. OUR FREEDOM In the accomplishment of its task Emmaus is not subordinate to any other ideal than that expressed in this Manifesto, or to any other authority than that established internally according to its own rules. It acts in conformity with the Declaration of Human Rights, adopted by the United Nations, and with the just laws of every society and nation, without political, racial, linguistic, spiritual or any other distinction. Nothing else is required of anyone wishing to participate in our action other than the acceptance of the content of this Manifesto.

8. OUR MEMBERS This Manifesto constitutes the simple and clearly defined foundation of the Emmaus Movement. It must be adopted and applied by any group wishing to become an active member of the Movement.

The author of the Universal Manifesto is not recorded, but it has been confirmed as Abbé Pierre by someone who knew him well. He also supplied two further insights into its composition. Abbé Pierre had told him that his personal thinking and experience had formed the basis of the Manifesto, but that he had also been influenced in its creation by the ideas of Lord Beveridge, architect of the British welfare state, with whom he had worked after the war on a European committee for social and housing issues; and, it can also be surmised that aspects of liberation theology may have rubbed off on Abbé Pierre, given the number of priests of that persuasion who had served Emmaus in France and South America during the 1950s and 1960s.[4]

During the Berne debate, nobody questioned the Manifesto's Christian links. The origin of the Emmaus name and the thinly veiled reference to the resurrection of Jesus Christ that opens the preamble were agreed without comment. Indeed, it was noted in the minutes that the first three sentences of the Manifesto contained all the general principles needed to guide the movement.

There was also no disagreement about the objectives. Everyone agreed that changing the structures of worldwide society was what Emmaus was about. Even though homelessness triggered the beginning of Emmaus and continues to this day as a major focus of the movement, the vision has always been much bigger. Sometime in the late afternoon of 24 May 1969, the Universal Manifesto was approved 'by acclamation', a dramatic moment in the history of the Emmaus movement.[5]

The development of Emmaus

Emmaus has grown in its home country to 18,000 people, operating in 284 groups. Of these 117 are intentional communities, 'working to live and give' in a similar way to the original communities. The other 167 groups are engaged in a variety of social actions. Social housing remains high on the Emmaus agenda.[6]

As the Emmaus movement developed in France, it split into half a dozen different streams, each with its own particular emphasis. Today, the diversity remains, but all the streams re-converged some ten years ago to form one federation. Abbé Pierre, who had never taken sides, happily lived to see the movement reunited. A recent visit to France to check on a couple of large communities confirmed them to be in good health and staying true to the Manifesto. For example, Lyon has around 100 Companions and operates an industrial scale 32,000 sq. ft. recycling and upcycling business with an annual turnover of 2.6 million Euros (see Figures 34.2 and 34.3) The retail space, 'antiqu'emmaus', is dedicated to the less valuable items, mainly small furniture and bric-a-brac, with the pricing controlled by the Companions. The small shop in Rue de Crequi reportedly makes more profit per square feet than any other Emmaus shop in France.

A similar pattern has been experienced in the UK. The movement reached these shores in 1991, and today has 29 operating communities and another 4 in formation. Britain now has a larger and faster growing Emmaus presence than any country other than France, and includes some sizeable operations. For example, the Emmaus Preston megastore at 47,000 sq. ft. is the largest shop in the UK Charity Retail Association (see Figure 34.4).

Figure 34.2 Lyon Emmaus: bargain department
Source: photo: John Webbe.

Figure 34.3 Lyon Emmaus: Rue de Crequi shop front
Source: photo: John Webbe.

Figure 34.4 Emmaus Preston megastore
Source: photo: Stephen Buchanan.

There appears to be no end in sight to homelessness in the UK, and the latest figures from Crisis, the UK national charity for single homeless people, indicate that 78,000 homeless families and individuals were placed in temporary accommodation last year, an increase of 60% since 2012 (Fitzpatrick et al. 2018). The upward trend in rough sleeping also continues, with the estimated national total up by 169% since 2010 (see Figure 34.5).

Figure 34.5 Rough sleeper, Tottenham High Road, London
Source: Wikipedia Commons.

In 2006, Emmaus in the UK split into two streams following disagreements about core values and practice, but was reunited six years later. Today, as neo-liberal austerity continues to diminish the welfare state, those Emmaus communities that stayed farthest from the culture of entitlement are helping those that did not. A strong focus on self-sufficiency is being developed, and the movement is beginning to explore other models of operation. The Universal Manifesto permits unlimited variations of service to humanity provided the key principles of Emmaus are faithfully observed and applied.

One line of development that particularly interests me is the potential for a social enterprise like Emmaus to offset the negative aspects of capitalism. If poverty is to be eliminated, then every adult of working age needs a job that enables them to 'live, have a place and be fulfilled' (Universal Manifesto 1969) by paying them a living wage or its equivalent in kind. However, a regular, profit-dependent business needs to be able to compete in an unforgiving marketplace, and that requires them to employ the most effective people they can afford. This will inevitably exclude some people from employment opportunities because of their lack of ability and skills or their personal support needs.

In contrast, a social enterprise like Emmaus has the reverse perspective. It starts by enrolling a person who has difficulty in finding and keeping a regular job and then sets about creating ways that he or she can use his or her talents to meet his or her own needs and help others. If capitalist and social enterprises were linked together in partnership, then unemployment could become a soluble problem at minimum cost to society, and we could move a step closer to the enlightened city.

Spiritual roots of postsecularism

The Universal Manifesto, then, envisaged a postsecular world nearly 30 years before that term was invented. Emmaus is not in the business of peddling any ideology, social, political, or religious. It is concerned to unite believers and non-believers, privileged people and marginalized persons in working together to meet the material, the emotional, and the everyday needs of the most distressed people on our planet. Abbé Pierre saw this clearly in the teaching of Jesus:

> There was a day when the disciples came and asked Our Lord concerning the Day of Judgment. What does Our Lord say? What he says is extraordinary. He says nothing about sacraments, nothing about commandments, nothing about practices, nothing about virtues, nothing even about prayers. He speaks only of one thing; He says: "On that day, the Son of Man will appear in all the power of his majesty, and he will say to some, 'I was hungry, I was cold, I was lonely, I was homeless, and you shared what you had with Me, you helped Me in your suffering brother. So, come.' And to others, 'You did not help Me, you did not share what you had, so go!'".
>
> *(See note 2)*

The Christian church for much of its history has ignored Jesus' teaching and mistakenly attempted to build the Kingdom of Heaven from the top down. But the Kingdom needs to be built from the bottom up. That is why Abbé Pierre's perception is so important. Believers and non-believers, while acknowledging their differences, must work together to serve those who suffer most.

In Emmaus, some see themselves as following in the footsteps of Jesus; others practise humanistic benevolence, seeing it as a fortuitous by-product of evolution. Can we know, in this life, which view is right? No. But does that matter if the hungry are being fed, the thirsty given drink, the stranger is being taken in, the naked clothed, the sick visited, and the prisoners cared for? And, according to Jesus, in the Final Analysis, that is all that will count.

Emmaus is neither atheist nor Christian. It does not deny or dumb down either of those worldviews, but affirms both as logically coherent even though mutually exclusive. Emmaus accepts and celebrates different beliefs and faiths and encourages people to cooperate at the fuzzy interface between the sacred and the secular where an uncomfortable, but highly creative, chemistry produces remarkable results.

Abbé Pierre felt this paradox keenly:

> The tension was difficult, but healthy and fertile, between the absolute respect of man and the full force of Life speaking within. One day a neighbour, a daily witness to our work, exclaimed, "I don't know if God exists, but what I do know is that if He exists He is in what you are doing".
>
> *(See note 2)*

While it is undoubtedly true that there have always been Christians serving people on the margins of society, it does seem that this has been on the increase for several decades. My own observation from many years of Christian discipleship and service is that this change is being driven by the large-scale re-emergence of individual spiritual experience. An important result of this has been a change in the focus of evangelistic outreach from 'belief-based conversion' to 'whole person salvation'. This helps to make Christians and non-believers more comfortable as colleagues in social action.

Before exploring reasons for the shift to the spiritual within Christianity, it is important to note that there has also been movement from the non-believing side of the divide. The savage tones of Richard Dawkins in *The God Delusion* (2006; 2016) embarrassed many of his fellow atheists, and more constructive contributions to the dialogue have started to be heard from the non-faith community, for example, Daniel Dennett's definition of atheistic spirituality:

> Let yourself go. If you can approach the world's complexities, both its glories and its horrors, with an attitude of humble curiosity, acknowledging that however deeply you have seen, you have only just scratched the surface, you will find worlds within worlds, beauties you could not heretofore imagine, and your own mundane preoccupations will shrink to proper size, not all that important in the greater scheme of things. Keeping that awestruck vision of the world ready to hand while dealing with the demands of daily living is no easy exercise, but it is definitely worth the effort, for if you can stay centered, and engaged, you will find the hard choices easier, the right words will come to you when you need them, and you will indeed be a better person. That is the secret to spirituality, and it has nothing at all to do with believing in an immortal soul, or in anything supernatural.
>
> *(Dennett 2007: 303)*

Dennett has narrowed an important gap between atheism and faith across which enlightened believers can construct a bridge. The materials to do so will come from many streams of Christianity, but one phenomenon in particular stands out: the worldwide Charismatic Movement.

The origins of this spiritual awakening can be traced back to Wesleyan Methodism in the eighteenth century, and onward through the nineteenth century Holiness movement and the Keswick Convention. However, the revival that began on 9 April 1906 in Azusa Street, Los Angeles, is generally considered to have been the start of the modern awakening. The first wave, which was largely rejected by the mainstream Christian denominations, resulted in the formation of many new Pentecostal churches across the world during the first half of the twentieth century. The second wave, which began around 1960, still continues. As with the first wave, new churches have been created but significantly all the historic denominations of Christianity, including Catholic, Orthodox, and Protestant traditions, have felt its effects. At the heart of this Christian renewal movement is an experience called the Baptism or Filling or Anointing of the Holy Spirit. There are two biblical models for this; the descent of the Holy Spirit upon Jesus that launched his public ministry; and the events of the Day of Pentecost when the Christian church came into existence. The defining characteristic of this experience is generally said to be the rediscovery and practice of certain spiritual gifts such as prophecy, healing, and speaking in tongues that were evident in the early church. These are undoubtedly important, but there are other less obvious aspects such as a greater awareness of the presence of God, a deepening of the spiritual life, an elevation in worship, an expansiveness of love, a sense of divine guidance, and an empowerment for service. I can personally testify to all of these.

The significant element for our present enquiry is that this spontaneous move of the Holy Spirit, for as such it is seen by Christians, has cut right across denominational boundaries, ignoring doctrinal and theological differences and drawing together in fellowship people who previously did not and possibly might not even have wanted to befriend each other. Expressed subjectively, it feels as if God has been saying to his people, 'Forget your treasured labels and your theological categories; just concentrate on getting your hearts together and we will sort your heads out later'. Now once a Christian has digested the idea that God is less interested in theology than in practical compassion, it is a short step to the next question, 'Is God actually fazed by people not even believing in him?' This has led some Christians back to the gospels to find out how Jesus might have related to an atheist if he had actually met one, and it turns out that Jesus was a whole lot more inclusive than his followers have sometimes been. So, the way is open and the motivation is churning, I would say, for spirit-filled Christians to get involved in postsecular humanitarian enterprises.

Conclusion

It is not difficult to see Abbé Pierre's spiritual journey as 'charismatic' even though, as far as we know, he had no direct contact with the actual movement. The 'violent and indescribable emotion' experienced at Assisi that awoke within him an intense adoration, a cocktail of love and worship that drove him to action, is typical of the baptism of the Holy Spirit; the extraordinary encounter with Georges Legay when, without premeditation, he found some words that not only saved a man's life but also tapped into a deep level of human motivation and triggered a worldwide humanitarian movement was, in Pentecostal terminology, the spiritual gift of a 'word of wisdom'; also, his continual challenging of the French collective conscience over more than 50 years marks him out clearly as a man with a prophetic calling and ministry. Yet, although he witnessed constantly to his faith, he did not proselytize. He said that he only distinguished between those who were prepared to tackle poverty and those who were not; what people believed did not matter to him. He drew up a Manifesto for the Emmaus movement that can now be seen as a working template for postsecular humanitarianism 40 years before its time. On 17 occasions, he topped the list of France's best-loved figures. His death in 2007 saw the political class united in their tributes. I think it is fair to say that Abbé Pierre prefigured postsecularity.

Notes

1 See 'Henri Grouès Dit L'Abbé Pierre (1912–2007)', Theophraste-la-Cinquieme Odysee. Avril 2001. An English version was made available by Cambridge Film and Television Productions in 2001 and distributed by Emmaus UK. See also www.youtube.com/watch?v=PnSVC9WLiyk accessed online 06-06-2018.
2 These quotes were taken from an Emmaus International pamphlet dated 1994 that was in circulation within the movement in the late 1990s.
3 Universal Manifesto (1969) taken from the Emmaus International website, see /www.emmaus-international.org/en/who-are-we/2014-12-03-10-21-24/founding-texts.html accessed online 06-06-2018
4 Selwyn D. Image CBE JP, an international marketing consultant, first encountered Emmaus as a student on a French work camp in 1960. He introduced Emmaus to the UK in 1991, when he founded Emmaus Cambridge and spoke with Abbe Pierre on several occasions.
5 Emmaus World Assembly (1969) Minutes. I obtained a copy of these minutes from Emmaus International some years ago and had a translation made.
6 See Emmaus France: http://emmaus-france.org/qui-sommes-nous/notre-organisation/accessed online 06-06-2018.

Further reading

Abbé Pierre (2005) *Jesus and the Quest for Secular Justice*, Niagara Falls, NY: Seraphim Editions.

Intended for the general reader, this book surveys major contemporary writings on the historical Jesus. Presenting a realistic image of flawed human nature, Abbé Pierre argues that religious principles of compassion offer the best hope for achieving genuine international justice.

Martin, D. and P. Mullen (eds.) (1984) *Strange Gifts? A guide to charismatic renewal*, Oxford: Basil Blackwell.

This volume brings together supporters and critics of the Charismatic Movement, one of the more dramatic and contested developments in Christianity.

Menzies, W. W. and R. P. Menzies (2000) *Spirit and Power: foundations of Pentecostal experience*, Grand Rapids, MI: Zondervan.

This book provides a cutting-edge look at Pentecostal theology from the standpoints of history, hermeneutics, and exegesis, and offering a continuing dialogue between Pentecostals and Evangelicals.

Repland, L. C. (1956) *Abbé Pierre Speaks: speeches collected by L. C. Repland*, translated into English, New York, NY: Sheed & Ward.

This book presents a series of translated speeches by Abbé Pierre, capturing him at his most potent in the spoken word.

References

Abbé Pierre and J. Rousseau (2015) *Pensees Inedites Pour Un Monde Plus Juste* ("Previously Unpublished Thoughts for a Fairer World"), Paris: Le Cherche Midi.

Beaumont, J. (2018) 'The enlightened city', *Political Geography*, under review.

Blond, P. (1997) *Post-Secular Philosophy: between philosophy and theology*, London/New York, NY: Routledge.

Brodiez-Dolino, A. (2013) *Emmaus and Abbé Pierre: an alternative model of enterprise, charity and society*, English translation: Alexandra Harwood, Paris: Les Presses de Sciences Po.

Cloke, P., Baker, C., Sutherland, C. and A. Williams (2019) *Geographies of Postsecularity: re-envisioning politics, subjectivity and ethics*, London/New York: Routledge.

Cloke, P. and J. Beaumont (2013) 'Geographies of postsecular rapprochement in the city', *Progress in Human Geography*, 37(1): 27–51.

Cloke, P., May, J. and S. Johnsen (2011) *Swept Up Lives?: re-envisioning the homeless city*, Oxford: Wiley-Blackwell.

Cloke, P., May, J. and A. Williams (2017) 'The geographies of food banks in the mean times', *Progress in Human Geography*, 41(6): 703–26.

Dawkins, R. (2006; 2016) *The God Delusion*, 10th anniversary edition with new material, London: Black Swan (Penguin).

Dennett, D. C. (2007) *Breaking the Spell: religion as a natural phenomenon*, London: Viking (Penguin).

Fitzpatrick, S., Pawson, H., Bramley, G., Wilcox, S., Watts, B. and J. Wood (2018) *Crisis: The Homelessness Monitor England 2018*, London: Crisis UK, accessed online 06-06-2018.

Habermas, J. (2008) 'Secularism's crisis of faith: notes on post-secular society', *New Perspectives Quarterly*, 25: 17–29.

Joas, H. (2008) *Do We Need Religion?*, trans. A. Skinner, Boulder, CO: Paradigm.

Kong, L. (2010) 'Global shifts, theoretical shifts: changing geographies of religion', *Progress in Human Geography*, 34(6): 755–76.

Ley, D. (2011) 'Preface: towards the postsecular city?'. In Beaumont, J. and C. Baker (eds.) *Postsecular Cities: space, theory and practice*, London/New York, NY: Continuum.

May, J. and P. Cloke (2014) 'Modes of attentiveness: reading for difference in geographies of homelessness', *Antipode*, 46(4): 894–920.

Reder, M. (2010) 'How far can faith and reason be distinguished?'. In Habermas, J., Brieskorn, N., Reder, M., Ricken, F. and J. Schmidt (eds.) *An Awareness of What Is Missing: faith and reason in a Postsecular Age*, Cambridge, MA: Polity Press, pp. 36–50.

Wilford, J. (2010) 'Sacred archipelagos: geographies of secularization', *Progress in Human Geography*, 34(3): 328–48.

Williams, A., Cloke, P., May, J. and M. Goodwin (2016) 'Contested space: the contradictory political dynamics of food banking in the UK', *Environment and Planning A*, 48(11): 2291–316.

Afterword

35

Reflexive secularization

Eduardo Mendieta and Justin Beaumont

A handbook, as every dictionary will inform you, is a book that gathers in an organized fashion basic factual information about a given subject, and that thus contains specific and targeted instructions on how to operate, sometimes machines, sometimes software, sometimes concepts. A handbook is a toolkit, as well as a lexicon. It should be useful to newcomers to the term, but also to experts who may want to have a broader and deeper understanding of the term. It should be ecumenical but also thorough.

The 35 chapters in this handbook, penned by a truly international group of scholars, offer basic information, in an organized manner, about postsecularism, postsecularity, and the postsecular, and many of its cognates, gathered under quite useful rubrics. The rubrics call to mind 'a certain Chinese encyclopedia called the *Heavenly Emporium of Benevolent Knowledge*' to which Jorge Luis Borges refers in his mesmerizing essay 'John Wilkins' Analytical Knowledge', which set out to remedy the blatant omission from the *Encyclopedia Britannica*, namely the absence of an entry of John Wilkins, who, among other things, devoted his life to speculating over whether a 'world language' would be possible and what would its semantics and grammar look like. Be that as it may, Borges tells us that the ancient Chinese encyclopedia had divided animals into: '(a) those that belong to the emperor; (b) embalmed ones; (c) those that are trained; (d) suckling pigs; (e) mermaids; (f) fabulous ones; (g) stray dogs; (h) those that are included in this classification; (i) those that tremble as if they were mad…', and so on and so forth. Then, Borges adds: 'there is no classification of the universe that is not arbitrary and speculative' (Borges 1999: 231).

In the case of this handbook, it is speculative but not entirely arbitrary. What's evident is that the concepts related to postsecularity have emerged in a given context, responding to specific histories, dynamics, traditions, and it is meant to name a distinct phenomenon that deciphers a specific or peculiar aspect of our contemporaneity. In this sense, all lexicons, dictionaries, encyclopaedias, and handbooks become portraits of our age, snap shots of whom we take ourselves to be, or have become, and, to this extent, they also are supererogatory acts of theory-making that may be resented by those that feel excluded. The category postsecularity, then, is provisional and contested. It can and should be used in the mode of the *als ob*, as if, until further notice.

Nonetheless, this volume, the brainchild and product of the industry and dogged persistence of Justin Beaumont, has gotten us to an amazing start. We must begin somewhere. The stage is now set. Laughter, even Borgeisan, maybe the shortest distance between two people. But it's not enough. One must attempt to make sense of our times.

It is undeniably tempting to relate the growing relevance of postsecularity to ongoing crises throughout the world: climate change, and the resultant severe weather that is creating climate refugees, fuelling resource wars, catalyzing massive migrations, which in turn have given rise to many humanitarian crises; we could also refer to Brexit, the Catalan referendum and Spanish constitutional crisis, the austerity measure imposed on Greece, and the rise of xenophobic, nationalistic, and, in many cases, racist populist movements that have turned many polities towards authoritarian governments that are imposing draconian measures and legislating policies that affect perceived foreigners, refugees, and migrants. It is difficult not to approach the terms under these darkening skies. On the other hand, the *Me Too* (#MeToo) *Movement*, as well as many protests, in the USA, against Trump, and the rise of the, lets us call them, in short, the Occupy and *Indignados* mobilizations in the West, and the many uprisings throughout the Middle East, are different, if not positive, signs of alternative forces at play. Absent sociological data on whether a *postsecular* orientation towards the world results in specific attitudes about climate injustice, gender, migrants, and so on, we can only speculate. Still, it is not difficult to acknowledge that critiques of the rapacious commodification and monetization of nature are entangled with appraisals of certain views of Christianity, God, the religious, and the sacred. Similarly, critiques of jingoistic and racist xenophobia are entangled with critiques of certain views of religious traditions that have fomented racist views of strangers. It bears asking: to what extent Brexit and the emergence of the Trump phenomenon are related to religious trends, or whether these two phenomena instead rather belie any such connections? In the USA, Republican Politics have been closely allied with Evangelicals and conservative Protestant denominations that advance conservative social agendas, at least since Richard Nixon, who exemplified what has been called the Southern Strategy. In the case of Trump, conservative Protestants wilfully turned a blind eye from his blatant sexism, racism, and evident lack of personal character, because he promised to appoint pro-life judges to the Supreme Court. The lack indeed of serious critique and call to accountability by religious groups was noteworthy and sobering.

These reflections aside, which localize partly our locus of enunciation, it is important to ask: in what sense is the term *postsecular* more than a moniker for an intellectual fashion, one that, despite the trend, captures both a shift in the way scholars of the social sciences and humanities ponder our global predicament and the role of religion in modern societies, and one that, more specifically, acts as a seismograph, registering tectonic shift in social reality. Alternatively, it can be asked: in what sense is *postsecular* a theoretically useful concept that illuminates, in enlightened and perspicacious ways, our now global social reality that would not otherwise be illuminated if we either lacked the concept or left it to the side? And, let's not forget, how would the concept contribute to a social-empirical research agenda.

In order to provide some preliminary answers to these questions, in what follows, we would like to take up the demand of what is or was the secular and secularization. After all, the term *postsecular* makes sense as either taking leave of the secular and the halting of secularization, or as a problematization of these two terms, that is, we no longer really recognize them *simpliciter, unconditionally*. We want to argue that the secular is what results from the process of secularization. Analogous here would be to say that race is the detritus of racism, and that concepts of race do not precede practices of racism. The secular is the name for what secularization has left in its wake.

To this extent, we can say, paraphrasing Bruno Latour's famous and ground-breaking book, 'we have never been secular' (see Chapters 1 and 2). We would add: 'enough and for the time being'. In order to make good on this argument, we'll rely, first, on a reconstruction of some uses of the terms *secular* and *secularization* with reference to José Casanova's (2013) synoptic essay 'Exploring the postsecular: three meanings of the "the secular" and their possible transcendence' and the entry on 'Säkularisation, Säkularisierung' in the historical lexicon *Geschichtliche Grundbegrieffe,* volume 5 (Brunner et al. 1984). Following that, we turn to the closely related question: in what senses can we speak of the *postsecular,* on the basis of what we have taken secularization to mean. We'll rely partly on James A. Beckford's SSSR presidential address entitled 'Public religions and the postsecular: critical reflections' (Beckford 2012). Drawing on Eduardo Mendieta's ideas, we articulate an alternative typology, or five rubrics of the postsecular, and conclude with thoughts for research futures.

Exploring the secular and postsecular

The noun *secular* has its roots in the Latin *saeculum,* which has some of its origins in the Etruscan ancient Italian civilization. The term is considered the longest life span of a human life. According to Leofranc Holford-Strevens, writing in the *Encyclopedia of Ancient History*:

> The Etruscans believed that cities were allotted a fated number of *saecula,* each ending with the death of the longest-living person born on its first day, but the Romans equated the *saeculum* with a hundred years; tradition made the twelve vultures seen by Romulus to promise the city twelve such *saecula* of existence. (The word is also used for "generation," "fashion of the age," and in Christian parlance "the world," whence English "secular" meaning "non-religious.")
>
> *(Holford-Strevens 2013: 6006–7)*

Later Latin uses of the term mean something like an indefinite span of time, as in *per saecula saeulorum* (unto the ages of the ages, i.e., infinity) (Casanova 2013: 28). Our present-day use of secular, however, is rooted in Saint Augustine's retooling of the noun. For him, *saeculum* means the temporal gap between the present and the eschatological *parousia,* the time of divine redemption, when all humans, both Pagan and Christian, would be reunited under God's grace (what Kant called the Kingdom of Ends). For Augustine, the *saeculum* names the time of the *city of man,* in contrast to the time of the *city of God,* the heavenly city (Markus 1988). Thus, in Augustine's distinction between the secular, that is, the human time of the terrestrial city, what can also be called profane time, and the divine time of eternity, the idea that the secular refers to the time of the political sphere is already operative (Momigliano 1963).

Three things need to be noted: on the one hand, *saeculum,* as profane time, is the time of heterogeneity, the coexistence of Pagans and Christians, and thus of a time that is commonly shared; on the other, it is also a time that implies contestation, of dissent and competition about how to draw the boundaries between the 'sacred' and the 'profane'. This second distinction will cut both horizontally and vertically, meaning that the secular as that which is distinct from the 'sacred' will not only be contested by non-Christians, but it will also be a subject of debate for Christians: as in the question—when is the time for the sacred, for the religious? To paraphrase Casanova: 'The Christian sacred was the pagan profane and vice versa', but the Christian sacred was also shattered by other Christians' profane (Casanova 2013: 28). The third note, or gloss, is as momentous, if not more. *Saeculum* in this third sense also means that *all* human history has unity, that there is something like a universal history, a common history of all humans. The secular, and secularization more specifically, is the imperative to make sense of human history, in general, as a common, shared, historical fate. *Saeculum* announces the commonality and unity of human agency in time, human time, even as the one God watches and seeks to fulfil her plans. *Saeculum* thus is also the name for world, global, human history.

If one takes Augustine to inaugurate Western medieval Christendom, then since the fourth century we have the dyad (the Pythagorean 'two' or 'otherness') religious/secular, one that will configure both temporality and spatiality of society. This duality, however, will become polysemic inasmuch as Christendom takes hold of Medieval Europe, divine time becomes the religious-spiritual-sacred, while the *saeculum* becomes the secular-temporal-profane. This leads to a superimposition. Whereas the religious-spiritual-sacred gets spatialized in the heavenly city, the secular-temporal-profane gets spatialized as the earthly urban realm. This necessarily draws a distinction between the 'religious' clergy that withdraws into monasteries and cloisters, and the 'secular' clergy which comes to serve and live in the secular realm with the laity, the people. One can see how the secular is also the time-space for the intervention of the religious and sacred.

During Medieval Christianity, this transposed duality gave rise to two dynamics: on the one hand, we have the intervention of Christianity in the *saeculum*. This intervention goes by the name of the 'spiritualization' of the secular through the introduction of the religious practices of the monasteries into the profane world. This dynamic, this immanent tendency to seek to spiritualize the world, will be source of the radical reform movements within Christianity, which culminated with the Protestant Reformation and the plurification of denominations both within Protestantism and Catholicism (as with the Dominicans, Jesuits, and so on.). The other dynamic will be the thrust towards laicization, which aimed to have the church withdraw from all spheres of the secular world. At times, this withdrawal was driven by the church itself, at others, it was driven by the rising state, which in Late Medieval times and early Renaissance, would uncouple from the Papacy and the Church—a detailed and complex story that cannot be engaged here (see Manent 1996). This counter-tendency gives rise to what in French is called *laïcité*, literally, secularity. But in contrast to the medieval conception of the secular, *laïcité*, means freed from religion, not beholden at all to the dyad, profane-sacred. Thus, if one tendency was for the introjection of the sacred into the profane, the other was the ejection of the religious from the common space of subjects. Religion, as it were, is exiled from the emergent civil society and public sphere.

This outline describes what Casanova called 'mere secularity'. There is another sense of secular, and this is what he calls 'self-contained secularity' in which agents behave and act as if 'God did not exist', and for whom, religion or the religious form of life is one option among many. Following Charles Taylor's (2007) *The Secular Age*, Casanova argues that secular refers to the phenomenological experience of agents orienting themselves to differentiated cosmic, social, aesthetic, and natural orders. Religion is among one of those orders. This phenomenological and existential experience is called by Taylor the 'immanent frame', that is, the experience that account of oneself, nature, and the social world can only be given with reference to intramundane events, dispensing with all references to transcendent or higher realms. Taylor and Casanova agree that this self-contained secularity began to emerge in the sixteenth century and came about because of the momentous revolutions in science, technology, the expansion of the world and cosmos in general, as well as the Reformation and the Peace of Westphalia. This self-contained secularity could be characterized as an epistemic orientation towards the spheres of the individual, the social, and the natural that put religion on the same level as any other way of seeing these domains.

Casanova also discusses a third sense of secular, one that he calls 'secularist secularity' or 'secularism as a stadial consciousness'. This third sense can be seen as responding to 'self-contained secularity', which appears as a naïve self-presentation of episodic, contingent, and non-causally related historical events. Secularist secularity aims to give a developmental account of the emergence of a secular and secularist self-understanding, one that sees itself as the result of a diversity of progressive and transformative social, political, cognitive, and moral developments that are based on insights into the logics of social differentiation. One can thus say that secularist secularity is the latest layer, or stage, to level in the phylogenesis (evolutionary development or diversification) of *Geist* (mind/spirit). This developmental, progressivist, and thus presentist way of thinking about the secular and secularity is one that orients most sociological theory since Wilfred Pareto, Émile Durkheim, and Max Weber, and that continues to inform most sociologists following in their footsteps, among those, of course, Talcott Parsons, Jürgen Habermas, and Anthony Giddens.

Now, returning to Casanova, we can speak of secularization in at least three ways (Casanova 2013: 34): *first*, as the differential fanning out of spheres of human activity that are no longer under the reigns of either the church, or the religions. This is what Weber referred to as the differentiation of value spheres (state, economy, science, law, art, politics, and, of course, private life). This process is seen as one of emancipation and liberation. This process of secularization as a differentiation of value spheres also means the production of the religious as its own legitimate value sphere. Secularization then can be seen as both a process of the emancipation of different human dimensions of human action, and as the legitimate constitution of the religious as a sphere within which religious commitments can be practiced and lived.

Second, secularization can also be understood as the inevitable process of the decline of the performance of religious beliefs and their declining importance in giving meaning to the lifeworld of modern subjects. In this sense, secularization means the slow extinction or cessation of the relevance and presence of both religious practices and beliefs in the modern world. Secularization in this meaning entails an attendant idea, namely that religion and religious practices have become atavistic and vestigial practices that are either in discord or are anachronistic—*unzeitgemässe* (untimely), in Nietzsche's use of the term.

There is a *third* way in which we have come to understand secularization and that is as the *privatization of religion*, which is considered to be a condition *sine qua non* for the rise of modern heterogeneous, liberal, rule of law, constitutional democracies. Casanova, and Klaus Eder, have called into question this sociological hypothesis, not only on empirical grounds (religion has not been exiled from the public sphere) but also on normative grounds (in reaction to the secularization of the public sphere, public religions have emerged that continue to have decisive impact on all other spheres of social life). It is fair to say that Casanova's numerous texts, as well as his classic work *Public Religions in the Modern World* (1994), have contributed to our deeper understanding of the expansion of the secular, with the simultaneous and contentious revitalization of the religious within the secular.

Still, we want to complement and expand Casanova's important differentiations. To do so, we'll now turn to the entry on 'Säkularisation, Säkularisierung' in the massive and important *Geschichtliche Grundbegriffe*, published in Germany, in seven volumes, from 1972 to 1990, and which was edited by Otto Brunner, Werner Conze, and Reinhart Koselleck. The full title of the volume is *Geschichtliche Grundbegriffe: Historisches Lexikon zur Politisch-sozialen Sprache in Deutschland* ('Basic Concepts in History: a dictionary on historical principles on political and social language in German') (Richter 2011: 2)—it should be noted we are not entirely happy with this translation, as a more literal one would be: 'Historical Basic

Concepts: historical lexicon to the political and social language in Germany'. This lexicon is without question the utmost exemplar and achievement of the distinct field of historical study, namely, the history of concepts. This field distinguishes from intellectual history, although the latter relies on the former. Related notions are historical semantics, social history, intellectual and social history, and history of crises and social transformation. As Koselleck (2011: 8) noted in his introduction to the *Geschichtliche Grundbegriffe*,

> The central problematic (*die leitende fragestellung*) of this lexicon is the dissolution of old society of orders and estates, and the development of the modern world. These twin processes are studied by tracking the history of how they were conceptually registered.

A fair question would be: why turn to a lexicon that is narrowly focused on the German political and social language? There are abundant reasons. A key justification is because Germans struggle with Latin, and the Romance languages in general, in order to develop their own language; this process of translation was no less important than the process of translation that took place between Greek and Latin, between, let's say, Plato and Cicero. An evident reason also is because the German state(s) also struggled with the process of secularization if only because they became the historical theatre on which the drama of the Reformation was played out. Another reason would be because Germany also produced some of the most important thinkers of Christianity, in particular, and religion, in general—from Luther through as recently as Habermas and Peter Sloterdijk (2009, 2017). A very important reason is because Germany, and the German language, both as synecdoches for the defence and preservation of a version of Christianity, also became the stage for one of the most disastrous crusades *in* European and world history about the meaning of the *saeculum*, namely, the struggle to purify it of Jewish contamination. Finally, and more to the point of our aims here, this monumental contribution illustrates how concepts have histories and how those histories are impacted by the transformation of language. Beaumont's collection, after all, is a handbook.

It is not the intention to reply on every move that is tracked in this fascinating entry. We'll restrict ourselves to highlighting eight senses or meanings of secularization that emerged in Germany during the last 400 years. Before that, however, it must be noted that initially the term *Säkularisierung* was used in order to translate *saecularisatio*, which is why another term was used in German as well, namely *Säkularisation*, which retained the Latin/Romance languages inflection of the process of turning social life to the *saeculum* as described earlier. Both terms, however, are German latinizations of secularization. Translations into German, however, were *Verweltlichung* and *Weltlichung,* which are substantive versions of the transitive verbs, *verweltlichen/weltlichen*, correspondingly. The transitive verbs mean to make worldly or to make into the world, or to make mundane. This translation is appealing because it signals that the process of secularization as a form of *Verweltlichung* and *Weltlichung* is not simply turning something over to an extant realm, but means to literally create that horizon, to open up that time-space. To secularize, then, means to world, to make a world, to constitute a world that is mundane. In this sense, we can see more clearly, what Habermas calls, as we will discuss later, the dialectics of secularization. On the one hand, the secular world is produced from within religion; on the other, the religious realm both expands and depends in its relation to the secular world, that with which it is in a dynamic and generative tension.

Herman Zabel, the author of the entry under discussion, covers in a sobering and exacting way the layering of the terms secularization in the German language, covering nearly 600 years. Among the many layers and permutations, one can highlight the following. *First*, secularization as the *transfer*, turning over into the world, the saeculum, as when priests leave the monasteries and go into the city, among the lay people. This process is seen as necessary and not in an invidious sense. *Second*, there is the transfer of both church property and formulation and translation of canon law that can coordinate the *Jus Gentium* ('law of nations') *and Ius Civilis* ('civil law'). Here secularization meant something like the *juridification* of the relationship between the church and extant legal orders inherited from the Roman Empire. Again, here secularization is still not seen pejoratively, but rather a logical necessity of having to coordinate two, or several, orders of law (see the important work by Schiavone 2012). *Third*, there is the violent *expropriation* of church property by the state, which was conceived as an assault on the church and the religious realm. Here secularization could be taken in a pejorative and violent sense, as the dismantling of a powerful and long-standing order of society. *Fourth*, secularization came to mean also the *spiritualization*, or extension of the religious attitude and worldview into the secular realm. This is how Protestantism was understood. In order to accomplish this spiritualization, *Vergeistigung*, of the social world, religion had to detach itself from the church, which now had come to be conceived as another worldly institution. *Fifth*, and we already discussed this earlier, secularization came to be discussed as the drawing of strict and clear demarcations between the religious realm and the profane, with the subsidiary clarification that this realm may belong entirely within the inner life of agents. So, here secularization came to mean something like individuation and subjectification, the deepening of an internal realm in which subjects could have a direct relationship to the God or the divine. *Sixth*, secularization came to mean the logical and historical fulfilment of Christianity; the actualization of the deepest and most fundamental principle, the principle of subjective freedom, which came to be enshrined in the state. Evidently, the key figure here is Hegel, who made these arguments in his lectures on the history of philosophy and the philosophy of history. Here secularization meant the fulfilment of Christianity itself in the secular realm. *Seventh*, secularization meant the incomplete processes of the detheologization and defetishization of basic concepts of humanity. In this sense, secularization, as something still underway, was the means for the liberation of humanity from its created Gods, which are conceived as alienated forms of its own potentiality. Evidently, the referents there are Ludwig Feuerbach and Karl Marx.

Departing now from Zabel's masterful reconstruction and archaeological work in historical semantics, we would add an *eighth* sense, namely, secularization as an inauthentic, self-deluding, and even illegitimate process of passing off key ideas in Christian and religious thought as if they were autochthonous (indigenous) and self-generated by modernity. Here the key names are Carl Schmitt, Carl Löwith, on one side, and Hans Blumenberg and Habermas, on the other (see Habermas 2017: 87–92). One could call this, to paraphrase Blumenberg, the problem of the *legitimacy of the secular*. Overall, we would argue that there is no secular realm without the process of secularization, but secularization has meant many things at different times. This means, at the very least, that neither category has been exhausted and thus both remain, for the time being, generative.

Public religions and the postsecular

Given this polysemia related to secular and secularization, in what sense can we use the terms postsecular and postsecularity? Here it is insightful to draw upon Jim Beckford's presidential address to the Society for the Scientific Study of Religion (SSSR) from 2012 because, at the very least, he has discovered some of the earliest references to the term 'postsecular' in the English literature. According to Beckford, the earliest uses in the English print correspond to Andrew Greely in an article from 1966, and to Richard John Neuhaus in a piece from 1982 (Beckford 2012: 2). It is after the 1990s, however, that Beckford registers that 'the concept quickly acquired diverse and divergent meanings' (*ibid*.: 2).

He then discusses the usages of *postsecular* under six different clustering subheadings. *First*, there is the 'secularization deniers and doubters', which not only doubt whether secularization is the appropriate term for what happened to religion in general, and Christianity in particular. The term (postsecular), then, is used as a corrective and a way to move beyond a mistake and 'mischievous' theorization of the modern religious condition. *Second*, there is the 'building on the secular', which does not take a contentious attitude towards secularity, but instead sees the postsecular as building on both achievement of secularity and the new religious sensibilities. Beckford clusters under this subheading the work of the RO thinkers, such as John Milbank and Phillip Blond, who have called for a re-Christianization of culture in the face of the excesses and failures of modernity and postmodernity. In this movement, accordingly, the postsecular and postmodern converge. *Third*, there is the 'reenchantment of culture' cluster of uses of the term that have emerged from studies of various art forms. The intersections and cross-fertilization of different currents in literary, philosophical, and theological thinking have led to innovative and distinctive valences of the term postsecular. These discussions are indeed interesting, as they look towards aesthetic sensibilities that signal new spiritual sensibilities that are not tethered to a traditional understanding of religiosity and/or membership in some sort of organized church. Some of this discussion points to the works of Don DeLillo, Toni Morrison, Alice Walker, and Thomas Pynchon, as exemplary of these postsecular religious sensibilities. Without doubt there are also European and non-European writers who could be brought into this cluster. *Fourth*, there is the 'public resurgence of religion' cluster. Here Beckford discusses Bosetti and Eder (2006), but also Clayton Crockett, as challenging the disappearance of religion from the public sphere, affirming its resurgence and vitality. At the core of the uses of the term postsecular in this cluster is contesting the dyad private/public and how it does not map onto the all too evident visibility of public religious acts, such as progressive social justice activism among faith-based organizations (e.g., Cloke and Beaumont 2013), and appeals to religious language and imaginary. *Fifth*, there is the 'politics, philosophy, and theology' cluster. Under this cluster, Beckford focuses exclusively on Habermas and confesses sharing with Hans Joas' 'astonishment' at Habermas' embrace of the term. According to Beckford, Habermas' use of the postsecular is different from those thus far discussed, as it seems to have a more positive valence, insofar as the term 'seems to refer to the unfolding of the latest phase of secularization in the sense of functional differentiation and subjectivization process' (Beckford 2012: 8). In any case, he summarizes Habermas' usage, and, for the most part, accurately:

> He [Habermas] strongly believes that the central institutions of the state should remain secular but that, at the same time, religious and spiritual arguments have a right to be heard in the public sphere of civil society. This recognition of the potential value of religion in public life—especially faith-based moral values—is relatively new for Habermas and this is probably why he felt the need to use the term "postsecular." But this use of it tells us more about his deep concern about the state of democratic politics and the lack of human solidarity than about any respect that he may have for religion.
>
> *(Beckford 2012: 10)*

Evidently, this last remark is off mark, given how extensively Habermas has engaged religious thinkers since his earliest writings on Jewish German thinkers and non-Jewish thinkers such as Karl Jaspers and Johannes Baptist Metz (see Habermas 2002; Mendieta 2013). *Sixth*, and finally, there is the 'a plague on all your houses' cluster, which essentially is critical and dismissive of the concept.

Five rubrics on the postsecular

As useful as Beckford's overview and clustering of the uses of the term postsecular might be, we would like to offer a more methodologically oriented typology, without attempting to attach specific scholars, schools, or trends. Alongside Beckford's approach, there are other valuable groupings with some overlaps with ours (see Chapters 9 and 22). We contend, in any case, that the postsecular could be deployed in at least five ways.

Naively: The term postsecular could be conceptualized in the sense that the postmodern may have been used in the past, reactively and without proper engagement with the phenomenon that it names as overcoming and having gone beyond. *Critically*: Here commentators see the term as a way to challenge the self-congratulating and presentist or Whiggish narratives that assume the we have indeed become secular, that secularity is equivalent to modernity, and that it has been concurrent with modernization and democratization. *Reflexively*: Scholars here see the concept as way to engage in a dual process of seeing what has been accomplished and also what remains yet to be done. Reflexive in this attitude also means dialectic. We are thinking here how the term reflexive modernity was used to complement and criticize postmodern critics of modernity. Postsecular, in this sense, then would be a form of 'reflexive secularity'. *Genealogically*: Used here in the triple sense of genealogical analyses offered by Colin Koopman (2013) and expanded by Amy Allen (2013), as either subversive, vindicatory, or problematizing. These three versions of genealogical analysis are implicit in Friedrich Nietzsche's (1994) *On the Genealogy of Morality* (1994), which not only aims to unearth the origins of our moral values but also to subvert and problematize their usefulness and relevance. 'Vindicatory' aligns itself with, what we called earlier, the Whiggish version of congratulating narratives about how we have become secular, thus modern and more 'mature'. 'Problematizing' aims to get us to take a meta-look at our obsession with the religious and our ceaseless attempts to both affirm and reject it within the same step. *Begriffgeschichtliche/history of concepts/historical semantics*: This final deployment or attitude, emphasizing intellectual and social history, as well as history of crises and social transformation, is partly illustrated earlier with reference to the *Geschichtliche Grundbegriffe* dictionary and the Beckford (2012) article.

As this Afterword—and volume as a whole—nears completion, one should note that the term postsecular by necessity relates reflexively to and is, in a sense, parasitic on the uses of the terms secular and secularization. The secular is the moraine left over by the multiple processes of secularization that have taken place over the last two and a half millennia—if we extend our analysis to the Axial Age as identified by Jaspers and the evolution of postsecular consciousness (Bellah and Joas 2012; Assmann 2018; Gordon forthcoming; Mendieta forthcoming). The secular is to secularization as the postsecular is to postsecularization. The challenge, now, is to begin to survey those processes that in their wake are clearing conceptual and social spaces for new sensibilities that allow us to review the complexity of secularization on the one hand, and the way in which we are always already secular as we claim to be postsecular on the other. Here, postsecularity is more than complex secularity. It concerns and attempts to foreground what we call *reflexive secularization*. Postsecularity concerns assemblages of antagonist processes unfolding through state secularization, and structures and practices that uphold respect for one's right to faith. Bridled with sights on new pathways for social evolution or transformation, our approach yields a new sense of conflicts and solidarities, complementarity of discourses, and confrontation of normativities—all pointing to new empirical challenges ahead. A thorough and multilayer reflexive secularization approach to the postsecular requires a combination of four of the five disaggregated attitudes mentioned earlier—the critical, the reflexive, the genealogical, and the *Begriffgeschichtliche*—as, arguably, Habermas does in this appropriation and elaboration of the term (Aguirre forthcoming; Mendieta forthcoming; see also Chapter 4).

One last analogy helps illustrate what we mean by reflexive secularization. The Peruvian sociologist Anibal Quijano, one of the most influential to have come from Latin America in the last half of the century, introduced the concept of 'coloniality of power' in order to name what can be called the 'colonial present'. By the concept 'coloniality of power', Quijano meant to challenge the naïve attitude of postcolonial and postmodern thinkers that assume that either coloniality and modernity could be overcome as if by a Munchausen trick. The project and experience of European colonialism required the creation and establishment of a matrix of intelligibility that secured and legitimated European power and superiority over its colonial others. This matrix called for the constructions of races, classes, genders, and, of course, forms of understanding cultures that sedimented invidious and hierarchical categories. Quijano argued, pointedly, that we live in the wake of the power unleashed by the modern/colonial matrix of intelligibility and arranging social relations. In tandem, thinkers like Nelson Maldonado-Torres, Santiago Castro-Gómez, and Walter Mignolo have called for a 'decolonial turn' rather than a postcolonial turn (see Moraña et al. 2008, for a sampling of these critiques and theoretical proposals; see also Mendieta 2010). By the decolonial, they mean having to deal with the colonial present in which we live, still, even after the decolonizing movements and struggles of the last half a century.

We would argue that something similar is at play with the concepts secular/postsecular. We live in a secular present, one that is not only implicated with colonialism and imperialism, which have taken many permuations, one of them being the 'evangelization' of colonial others, but also the dual secularization/desecularization of civil society in Europe itself. If we live in a colonial present, we also live in a secular present. Decolonial is to coloniality as postsecular is to secular, that is, immanent, reflexive, critical postures that aim to elucidate the co-determination of the colonial/decolonial, secular/postsecular dialectical dyads. If we have never been secular, we can also never become postsecular enough.

At least two lines of inquiry emerge from this critical and reflexive reading of secularization and postsecularity, and the contributions to the volume as a whole. *First*, new work could focus on the spatialities as well as temporalities of postsecularity. New theorizations of *urbanization of the postsecular and the enlightened city* would be a possibility. The notion of the 'postsecular city' (Molendijk et al. 2010; Beaumont and Baker 2011; see also Chapters 25–27) ought to broaden to what Beaumont (2018) calls the 'enlightened city'—the conceptualization of urbanization as overlapping ethical and political imperatives that reconcile radical difference and confront injustices. Klaus Eder's empirical interest in structurally equivalent positions in narrative networks, if spatialized, would be a valuable critical-theoretical contribution to understanding postsecular urban society. Larger-scale scholarship could explore the immanentist and postmetaphysical connotations of urbanization of postsecularity, which might reach out to transcendent, monist, and pantheist modes of thought. Eduardo Mendieta's work linking decoloniality, urbanism, and the Anthropocene ties into this agenda. *Second*, new inquiries could attend to the growing recognition of multiple postsecularisms, critical postsecular consciousness, and moves to decolonize the secular. New scholarship could raise and address important questions about the *universal and authentic human condition*, through respect for context-specific postsecularisms outside the Occident including Russia, India, and Asia.

References

Allen, A. (2013) 'Having one's cake and eating it too: Habermas' genealogy of postsecular reason'. In Calhoun, C., Mendieta, E. and J. VanAntwerpen (eds.) *Habermas and Religion*, Cambridge, MA: Polity Press.

Aguirre, J. (Forthcoming) 'Secularization/Postsecularism'. In Allen, A. and E. Mendieta (eds.) *The Cambridge Habermas Lexicon*, Cambridge: Cambridge University Press.

Assmann, J. (2018) *The Invention of Religion: faith and covenant in the Book of Exodus*, Princeton, NJ: Princeton University Press.

Beaumont, J. (2018) 'The enlightened city', submitted to *Political Geography.*

Beaumont, J. and C. Baker (eds.) (2011) *Postsecular Cities: space, theory and practice*, London: Continuum.

Beckford, J. A. (2012) 'SSSR presidential address: Public religions and the postsecular: critical reflections', *Journal for the Scientific Study of Religion*, 51(1): 1–19.

Bellah, R. N. and H. Joas (eds.) (2012) *The Axial Age and its Consequences*, Cambridge, MA: The Belknap Press of Harvard University Press.

Borges, J. L. (1999) *The Total Library: non-fiction 1922–1986*, New York, NY: Penguin Books.

Bosetti, G. and K. Eder (2006) 'Post-secularism: a return to the public sphere', interview with G. Bosetti, *Eurozine*, 17 August. www.eurozine.com/post-secularism-a-return-to-the-public-sphere/

Brunner, O., Conze, W. and R. Koselleck (eds.) (1984) *Geschichtliche Grundbegrieffe: historisches lexikon zur politisch-sozialen prace in Deutschland*, Stuttgart: Klett-Cotta.

Casanova, J. (1994) *Public Religions in the Modern World*, Chicago, IL: Chicago University Press.

——— (2013) 'Exploring the postsecular: three meanings of the "the secular" and their possible transcendence'. In Calhoun, C., Mendieta, E. and J. VanAntwerpen (eds.) *Habermas and Religion*, Cambridge, MA: Polity Press, pp. 27–48.

Cloke, P. and J. Beaumont (2013) 'Geographies of postsecular rapprochement in the city', *Progress in Human Geography*, 37(1): 27–51.

Gordon, P. E. (Forthcoming) 'Axial age ("Achenzeit")'. In Allen, A. and E. Mendieta (eds.) *The Cambridge Habermas Lexicon*, Cambridge: Cambridge University Press.

Habermas, J. (2002) *Religion and Rationality: essays on reason, God and modernity*, Cambridge, MA: MIT Press.

——— (2017) *Postmetaphysical Thinking II. Essays and Replies*, Cambridge, MA: Polity Press.

Holford-Strevens, L. (2013) 'Saeculum'. In Bagnall, R. S., Brodersen, K., Champion, C. B., Erskine, A. and S. R. Huebner (eds.) *The Encyclopedia of Ancient History*, Lanham, MD: Blackwell Publishing Ltd., pp. 6006–7.

Koopman, C. (2013) *Genealogy as Critique: Foucault and the problems of modernity*, Bloomington: Indiana University Press.

Koselleck, R. (2011) 'Introduction and prefaces to the *Geschichtliche Grundbegrieffe*', *Contributions to the History of Concepts*, 6(1): 1–37.

Manent, P. (1996) *An Intellectual History of Liberalism*, trans. Rebecca Balinski, with foreword by Jerrold E. Seigel, reprint edition, Princeton, NJ: Princeton University Press.

Markus, R. A. (1988) *Saeculum: history and society in the theology of St. Augustine*, Cambridge: Cambridge University Press.

Mendieta, E. (2010) 'Postcolonialism, postorientalism, postoccidentalism: the past that never went away and the future that never arrived'. In May, T. (ed.) *Emerging Trends in Continental Philosophy, Volume 8: history of continental philosophy*, General Editor Alan D. Schrift. Durham: Acumen, pp. 149–71.

——— (2013) 'Appendix: religion in Habermas's Work. In Calhoun, C., Mendieta, E. and J. VanAntwerpen (eds.) *Habermas and Religion*, Cambridge, MA: Polity Press, pp. 391–407.

——— (Forthcoming) 'The axial age, social evolution, and postsecular consciousness', *Critical Research on Religion.*

Molendijk, A. L., Beaumont, J. and C. Jedan (2010) *Exploring the Postsecular: the religious, the political and the urban*, Leiden/Boston, MA: Brill.

Momigliano, A. (1963) 'Pagan and Christian historiography in the fourth century AD'. In Momigliano, A. (ed.) (1963) *The Conflict between Paganism and Christianity in the fourth century*, Oxford: Oxford University Press, pp. 79–99.

Moraña, M., Dussel, E. and C. A. Jáuregui (eds.) (2008) *Coloniality at Large: Latin America and the postcolonial debate*, Durham, NC: Duke University Press.

Nietzsche, F. (1994) *On the Genealogy of Morality*, Cambridge: Cambridge University Press.
Richter, M. (2011) 'Introduction and prefaces to the *Geschichtliche Grundbegrieffe*: a note on the translations', *Contributions to the History of Concepts*, 6(1): 1–37.
Schiavone, A. (2012) *The Invention of Law in the West*, Cambridge, MA: Harvard University Press.
Sloterdijk, P. (2009) *God's Zeal: the battle of the three monotheisms*, Cambridge, MA: Polity Press.
——— (2017) *Nach Gott: die konsequenzen aus dem satz 'Gott is Tot'*, Frankfurt am Main: Suhrkamp Verlag.
Taylor, C. (2007) *A Secular Age*, Cambridge, MA: Harvard University Press.

Index

Note: Boldface page numbers refer to tables; italic page numbers refer to figures.

For Product Safety Concerns and Information please contact our EU
representative GPSR@taylorandfrancis.com
Taylor & Francis Verlag GmbH, Kaufingerstraße 24, 80331 München, Germany

www.ingramcontent.com/pod-product-compliance
Ingram Content Group UK Ltd.
Pitfield, Milton Keynes, MK11 3LW, UK
UKHW050232200926
475702UK00020B/2979
* 9 7 8 0 3 6 7 7 3 2 6 4 6 *